ESSENTIALS OF ROBUST CONTROL

ESSENTIALS OF ROBUST CONTROL

KEMIN ZHOU
Louisiana State University

with John C. Doyle
California Institute of Technology

Prentice Hall
Upper Saddle River, New Jersey 07458

Library of Congress Catalog in Publicaton Data
Zhou, Kemin.
        Essentials of robust control / Kemin Zhou
                p. cm.
        Includes bibliographical references and index.
        ISBN 0-13-525833-2
        1. Control theory. 2.H^∞ . I. Title.
QA402.3Z475 1998   97-29568
629.8--dc21   CIP

Publisher: Tom Robbins
Associate editor: Alice Dworkin
Editor-in-chief: Marcia Horton
Production editor: AnnMarie Longobardo
Copy editor: Patricia M. Daly
Director of production and manufacturing: David W. Riccardi
Art director: Jayne Conte
Managing editor: Bayani Mendoza de Leon
Cover designer: Bruce Kenselaar
Manufacturing buyer: Julia Meehan
Editorial assistant: Nancy Garcia

©1998 by Prentice-Hall, Inc.
Upper Saddle River, New Jersey 07458

All rights reserved. No part of this book may be reproduced, in any form or by any means, without permission in writing from the publisher.

The author and publisher of this book have used their best efforts in preparing this book. These efforts include the development, research, and testing of the theories and programs to determine their effectiveness. The author and publisher make no warranty of any kind, expressed or implied, with regard to these programs or the documentation contained in this book. The author and publisher shall not be liable in any event for incidental or consequential damages in connection with, or arising out of, the furnishing, performance, or use of these programs.

Printed in the United States of America

10 9 8 7 6 5 4 3 2

ISBN 0-13-525833-2

Prentice-Hall International (UK) Limited, *London*
Prentice-Hall of Australia Pty. Limited, *Sydney*
Prentice-Hall Canada Inc., *Toronto*
Prentice-Hall Hispanoamericana, S.A., *Mexico*
Prentice-Hall of India Private Limited, *New Delhi*
Prentice-Hall of Japan, Inc., *Tokyo*
Pearson Education Asia Pte. Ltd., *Singapore*
Editoria Prentice-Hall do Brasil, Ltda., *Rio De Janeiro*

To

my wife, my son,

my sister, and my brothers

Preface

Robustness of control systems to disturbances and uncertainties has always been the central issue in feedback control. Feedback would not be needed for most control systems if there were no disturbances and uncertainties. Developing multivariable robust control methods has been the focal point in the last two decades in the control community. The state-of-the-art $\mathcal{H}_\infty$ robust control theory is the result of this effort.

This book introduces some essentials of robust and $\mathcal{H}_\infty$ control theory. It grew from another book by this author, John C. Doyle, and Keith Glover, entitled *Robust and Optimal Control*, which has been extensively class-tested in many universities around the world. Unlike that book, which is intended primarily as a comprehensive reference of robust and $\mathcal{H}_\infty$ control theory, this book is intended to be a text for a graduate course in multivariable control. It is also intended to be a reference for practicing control engineers who are interested in applying the state-of-the-art robust control techniques in their applications. With this objective in mind, I have streamlined the presentation, added more than 50 illustrative examples, included many related MATLAB® commands[1] and more than 150 exercise problems, and added some recent developments in the area of robust control such as gap metric, ν-gap metric, model validation, and mixed μ problem. In addition, many proofs are completely rewritten and some advanced topics are either deleted completely or do not get an in-depth treatment.

The prerequisite for reading this book is some basic knowledge of classical control theory and state-space theory. The text contains more material than could be covered in detail in a one-semester or a one-quarter course. Chapter 1 gives a chapter-by-chapter summary of the main results presented in the book, which could be used as a guide for the selection of topics for a specific course. Chapters 2 and 3 can be used as a refresher for some linear algebra facts and some standard linear system theory. A course focusing on $\mathcal{H}_\infty$ control should cover at least most parts of Chapters 4–6, 8, 9, 11–13, and Sections 14.1 and 14.2. An advanced $\mathcal{H}_\infty$ control course should also include the rest of Chapter 14, Chapter 16, and possibly Chapters 10, 7, and 15. A course focusing on robustness and model uncertainty should cover at least Chapters 4, 5, and 8–10. Chapters 17 and 18 can be added to any advanced robust and $\mathcal{H}_\infty$ control course if time permits.

I have tried hard to eliminate obvious mistakes. It is, however, impossible for me to make the book perfect. Readers are encouraged to send corrections, comments, and

---

[1] MATLAB is a registered trademark of The MathWorks, Inc.

suggestions to me, preferably by electronic mail, at

kemin@ee.lsu.edu

I am also planning to put any corrections, modifications, and extensions on the Internet so that they can be obtained either from the following anonymous ftp:

ftp gate.ee.lsu.edu   cd pub/kemin/books/essentials/

or from the author's home page:

http://kilo.ee.lsu.edu/kemin/books/essentials/

This book would not be possible without the work done jointly for the previous book with Professor John C. Doyle and Professor Keith Glover. I thank them for their influence on my research and on this book. Their serious attitudes toward scientific research have been reference models for me. I am especially grateful to John for having me as a research fellow in Caltech, where I had two very enjoyable years and had opportunities to catch a glimpse of his "BIG PICTURE" of control.

I want to thank my editor from Prentice Hall, Tom Robbins, who originally proposed the idea for this book and has been a constant source of support for me while writing it. Without his support and encouragement, this project would have been a difficult one. It has been my great pleasure to work with him.

I would like to express my sincere gratitude to Professor Bruce A. Francis for giving me many helpful comments and suggestions on this book. Professor Francis has also kindly provided many exercises in the book. I am also grateful to Professor Kang-Zhi Liu and Professor Zheng-Hua Luo, who have made many useful comments and suggestions. I want to thank Professor Glen Vinnicombe for his generous help in the preparation of Chapters 16 and 17. Special thanks go to Professor Jianqing Mao for providing me the opportunity to present much of this material in a series of lectures at Beijing University of Aeronautics and Astronautics in the summer of 1996.

In addition, I would like to thank all those who have helped in many ways in making this book possible, especially Professor Pramod P. Khargonekar, Professor André Tits, Professor Andrew Packard, Professor Jie Chen, Professor Jakob Stoustrup, Professor Hans Henrik Niemann, Professor Malcolm Smith, Professor Tryphon Georgiou, Professor Tongwen Chen, Professor Hitay Özbay, Professor Gary Balas, Professor Carolyn Beck, Professor Dennis S. Bernstein, Professor Mohamed Darouach, Dr. Bobby Bodenheimer, Professor Guoxiang Gu, Dr. Weimin Lu, Dr. John Morris, Dr. Matt Newlin, Professor Li Qiu, Professor Hector P. Rotstein, Professor Andrew Teel, Professor Jagannathan Ramanujam, Dr. Linda G. Bushnell, Xiang Chen, Greg Salomon, Pablo A. Parrilo, and many other people.

I would also like to thank the following agencies for supporting my research: National Science Foundation, Army Research Office (ARO), Air Force of Scientific Research, and the Board of Regents in the State of Louisiana.

Finally, I would like to thank my wife, Jing, and my son, Eric, for their generous support, understanding, and patience during the writing of this book.

Kemin Zhou

PREFACE

Here is how $\mathcal{H}_\infty$ is pronounced in Chinese:

It means "The joy of love is endless."

Contents

Preface vii

Notation and Symbols xv

List of Acronyms xvii

1 Introduction 1
   1.1 What Is This Book About? 1
   1.2 Highlights of This Book 3
   1.3 Notes and References 9
   1.4 Problems 10

2 Linear Algebra 11
   2.1 Linear Subspaces 11
   2.2 Eigenvalues and Eigenvectors 12
   2.3 Matrix Inversion Formulas 13
   2.4 Invariant Subspaces 15
   2.5 Vector Norms and Matrix Norms 16
   2.6 Singular Value Decomposition 19
   2.7 Semidefinite Matrices 23
   2.8 Notes and References 24
   2.9 Problems 24

3 Linear Systems 27
   3.1 Descriptions of Linear Dynamical Systems 27
   3.2 Controllability and Observability 28
   3.3 Observers and Observer-Based Controllers 31
   3.4 Operations on Systems 34
   3.5 State-Space Realizations for Transfer Matrices 35
   3.6 Multivariable System Poles and Zeros 38
   3.7 Notes and References 41
   3.8 Problems 42

4 $\mathcal{H}_2$ and $\mathcal{H}_\infty$ Spaces — 45
- 4.1 Hilbert Spaces 45
- 4.2 $\mathcal{H}_2$ and $\mathcal{H}_\infty$ Spaces 47
- 4.3 Computing $\mathcal{L}_2$ and $\mathcal{H}_2$ Norms 53
- 4.4 Computing $\mathcal{L}_\infty$ and $\mathcal{H}_\infty$ Norms 55
- 4.5 Notes and References 61
- 4.6 Problems 62

5 Internal Stability — 65
- 5.1 Feedback Structure 65
- 5.2 Well-Posedness of Feedback Loop 66
- 5.3 Internal Stability 68
- 5.4 Coprime Factorization over $\mathcal{RH}_\infty$ 71
- 5.5 Notes and References 77
- 5.6 Problems 77

6 Performance Specifications and Limitations — 81
- 6.1 Feedback Properties 81
- 6.2 Weighted $\mathcal{H}_2$ and $\mathcal{H}_\infty$ Performance 85
- 6.3 Selection of Weighting Functions 89
- 6.4 Bode's Gain and Phase Relation 94
- 6.5 Bode's Sensitivity Integral 98
- 6.6 Analyticity Constraints 100
- 6.7 Notes and References 102
- 6.8 Problems 102

7 Balanced Model Reduction — 105
- 7.1 Lyapunov Equations 106
- 7.2 Balanced Realizations 107
- 7.3 Model Reduction by Balanced Truncation 117
- 7.4 Frequency-Weighted Balanced Model Reduction 124
- 7.5 Notes and References 126
- 7.6 Problems 127

8 Uncertainty and Robustness — 129
- 8.1 Model Uncertainty 129
- 8.2 Small Gain Theorem 137
- 8.3 Stability under Unstructured Uncertainties 141
- 8.4 Robust Performance 147
- 8.5 Skewed Specifications 150
- 8.6 Classical Control for MIMO Systems 154
- 8.7 Notes and References 157
- 8.8 Problems 158

9 Linear Fractional Transformation — 165
- 9.1 Linear Fractional Transformations — 165
- 9.2 Basic Principle — 173
- 9.3 Redheffer Star Products — 178
- 9.4 Notes and References — 180
- 9.5 Problems — 181

10 μ and μ Synthesis — 183
- 10.1 General Framework for System Robustness — 184
- 10.2 Structured Singular Value — 187
- 10.3 Structured Robust Stability and Performance — 200
- 10.4 Overview of μ Synthesis — 213
- 10.5 Notes and References — 216
- 10.6 Problems — 217

11 Controller Parameterization — 221
- 11.1 Existence of Stabilizing Controllers — 222
- 11.2 Parameterization of All Stabilizing Controllers — 224
- 11.3 Coprime Factorization Approach — 228
- 11.4 Notes and References — 231
- 11.5 Problems — 231

12 Algebraic Riccati Equations — 233
- 12.1 Stabilizing Solution and Riccati Operator — 234
- 12.2 Inner Functions — 245
- 12.3 Notes and References — 246
- 12.4 Problems — 246

13 $\mathcal{H}_2$ Optimal Control — 253
- 13.1 Introduction to Regulator Problem — 253
- 13.2 Standard LQR Problem — 255
- 13.3 Extended LQR Problem — 258
- 13.4 Guaranteed Stability Margins of LQR — 259
- 13.5 Standard $\mathcal{H}_2$ Problem — 261
- 13.6 Stability Margins of $\mathcal{H}_2$ Controllers — 265
- 13.7 Notes and References — 267
- 13.8 Problems — 267

14 $\mathcal{H}_\infty$ Control — 269
- 14.1 Problem Formulation — 269
- 14.2 A Simplified $\mathcal{H}_\infty$ Control Problem — 270
- 14.3 Optimality and Limiting Behavior — 282
- 14.4 Minimum Entropy Controller — 286
- 14.5 An Optimal Controller — 286

14.6 General $\mathcal{H}_\infty$ Solutions 288
14.7 Relaxing Assumptions 291
14.8 $\mathcal{H}_2$ and $\mathcal{H}_\infty$ Integral Control 294
14.9 $\mathcal{H}_\infty$ Filtering 297
14.10 Notes and References 299
14.11 Problems 300

15 Controller Reduction   305
15.1 $\mathcal{H}_\infty$ Controller Reductions 306
15.2 Notes and References 312
15.3 Problems 313

16 $\mathcal{H}_\infty$ Loop Shaping   315
16.1 Robust Stabilization of Coprime Factors 315
16.2 Loop-Shaping Design 325
16.3 Justification for $\mathcal{H}_\infty$ Loop Shaping 328
16.4 Further Guidelines for Loop Shaping 334
16.5 Notes and References 341
16.6 Problems 342

17 Gap Metric and ν-Gap Metric   349
17.1 Gap Metric 350
17.2 ν-Gap Metric 357
17.3 Geometric Interpretation of ν-Gap Metric 370
17.4 Extended Loop-Shaping Design 373
17.5 Controller Order Reduction 375
17.6 Notes and References 375
17.7 Problems 375

18 Miscellaneous Topics   377
18.1 Model Validation 377
18.2 Mixed μ Analysis and Synthesis 381
18.3 Notes and References 389
18.4 Problems 390

Bibliography   391

Index   407

Notation and Symbols

$\mathbb{R}$ and $\mathbb{C}$	fields of real and complex numbers
$\mathbb{F}$	field, either $\mathbb{R}$ or $\mathbb{C}$
$\mathbb{C}_-$ and $\overline{\mathbb{C}}_-$	open and closed left-half plane
$\mathbb{C}_+$ and $\overline{\mathbb{C}}_+$	open and closed right-half plane
$j\mathbb{R}$	imaginary axis
$\in$	belong to
$\subset$	subset
$\cup$	union
$\cap$	intersection
$\square$	end of proof
$\diamond$	end of remark
$:=$	defined as
$\gtrapprox$ and $\lessapprox$	asymptotically greater and less than
$\gg$ and $\ll$	much greater and less than
$\overline{\alpha}$	complex conjugate of $\alpha \in \mathbb{C}$
$|\alpha|$	absolute value of $\alpha \in \mathbb{C}$
$Re(\alpha)$	real part of $\alpha \in \mathbb{C}$
I_n	$n \times n$ identity matrix
$[a_{ij}]$	a matrix with a_{ij} as its ith row and jth column element
$\mathrm{diag}(a_1, \ldots, a_n)$	an $n \times n$ diagonal matrix with a_i as its ith diagonal element
A^T and A^*	transpose and complex conjugate transpose of A
A^{-1} and A^+	inverse and pseudoinverse of A
A^{-*}	shorthand for $(A^{-1})^*$
$\det(A)$	determinant of A
$\mathrm{trace}(A)$	trace of A

$\lambda(A)$	eigenvalue of A
$\rho(A)$	spectral radius of A
$\rho_R(A)$	real spectrum radius of A
$\overline{\sigma}(A)$ and $\underline{\sigma}(A)$	the largest and the smallest singular values of A
$\sigma_i(A)$	ith singular value of A
$\kappa(A)$	condition number of A
$\|A\|$	spectral norm of A: $\|A\| = \overline{\sigma}(A)$
Im(A), R(A)	image (or range) space of A
Ker(A), N(A)	kernel (or null) space of A
$\mathcal{X}_-(A)$	stable invariant subspace of A
Ric(H)	the stabilizing solution of an ARE
$g * f$	convolution of g and f
$\angle$	angle
$\langle , \rangle$	inner product
$x \perp y$	orthogonal, $\langle x, y \rangle = 0$
$D_\perp$	orthogonal complement of D
$S^\perp$	orthogonal complement of subspace S, e.g., $\mathcal{H}_2^\perp$
$\mathcal{L}_2(-\infty, \infty)$	time domain square integrable functions
$\mathcal{L}_{2+} := \mathcal{L}_2[0, \infty)$	subspace of $\mathcal{L}_2(-\infty, \infty)$ with functions zero for $t < 0$
$\mathcal{L}_{2-} := \mathcal{L}_2(-\infty, 0]$	subspace of $\mathcal{L}_2(-\infty, \infty)$ with functions zero for $t > 0$
$\mathcal{L}_2(j\mathbb{R})$	square integrable functions on $\mathbb{C}_0$ including at ∞
$\mathcal{H}_2$	subspace of $\mathcal{L}_2(j\mathbb{R})$ with functions analytic in Re(s) > 0
$\mathcal{H}_2^\perp$	subspace of $\mathcal{L}_2(j\mathbb{R})$ with functions analytic in Re(s) < 0
$\mathcal{L}_\infty(j\mathbb{R})$	functions bounded on Re(s) = 0 including at ∞
$\mathcal{H}_\infty$	the set of $\mathcal{L}_\infty(j\mathbb{R})$ functions analytic in Re(s) > 0
$\mathcal{H}_\infty^-$	the set of $\mathcal{L}_\infty(j\mathbb{R})$ functions analytic in Re(s) < 0
prefix **B** and **B**^{o}	*closed* and *open* unit ball, e.g. **B**Δ and **B**$^o\Delta$
prefix $\mathcal{R}$	real rational, e.g., $\mathcal{R}\mathcal{H}_\infty$ and $\mathcal{R}\mathcal{H}_2$, etc.
$\mathcal{R}_p(s)$	rational proper transfer matrices
$G^\sim(s)$	shorthand for $G^T(-s)$
$\left[\begin{array}{c\|c} A & B \\ \hline C & D \end{array}\right]$	shorthand for state space realization $C(sI - A)^{-1}B + D$
$\eta(G(s))$	number of right-half plane poles
$\eta_0(G(s))$	number of imaginary axis poles
$wno(G)$	winding number
$\mathcal{F}_\ell(M, Q)$	lower LFT
$\mathcal{F}_u(M, Q)$	upper LFT
$M \star N$	star product

List of Acronyms

ARE	algebraic Riccati equation
FDLTI	finite dimensional linear time invariant
iff	if and only if
lcf	left coprime factorization
LFT	linear fractional transformation
lhp or LHP	left-half plane $\operatorname{Re}(s) < 0$
LQG	linear quadratic Gaussian
LTI	linear time invariant
MIMO	multi-input multioutput
nlcf	normalized left coprime factorization
NP	nominal performance
nrcf	normalized right coprime factorization
NS	nominal stability
rcf	right coprime factorization
rhp or RHP	right-half plane $\operatorname{Re}(s) > 0$
RP	robust performance
RS	robust stability
SISO	single-input single-output
SSV	structured singular value (μ)
SVD	singular value decomposition

Chapter 1

Introduction

This chapter gives a brief description of the problems considered in this book and the key results presented in each chapter.

1.1 What Is This Book About?

This book is about basic robust and $\mathcal{H}_\infty$ control theory. We consider a control system with possibly multiple sources of uncertainties, noises, and disturbances as shown in Figure 1.1.

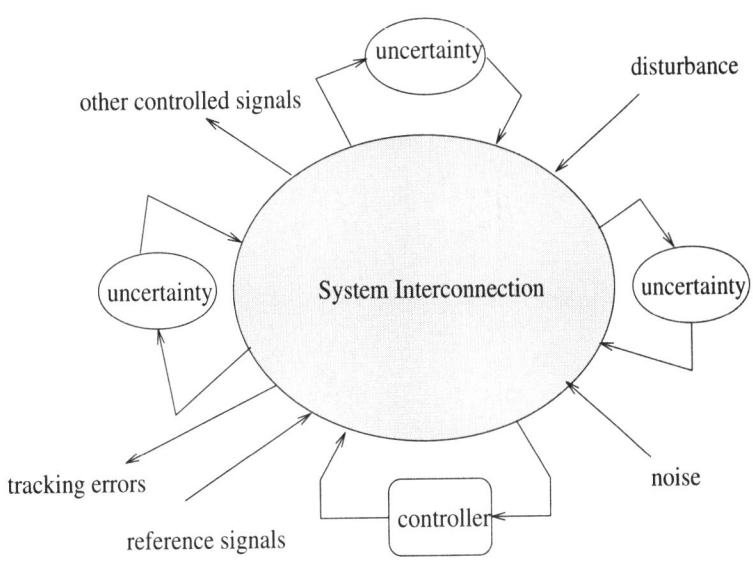

Figure 1.1: General system interconnection

We consider mainly two types of problems:

- Analysis problems: Given a controller, determine if the controlled signals (including tracking errors, control signals, etc.) satisfy the desired properties for all admissible noises, disturbances, and model uncertainties.

- Synthesis problems: Design a controller so that the controlled signals satisfy the desired properties for all admissible noises, disturbances, and model uncertainties.

Most of our analysis and synthesis will be done on a unified linear fractional transformation (LFT) framework. To that end, we shall show that the system shown in Figure 1.1 can be put in the general diagram in Figure 1.2, where P is the interconnection matrix, K is the controller, Δ is the set of all possible uncertainty, w is a vector signal including noises, disturbances, and reference signals, z is a vector signal including all controlled signals and tracking errors, u is the control signal, and y is the measurement.

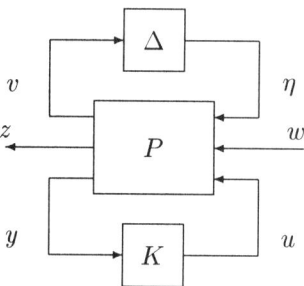

Figure 1.2: General LFT framework

The block diagram in Figure 1.2 represents the following equations:

$$\begin{bmatrix} v \\ z \\ y \end{bmatrix} = P \begin{bmatrix} \eta \\ w \\ u \end{bmatrix}$$
$$\eta = \Delta v$$
$$u = Ky.$$

Let the transfer matrix from w to z be denoted by T_{zw} and assume that the admissible uncertainty Δ satisfies $\bar{\sigma}(\Delta) < 1/\gamma_u$ for some $\gamma_u > 0$. Then our analysis problem is to answer if the closed-loop system is stable for all admissible Δ and $\|T_{zw}\|_\infty \leq \gamma_p$ for some prespecified $\gamma_p > 0$, where $\|T_{zw}\|_\infty$ is the $\mathcal{H}_\infty$ norm defined as $\|T_{zw}\|_\infty = \sup_\omega \bar{\sigma}(T_{zw}(j\omega))$. The synthesis problem is to design a controller K so that the aforementioned robust stability and performance conditions are satisfied.

In the simplest form, we have either $\Delta = 0$ or $w = 0$. The former becomes the well-known $\mathcal{H}_\infty$ control problem and the later becomes the robust stability problem. The two

1.2. Highlights of This Book

problems are equivalent when Δ is a single-block unstructured uncertainty through the application of the small gain theorem (see Chapter 8). This robust stability consequence was probably the main motivation for the development of $\mathcal{H}_\infty$ methods.

The analysis and synthesis for systems with multiple-block Δ can be reduced in most cases to an equivalent $\mathcal{H}_\infty$ problem with suitable scalings. Thus a solution to the $\mathcal{H}_\infty$ control problem is the key to all robustness problems considered in this book. In the next section, we shall give a chapter-by-chapter summary of the main results presented in this book.

We refer readers to the book *Robust and Optimal Control* by K. Zhou, J. C. Doyle, and K. Glover [1996] for a brief historical review of $\mathcal{H}_\infty$ and robust control and for some detailed treatment of some advanced topics.

1.2 Highlights of This Book

The key results in each chapter are highlighted in this section. Readers should consult the corresponding chapters for the exact statements and conditions.

Chapter 2 reviews some basic linear algebra facts.

Chapter 3 reviews system theoretical concepts: controllability, observability, stabilizability, detectability, pole placement, observer theory, system poles and zeros, and state-space realizations.

Chapter 4 introduces the $\mathcal{H}_2$ spaces and the $\mathcal{H}_\infty$ spaces. State-space methods of computing real rational $\mathcal{H}_2$ and $\mathcal{H}_\infty$ transfer matrix norms are presented. For example, let

$$G(s) = \left[\begin{array}{c|c} A & B \\ \hline C & 0 \end{array}\right] \in \mathcal{RH}_\infty.$$

Then

$$\|G\|_2^2 = \operatorname{trace}(B^*QB) = \operatorname{trace}(CPC^*)$$

and

$$\|G\|_\infty = \max\{\gamma : H \text{ has an eigenvalue on the imaginary axis}\},$$

where P and Q are the controllability and observability Gramians and

$$H = \begin{bmatrix} A & BB^*/\gamma^2 \\ -C^*C & -A^* \end{bmatrix}.$$

Chapter 5 introduces the feedback structure and discusses its stability.

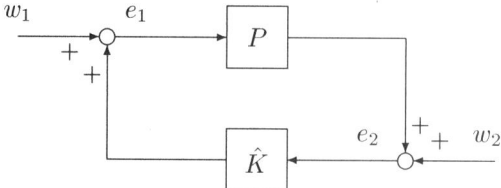

We define that the above closed-loop system is internally stable if and only if

$$\begin{bmatrix} I & -\hat{K} \\ -P & I \end{bmatrix}^{-1} = \begin{bmatrix} (I - \hat{K}P)^{-1} & \hat{K}(I - P\hat{K})^{-1} \\ P(I - \hat{K}P)^{-1} & (I - P\hat{K})^{-1} \end{bmatrix} \in \mathcal{RH}_\infty.$$

Alternative characterizations of internal stability using coprime factorizations are also presented.

Chapter 6 considers the feedback system properties and design limitations. The formulations of optimal $\mathcal{H}_2$ and $\mathcal{H}_\infty$ control problems and the selection of weighting functions are also considered in this chapter.

Chapter 7 considers the problem of reducing the order of a linear multivariable dynamical system using the balanced truncation method. Suppose

$$G(s) = \left[\begin{array}{cc|c} A_{11} & A_{12} & B_1 \\ A_{21} & A_{22} & B_2 \\ \hline C_1 & C_2 & D \end{array} \right] \in \mathcal{RH}_\infty$$

is a balanced realization with controllability and observability Gramians $P = Q = \Sigma = \text{diag}(\Sigma_1, \Sigma_2)$

$$\begin{aligned} \Sigma_1 &= \text{diag}(\sigma_1 I_{s_1}, \sigma_2 I_{s_2}, \ldots, \sigma_r I_{s_r}) \\ \Sigma_2 &= \text{diag}(\sigma_{r+1} I_{s_{r+1}}, \sigma_{r+2} I_{s_{r+2}}, \ldots, \sigma_N I_{s_N}). \end{aligned}$$

Then the truncated system $G_r(s) = \left[\begin{array}{c|c} A_{11} & B_1 \\ \hline C_1 & D \end{array} \right]$ is stable and satisfies an additive error bound:

$$\|G(s) - G_r(s)\|_\infty \leq 2 \sum_{i=r+1}^{N} \sigma_i.$$

Frequency-weighted balanced truncation method is also discussed.

Chapter 8 derives robust stability tests for systems under various modeling assumptions through the use of the small gain theorem. In particular, we show that a system, shown at the top of the following page, with an unstructured uncertainty $\Delta \in \mathcal{RH}_\infty$

1.2. Highlights of This Book

with $\|\Delta\|_\infty < 1$ is robustly stable if and only if $\|T_{zw}\|_\infty \leq 1$, where T_{zw} is the matrix transfer function from w to z.

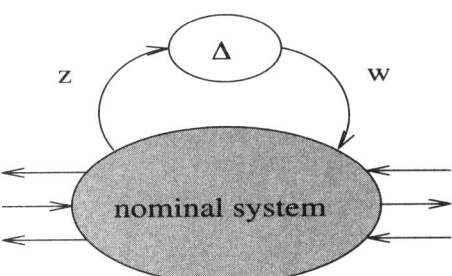

Chapter 9 introduces the LFT in detail. We show that many control problems can be formulated and treated in the LFT framework. In particular, we show that every analysis problem can be put in an LFT form with some structured $\Delta(s)$ and some interconnection matrix $M(s)$ and every synthesis problem can be put in an LFT form with a generalized plant $G(s)$ and a controller $K(s)$ to be designed.

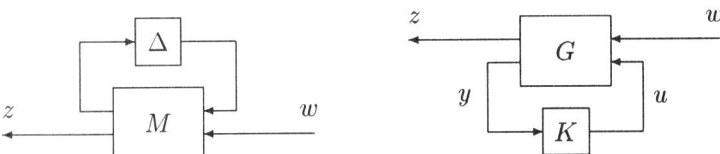

Chapter 10 considers robust stability and performance for systems with multiple sources of uncertainties. We show that an uncertain system is robustly stable and satisfies some $\mathcal{H}_\infty$ performance criterion for all $\Delta_i \in \mathcal{RH}_\infty$ with $\|\Delta_i\|_\infty < 1$ if and only if the structured singular value (μ) of the corresponding interconnection model is no greater than 1.

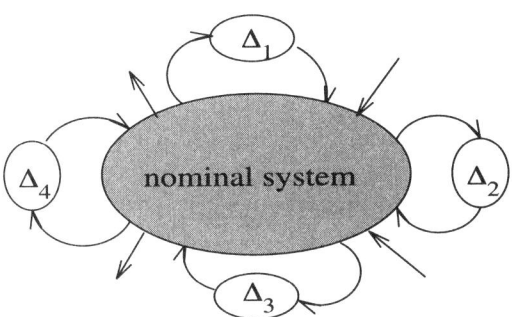

Chapter 11 characterizes in state-space all controllers that stabilize a given dynamical system $G(s)$. For a given generalized plant

$$G(s) = \begin{bmatrix} G_{11}(s) & G_{12}(s) \\ G_{21}(s) & G_{22}(s) \end{bmatrix} = \left[\begin{array}{c|cc} A & B_1 & B_2 \\ \hline C_1 & D_{11} & D_{12} \\ C_2 & D_{21} & D_{22} \end{array}\right]$$

we show that all stabilizing controllers can be parameterized as the transfer matrix from y to u below where F and L are such that $A + LC_2$ and $A + B_2F$ are stable.

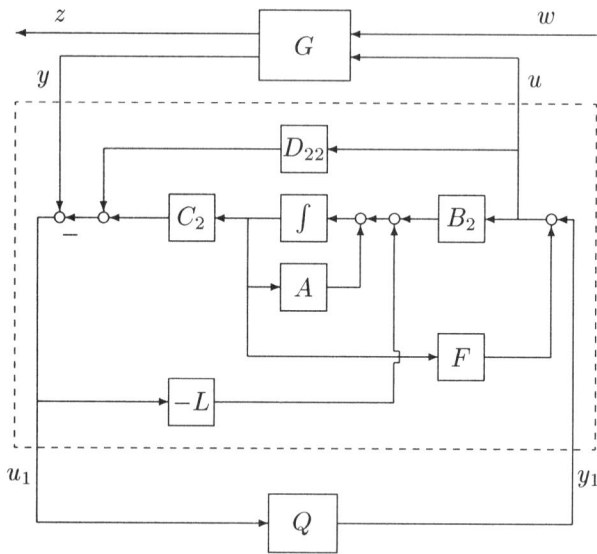

Chapter 12 studies the stabilizing solution to an *algebraic Riccati equation* (ARE). A solution to the following ARE

$$A^*X + XA + XRX + Q = 0$$

is said to be a stabilizing solution if $A + RX$ is stable. Now let

$$H := \begin{bmatrix} A & R \\ -Q & -A^* \end{bmatrix}$$

and let $\mathcal{X}_-(H)$ be the stable H invariant subspace and

$$\mathcal{X}_-(H) = \operatorname{Im} \begin{bmatrix} X_1 \\ X_2 \end{bmatrix},$$

where $X_1, X_2 \in \mathbb{C}^{n \times n}$. If X_1 is nonsingular, then $X := X_2 X_1^{-1}$ is uniquely determined by H, denoted by $X = \operatorname{Ric}(H)$. A key result of this chapter is the so-called bounded

1.2. Highlights of This Book

real lemma, which states that a stable transfer matrix $G(s)$ satisfies $\|G(s)\|_\infty < \gamma$ if and only if there exists an X such that $A + BB^*X/\gamma^2$ is stable and

$$XA + A^*X + XBB^*X/\gamma^2 + C^*C = 0.$$

The $\mathcal{H}_\infty$ control theory in Chapter 14 will be derived based on this lemma.

Chapter 13 treats the optimal control of linear time-invariant systems with quadratic performance criteria (i.e., $\mathcal{H}_2$ problems). We consider a dynamical system described by an LFT with

$$G(s) = \left[\begin{array}{c|cc} A & B_1 & B_2 \\ \hline C_1 & 0 & D_{12} \\ C_2 & D_{21} & 0 \end{array}\right].$$

Define

$$R_1 = D_{12}^*D_{12} > 0, \quad R_2 = D_{21}D_{21}^* > 0$$

$$H_2 := \begin{bmatrix} A - B_2 R_1^{-1} D_{12}^* C_1 & -B_2 R_1^{-1} B_2^* \\ -C_1^*(I - D_{12}R_1^{-1}D_{12}^*)C_1 & -(A - B_2 R_1^{-1} D_{12}^* C_1)^* \end{bmatrix}$$

$$J_2 := \begin{bmatrix} (A - B_1 D_{21}^* R_2^{-1} C_2)^* & -C_2^* R_2^{-1} C_2 \\ -B_1(I - D_{21}^* R_2^{-1} D_{21})B_1^* & -(A - B_1 D_{21}^* R_2^{-1} C_2) \end{bmatrix}$$

$$X_2 := \text{Ric}(H_2) \geq 0, \quad Y_2 := \text{Ric}(J_2) \geq 0$$

$$F_2 := -R_1^{-1}(B_2^* X_2 + D_{12}^* C_1), \quad L_2 := -(Y_2 C_2^* + B_1 D_{21}^*)R_2^{-1}.$$

Then the $\mathcal{H}_2$ optimal controller (i.e., the controller that minimizes $\|T_{zw}\|_2$) is given by

$$K_{\text{opt}}(s) := \left[\begin{array}{c|c} A + B_2 F_2 + L_2 C_2 & -L_2 \\ \hline F_2 & 0 \end{array}\right].$$

Chapter 14 first considers an $\mathcal{H}_\infty$ control problem with the generalized plant $G(s)$ as given in Chapter 13 but with some additional simplifications: $R_1 = I$, $R_2 = I$, $D_{12}^* C_1 = 0$, and $B_1 D_{21}^* = 0$. We show that there exists an admissible controller such that $\|T_{zw}\|_\infty < \gamma$ if and only if the following three conditions hold:

(i) $H_\infty \in \text{dom}(\text{Ric})$ and $X_\infty := \text{Ric}(H_\infty) \geq 0$, where

$$H_\infty := \begin{bmatrix} A & \gamma^{-2} B_1 B_1^* - B_2 B_2^* \\ -C_1^* C_1 & -A^* \end{bmatrix};$$

(ii) $J_\infty \in \mathrm{dom}(\mathrm{Ric})$ and $Y_\infty := \mathrm{Ric}(J_\infty) > 0$, where

$$J_\infty := \begin{bmatrix} A^* & \gamma^{-2}C_1^*C_1 - C_2^*C_2 \\ -B_1B_1^* & -A \end{bmatrix};$$

(iii) $\rho(X_\infty Y_\infty) < \gamma^2$.

Moreover, an admissible controller such that $\|T_{zw}\|_\infty < \gamma$ is given by

$$K_{\mathrm{sub}} = \left[\begin{array}{c|c} \hat{A}_\infty & -Z_\infty L_\infty \\ \hline F_\infty & 0 \end{array} \right]$$

where

$$\hat{A}_\infty := A + \gamma^{-2}B_1B_1^*X_\infty + B_2F_\infty + Z_\infty L_\infty C_2$$
$$F_\infty := -B_2^*X_\infty, \quad L_\infty := -Y_\infty C_2^*, \quad Z_\infty := (I - \gamma^{-2}Y_\infty X_\infty)^{-1}.$$

We then consider further the general $\mathcal{H}_\infty$ control problem. We indicate how various assumptions can be relaxed to accommodate other more complicated problems, such as singular control problems. We also consider the integral control in the $\mathcal{H}_2$ and $\mathcal{H}_\infty$ theory and show how the general $\mathcal{H}_\infty$ solution can be used to solve the $\mathcal{H}_\infty$ filtering problem.

Chapter 15 considers the design of reduced-order controllers by means of controller reduction. Special attention is paid to the controller reduction methods that preserve the closed-loop stability and performance. Methods are presented that give sufficient conditions in terms of frequency-weighted model reduction.

Chapter 16 first solves a special $\mathcal{H}_\infty$ minimization problem. Let $P = \tilde{M}^{-1}\tilde{N}$ be a normalized left coprime factorization. Then we show that

$$\inf_{K \text{ stabilizing}} \left\| \begin{bmatrix} K \\ I \end{bmatrix} (I + PK)^{-1} \begin{bmatrix} I & P \end{bmatrix} \right\|_\infty$$

$$= \inf_{K \text{ stabilizing}} \left\| \begin{bmatrix} K \\ I \end{bmatrix} (I + PK)^{-1} \tilde{M}^{-1} \right\|_\infty = \left(\sqrt{1 - \left\| \begin{bmatrix} \tilde{N} & \tilde{M} \end{bmatrix} \right\|_H^2} \right)^{-1}.$$

This implies that there is a robustly stabilizing controller for

$$P_\Delta = (\tilde{M} + \tilde{\Delta}_M)^{-1}(\tilde{N} + \tilde{\Delta}_N)$$

with

$$\left\| \begin{bmatrix} \tilde{\Delta}_N & \tilde{\Delta}_M \end{bmatrix} \right\|_\infty < \epsilon$$

if and only if

$$\epsilon \leq \sqrt{1 - \left\| \begin{bmatrix} \tilde{N} & \tilde{M} \end{bmatrix} \right\|_H^2}.$$

Using this stabilization result, a loop-shaping design technique is proposed. The proposed technique uses only the basic concept of loop-shaping methods, and then a robust stabilization controller for the normalized coprime factor perturbed system is used to construct the final controller.

Chapter 17 introduces the gap metric and the ν-gap metric. The frequency domain interpretation and applications of the ν-gap metric are discussed. The controller order reduction in the gap or ν-gap metric framework is also considered.

Chapter 18 considers briefly the problems of model validation and the mixed real and complex μ analysis and synthesis.

Most computations and examples in this book are done using MATLAB. Since we shall use MATLAB as a major computational tool, it is assumed that readers have some basic working knowledge of the MATLAB operations (for example, how to input vectors and matrices). We have also included in this book some brief explanations of MATLAB, SIMULINK®, Control System Toolbox, and μ Analysis and Synthesis Toolbox[1] commands. In particular, this book is written consistently with the μ Analysis and Synthesis Toolbox. (Robust Control Toolbox, LMI Control Toolbox, and other software packages may equally be used with this book.) Thus it is helpful for readers to have access to this toolbox. It is suggested at this point to try the following demo programs from this toolbox.

≫ **msdemo1**

≫ **msdemo2**

We shall introduce many more MATLAB commands in the subsequent chapters.

1.3 Notes and References

The original formulation of the $\mathcal{H}_\infty$ control problem can be found in Zames [1981]. Relations between $\mathcal{H}_\infty$ have now been established with many other topics in control: for example, risk-sensitive control of Whittle [1990]; differential games (see Başar and Bernhard [1991], Limebeer, Anderson, Khargonekar, and Green [1992]; Green and Limebeer [1995]); chain-scattering representation, and J-lossless factorization (Green [1992] and Kimura [1997]). See also Zhou, Doyle, and Glover [1996] for additional discussions and references. The state-space theory of $\mathcal{H}_\infty$ has also been carried much further, by generalizing time invariant to time varying, infinite horizon to finite horizon, and finite dimensional to infinite dimensional, and even to some nonlinear settings.

---

[1]SIMULINK is a registered trademark of The MathWorks, Inc.; μ-Analysis and Synthesis is a trademark of The MathWorks, Inc. and MUSYN Inc.; Control System Toolbox, Robust Control Toolbox, and LMI Control Toolbox are trademarks of The MathWorks, Inc.

1.4 Problems

Problem 1.1 We shall solve an easy problem first. When you read a paper or a book, you often come across a statement like this "It is easy ...". What the author really meant was one of the following: (a) it is really easy; (b) it seems to be easy; (c) it is easy for an expert; (d) the author does not know how to show it but he or she thinks it is correct. Now prove that when I say "It is easy" in this book, I mean it is really easy. (Hint: If you can prove it after you read the whole book, ask your boss for a promotion. If you cannot prove it after you read the whole book, trash the book and write a book yourself. Remember use something like "it is easy ..." if you are not sure what you are talking about.)

Chapter 2

Linear Algebra

Some basic linear algebra facts will be reviewed in this chapter. The detailed treatment of this topic can be found in the references listed at the end of the chapter. Hence we shall omit most proofs and provide proofs only for those results that either cannot be easily found in the standard linear algebra textbooks or are insightful to the understanding of some related problems.

2.1 Linear Subspaces

Let $\mathbb{R}$ denote the real scalar field and $\mathbb{C}$ the complex scalar field. For the interest of this chapter, let $\mathbb{F}$ be either $\mathbb{R}$ or $\mathbb{C}$ and let $\mathbb{F}^n$ be the vector space over $\mathbb{F}$ (i.e., $\mathbb{F}^n$ is either $\mathbb{R}^n$ or $\mathbb{C}^n$). Now let $x_1, x_2, \ldots, x_k \in \mathbb{F}^n$. Then an element of the form $\alpha_1 x_1 + \ldots + \alpha_k x_k$ with $\alpha_i \in \mathbb{F}$ is a *linear combination* over $\mathbb{F}$ of $x_1, \ldots, x_k$. The set of all linear combinations of $x_1, x_2, \ldots, x_k \in \mathbb{F}^n$ is a subspace called the *span* of $x_1, x_2, \ldots, x_k$, denoted by

$$\mathrm{span}\{x_1, x_2, \ldots, x_k\} := \{x = \alpha_1 x_1 + \ldots + \alpha_k x_k : \alpha_i \in \mathbb{F}\}.$$

A set of vectors $x_1, x_2, \ldots, x_k \in \mathbb{F}^n$ is said to be *linearly dependent* over $\mathbb{F}$ if there exists $\alpha_1, \ldots, \alpha_k \in \mathbb{F}$ not all zero such that $\alpha_1 x_2 + \ldots + \alpha_k x_k = 0$; otherwise the vectors are said to be *linearly independent*.

Let S be a subspace of $\mathbb{F}^n$, then a set of vectors $\{x_1, x_2, \ldots, x_k\} \in S$ is called a *basis* for S if $x_1, x_2, \ldots, x_k$ are linearly independent and $S = \mathrm{span}\{x_1, x_2, \ldots, x_k\}$. However, such a basis for a subspace S is not unique but all bases for S have the same number of elements. This number is called the *dimension* of S, denoted by $\dim(S)$.

A set of vectors $\{x_1, x_2, \ldots, x_k\}$ in $\mathbb{F}^n$ is mutually *orthogonal* if $x_i^* x_j = 0$ for all $i \neq j$ and *orthonormal* if $x_i^* x_j = \delta_{ij}$, where the superscript $*$ denotes complex conjugate transpose and δ_{ij} is the Kronecker delta function with $\delta_{ij} = 1$ for $i = j$ and $\delta_{ij} = 0$ for $i \neq j$. More generally, a collection of subspaces $S_1, S_2, \ldots, S_k$ of $\mathbb{F}^n$ is mutually orthogonal if $x^* y = 0$ whenever $x \in S_i$ and $y \in S_j$ for $i \neq j$.

The *orthogonal complement* of a subspace $S \subset \mathbb{F}^n$ is defined by

$$S^\perp := \{y \in \mathbb{F}^n : y^*x = 0 \text{ for all } x \in S\}.$$

We call a set of vectors $\{u_1, u_2, \ldots, u_k\}$ an *orthonormal basis* for a subspace $S \in \mathbb{F}^n$ if the vectors form a basis of S and are orthonormal. It is always possible to extend such a basis to a full orthonormal basis $\{u_1, u_2, \ldots, u_n\}$ for $\mathbb{F}^n$. Note that in this case

$$S^\perp = \text{span}\{u_{k+1}, \ldots, u_n\},$$

and $\{u_{k+1}, \ldots, u_n\}$ is called an *orthonormal completion* of $\{u_1, u_2, \ldots, u_k\}$.

Let $A \in \mathbb{F}^{m \times n}$ be a linear transformation from $\mathbb{F}^n$ to $\mathbb{F}^m$; that is,

$$A : \mathbb{F}^n \longmapsto \mathbb{F}^m.$$

Then the *kernel* or *null space* of the linear transformation A is defined by

$$\text{Ker} A = N(A) := \{x \in \mathbb{F}^n : Ax = 0\},$$

and the *image* or *range* of A is

$$\text{Im} A = R(A) := \{y \in \mathbb{F}^m : y = Ax, \ x \in \mathbb{F}^n\}.$$

Let a_i, $i = 1, 2, \ldots, n$ denote the columns of a matrix $A \in \mathbb{F}^{m \times n}$; then

$$\text{Im} A = \text{span}\{a_1, a_2, \ldots, a_n\}.$$

A square matrix $U \in F^{n \times n}$ whose columns form an orthonormal basis for $\mathbb{F}^n$ is called a *unitary matrix* (or *orthogonal matrix* if $\mathbb{F} = \mathbb{R}$), and it satisfies $U^*U = I = UU^*$.

Now let $A = [a_{ij}] \in \mathbb{C}^{n \times n}$; then the *trace* of A is defined as

$$\text{trace}(A) := \sum_{i=1}^{n} a_{ii}.$$

Illustrative MATLAB Commands:

≫ **basis_of_KerA = null(A); basis_of_ImA = orth(A); rank_of_A = rank(A);**

2.2 Eigenvalues and Eigenvectors

Let $A \in \mathbb{C}^{n \times n}$; then the *eigenvalues* of A are the n roots of its characteristic polynomial $p(\lambda) = \det(\lambda I - A)$. The maximal modulus of the eigenvalues is called the *spectral radius*, denoted by

$$\rho(A) := \max_{1 \leq i \leq n} |\lambda_i|$$

if λ_i is a root of $p(\lambda)$, where, as usual, $|\cdot|$ denotes the magnitude. The real spectral radius of a matrix A, denoted by $\rho_R(A)$, is the maximum modulus of the real eigenvalues of A; that is, $\rho_R(A) := \max_{\lambda_i \in \mathbb{R}} |\lambda_i|$. A nonzero vector $x \in \mathbb{C}^n$ that satisfies

$$Ax = \lambda x$$

is referred to as a *right eigenvector* of A. Dually, a nonzero vector y is called a *left eigenvector* of A if

$$y^* A = \lambda y^*.$$

In general, eigenvalues need not be real, and neither do their corresponding eigenvectors. However, if A is real and λ is a real eigenvalue of A, then there is a real eigenvector corresponding to λ. In the case that all eigenvalues of a matrix A are real, we will denote $\lambda_{\max}(A)$ for the largest eigenvalue of A and $\lambda_{\min}(A)$ for the smallest eigenvalue. In particular, if A is a Hermitian matrix (i.e., $A = A^*$), then there exist a unitary matrix U and a real diagonal matrix Λ such that $A = U\Lambda U^*$, where the diagonal elements of Λ are the eigenvalues of A and the columns of U are the eigenvectors of A.

Lemma 2.1 *Consider the Sylvester equation*

$$AX + XB = C, \tag{2.1}$$

where $A \in \mathbb{F}^{n \times n}$, $B \in \mathbb{F}^{m \times m}$, and $C \in \mathbb{F}^{n \times m}$ are given matrices. There exists a unique solution $X \in \mathbb{F}^{n \times m}$ if and only if $\lambda_i(A) + \lambda_j(B) \neq 0$, $\forall i = 1, 2, \ldots, n$, and $j = 1, 2, \ldots, m$.

In particular, if $B = A^*$, equation (2.1) is called the Lyapunov equation; and the necessary and sufficient condition for the existence of a unique solution is that $\lambda_i(A) + \bar{\lambda}_j(A) \neq 0$, $\forall i, j = 1, 2, \ldots, n$.

Illustrative MATLAB Commands:

≫ [V, D] = eig(A)   % $AV = VD$

≫ X=lyap(A,B,-C)   % solving Sylvester equation.

2.3 Matrix Inversion Formulas

Let A be a square matrix partitioned as follows:

$$A := \begin{bmatrix} A_{11} & A_{12} \\ A_{21} & A_{22} \end{bmatrix},$$

where A_{11} and A_{22} are also square matrices. Now suppose A_{11} is nonsingular; then A has the following decomposition:

$$\begin{bmatrix} A_{11} & A_{12} \\ A_{21} & A_{22} \end{bmatrix} = \begin{bmatrix} I & 0 \\ A_{21}A_{11}^{-1} & I \end{bmatrix} \begin{bmatrix} A_{11} & 0 \\ 0 & \Delta \end{bmatrix} \begin{bmatrix} I & A_{11}^{-1}A_{12} \\ 0 & I \end{bmatrix}$$

with $\Delta := A_{22} - A_{21}A_{11}^{-1}A_{12}$, and A is nonsingular iff Δ is nonsingular. Dually, if A_{22} is nonsingular, then

$$\begin{bmatrix} A_{11} & A_{12} \\ A_{21} & A_{22} \end{bmatrix} = \begin{bmatrix} I & A_{12}A_{22}^{-1} \\ 0 & I \end{bmatrix} \begin{bmatrix} \hat{\Delta} & 0 \\ 0 & A_{22} \end{bmatrix} \begin{bmatrix} I & 0 \\ A_{22}^{-1}A_{21} & I \end{bmatrix}$$

with $\hat{\Delta} := A_{11} - A_{12}A_{22}^{-1}A_{21}$, and A is nonsingular iff $\hat{\Delta}$ is nonsingular. The matrix Δ ($\hat{\Delta}$) is called the *Schur complement* of A_{11} (A_{22}) in A.

Moreover, if A is nonsingular, then

$$\begin{bmatrix} A_{11} & A_{12} \\ A_{21} & A_{22} \end{bmatrix}^{-1} = \begin{bmatrix} A_{11}^{-1} + A_{11}^{-1}A_{12}\Delta^{-1}A_{21}A_{11}^{-1} & -A_{11}^{-1}A_{12}\Delta^{-1} \\ -\Delta^{-1}A_{21}A_{11}^{-1} & \Delta^{-1} \end{bmatrix}$$

and

$$\begin{bmatrix} A_{11} & A_{12} \\ A_{21} & A_{22} \end{bmatrix}^{-1} = \begin{bmatrix} \hat{\Delta}^{-1} & -\hat{\Delta}^{-1}A_{12}A_{22}^{-1} \\ -A_{22}^{-1}A_{21}\hat{\Delta}^{-1} & A_{22}^{-1} + A_{22}^{-1}A_{21}\hat{\Delta}^{-1}A_{12}A_{22}^{-1} \end{bmatrix}.$$

The preceding matrix inversion formulas are particularly simple if A is block triangular:

$$\begin{bmatrix} A_{11} & 0 \\ A_{21} & A_{22} \end{bmatrix}^{-1} = \begin{bmatrix} A_{11}^{-1} & 0 \\ -A_{22}^{-1}A_{21}A_{11}^{-1} & A_{22}^{-1} \end{bmatrix}$$

$$\begin{bmatrix} A_{11} & A_{12} \\ 0 & A_{22} \end{bmatrix}^{-1} = \begin{bmatrix} A_{11}^{-1} & -A_{11}^{-1}A_{12}A_{22}^{-1} \\ 0 & A_{22}^{-1} \end{bmatrix}.$$

The following identity is also very useful. Suppose A_{11} and A_{22} are both nonsingular matrices; then

$$(A_{11} - A_{12}A_{22}^{-1}A_{21})^{-1} = A_{11}^{-1} + A_{11}^{-1}A_{12}(A_{22} - A_{21}A_{11}^{-1}A_{12})^{-1}A_{21}A_{11}^{-1}.$$

As a consequence of the matrix decomposition formulas mentioned previously, we can calculate the determinant of a matrix by using its submatrices. Suppose A_{11} is nonsingular; then

$$\det A = \det A_{11} \det(A_{22} - A_{21}A_{11}^{-1}A_{12}).$$

On the other hand, if A_{22} is nonsingular, then

$$\det A = \det A_{22} \det(A_{11} - A_{12}A_{22}^{-1}A_{21}).$$

In particular, for any $B \in \mathbb{C}^{m \times n}$ and $C \in \mathbb{C}^{n \times m}$, we have

$$\det \begin{bmatrix} I_m & B \\ -C & I_n \end{bmatrix} = \det(I_n + CB) = \det(I_m + BC)$$

and for $x, y \in \mathbb{C}^n$

$$\det(I_n + xy^*) = 1 + y^*x.$$

Related MATLAB Commands: inv, det

2.4 Invariant Subspaces

Let $A : \mathbb{C}^n \longmapsto \mathbb{C}^n$ be a linear transformation, λ be an eigenvalue of A, and x be a corresponding eigenvector, respectively. Then $Ax = \lambda x$ and $A(\alpha x) = \lambda(\alpha x)$ for any $\alpha \in \mathbb{C}$. Clearly, the eigenvector x defines a one-dimensional subspace that is invariant with respect to premultiplication by A since $A^k x = \lambda^k x, \forall k$. In general, a subspace $S \subset \mathbb{C}^n$ is called *invariant* for the transformation A, or *A-invariant*, if $Ax \in S$ for every $x \in S$. In other words, that S is invariant for A means that the image of S under A is contained in S: $AS \subset S$. For example, $\{0\}$, $\mathbb{C}^n$, Ker$A$, and ImA are all A-invariant subspaces.

As a generalization of the one-dimensional invariant subspace induced by an eigenvector, let $\lambda_1, \ldots, \lambda_k$ be eigenvalues of A (not necessarily distinct), and let x_i be the corresponding eigenvectors and the generalized eigenvectors. Then $S = \text{span}\{x_1, \ldots, x_k\}$ is an A-invariant subspace provided that all the lower-rank generalized eigenvectors are included. More specifically, let $\lambda_1 = \lambda_2 = \cdots = \lambda_l$ be eigenvalues of A, and let $x_1, x_2, \ldots, x_l$ be the corresponding eigenvector and the generalized eigenvectors obtained through the following equations:

$$\begin{aligned}
(A - \lambda_1 I)x_1 &= 0 \\
(A - \lambda_1 I)x_2 &= x_1 \\
&\vdots \\
(A - \lambda_1 I)x_l &= x_{l-1}.
\end{aligned}$$

Then a subspace S with $x_t \in S$ for some $t \leq l$ is an A-invariant subspace only if all lower-rank eigenvectors and generalized eigenvectors of x_t are in S (i.e., $x_i \in S$, $\forall 1 \leq i \leq t$). This will be further illustrated in Example 2.1.

On the other hand, if S is a nontrivial subspace[1] and is A-invariant, then there is $x \in S$ and λ such that $Ax = \lambda x$.

An A-invariant subspace $S \subset \mathbb{C}^n$ is called a *stable invariant subspace* if all the eigenvalues of A constrained to S have negative real parts. Stable invariant subspaces will play an important role in computing the stabilizing solutions to the algebraic Riccati equations in Chapter 12.

Example 2.1 Suppose a matrix A has the following Jordan canonical form:

$$A \begin{bmatrix} x_1 & x_2 & x_3 & x_4 \end{bmatrix} = \begin{bmatrix} x_1 & x_2 & x_3 & x_4 \end{bmatrix} \begin{bmatrix} \lambda_1 & 1 & & \\ & \lambda_1 & & \\ & & \lambda_3 & \\ & & & \lambda_4 \end{bmatrix}$$

---
[1] We will say subspace S is trivial if $S = \{0\}$.

with $\text{Re}\lambda_1 < 0$, $\lambda_3 < 0$, and $\lambda_4 > 0$. Then it is easy to verify that

$$
\begin{array}{llllll}
S_1 & = \text{span}\{x_1\} & S_{12} & = \text{span}\{x_1, x_2\} & S_{123} & = \text{span}\{x_1, x_2, x_3\} \\
S_3 & = \text{span}\{x_3\} & S_{13} & = \text{span}\{x_1, x_3\} & S_{124} & = \text{span}\{x_1, x_2, x_4\} \\
S_4 & = \text{span}\{x_4\} & S_{14} & = \text{span}\{x_1, x_4\} & S_{34} & = \text{span}\{x_3, x_4\}
\end{array}
$$

are all A-invariant subspaces. Moreover, S_1, S_3, S_{12}, S_{13}, and S_{123} are stable A-invariant subspaces. The subspaces $S_2 = \text{span}\{x_2\}$, $S_{23} = \text{span}\{x_2, x_3\}$, $S_{24} = \text{span}\{x_2, x_4\}$, and $S_{234} = \text{span}\{x_2, x_3, x_4\}$ are, however, not A-invariant subspaces since the lower-rank eigenvector x_1 is not in these subspaces. To illustrate, consider the subspace S_{23}. Then by definition, $Ax_2 \in S_{23}$ if it is an A-invariant subspace. Since

$$Ax_2 = \lambda x_2 + x_1,$$

$Ax_2 \in S_{23}$ would require that x_1 be a linear combination of x_2 and x_3, but this is impossible since x_1 is independent of x_2 and x_3.

2.5 Vector Norms and Matrix Norms

In this section, we shall define vector and matrix norms. Let X be a vector space. A real-valued function $\|\cdot\|$ defined on X is said to be a *norm* on X if it satisfies the following properties:

(i) $\|x\| \geq 0$ (positivity);

(ii) $\|x\| = 0$ if and only if $x = 0$ (positive definiteness);

(iii) $\|\alpha x\| = |\alpha| \|x\|$, for any scalar α (homogeneity);

(iv) $\|x + y\| \leq \|x\| + \|y\|$ (triangle inequality)

for any $x \in X$ and $y \in X$.

Let $x \in \mathbb{C}^n$. Then we define the vector p-norm of x as

$$\|x\|_p := \left(\sum_{i=1}^n |x_i|^p\right)^{1/p}, \text{ for } 1 \leq p \leq \infty.$$

In particular, when $p = 1, 2, \infty$ we have

$$\|x\|_1 := \sum_{i=1}^n |x_i|;$$

$$\|x\|_2 := \sqrt{\sum_{i=1}^n |x_i|^2};$$

2.5. Vector Norms and Matrix Norms

$$\|x\|_\infty := \max_{1\leq i\leq n} |x_i|.$$

Clearly, norm is an abstraction and extension of our usual concept of length in three-dimensional Euclidean space. So a norm of a vector is a measure of the vector "length" (for example, $\|x\|_2$ is the Euclidean distance of the vector x from the origin). Similarly, we can introduce some kind of measure for a matrix.

Let $A = [a_{ij}] \in \mathbb{C}^{m\times n}$; then the matrix norm *induced* by a vector p-norm is defined as

$$\|A\|_p := \sup_{x\neq 0} \frac{\|Ax\|_p}{\|x\|_p}.$$

The matrix norms induced by vector p-norms are sometimes called *induced p-norms*. This is because $\|A\|_p$ is defined by or induced from a vector p-norm. In fact, A can be viewed as a mapping from a vector space $\mathbb{C}^n$ equipped with a vector norm $\|\cdot\|_p$ to another vector space $\mathbb{C}^m$ equipped with a vector norm $\|\cdot\|_p$. So from a system theoretical point of view, the induced norms have the interpretation of input/output amplification gains.

In particular, the induced matrix 2-norm can be computed as

$$\|A\|_2 = \sqrt{\lambda_{\max}(A^*A)}.$$

We shall adopt the following convention throughout this book for the vector and matrix norms unless specified otherwise: Let $x \in \mathbb{C}^n$ and $A \in \mathbb{C}^{m\times n}$; then we shall denote the Euclidean 2-norm of x simply by

$$\|x\| := \|x\|_2$$

and the induced 2-norm of A by

$$\|A\| := \|A\|_2.$$

The Euclidean 2-norm has some very nice properties:

Lemma 2.2 *Let $x \in \mathbb{F}^n$ and $y \in \mathbb{F}^m$.*

1. *Suppose $n \geq m$. Then $\|x\| = \|y\|$ iff there is a matrix $U \in \mathbb{F}^{n\times m}$ such that $x = Uy$ and $U^*U = I$.*

2. *Suppose $n = m$. Then $|x^*y| \leq \|x\|\|y\|$. Moreover, the equality holds iff $x = \alpha y$ for some $\alpha \in \mathbb{F}$ or $y = 0$.*

3. *$\|x\| \leq \|y\|$ iff there is a matrix $\Delta \in \mathbb{F}^{n\times m}$ with $\|\Delta\| \leq 1$ such that $x = \Delta y$. Furthermore, $\|x\| < \|y\|$ iff $\|\Delta\| < 1$.*

4. *$\|Ux\| = \|x\|$ for any appropriately dimensioned unitary matrices U.*

Another often used matrix norm is the so called *Frobenius norm*. It is defined as

$$\|A\|_F := \sqrt{\operatorname{trace}(A^*A)} = \sqrt{\sum_{i=1}^{m}\sum_{j=1}^{n}|a_{ij}|^2}\ .$$

However, the Frobenius norm is not an induced norm.

The following properties of matrix norms are easy to show:

Lemma 2.3 *Let A and B be any matrices with appropriate dimensions. Then*

1. $\rho(A) \leq \|A\|$ *(this is also true for the F-norm and any induced matrix norm).*

2. $\|AB\| \leq \|A\|\|B\|$. *In particular, this gives* $\|A^{-1}\| \geq \|A\|^{-1}$ *if A is invertible. (This is also true for any induced matrix norm.)*

3. $\|UAV\| = \|A\|$, *and* $\|UAV\|_F = \|A\|_F$, *for any appropriately dimensioned unitary matrices U and V.*

4. $\|AB\|_F \leq \|A\|\|B\|_F$ *and* $\|AB\|_F \leq \|B\|\|A\|_F$.

Note that although premultiplication or postmultiplication of a unitary matrix on a matrix does not change its induced 2-norm and F-norm, it does change its eigenvalues. For example, let

$$A = \begin{bmatrix} 1 & 0 \\ 1 & 0 \end{bmatrix}.$$

Then $\lambda_1(A) = 1, \lambda_2(A) = 0$. Now let

$$U = \begin{bmatrix} \frac{1}{\sqrt{2}} & \frac{1}{\sqrt{2}} \\ -\frac{1}{\sqrt{2}} & \frac{1}{\sqrt{2}} \end{bmatrix};$$

then U is a unitary matrix and

$$UA = \begin{bmatrix} \sqrt{2} & 0 \\ 0 & 0 \end{bmatrix}$$

with $\lambda_1(UA) = \sqrt{2}$, $\lambda_2(UA) = 0$. This property is useful in some matrix perturbation problems, particularly in the computation of bounds for structured singular values, which will be studied in Chapter 9.

Related MATLAB Commands: norm, normest

2.6 Singular Value Decomposition

A very useful tool in matrix analysis is *singular value decomposition (SVD)*. It will be seen that singular values of a matrix are good measures of the "size" of the matrix and that the corresponding singular vectors are good indications of strong/weak input or output directions.

Theorem 2.4 *Let $A \in \mathbb{F}^{m \times n}$. There exist unitary matrices*

$$U = [u_1, u_2, \ldots, u_m] \in \mathbb{F}^{m \times m}$$
$$V = [v_1, v_2, \ldots, v_n] \in \mathbb{F}^{n \times n}$$

such that

$$A = U\Sigma V^*, \quad \Sigma = \begin{bmatrix} \Sigma_1 & 0 \\ 0 & 0 \end{bmatrix},$$

where

$$\Sigma_1 = \begin{bmatrix} \sigma_1 & 0 & \cdots & 0 \\ 0 & \sigma_2 & \cdots & 0 \\ \vdots & \vdots & \ddots & \vdots \\ 0 & 0 & \cdots & \sigma_p \end{bmatrix}$$

and

$$\sigma_1 \geq \sigma_2 \geq \cdots \geq \sigma_p \geq 0, \quad p = \min\{m, n\}.$$

Proof. Let $\sigma = \|A\|$ and without loss of generality assume $m \geq n$. Then, from the definition of $\|A\|$, there exists a $z \in \mathbb{F}^n$ such that

$$\|Az\| = \sigma \|z\|.$$

By Lemma 2.2, there is a matrix $\tilde{U} \in F^{m \times n}$ such that $\tilde{U}^* \tilde{U} = I$ and

$$Az = \sigma \tilde{U} z.$$

Now let

$$x = \frac{z}{\|z\|} \in \mathbb{F}^n, \quad y = \frac{\tilde{U}z}{\|\tilde{U}z\|} \in \mathbb{F}^m.$$

We have $Ax = \sigma y$. Let

$$V = \begin{bmatrix} x & V_1 \end{bmatrix} \in \mathbb{F}^{n \times n}$$

and

$$U = \begin{bmatrix} y & U_1 \end{bmatrix} \in \mathbb{F}^{m \times m}$$

be unitary.[2] Consequently, $U^* AV$ has the following structure:

$$A_1 := U^*AV = \begin{bmatrix} y^*Ax & y^*AV_1 \\ U_1^*Ax & U_1^*AV_1 \end{bmatrix} = \begin{bmatrix} \sigma y^*y & y^*AV_1 \\ \sigma U_1^*y & U_1^*AV_1 \end{bmatrix} = \begin{bmatrix} \sigma & w^* \\ 0 & B \end{bmatrix},$$

---

[2] Recall that it is always possible to extend an orthonormal set of vectors to an orthonormal basis for the whole space.

where $w := V_1^* A^* y \in \mathbb{F}^{n-1}$ and $B := U_1^* A V_1 \in \mathbb{F}^{(m-1)\times(n-1)}$.
Since
$$\left\| A_1^* \begin{bmatrix} 1 \\ 0 \\ \vdots \\ 0 \end{bmatrix} \right\|_2^2 = (\sigma^2 + w^* w),$$
it follows that $\|A_1\|^2 \geq \sigma^2 + w^* w$. But since $\sigma = \|A\| = \|A_1\|$, we must have $w = 0$. An obvious induction argument gives
$$U^* A V = \Sigma.$$
This completes the proof. □

The σ_i is the ith *singular value* of A, and the vectors u_i and v_j are, respectively, the ith *left singular vector* and the jth *right singular vector*. It is easy to verify that
$$\begin{aligned} A v_i &= \sigma_i u_i \\ A^* u_i &= \sigma_i v_i. \end{aligned}$$
The preceding equations can also be written as
$$\begin{aligned} A^* A v_i &= \sigma_i^2 v_i \\ A A^* u_i &= \sigma_i^2 u_i. \end{aligned}$$
Hence σ_i^2 is an eigenvalue of AA^* or A^*A, u_i is an eigenvector of AA^*, and v_i is an eigenvector of A^*A.

The following notations for singular values are often adopted:
$$\overline{\sigma}(A) = \sigma_{\max}(A) = \sigma_1 = \text{the largest singular value of } A;$$
and
$$\underline{\sigma}(A) = \sigma_{\min}(A) = \sigma_p = \text{the smallest singular value of } A.$$
Geometrically, the singular values of a matrix A are precisely the lengths of the semi-axes of the hyperellipsoid E defined by
$$E = \{y : y = Ax, \; x \in \mathbb{C}^n, \; \|x\| = 1\}.$$
Thus v_1 is the direction in which $\|y\|$ is largest for all $\|x\| = 1$; while v_n is the direction in which $\|y\|$ is smallest for all $\|x\| = 1$. From the input/output point of view, v_1 (v_n) is the *highest (lowest) gain input (or control) direction*, while u_1 (u_m) is the *highest (lowest) gain output (or observing) direction*. This can be illustrated by the following 2×2 matrix:
$$A = \begin{bmatrix} \cos\theta_1 & -\sin\theta_1 \\ \sin\theta_1 & \cos\theta_1 \end{bmatrix} \begin{bmatrix} \sigma_1 & \\ & \sigma_2 \end{bmatrix} \begin{bmatrix} \cos\theta_2 & -\sin\theta_2 \\ \sin\theta_2 & \cos\theta_2 \end{bmatrix}.$$

2.6. Singular Value Decomposition

It is easy to see that A maps a unit circle to an ellipsoid with semiaxes of σ_1 and σ_2.

Hence it is often convenient to introduce the following alternative definitions for the largest singular value $\bar{\sigma}$:
$$\bar{\sigma}(A) := \max_{\|x\|=1} \|Ax\|$$
and for the smallest singular value $\underline{\sigma}$ of a *tall matrix*:
$$\underline{\sigma}(A) := \min_{\|x\|=1} \|Ax\|.$$

Lemma 2.5 *Suppose A and Δ are square matrices. Then*

(i) $|\underline{\sigma}(A + \Delta) - \underline{\sigma}(A)| \leq \bar{\sigma}(\Delta)$;

(ii) $\underline{\sigma}(A\Delta) \geq \underline{\sigma}(A)\underline{\sigma}(\Delta)$;

(iii) $\bar{\sigma}(A^{-1}) = \dfrac{1}{\underline{\sigma}(A)}$ *if A is invertible.*

Proof.

(i) By definition
$$\begin{aligned}\underline{\sigma}(A + \Delta) &:= \min_{\|x\|=1} \|(A + \Delta)x\| \geq \min_{\|x\|=1} \{\|Ax\| - \|\Delta x\|\} \\ &\geq \min_{\|x\|=1} \|Ax\| - \max_{\|x\|=1} \|\Delta x\| \\ &= \underline{\sigma}(A) - \bar{\sigma}(\Delta).\end{aligned}$$

Hence $-\bar{\sigma}(\Delta) \leq \underline{\sigma}(A + \Delta) - \underline{\sigma}(A)$. The other inequality $\underline{\sigma}(A + \Delta) - \underline{\sigma}(A) \leq \bar{\sigma}(\Delta)$ follows by replacing A by $A + \Delta$ and Δ by $-\Delta$ in the preceding proof.

(ii) This follows by noting that
$$\begin{aligned}\underline{\sigma}(A\Delta) &:= \min_{\|x\|=1} \|A\Delta x\| \\ &= \sqrt{\min_{\|x\|=1} x^* \Delta^* A^* A \Delta x} \\ &\geq \underline{\sigma}(A) \min_{\|x\|=1} \|\Delta x\| = \underline{\sigma}(A)\underline{\sigma}(\Delta).\end{aligned}$$

(iii) Let the singular value decomposition of A be $A = U\Sigma V^*$; then $A^{-1} = V\Sigma^{-1}U^*$. Hence $\bar{\sigma}(A^{-1}) = \bar{\sigma}(\Sigma^{-1}) = 1/\underline{\sigma}(\Sigma) = 1/\underline{\sigma}(A)$.

□

Note that (*ii*) may not be true if A and Δ are not square matrices. For example, consider $A = \begin{bmatrix} 1 \\ 2 \end{bmatrix}$ and $\Delta = \begin{bmatrix} 3 & 4 \end{bmatrix}$; then $\underline{\sigma}(A\Delta) = 0$ but $\underline{\sigma}(A) = \sqrt{5}$ and $\underline{\sigma}(\Delta) = 5$.

Some useful properties of SVD are collected in the following lemma.

Lemma 2.6 *Let $A \in \mathbb{F}^{m \times n}$ and*
$$\sigma_1 \geq \sigma_2 \geq \cdots \geq \sigma_r > \sigma_{r+1} = \cdots = 0, \ r \leq \min\{m, n\}.$$
Then

1. $\text{rank}(A) = r$;

2. $\text{Ker}A = \text{span}\{v_{r+1}, \ldots, v_n\}$ and $(\text{Ker}A)^\perp = \text{span}\{v_1, \ldots, v_r\}$;

3. $\text{Im}A = \text{span}\{u_1, \ldots, u_r\}$ and $(\text{Im}A)^\perp = \text{span}\{u_{r+1}, \ldots, u_m\}$;

4. $A \in \mathbb{F}^{m \times n}$ has a dyadic expansion:
$$A = \sum_{i=1}^{r} \sigma_i u_i v_i^* = U_r \Sigma_r V_r^*,$$
where $U_r = [u_1, \ldots, u_r]$, $V_r = [v_1, \ldots, v_r]$, and $\Sigma_r = \text{diag}\,(\sigma_1, \ldots, \sigma_r)$;

5. $\|A\|_F^2 = \sigma_1^2 + \sigma_2^2 + \cdots + \sigma_r^2$;

6. $\|A\| = \sigma_1$;

7. $\sigma_i(U_0 A V_0) = \sigma_i(A)$, $i = 1, \ldots, p$ *for any appropriately dimensioned unitary matrices U_0 and V_0;*

8. *Let $k < r = \text{rank}(A)$ and $A_k := \sum_{i=1}^{k} \sigma_i u_i v_i^*$; then*
$$\min_{\text{rank}(B) \leq k} \|A - B\| = \|A - A_k\| = \sigma_{k+1}.$$

Proof. We shall only give a proof for part 8. It is easy to see that $\text{rank}(A_k) \leq k$ and $\|A - A_k\| = \sigma_{k+1}$. Hence, we only need show that $\min_{\text{rank}(B) \leq k} \|A - B\| \geq \sigma_{k+1}$. Let B be any matrix such that $\text{rank}(B) \leq k$. Then

$$\begin{aligned} \|A - B\| &= \|U\Sigma V^* - B\| = \|\Sigma - U^* BV\| \\ &\geq \left\| \begin{bmatrix} I_{k+1} & 0 \end{bmatrix} (\Sigma - U^* BV) \begin{bmatrix} I_{k+1} \\ 0 \end{bmatrix} \right\| = \left\| \Sigma_{k+1} - \hat{B} \right\|, \end{aligned}$$

where $\hat{B} = \begin{bmatrix} I_{k+1} & 0 \end{bmatrix} U^* BV \begin{bmatrix} I_{k+1} \\ 0 \end{bmatrix} \in \mathbb{F}^{(k+1) \times (k+1)}$ and $\text{rank}(\hat{B}) \leq k$. Let $x \in \mathbb{F}^{k+1}$ be such that $\hat{B}x = 0$ and $\|x\| = 1$. Then

$$\|A - B\| \geq \left\| \Sigma_{k+1} - \hat{B} \right\| \geq \left\| (\Sigma_{k+1} - \hat{B})x \right\| = \|\Sigma_{k+1} x\| \geq \sigma_{k+1}.$$

Since B is arbitrary, the conclusion follows. □

2.7. Semidefinite Matrices

Illustrative MATLAB Commands:

≫ $[\mathbf{U}, \mathbf{\Sigma}, \mathbf{V}] = \mathbf{svd}(\mathbf{A})$ % $A = U\Sigma V^*$

Related MATLAB Commands: cond, condest

2.7 Semidefinite Matrices

A square Hermitian matrix $A = A^*$ is said to be *positive definite (semidefinite)*, denoted by $A > 0$ (≥ 0), if $x^* A x > 0$ (≥ 0) for all $x \neq 0$. Suppose $A \in \mathbb{F}^{n \times n}$ and $A = A^* \geq 0$; then there exists a $B \in \mathbb{F}^{n \times r}$ with $r \geq \text{rank}(A)$ such that $A = BB^*$.

Lemma 2.7 *Let $B \in \mathbb{F}^{m \times n}$ and $C \in \mathbb{F}^{k \times n}$. Suppose $m \geq k$ and $B^* B = C^* C$. Then there exists a matrix $U \in \mathbb{F}^{m \times k}$ such that $U^* U = I$ and $B = UC$.*

Proof. Let V_1 and V_2 be unitary matrices such that

$$B = V_1 \begin{bmatrix} B_1 \\ 0 \end{bmatrix}, \quad C = V_2 \begin{bmatrix} C_1 \\ 0 \end{bmatrix},$$

where B_1 and C_1 are full-row rank. Then B_1 and C_1 have the same number of rows and $V_3 := B_1 C_1^* (C_1 C_1^*)^{-1}$ satisfies $V_3^* V_3 = I$ since $B^* B = C^* C$. Hence V_3 is a unitary matrix and $V_3^* B_1 = C_1$. Finally, let

$$U = V_1 \begin{bmatrix} V_3 & 0 \\ 0 & V_4 \end{bmatrix} V_2^*$$

for any suitably dimensioned V_4 such that $V_4^* V_4 = I$.   □

We can define square root for a positive semidefinite matrix A, $A^{1/2} = (A^{1/2})^* \geq 0$, by

$$A = A^{1/2} A^{1/2}.$$

Clearly, $A^{1/2}$ can be computed by using spectral decomposition or SVD: Let $A = U\Lambda U^*$; then

$$A^{1/2} = U\Lambda^{1/2} U^*,$$

where

$$\Lambda = \text{diag}\{\lambda_1, \ldots, \lambda_n\}, \quad \Lambda^{1/2} = \text{diag}\{\sqrt{\lambda_1}, \ldots, \sqrt{\lambda_n}\}.$$

Lemma 2.8 *Suppose $A = A^* > 0$ and $B = B^* \geq 0$. Then $A > B$ iff $\rho(BA^{-1}) < 1$.*

Proof. Since $A > 0$, we have $A > B$ iff
$$0 < I - A^{-1/2}BA^{-1/2} = I - A^{-1/2}(BA^{-1})A^{1/2}.$$
However, $A^{-1/2}BA^{-1/2}$ and BA^{-1} are similar, hence $\lambda_i(BA^{-1}) = \lambda_i(A^{-1/2}BA^{-1/2})$. Therefore, the conclusion follows by the fact that
$$0 < I - A^{-1/2}BA^{-1/2}$$
iff $\rho(A^{-1/2}BA^{-1/2}) < 1$ iff $\rho(BA^{-1}) < 1$. □

2.8 Notes and References

A very extensive treatment of most topics in this chapter can be found in Brogan [1991], Horn and Johnson [1990, 1991] and Lancaster and Tismenetsky [1985]. Golub and Van Loan's book [1983] contains many numerical algorithms for solving most of the problems in this chapter.

2.9 Problems

Problem 2.1 Let
$$A = \begin{pmatrix} 1 & 1 & 0 \\ 1 & 0 & 1 \\ 2 & 1 & 1 \\ 1 & 0 & 1 \\ 2 & 0 & 2 \end{pmatrix}.$$
Determine the row and column rank of A and find bases for $\text{Im}(A)$, $\text{Im}(A^*)$, and $\text{Ker}(A)$.

Problem 2.2 Let $D_0 = \begin{bmatrix} 1 & 4 \\ 2 & 5 \\ 3 & 6 \end{bmatrix}$. Find a D such that $D^*D = I$ and $\text{Im} D = \text{Im} D_0$. Furthermore, find a $D_\perp$ such that $\begin{bmatrix} D & D_\perp \end{bmatrix}$ is a unitary matrix.

Problem 2.3 Let A be a nonsingular matrix and $x, y \in \mathbb{C}^n$. Show
$$(A^{-1} + xy^*)^{-1} = A - \frac{Axy^*A}{1 + y^*Ax}$$
and
$$\det(A^{-1} + xy^*)^{-1} = \frac{\det A}{1 + y^*Ax}.$$

2.9. Problems

Problem 2.4 Let A and B be compatible matrices. Show
$$B(I+AB)^{-1} = (I+BA)^{-1}B, \quad (I+A)^{-1} = I - A(I+A)^{-1}.$$

Problem 2.5 Find a basis for the maximum dimensional stable invariant subspace of
$H = \begin{bmatrix} A & R \\ -Q & -A^* \end{bmatrix}$ with

1. $A = \begin{bmatrix} -1 & 2 \\ 3 & 0 \end{bmatrix}$, $R = \begin{bmatrix} -1 & -1 \\ -1 & -1 \end{bmatrix}$, and $Q = 0$

2. $A = \begin{bmatrix} 0 & 1 \\ 0 & 2 \end{bmatrix}$, $R = \begin{bmatrix} 0 & 0 \\ 0 & -1 \end{bmatrix}$, and $Q = \begin{bmatrix} 1 & 2 \\ 2 & 4 \end{bmatrix}$

3. $A = 0$, $R = \begin{bmatrix} 1 & 2 \\ 2 & 5 \end{bmatrix}$, and $Q = I_2$.

Problem 2.6 Let $A = [a_{ij}]$. Show that $\alpha(A) := \max_{i,j} |a_{ij}|$ defines a matrix norm. Give examples so that $\alpha(A) < \rho(A)$ and $\alpha(AB) > \alpha(A)\alpha(B)$.

Problem 2.7 Let $A = \begin{pmatrix} 1 & 2 & 3 \\ 4 & 1 & -1 \end{pmatrix}$ and $B = \begin{pmatrix} 0 \\ 1 \end{pmatrix}$. (a) Find all x such that $Ax = B$.
(b) Find the minimal norm solution x: $\min\{\|x\| : Ax = B\}$.

Problem 2.8 Let $A = \begin{pmatrix} 1 & 2 \\ -2 & -5 \\ 0 & 1 \end{pmatrix}$ and $B = \begin{pmatrix} 3 \\ 4 \\ 5 \end{pmatrix}$. Find an x such that $\|Ax - B\|$ is minimized.

Problem 2.9 Let $\|A\| < 1$. Show

1. $(I-A)^{-1} = I + A + A^2 + \cdots$.

2. $\|(I-A)^{-1}\| \leq 1 + \|A\| + \|A\|^2 + \cdots = \frac{1}{1-\|A\|}$.

3. $\|(I-A)^{-1}\| \geq \frac{1}{1+\|A\|}$.

Problem 2.10 Let $A \in \mathbb{C}^{m \times n}$. Show that
$$\frac{1}{\sqrt{m}} \|A\|_2 \leq \|A\|_\infty \leq \sqrt{n} \|A\|_2 ;$$
$$\frac{1}{\sqrt{n}} \|A\|_2 \leq \|A\|_1 \leq \sqrt{m} \|A\|_2 ;$$
$$\frac{1}{n} \|A\|_\infty \leq \|A\|_1 \leq m \|A\|_\infty .$$

Problem 2.11 Let $A = xy^*$ and $x, y \in \mathbb{C}^n$. Show that $\|A\|_2 = \|A\|_F = \|x\| \|y\|$.

Problem 2.12 Let $A = \begin{bmatrix} 1 & 1+j \\ 1-j & 2 \end{bmatrix}$. Find $A^{\frac{1}{2}}$ and a $B \in \mathbb{C}^2$ such that $A = BB^*$.

Problem 2.13 Let $P = P^* = \begin{bmatrix} P_{11} & P_{12} \\ P_{12}^* & P_{22} \end{bmatrix} \geq 0$ with $P_{11} \in \mathbb{C}^{k \times k}$. Show $\lambda_i(P) \geq \lambda_i(P_{11})$, $\forall\, 1 \leq i \leq k$.

Problem 2.14 Let $X = X^* \geq 0$ be partitioned as $X = \begin{bmatrix} X_{11} & X_{12} \\ X_{12}^* & X_{22} \end{bmatrix}$. (a) Show $\text{Ker} X_{22} \subset \text{Ker} X_{12}$; (b) let $X_{22} = U_2 \text{diag}(\Lambda_1, 0) U_2^*$ be such that Λ_1 is nonsingular and define $X_{22}^+ := U_2 \text{diag}(\Lambda_1^{-1}, 0) U_2^*$ (the pseudoinverse of X_{22}); then show that $Y = X_{12} X_{22}^+$ solves $Y X_{22} = X_{12}$; and (c) show that

$$\begin{bmatrix} X_{11} & X_{12} \\ X_{12}^* & X_{22} \end{bmatrix} = \begin{bmatrix} I & X_{12} X_{22}^+ \\ 0 & I \end{bmatrix} \begin{bmatrix} X_{11} - X_{12} X_{22}^+ X_{12}^* & 0 \\ 0 & X_{22} \end{bmatrix} \begin{bmatrix} I & 0 \\ X_{22}^+ X_{12}^* & I \end{bmatrix}.$$

Chapter 3

Linear Systems

This chapter reviews some basic system theoretical concepts. The notions of controllability, observability, stabilizability, and detectability are defined and various algebraic and geometric characterizations of these notions are summarized. Observer theory is then introduced. System interconnections and realizations are studied. Finally, the concepts of system poles and zeros are introduced.

3.1 Descriptions of Linear Dynamical Systems

Let a finite dimensional linear time invariant (FDLTI) dynamical system be described by the following linear constant coefficient differential equations:

$$\dot{x} = Ax + Bu, \quad x(t_0) = x_0 \quad (3.1)$$
$$y = Cx + Du, \quad (3.2)$$

where $x(t) \in \mathbb{R}^n$ is called the system *state*, $x(t_0)$ is called the *initial condition* of the system, $u(t) \in \mathbb{R}^m$ is called the system *input*, and $y(t) \in \mathbb{R}^p$ is the system *output*. The $A, B, C,$ and D are appropriately dimensioned real constant matrices. A dynamical system with single-input ($m = 1$) and single-output ($p = 1$) is called a SISO (single-input and single-output) system; otherwise it is called a MIMO (multiple-input and multiple-output) system. The corresponding transfer matrix from u to y is defined as

$$Y(s) = G(s)U(s),$$

where $U(s)$ and $Y(s)$ are the Laplace transforms of $u(t)$ and $y(t)$ with zero initial condition ($x(0) = 0$). Hence, we have

$$G(s) = C(sI - A)^{-1}B + D.$$

Note that the system equations (3.1) and (3.2) can be written in a more compact matrix form:

$$\begin{bmatrix} \dot{x} \\ y \end{bmatrix} = \begin{bmatrix} A & B \\ C & D \end{bmatrix} \begin{bmatrix} x \\ u \end{bmatrix}.$$

To expedite calculations involving transfer matrices, we shall use the following notation:

$$\left[\begin{array}{c|c} A & B \\ \hline C & D \end{array}\right] := C(sI - A)^{-1}B + D.$$

In MATLAB the system can also be written in the packed form using the command

≫ **G=pck(A, B, C, D)** % pack the realization in partitioned form

≫ **seesys(G)** % display G in partitioned format

≫ **[A, B, C, D]=unpck(G)** % unpack the system matrix

Note that
$$\left[\begin{array}{cc} A & B \\ C & D \end{array}\right]$$
is a real block matrix, not a transfer function.

Illustrative MATLAB Commands:

≫ **G=pck([], [], [], 10)** % create a constant system matrix

≫ **[y, x, t]=step(A, B, C, D, Iu)** % Iu=i (step response of the ith channel)

≫ **[y, x, t]=initial(A, B, C, D, x₀)** % initial response with initial condition x_0

≫ **[y, x, t]=impulse(A, B, C, D, Iu)** % impulse response of the Iuth channel

≫ **[y,x]=lsim(A,B,C,D,U,T)** % U is a length(T) × column(B) matrix input; T is the sampling points.

Related MATLAB Commands: minfo, trsp, cos_tr, sin_tr, siggen

3.2 Controllability and Observability

We now turn to some very important concepts in linear system theory.

Definition 3.1 The dynamical system described by equation (3.1) or the pair (A, B) is said to be *controllable* if, for any initial state $x(0) = x_0$, $t_1 > 0$ and final state x_1, there exists a (piecewise continuous) input $u(\cdot)$ such that the solution of equation (3.1) satisfies $x(t_1) = x_1$. Otherwise, the system or the pair (A, B) is said to be *uncontrollable*.

The controllability (and the observability introduced next) of a system can be verified through some algebraic or geometric criteria.

3.2. Controllability and Observability

Theorem 3.1 *The following are equivalent:*

(i) (A, B) *is controllable.*

(ii) The matrix

$$W_c(t) := \int_0^t e^{A\tau} BB^* e^{A^*\tau} d\tau$$

is positive definite for any $t > 0$.

(iii) The controllability matrix

$$\mathcal{C} = \begin{bmatrix} B & AB & A^2B & \cdots & A^{n-1}B \end{bmatrix}$$

has full-row rank or, in other words, $\langle A\,|\,\text{Im}\,B\rangle := \sum_{i=1}^{n} \text{Im}(A^{i-1}B) = \mathbb{R}^n$.

(iv) The matrix $[A - \lambda I, B]$ has full-row rank for all λ in $\mathbb{C}$.

(v) Let λ and x be any eigenvalue and any corresponding left eigenvector of A (i.e., $x^*A = x^*\lambda$); then $x^*B \neq 0$.

(vi) The eigenvalues of $A+BF$ can be freely assigned (with the restriction that complex eigenvalues are in conjugate pairs) by a suitable choice of F.

Example 3.1 Let $A = \begin{bmatrix} 2 & 0 \\ 0 & 2 \end{bmatrix}$ and $B = \begin{bmatrix} 1 \\ 1 \end{bmatrix}$. Then $x_1 = \begin{bmatrix} 1 \\ 0 \end{bmatrix}$ and $x_2 = \begin{bmatrix} 0 \\ 1 \end{bmatrix}$ are independent eigenvectors of A and $x_i^*B \neq 0$, $i = 1, 2$. However, this should not lead one to conclude that (A, B) is controllable. In fact, $x = x_1 - x_2$ is also an eigenvector of A and $x^*B = 0$, which implies that (A, B) is not controllable. Hence one must check for all possible eigenvectors in using criterion (v).

Definition 3.2 An unforced dynamical system $\dot{x} = Ax$ is said to be *stable* if all the eigenvalues of A are in the open left half plane; that is, $\text{Re}\,\lambda(A) < 0$. A matrix A with such a property is said to be stable or Hurwitz.

Definition 3.3 The dynamical system of equation (3.1), or the pair (A, B), is said to be *stabilizable* if there exists a state feedback $u = Fx$ such that the system is stable (i.e., $A + BF$ is stable).

It is more appropriate to call this stabilizability the *state feedback stabilizability* to differentiate it from the *output feedback stabilizability* defined later.

The following theorem is a consequence of Theorem 3.1.

Theorem 3.2 *The following are equivalent:*

(i) (A, B) *is stabilizable.*

(ii) *The matrix* $[A - \lambda I, B]$ *has full-row rank for all* $\operatorname{Re}\lambda \geq 0$.

(iii) *For all* λ *and* x *such that* $x^*A = x^*\lambda$ *and* $\operatorname{Re}\lambda \geq 0$, $x^*B \neq 0$.

(iv) *There exists a matrix* F *such that* $A + BF$ *is Hurwitz.*

We now consider the dual notions: observability and detectability of the system described by equations (3.1) and (3.2).

Definition 3.4 *The dynamical system described by equations (3.1) and (3.2) or by the pair* (C, A) *is said to be* observable *if, for any* $t_1 > 0$, *the initial state* $x(0) = x_0$ *can be determined from the time history of the input* $u(t)$ *and the output* $y(t)$ *in the interval of* $[0, t_1]$. *Otherwise, the system, or* (C, A), *is said to be* unobservable.

Theorem 3.3 *The following are equivalent:*

(i) (C, A) *is observable.*

(ii) *The matrix*
$$W_o(t) := \int_0^t e^{A^*\tau} C^* C e^{A\tau} d\tau$$
is positive definite for any $t > 0$.

(iii) *The observability matrix*
$$\mathcal{O} = \begin{bmatrix} C \\ CA \\ CA^2 \\ \vdots \\ CA^{n-1} \end{bmatrix}$$
has full-column rank or $\bigcap_{i=1}^n \operatorname{Ker}(CA^{i-1}) = 0$.

(iv) *The matrix* $\begin{bmatrix} A - \lambda I \\ C \end{bmatrix}$ *has full-column rank for all* λ *in* $\mathbb{C}$.

(v) *Let* λ *and* y *be any eigenvalue and any corresponding right eigenvector of* A *(i.e.,* $Ay = \lambda y$); *then* $Cy \neq 0$.

(vi) *The eigenvalues of* $A + LC$ *can be freely assigned (with the restriction that complex eigenvalues are in conjugate pairs) by a suitable choice of* L.

(vii) (A^*, C^*) *is controllable.*

Definition 3.5 The system, or the pair (C, A), is *detectable* if $A + LC$ is stable for some L.

Theorem 3.4 *The following are equivalent:*

(i) (C, A) *is detectable.*

(ii) The matrix $\begin{bmatrix} A - \lambda I \\ C \end{bmatrix}$ *has full-column rank for all* $\operatorname{Re}\lambda \geq 0$.

(iii) For all λ *and* x *such that* $Ax = \lambda x$ *and* $\operatorname{Re}\lambda \geq 0$, $Cx \neq 0$.

(iv) There exists a matrix L *such that* $A + LC$ *is Hurwitz.*

(v) (A^*, C^*) *is stabilizable.*

The conditions (iv) and (v) of Theorems 3.1 and 3.3 and the conditions (ii) and (iii) of Theorems 3.2 and 3.4 are often called Popov-Belevitch-Hautus (PBH) tests. In particular, the following definitions of modal controllability and observability are often useful.

Definition 3.6 Let λ be an eigenvalue of A or, equivalently, a mode of the system. Then the mode λ is said to be controllable (observable) if $x^*B \neq 0$ ($Cx \neq 0$) for *all* left (right) eigenvectors of A associated with λ; that is, $x^*A = \lambda x^*$ ($Ax = \lambda x$) and $0 \neq x \in \mathbb{C}^n$. Otherwise, the mode is said to be uncontrollable (unobservable).

It follows that a system is controllable (observable) if and only if every mode is controllable (observable). Similarly, a system is stabilizable (detectable) if and only if every unstable mode is controllable (observable).

Illustrative MATLAB Commands:

≫ $\mathcal{C}$= **ctrb(A, B);** $\mathcal{O}$= **obsv(A, C);**

≫ **W**$_c(\infty)$=**gram(A, B);** % if A is stable.

≫ **F=-place(A, B, P)** % P is a vector of desired eigenvalues.

Related MATLAB Commands: ctrbf, obsvf, canon, strans, acker

3.3 Observers and Observer-Based Controllers

It is clear that if a system is controllable and the system states are available for feedback, then the system closed-loop poles can be assigned arbitrarily through a constant feedback. However, in most practical applications, the system states are not completely accessible and all the designer knows are the output y and input u. Hence, the estimation of the system states from the given output information y and input u is often necessary to realize some specific design objectives. In this section, we consider such an estimation problem and the application of this state estimation in feedback control.

Consider a plant modeled by equations (3.1) and (3.2). An observer is a dynamical system with input (u, y) and output (say, $\hat{x}$), that asymptotically estimates the state x, that is, $\hat{x}(t) - x(t) \to 0$ as $t \to \infty$ for all initial states and for every input.

Theorem 3.5 *An observer exists iff (C, A) is detectable. Further, if (C, A) is detectable, then a full-order Luenberger observer is given by*

$$\dot{q} = Aq + Bu + L(Cq + Du - y) \qquad (3.3)$$
$$\hat{x} = q, \qquad (3.4)$$

where L is any matrix such that $A + LC$ is stable.

Recall that, for a dynamical system described by the equations (3.1) and (3.2), if (A, B) is controllable and state x is available for feedback, then there is a state feedback $u = Fx$ such that the closed-loop poles of the system can be arbitrarily assigned. Similarly, if (C, A) is observable, then the system observer poles can be arbitrarily placed so that the state estimator $\hat{x}$ can be made to approach x arbitrarily fast. Now let us consider what will happen if the system states are not available for feedback so that the estimated state has to be used. Hence, the controller has the following dynamics:

$$\dot{\hat{x}} = (A + LC)\hat{x} + Bu + LDu - Ly$$
$$u = F\hat{x}.$$

Then the total system state equations are given by

$$\begin{bmatrix} \dot{x} \\ \dot{\hat{x}} \end{bmatrix} = \begin{bmatrix} A & BF \\ -LC & A + BF + LC \end{bmatrix} \begin{bmatrix} x \\ \hat{x} \end{bmatrix}.$$

Let $e := x - \hat{x}$; then the system equation becomes

$$\begin{bmatrix} \dot{e} \\ \dot{\hat{x}} \end{bmatrix} = \begin{bmatrix} A + LC & 0 \\ -LC & A + BF \end{bmatrix} \begin{bmatrix} e \\ \hat{x} \end{bmatrix}$$

and the closed-loop poles consist of two parts: the poles resulting from state feedback $\lambda_i(A + BF)$ and the poles resulting from the state estimation $\lambda_j(A + LC)$. Now if (A, B) is controllable and (C, A) is observable, then there exist F and L such that the eigenvalues of $A + BF$ and $A + LC$ can be arbitrarily assigned. In particular, they can be made to be stable. Note that a slightly weaker result can also result even if (A, B) and (C, A) are only stabilizable and detectable.

The controller given above is called an observer-based controller and is denoted as

$$u = K(s)y$$

and

$$K(s) = \left[\begin{array}{c|c} A + BF + LC + LDF & -L \\ \hline F & 0 \end{array} \right].$$

3.3. Observers and Observer-Based Controllers

Now denote the open-loop plant by

$$G = \left[\begin{array}{c|c} A & B \\ \hline C & D \end{array}\right];$$

then the closed-loop feedback system is as shown below:

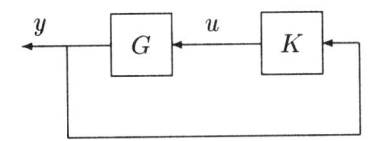

```
  y         u
◄──── G ◄──── K ◄──┐
      │            │
      └────────────┘
```

In general, if a system is stabilizable through feeding back the output y, then it is said to be *output feedback stabilizable*. It is clear from the above construction that a system is output feedback stabilizable if and only if (A, B) is stabilizable and (C, A) is detectable.

Example 3.2 Let $A = \begin{bmatrix} 1 & 2 \\ 1 & 0 \end{bmatrix}$, $B = \begin{bmatrix} 1 \\ 0 \end{bmatrix}$, and $C = \begin{bmatrix} 1 & 0 \end{bmatrix}$. We shall design a state feedback $u = Fx$ such that the closed-loop poles are at $\{-2, -3\}$. This can be done by choosing $F = \begin{bmatrix} -6 & -8 \end{bmatrix}$ using

$$\gg F = -\text{place}(A, B, [-2, -3]).$$

Now suppose the states are not available for feedback and we want to construct an observer so that the observer poles are at $\{-10, -10\}$. Then $L = \begin{bmatrix} -21 \\ -51 \end{bmatrix}$ can be obtained by using

$$\gg L = -\text{acker}(A', C', [-10, -10])'$$

and the observer-based controller is given by

$$K(s) = \frac{-534(s + 0.6966)}{(s + 34.6564)(s - 8.6564)}.$$

Note that the stabilizing controller itself is unstable. Of course, this may not be desirable in practice.

3.4 Operations on Systems

In this section, we present some facts about system interconnection. Since these proofs are straightforward, we will leave the details to the reader.

Suppose that G_1 and G_2 are two subsystems with state-space representations:

$$G_1 = \left[\begin{array}{c|c} A_1 & B_1 \\ \hline C_1 & D_1 \end{array}\right], \quad G_2 = \left[\begin{array}{c|c} A_2 & B_2 \\ \hline C_2 & D_2 \end{array}\right].$$

Then the series or cascade connection of these two subsystems is a system with the output of the second subsystem as the input of the first subsystem, as shown in the following diagram:

$$\longleftarrow \boxed{G_1} \longleftarrow \boxed{G_2} \longleftarrow$$

This operation in terms of the transfer matrices of the two subsystems is essentially the product of two transfer matrices. Hence, a representation for the cascaded system can be obtained as

$$\begin{aligned} G_1 G_2 &= \left[\begin{array}{c|c} A_1 & B_1 \\ \hline C_1 & D_1 \end{array}\right] \left[\begin{array}{c|c} A_2 & B_2 \\ \hline C_2 & D_2 \end{array}\right] \\ &= \left[\begin{array}{cc|c} A_1 & B_1 C_2 & B_1 D_2 \\ 0 & A_2 & B_2 \\ \hline C_1 & D_1 C_2 & D_1 D_2 \end{array}\right] = \left[\begin{array}{cc|c} A_2 & 0 & B_2 \\ B_1 C_2 & A_1 & B_1 D_2 \\ \hline D_1 C_2 & C_1 & D_1 D_2 \end{array}\right]. \end{aligned}$$

Similarly, the parallel connection or the addition of G_1 and G_2 can be obtained as

$$G_1 + G_2 = \left[\begin{array}{c|c} A_1 & B_1 \\ \hline C_1 & D_1 \end{array}\right] + \left[\begin{array}{c|c} A_2 & B_2 \\ \hline C_2 & D_2 \end{array}\right] = \left[\begin{array}{cc|c} A_1 & 0 & B_1 \\ 0 & A_2 & B_2 \\ \hline C_1 & C_2 & D_1 + D_2 \end{array}\right].$$

For future reference, we shall also introduce the following definitions.

Definition 3.7 The *transpose* of a transfer matrix $G(s)$ or the *dual system* is defined as

$$G \longmapsto G^T(s) = B^*(sI - A^*)^{-1} C^* + D^*$$

or, equivalently,

$$\left[\begin{array}{c|c} A & B \\ \hline C & D \end{array}\right] \longmapsto \left[\begin{array}{c|c} A^* & C^* \\ \hline B^* & D^* \end{array}\right].$$

Definition 3.8 The *conjugate* system of $G(s)$ is defined as

$$G \longmapsto G^\sim(s) := G^T(-s) = B^*(-sI - A^*)^{-1} C^* + D^*$$

or, equivalently,

$$\left[\begin{array}{c|c} A & B \\ \hline C & D \end{array}\right] \longmapsto \left[\begin{array}{c|c} -A^* & -C^* \\ \hline B^* & D^* \end{array}\right].$$

3.5. State-Space Realizations for Transfer Matrices

In particular, we have $G^*(j\omega) := [G(j\omega)]^* = G^\sim(j\omega)$.

A real rational matrix $\hat{G}(s)$ is called an *inverse* of a transfer matrix $G(s)$ if $G(s)\hat{G}(s) = \hat{G}(s)G(s) = I$. Suppose $G(s)$ is square and D is invertible. Then

$$G^{-1} = \left[\begin{array}{c|c} A - BD^{-1}C & -BD^{-1} \\ \hline D^{-1}C & D^{-1} \end{array}\right].$$

The corresponding MATLAB commands are

$$G_1 G_2 \iff \mathbf{mmult}(\mathbf{G_1}, \mathbf{G_2}), \quad \begin{bmatrix} G_1 & G_2 \end{bmatrix} \iff \mathbf{sbs}(\mathbf{G_1}, \mathbf{G_2})$$

$$G_1 + G_2 \iff \mathbf{madd}(\mathbf{G_1}, \mathbf{G_2}), \quad G_1 - G_2 \iff \mathbf{msub}(\mathbf{G_1}, \mathbf{G_2})$$

$$\begin{bmatrix} G_1 \\ G_2 \end{bmatrix} \iff \mathbf{abv}(\mathbf{G_1}, \mathbf{G_2}), \quad \begin{bmatrix} G_1 & \\ & G_2 \end{bmatrix} \iff \mathbf{daug}(\mathbf{G_1}, \mathbf{G_2}),$$

$$G^T(s) \iff \mathbf{transp}(\mathbf{G}), \quad G^\sim(s) \iff \mathbf{cjt}(\mathbf{G}), \quad G^{-1}(s) \iff \mathbf{minv}(\mathbf{G})$$

$$\alpha\, G(s) \iff \mathbf{mscl}(\mathbf{G}, \alpha), \quad \alpha \text{ is a scalar.}$$

Related MATLAB Commands: append, parallel, feedback, series, cloop, sclin, sclout, sel

3.5 State-Space Realizations for Transfer Matrices

In some cases, the natural or convenient description for a dynamical system is in terms of matrix transfer function. This occurs, for example, in some highly complex systems for which the analytic differential equations are too hard or too complex to write down. Hence certain engineering approximation or identification has to be carried out; for example, input and output frequency responses are obtained from experiments so that some transfer matrix approximating the system dynamics can be obtained. Since the state-space computation is most convenient to implement on the computer, some appropriate state-space representation for the resulting transfer matrix is necessary.

In general, assume that $G(s)$ is a real rational transfer matrix that is *proper*. Then we call a state-space model (A, B, C, D) such that

$$G(s) = \left[\begin{array}{c|c} A & B \\ \hline C & D \end{array}\right]$$

a *realization* of $G(s)$.

Definition 3.9 A state-space realization (A, B, C, D) of $G(s)$ is said to be a *minimal realization* of $G(s)$ if A has the smallest possible dimension.

Theorem 3.6 *A state-space realization (A, B, C, D) of $G(s)$ is minimal if and only if (A, B) is controllable and (C, A) is observable.*

We now describe several ways to obtain a state-space realization for a given multiple-input and multiple-output transfer matrix $G(s)$. We shall first consider SIMO (single-input and multiple-output) and MISO (multiple-input and single-output) systems.

Let $G(s)$ be a column vector of transfer function with p outputs:

$$G(s) = \frac{\beta_1 s^{n-1} + \beta_2 s^{n-2} + \cdots + \beta_{n-1} s + \beta_n}{s^n + a_1 s^{n-1} + \cdots + a_{n-1} s + a_n} + d, \quad \beta_i \in \mathbb{R}^p, \; d \in \mathbb{R}^p.$$

Then $G(s) = \left[\begin{array}{c|c} A & b \\ \hline C & d \end{array}\right]$ with

$$A := \begin{bmatrix} -a_1 & -a_2 & \cdots & -a_{n-1} & -a_n \\ 1 & 0 & \cdots & 0 & 0 \\ 0 & 1 & \cdots & 0 & 0 \\ \vdots & \vdots & & \vdots & \vdots \\ 0 & 0 & \cdots & 1 & 0 \end{bmatrix} \quad b := \begin{bmatrix} 1 \\ 0 \\ 0 \\ \vdots \\ 0 \end{bmatrix}$$

$$C = \begin{bmatrix} \beta_1 & \beta_2 & \cdots & \beta_{n-1} & \beta_n \end{bmatrix}$$

is a so-called *controllable canonical form* or *controller canonical form*.

Dually, consider a multiple-input and single-output system

$$G(s) = \frac{\eta_1 s^{n-1} + \eta_2 s^{n-2} + \cdots + \eta_{n-1} s + \eta_n}{s^n + a_1 s^{n-1} + \cdots + a_{n-1} s + a_n} + d, \quad \eta_i^* \in \mathbb{R}^m, d^* \in \mathbb{R}^m.$$

Then

$$G(s) = \left[\begin{array}{c|c} A & B \\ \hline c & d \end{array}\right] = \left[\begin{array}{ccccc|c} -a_1 & 1 & 0 & \cdots & 0 & \eta_1 \\ -a_2 & 0 & 1 & \cdots & 0 & \eta_2 \\ \vdots & \vdots & \vdots & & \vdots & \vdots \\ -a_{n-1} & 0 & 0 & \cdots & 1 & \eta_{n-1} \\ -a_n & 0 & 0 & \cdots & 0 & \eta_n \\ \hline 1 & 0 & 0 & \cdots & 0 & d \end{array}\right]$$

is an *observable canonical form* or *observer canonical form*.

For a MIMO system, the simplest and most straightforward way to obtain a realization is by realizing each element of the matrix $G(s)$ and then combining all these individual realizations to form a realization for $G(s)$. To illustrate, let us consider a 2×2 (block) transfer matrix such as

$$G(s) = \begin{bmatrix} G_1(s) & G_2(s) \\ G_3(s) & G_4(s) \end{bmatrix}$$

and assume that $G_i(s)$ has a state-space realization of the form

$$G_i(s) = \left[\begin{array}{c|c} A_i & B_i \\ \hline C_i & D_i \end{array}\right], \quad i = 1, \ldots, 4.$$

3.5. State-Space Realizations for Transfer Matrices

Then a realization for $G(s)$ can be obtained as $(\mathbf{G} = \mathbf{abv}(\mathbf{sbs}(\mathbf{G_1}, \mathbf{G_2}), \mathbf{sbs}(\mathbf{G_3}, \mathbf{G_4})))$:

$$G(s) = \left[\begin{array}{cccc|cc} A_1 & 0 & 0 & 0 & B_1 & 0 \\ 0 & A_2 & 0 & 0 & 0 & B_2 \\ 0 & 0 & A_3 & 0 & B_3 & 0 \\ 0 & 0 & 0 & A_4 & 0 & B_4 \\ \hline C_1 & C_2 & 0 & 0 & D_1 & D_2 \\ 0 & 0 & C_3 & C_4 & D_3 & D_4 \end{array}\right].$$

Alternatively, if the transfer matrix $G(s)$ can be factored into the product and/or the sum of several simply realized transfer matrices, then a realization for G can be obtained by using the cascade or addition formulas given in the preceding section.

A problem inherited with these kinds of realization procedures is that a realization thus obtained will generally not be minimal. To obtain a minimal realization, a Kalman controllability and observability decomposition has to be performed to eliminate the uncontrollable and/or unobservable states. (An alternative numerically reliable method to eliminate uncontrollable and/or unobservable states is the *balanced realization* method, which will be discussed later.)

We shall now describe a procedure that does result in a minimal realization by using partial fractional expansion (the resulting realization is sometimes called *Gilbert's realization* due to E. G. Gilbert).

Let $G(s)$ be a $p \times m$ transfer matrix and write it in the following form:

$$G(s) = \frac{N(s)}{d(s)}$$

with $d(s)$ a scalar polynomial. For simplicity, we shall assume that $d(s)$ has only real and distinct roots $\lambda_i \neq \lambda_j$ if $i \neq j$ and

$$d(s) = (s - \lambda_1)(s - \lambda_2) \cdots (s - \lambda_r).$$

Then $G(s)$ has the following partial fractional expansion:

$$G(s) = D + \sum_{i=1}^{r} \frac{W_i}{s - \lambda_i}.$$

Suppose

$$\text{rank } W_i = k_i$$

and let $B_i \in \mathbb{R}^{k_i \times m}$ and $C_i \in \mathbb{R}^{p \times k_i}$ be two constant matrices such that

$$W_i = C_i B_i.$$

Then a realization for $G(s)$ is given by

$$G(s) = \left[\begin{array}{ccc|c} \lambda_1 I_{k_1} & & & B_1 \\ & \ddots & & \vdots \\ & & \lambda_r I_{k_r} & B_r \\ \hline C_1 & \cdots & C_r & D \end{array}\right].$$

It follows immediately from PBH tests that this realization is controllable and observable, and thus it is minimal.

An immediate consequence of this minimal realization is that a transfer matrix with an rth order polynomial denominator does not necessarily have an rth order state-space realization unless W_i for each i is a rank one matrix.

This approach can, in fact, be generalized to more complicated cases where $d(s)$ may have complex and/or repeated roots. Readers may convince themselves by trying some simple examples.

Illustrative MATLAB Commands:

≫ **G=nd2sys(num, den, gain); G=zp2sys(zeros, poles, gain);**

Related MATLAB Commands: ss2tf, ss2zp, zp2ss, tf2ss, residue, minreal

3.6 Multivariable System Poles and Zeros

A matrix is called a polynomial matrix of a variable if every element of the matrix is a polynomial of the variable.

Definition 3.10 Let $Q(s)$ be a $(p \times m)$ polynomial matrix (or a transfer matrix) of s. Then the *normal rank* of $Q(s)$, denoted normalrank $(Q(s))$, is the maximally possible rank of $Q(s)$ for at lease one $s \in \mathbb{C}$.

To show the difference between the normal rank of a polynomial matrix and the rank of the polynomial matrix evaluated at a certain point, consider

$$Q(s) = \begin{bmatrix} s & 1 \\ s^2 & 1 \\ s & 1 \end{bmatrix}.$$

Then $Q(s)$ has normal rank 2 since rank $Q(3) = 2$. However, $Q(0)$ has rank 1.

The poles and zeros of a transfer matrix can be characterized in terms of its state-space realizations. Let

$$\left[\begin{array}{c|c} A & B \\ \hline C & D \end{array}\right]$$

be a state-space realization of $G(s)$.

3.6. Multivariable System Poles and Zeros

Definition 3.11 The eigenvalues of A are called the *poles* of the realization of $G(s)$.

To define zeros, let us consider the following system matrix:

$$Q(s) = \begin{bmatrix} A - sI & B \\ C & D \end{bmatrix}.$$

Definition 3.12 A complex number $z_0 \in \mathbb{C}$ is called an *invariant zero* of the system realization if it satisfies

$$\mathrm{rank} \begin{bmatrix} A - z_0 I & B \\ C & D \end{bmatrix} < \mathrm{normalrank} \begin{bmatrix} A - sI & B \\ C & D \end{bmatrix}.$$

The invariant zeros are not changed by constant state feedback since

$$\mathrm{rank} \begin{bmatrix} A + BF - z_0 I & B \\ C + DF & D \end{bmatrix} = \mathrm{rank} \begin{bmatrix} A - z_0 I & B \\ C & D \end{bmatrix} \begin{bmatrix} I & 0 \\ F & I \end{bmatrix}$$

$$= \mathrm{rank} \begin{bmatrix} A - z_0 I & B \\ C & D \end{bmatrix}.$$

It is also clear that invariant zeros are not changed under similarity transformation.

The following lemma is obvious.

Lemma 3.7 *Suppose* $\begin{bmatrix} A - sI & B \\ C & D \end{bmatrix}$ *has full-column normal rank. Then* $z_0 \in \mathbb{C}$ *is an invariant zero of a realization* (A, B, C, D) *if and only if there exist* $0 \neq x \in \mathbb{C}^n$ *and* $u \in \mathbb{C}^m$ *such that*

$$\begin{bmatrix} A - z_0 I & B \\ C & D \end{bmatrix} \begin{bmatrix} x \\ u \end{bmatrix} = 0.$$

Moreover, if $u = 0$, *then* z_0 *is also a nonobservable mode.*

Proof. By definition, z_0 is an invariant zero if there is a vector $\begin{bmatrix} x \\ u \end{bmatrix} \neq 0$ such that

$$\begin{bmatrix} A - z_0 I & B \\ C & D \end{bmatrix} \begin{bmatrix} x \\ u \end{bmatrix} = 0$$

since $\begin{bmatrix} A - sI & B \\ C & D \end{bmatrix}$ has full-column normal rank.

On the other hand, suppose z_0 is an invariant zero; then there is a vector $\begin{bmatrix} x \\ u \end{bmatrix} \neq 0$ such that

$$\begin{bmatrix} A - z_0 I & B \\ C & D \end{bmatrix} \begin{bmatrix} x \\ u \end{bmatrix} = 0.$$

We claim that $x \neq 0$. Otherwise, $\begin{bmatrix} B \\ D \end{bmatrix} u = 0$ or $u = 0$ since $\begin{bmatrix} A - sI & B \\ C & D \end{bmatrix}$ has full-column normal rank (i.e., $\begin{bmatrix} x \\ u \end{bmatrix} = 0$), which is a contradiction.

Finally, note that if $u = 0$, then
$$\begin{bmatrix} A - z_0 I \\ C \end{bmatrix} x = 0$$
and z_0 is a nonobservable mode by PBH test. □

When the system is square, the invariant zeros can be computed by solving a generalized eigenvalue problem:
$$\underbrace{\begin{bmatrix} A & B \\ C & D \end{bmatrix}}_{M} \begin{bmatrix} x \\ u \end{bmatrix} = z_0 \underbrace{\begin{bmatrix} I & 0 \\ 0 & 0 \end{bmatrix}}_{N} \begin{bmatrix} x \\ u \end{bmatrix}$$

using a MATLAB command: **eig(M, N)**.

Lemma 3.8 *Suppose $\begin{bmatrix} A - sI & B \\ C & D \end{bmatrix}$ has full-row normal rank. Then $z_0 \in \mathbb{C}$ is an invariant zero of a realization (A, B, C, D) if and only if there exist $0 \neq y \in \mathbb{C}^n$ and $v \in \mathbb{C}^p$ such that*
$$\begin{bmatrix} y^* & v^* \end{bmatrix} \begin{bmatrix} A - z_0 I & B \\ C & D \end{bmatrix} = 0.$$
Moreover, if $v = 0$, then z_0 is also a noncontrollable mode.

Lemma 3.9 *$G(s)$ has full-column (row) normal rank if and only if $\begin{bmatrix} A - sI & B \\ C & D \end{bmatrix}$ has full-column (row) normal rank.*

Proof. This follows by noting that
$$\begin{bmatrix} A - sI & B \\ C & D \end{bmatrix} = \begin{bmatrix} I & 0 \\ C(A - sI)^{-1} & I \end{bmatrix} \begin{bmatrix} A - sI & B \\ 0 & G(s) \end{bmatrix}$$
and
$$\text{normalrank} \begin{bmatrix} A - sI & B \\ C & D \end{bmatrix} = n + \text{normalrank}(G(s)).$$
□

Lemma 3.10 *Let $G(s) \in \mathcal{R}_p(s)$ be a $p \times m$ transfer matrix and let (A, B, C, D) be a minimal realization. If the system input is of the form $u(t) = u_0 e^{\lambda t}$, where $\lambda \in \mathbb{C}$ is not a pole of $G(s)$ and $u_0 \in \mathbb{C}^m$ is an arbitrary constant vector, then the output due to the input $u(t)$ and the initial state $x(0) = (\lambda I - A)^{-1} B u_0$ is $y(t) = G(\lambda) u_0 e^{\lambda t}$, $\forall t \geq 0$. In particular, if λ is a zero of $G(s)$, then $y(t) = 0$.*

3.7. Notes and References

Proof. The system response with respect to the input $u(t) = u_0 e^{\lambda t}$ and the initial condition $x(0) = (\lambda I - A)^{-1} B u_0$ is (in terms of the Laplace transform)

$$\begin{aligned}
Y(s) &= C(sI-A)^{-1}x(0) + C(sI-A)^{-1}BU(s) + DU(s) \\
&= C(sI-A)^{-1}x(0) + C(sI-A)^{-1}Bu_0(s-\lambda)^{-1} + Du_0(s-\lambda)^{-1} \\
&= C(sI-A)^{-1}x(0) + C\left[(sI-A)^{-1} - (\lambda I - A)^{-1}\right]Bu_0(s-\lambda)^{-1} \\
&\quad + C(\lambda I - A)^{-1}Bu_0(s-\lambda)^{-1} + Du_0(s-\lambda)^{-1} \\
&= C(sI-A)^{-1}(x(0) - (\lambda I - A)^{-1}Bu_0) + G(\lambda)u_0(s-\lambda)^{-1} \\
&= G(\lambda)u_0(s-\lambda)^{-1}.
\end{aligned}$$

Hence $y(t) = G(\lambda)u_0 e^{\lambda t}$. □

Example 3.3 Let

$$G(s) = \left[\begin{array}{c|c} A & B \\ \hline C & D \end{array}\right] = \left[\begin{array}{ccc|ccc} -1 & -2 & 1 & 1 & 2 & 3 \\ 0 & 2 & -1 & 3 & 2 & 1 \\ -4 & -3 & -2 & 1 & 1 & 1 \\ \hline 1 & 1 & 1 & 0 & 0 & 0 \\ 2 & 3 & 4 & 0 & 0 & 0 \end{array}\right].$$

Then the invariant zeros of the system can be found using the MATLAB command

» **G=pck(A, B, C, D),** z_0 = **szeros(G)**, % or

» z_0 = **tzero(A, B, C, D)**

which gives $z_0 = 0.2$. Since $G(s)$ is full-row rank, we can find y and v such that

$$\begin{bmatrix} y^* & v^* \end{bmatrix} \begin{bmatrix} A - z_0 I & B \\ C & D \end{bmatrix} = 0,$$

which can again be computed using a MATLAB command:

» **null([A − z_0 ∗ eye(3), B; C, D]')** $\implies \begin{bmatrix} y \\ v \end{bmatrix} = \begin{bmatrix} 0.0466 \\ 0.0466 \\ -0.1866 \\ \hdashline -0.9702 \\ 0.1399 \end{bmatrix}$.

Related MATLAB Commands: spoles, rifd

3.7 Notes and References

Readers are referred to Brogan [1991], Chen [1984], Kailath [1980], and Wonham [1985] for extensive treatment of standard linear system theory.

3.8 Problems

Problem 3.1 Let $A \in \mathbb{C}^{n \times n}$ and $B \in \mathbb{C}^{m \times m}$. Show that $X(t) = e^{At}X(0)e^{Bt}$ is the solution to
$$\dot{X} = AX + XB.$$

Problem 3.2 Given the impulse response, $h_1(t)$, of a linear time-invariant system with model
$$\ddot{y} + a_1\dot{y} + a_2 y = u$$
find the impulse response, $h(t)$, of the system
$$\ddot{y} + a_1\dot{y} + a_2 y = b_0\ddot{u} + b_1\dot{u} + b_2 u.$$

Justify your answer and generalize your result to nth order systems.

Problem 3.3 Suppose a second order system is given by $\dot{x} = Ax$. Suppose it is known that $x(1) = \begin{bmatrix} 4 \\ -2 \end{bmatrix}$ for $x(0) = \begin{bmatrix} 1 \\ 1 \end{bmatrix}$ and $x(1) = \begin{bmatrix} 5 \\ -2 \end{bmatrix}$ for $x(0) = \begin{bmatrix} 1 \\ 2 \end{bmatrix}$. Find $x(n)$ for $x(0) = \begin{bmatrix} 1 \\ 0 \end{bmatrix}$. Can you determine A?

Problem 3.4 Assume (A, B) is controllable. Show that (F, G) with
$$F = \begin{bmatrix} A & 0 \\ C & 0 \end{bmatrix}, \quad G = \begin{bmatrix} B \\ 0 \end{bmatrix}$$
is controllable if and only if
$$\begin{bmatrix} A & B \\ C & 0 \end{bmatrix}$$
is a full-row rank matrix.

Problem 3.5 Let $\lambda_i, i = 1, 2, \ldots, n$ be n distinct eigenvalues of a matrix $A \in \mathbb{C}^{n \times n}$ and let x_i and y_i be the corresponding right- and left-unit eigenvectors such that $y_i^* x_i = 1$. Show that
$$y_i^* x_j = \delta_{ij}, \quad A = \sum_{i=1}^{n} \lambda_i x_i y_i^*$$
and
$$C(sI - A)^{-1}B = \sum_{i=1}^{n} \frac{(Cx_i)(y_i^* B)}{s - \lambda_i}.$$

Furthermore, show that the mode λ_i is controllable iff $y_i^* B \neq 0$ and the mode λ_i is observable iff $Cx_i \neq 0$.

3.8. Problems

Problem 3.6 Let (A, b, c) be a realization with $A \in \mathbb{R}^{n \times n}$, $b \in R^n$, and $c^* \in \mathbb{R}^n$. Assume that $\lambda_i(A) + \lambda_j(A) \neq 0$ for all i, j. This assumption ensures the existence of $X \in \mathbb{R}^{n \times n}$ such that
$$AX + XA + bc = 0$$
Show that X is nonsingular if and only if (A, b) is controllable and (c, A) is observable.

Problem 3.7 Compute the system zeros and the corresponding zero directions of the following transfer functions

$$G_1(s) = \left[\begin{array}{cc|cc} 1 & 2 & 2 & 2 \\ 1 & 0 & 3 & 4 \\ \hline 0 & 1 & 1 & 2 \\ 1 & 0 & 2 & 0 \end{array}\right], \quad G_2(s) = \left[\begin{array}{cc|cc} -1 & -2 & 1 & 2 \\ 0 & 1 & 2 & 1 \\ \hline 1 & 1 & 0 & 0 \\ 1 & 1 & 0 & 0 \end{array}\right],$$

$$G_3(s) = \left[\begin{array}{cc} \dfrac{2(s+1)(s+2)}{s(s+3)(s+4)} & \dfrac{s+2}{(s+1)(s+3)} \end{array}\right], \quad G_4(s) = \left[\begin{array}{cc} \dfrac{1}{s+1} & \dfrac{s+3}{(s+1)(s-2)} \\ \dfrac{10}{s-2} & \dfrac{5}{s+3} \end{array}\right]$$

Also find the vectors x and u whenever appropriate so that either

$$\left[\begin{array}{cc} A - zI & B \\ C & D \end{array}\right] \left[\begin{array}{c} x \\ u \end{array}\right] = 0 \quad \text{or} \quad \left[\begin{array}{cc} x^* & u^* \end{array}\right] \left[\begin{array}{cc} A - zI & B \\ C & D \end{array}\right] = 0.$$

Chapter 4

$\mathcal{H}_2$ and $\mathcal{H}_\infty$ Spaces

The most important objective of a control system is to achieve certain performance specifications in addition to providing internal stability. One way to describe the performance specifications of a control system is in terms of the size of certain signals of interest. For example, the performance of a tracking system could be measured by the size of the tracking error signal. In this chapter, we look at several ways of defining a signal's size (i.e., at several norms for signals). Of course, which norm is most appropriate depends on the situation at hand. For that purpose, we shall first introduce the Hardy spaces $\mathcal{H}_2$ and $\mathcal{H}_\infty$. Some state-space methods of computing real rational $\mathcal{H}_2$ and $\mathcal{H}_\infty$ transfer matrix norms are also presented.

4.1 Hilbert Spaces

Recall the inner product of vectors defined on a Euclidean space $\mathbb{C}^n$:

$$\langle x, y \rangle := x^* y = \sum_{i=1}^{n} \bar{x}_i y_i \quad \forall x = \begin{bmatrix} x_1 \\ \vdots \\ x_n \end{bmatrix}, y = \begin{bmatrix} y_1 \\ \vdots \\ y_n \end{bmatrix} \in \mathbb{C}^n.$$

Note that many important metric notions and geometrical properties, such as length, distance, angle, and the energy of physical systems, can be deduced from this inner product. For instance, the length of a vector $x \in \mathbb{C}^n$ is defined as

$$\|x\| := \sqrt{\langle x, x \rangle}$$

and the angle between two vectors $x, y \in \mathbb{C}^n$ can be computed from

$$\cos \angle(x, y) = \frac{\langle x, y \rangle}{\|x\| \, \|y\|}, \quad \angle(x, y) \in [0, \pi].$$

The two vectors are said to be *orthogonal* if $\angle(x, y) = \frac{\pi}{2}$.

We now consider a natural generalization of the inner product on $\mathbb{C}^n$ to more general (possibly infinite dimensional) vector spaces.

Definition 4.1 Let V be a vector space over $\mathbb{C}$. An *inner product*[1] on V is a complex-valued function,
$$\langle \cdot, \cdot \rangle : V \times V \longmapsto \mathbb{C}$$
such that for any $x, y, z \in V$ and $\alpha, \beta \in \mathbb{C}$

(i) $\langle x, \alpha y + \beta z \rangle = \alpha \langle x, y \rangle + \beta \langle x, z \rangle$

(ii) $\langle x, y \rangle = \overline{\langle y, x \rangle}$

(iii) $\langle x, x \rangle > 0$ if $x \neq 0$.

A vector space V with an inner product is called an *inner product space*.

It is clear that the inner product defined above induces a norm $\|x\| := \sqrt{\langle x, x \rangle}$, so that the norm conditions in Chapter 2 are satisfied. In particular, the distance between vectors x and y is $d(x, y) = \|x - y\|$.

Two vectors x and y in an inner product space V are said to be *orthogonal* if $\langle x, y \rangle = 0$, denoted $x \perp y$. More generally, a vector x is said to be orthogonal to a set $S \subset V$, denoted by $x \perp S$, if $x \perp y$ for all $y \in S$.

The inner product and the inner product induced norm have the following familiar properties.

Theorem 4.1 *Let V be an inner product space and let $x, y \in V$. Then*

(i) $|\langle x, y \rangle| \leq \|x\| \|y\|$ *(Cauchy-Schwarz inequality). Moreover, the equality holds if and only if $x = \alpha y$ for some constant α or $y = 0$.*

(ii) $\|x + y\|^2 + \|x - y\|^2 = 2 \|x\|^2 + 2 \|y\|^2$ *(Parallelogram law).*

(iii) $\|x + y\|^2 = \|x\|^2 + \|y\|^2$ *if $x \perp y$.*

A *Hilbert space* is a complete inner product space with the norm induced by its inner product. For example, $\mathbb{C}^n$ with the usual inner product is a (finite dimensional) Hilbert space. More generally, it is straightforward to verify that $\mathbb{C}^{n \times m}$ with the inner product defined as
$$\langle A, B \rangle := \operatorname{trace} A^* B = \sum_{i=1}^{n} \sum_{j=1}^{m} \bar{a}_{ij} b_{ij} \quad \forall A, B \in \mathbb{C}^{n \times m}$$
is also a (finite dimensional) Hilbert space.

A well-know infinite dimensional Hilbert space is $\mathcal{L}_2[a, b]$, which consists of all square integrable and Lebesgue measurable functions defined on an interval $[a, b]$ with the inner product defined as
$$\langle f, g \rangle := \int_a^b f(t)^* g(t) dt$$

---

[1] The property (i) in the following list is the other way around to the usual mathematical convention since we want to have $\langle x, y \rangle = x^* y$ rather than $y^* x$ for $x, y \in \mathbb{C}^n$.

for $f, g \in \mathcal{L}_2[a, b]$. Similarly, if the functions are vector or matrix-valued, the inner product is defined correspondingly as

$$\langle f, g \rangle := \int_a^b \text{trace}\,[f(t)^* g(t)]\, dt.$$

Some spaces used often in this book are $\mathcal{L}_2[0, \infty), \mathcal{L}_2(-\infty, 0], \mathcal{L}_2(-\infty, \infty)$. More precisely, they are defined as

$\mathcal{L}_2 = \mathcal{L}_2(-\infty, \infty)$: Hilbert space of matrix-valued functions on $\mathbb{R}$, with inner product

$$\langle f, g \rangle := \int_{-\infty}^{\infty} \text{trace}\,[f(t)^* g(t)]\, dt.$$

$\mathcal{L}_{2+} = \mathcal{L}_2[0, \infty)$: subspace of $\mathcal{L}_2(-\infty, \infty)$ with functions zero for $t < 0$.

$\mathcal{L}_{2-} = \mathcal{L}_2(-\infty, 0]$: subspace of $\mathcal{L}_2(-\infty, \infty)$ with functions zero for $t > 0$.

4.2 $\mathcal{H}_2$ and $\mathcal{H}_\infty$ Spaces

Let $S \subset \mathbb{C}$ be an open set, and let $f(s)$ be a complex-valued function defined on S:

$$f(s) : S \longmapsto \mathbb{C}.$$

Then $f(s)$ is said to be *analytic at a point* z_0 in S if it is differentiable at z_0 and also at each point in some neighborhood of z_0. It is a fact that if $f(s)$ is analytic at z_0 then f has continuous derivatives of all orders at z_0. Hence, a function analytic at z_0 has a power series representation at z_0. The converse is also true (i.e., if a function has a power series at z_0, then it is analytic at z_0). A function $f(s)$ is said to be *analytic in S* if it has a derivative or is analytic at each point of S. A matrix-valued function is analytic in S if every element of the matrix is analytic in S. For example, all real rational stable transfer matrices are analytic in the right-half plane and e^{-s} is analytic everywhere.

A well-know property of the analytic functions is the so-called *maximum modulus theorem*.

Theorem 4.2 *If $f(s)$ is defined and continuous on a closed-bounded set S and analytic on the interior of S, then $|f(s)|$ cannot attain the maximum in the interior of S unless $f(s)$ is a constant.*

The theorem implies that $|f(s)|$ can only achieve its maximum on the boundary of S; that is,

$$\max_{s \in S} |f(s)| = \max_{s \in \partial S} |f(s)|$$

where ∂S denotes the boundary of S. Next we consider some frequently used complex (matrix) function spaces.

$\mathcal{L}_2(j\mathbb{R})$ Space

$\mathcal{L}_2(j\mathbb{R})$ or simply $\mathcal{L}_2$ is a Hilbert space of matrix-valued (or scalar-valued) functions on $j\mathbb{R}$ and consists of all complex matrix functions F such that the following integral is bounded:

$$\int_{-\infty}^{\infty} \text{trace}\,[F^*(j\omega)F(j\omega)]\,d\omega < \infty.$$

The inner product for this Hilbert space is defined as

$$\langle F, G \rangle := \frac{1}{2\pi} \int_{-\infty}^{\infty} \text{trace}\,[F^*(j\omega)G(j\omega)]\,d\omega$$

for $F, G \in \mathcal{L}_2$, and the inner product induced norm is given by

$$\|F\|_2 := \sqrt{\langle F, F \rangle}.$$

For example, all real rational strictly proper transfer matrices with no poles on the imaginary axis form a subspace (not closed) of $\mathcal{L}_2(j\mathbb{R})$ that is denoted by $\mathcal{RL}_2(j\mathbb{R})$ or simply $\mathcal{RL}_2$.

$\mathcal{H}_2$ Space[2]

$\mathcal{H}_2$ is a (closed) subspace of $\mathcal{L}_2(j\mathbb{R})$ with matrix functions $F(s)$ analytic in $\text{Re}(s) > 0$ (open right-half plane). The corresponding norm is defined as

$$\|F\|_2^2 := \sup_{\sigma > 0} \left\{ \frac{1}{2\pi} \int_{-\infty}^{\infty} \text{trace}\,[F^*(\sigma + j\omega)F(\sigma + j\omega)]\,d\omega \right\}.$$

It can be shown[3] that

$$\|F\|_2^2 = \frac{1}{2\pi} \int_{-\infty}^{\infty} \text{trace}\,[F^*(j\omega)F(j\omega)]\,d\omega.$$

Hence, we can compute the norm for $\mathcal{H}_2$ just as we do for $\mathcal{L}_2$. The real rational subspace of $\mathcal{H}_2$, which consists of all strictly proper and real rational stable transfer matrices, is denoted by $\mathcal{RH}_2$.

$\mathcal{H}_2^\perp$ Space

$\mathcal{H}_2^\perp$ is the orthogonal complement of $\mathcal{H}_2$ in $\mathcal{L}_2$; that is, the (closed) subspace of functions in $\mathcal{L}_2$ that are analytic in the open left-half plane. The real rational subspace of $\mathcal{H}_2^\perp$, which consists of all strictly proper rational transfer matrices with all poles in the open right-half plane, will be denoted by $\mathcal{RH}_2^\perp$. It is easy to see that if G is a strictly proper, stable, and real rational transfer matrix, then $G \in \mathcal{H}_2$ and $G^\sim \in \mathcal{H}_2^\perp$. Most of our study in this book will be focused on the real rational case.

---

[2] The $\mathcal{H}_2$ space and $\mathcal{H}_\infty$ space defined in this subsection together with the $\mathcal{H}_p$ spaces, $p \geq 1$, which will not be introduced in this book, are usually called Hardy spaces and are named after the mathematician G. H. Hardy (hence the notation of $\mathcal{H}$).

[3] See Francis [1987].

4.2. $\mathcal{H}_2$ and $\mathcal{H}_\infty$ Spaces

The $\mathcal{L}_2$ spaces defined previously in the frequency domain can be related to the $\mathcal{L}_2$ spaces defined in the time domain. Recall the fact that a function in $\mathcal{L}_2$ space in the time domain admits a bilateral Laplace (or Fourier) transform. In fact, it can be shown that this bilateral Laplace transform yields an isometric isomorphism between the $\mathcal{L}_2$ spaces in the time domain and the $\mathcal{L}_2$ spaces in the frequency domain (this is what is called *Parseval's relations*):

$$\mathcal{L}_2(-\infty, \infty) \cong \mathcal{L}_2(j\mathbb{R})$$
$$\mathcal{L}_2[0, \infty) \cong \mathcal{H}_2$$
$$\mathcal{L}_2(-\infty, 0] \cong \mathcal{H}_2^\perp.$$

As a result, if $g(t) \in \mathcal{L}_2(-\infty, \infty)$ and if its bilateral Laplace transform is $G(s) \in \mathcal{L}_2(j\mathbb{R})$, then

$$\|G\|_2 = \|g\|_2.$$

Hence, whenever there is no confusion, the notation for functions in the time domain and in the frequency domain will be used interchangeably.

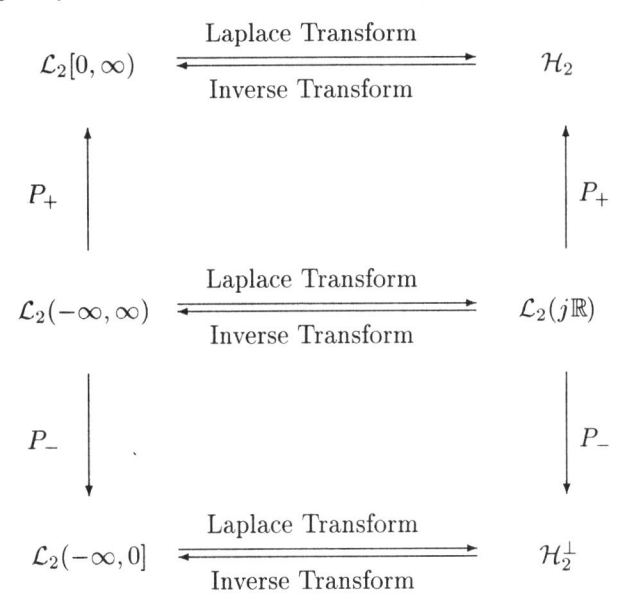

Figure 4.1: Relationships among function spaces

Define an orthogonal projection

$$P_+ : \mathcal{L}_2(-\infty, \infty) \longmapsto \mathcal{L}_2[0, \infty)$$

such that, for any function $f(t) \in \mathcal{L}_2(-\infty, \infty)$, we have $g(t) = P_+ f(t)$ with

$$g(t) := \begin{cases} f(t), & \text{for } t \geq 0; \\ 0, & \text{for } t < 0. \end{cases}$$

In this book, P_+ will also be used to denote the projection from $\mathcal{L}_2(j\mathbb{R})$ onto $\mathcal{H}_2$. Similarly, define P_- as another orthogonal projection from $\mathcal{L}_2(-\infty, \infty)$ onto $\mathcal{L}_2(-\infty, 0]$ (or $\mathcal{L}_2(j\mathbb{R})$ onto $\mathcal{H}_2^\perp$). Then the relationships between $\mathcal{L}_2$ spaces and $\mathcal{H}_2$ spaces can be shown as in Figure 4.1.

Other classes of important complex matrix functions used in this book are those bounded on the imaginary axis.

$\mathcal{L}_\infty(j\mathbb{R})$ Space

$\mathcal{L}_\infty(j\mathbb{R})$ or simply $\mathcal{L}_\infty$ is a Banach space of matrix-valued (or scalar-valued) functions that are (essentially) bounded on $j\mathbb{R}$, with norm

$$\|F\|_\infty := \operatorname*{ess\,sup}_{\omega \in \mathbb{R}} \overline{\sigma}\,[F(j\omega)].$$

The rational subspace of $\mathcal{L}_\infty$, denoted by $\mathcal{RL}_\infty(j\mathbb{R})$ or simply $\mathcal{RL}_\infty$, consists of all proper and real rational transfer matrices with no poles on the imaginary axis.

$\mathcal{H}_\infty$ Space

$\mathcal{H}_\infty$ is a (closed) subspace of $\mathcal{L}_\infty$ with functions that are analytic and bounded in the open right-half plane. The $\mathcal{H}_\infty$ norm is defined as

$$\|F\|_\infty := \sup_{\operatorname{Re}(s)>0} \overline{\sigma}\,[F(s)] = \sup_{\omega \in \mathbb{R}} \overline{\sigma}\,[F(j\omega)].$$

The second equality can be regarded as a generalization of the maximum modulus theorem for matrix functions. (See Boyd and Desoer [1985] for a proof.) The real rational subspace of $\mathcal{H}_\infty$ is denoted by $\mathcal{RH}_\infty$, which consists of all proper and real rational stable transfer matrices.

$\mathcal{H}_\infty^-$ Space

$\mathcal{H}_\infty^-$ is a (closed) subspace of $\mathcal{L}_\infty$ with functions that are analytic and bounded in the open left-half plane. The $\mathcal{H}_\infty^-$ norm is defined as

$$\|F\|_\infty := \sup_{\operatorname{Re}(s)<0} \overline{\sigma}\,[F(s)] = \sup_{\omega \in \mathbb{R}} \overline{\sigma}\,[F(j\omega)].$$

The real rational subspace of $\mathcal{H}_\infty^-$ is denoted by $\mathcal{RH}_\infty^-$, which consists of all proper, real rational, antistable transfer matrices (i.e., functions with all poles in the open right-half plane).

Let $G(s) \in \mathcal{L}_\infty$ be a $p \times q$ transfer matrix. Then a multiplication operator is defined as

$$M_G : \mathcal{L}_2 \longmapsto \mathcal{L}_2$$

$$M_G f := Gf.$$

4.2. $\mathcal{H}_2$ and $\mathcal{H}_\infty$ Spaces

In writing the preceding mapping, we have assumed that f has a compatible dimension. A more accurate description of the foregoing operator should be

$$M_G : \mathcal{L}_2^q \longmapsto \mathcal{L}_2^p.$$

That is, f is a q-dimensional vector function with each component in $\mathcal{L}_2$. However, we shall suppress all dimensions in this book and assume that all objects have compatible dimensions.

A useful fact about the multiplication operator is that the norm of a matrix G in $\mathcal{L}_\infty$ equals the norm of the corresponding multiplication operator.

Theorem 4.3 *Let $G \in \mathcal{L}_\infty$ be a $p \times q$ transfer matrix. Then $\|M_G\| = \|G\|_\infty$.*

Remark 4.1 It is also true that this operator norm equals the norm of the operator restricted to $\mathcal{H}_2$ (or $\mathcal{H}_2^\perp$); that is,

$$\|M_G\| = \|M_G|_{\mathcal{H}_2}\| := \sup\left\{\|Gf\|_2 : f \in \mathcal{H}_2, \|f\|_2 \leq 1\right\}.$$

This will be clear in the proof where an $f \in \mathcal{H}_2$ is constructed. ◇

Proof. By definition, we have

$$\|M_G\| = \sup\left\{\|Gf\|_2 : f \in \mathcal{L}_2, \|f\|_2 \leq 1\right\}.$$

First we see that $\|G\|_\infty$ is an upper bound for the operator norm:

$$\begin{aligned}
\|Gf\|_2^2 &= \frac{1}{2\pi} \int_{-\infty}^{\infty} f^*(j\omega) G^*(j\omega) G(j\omega) f(j\omega)\, d\omega \\
&\leq \|G\|_\infty^2 \frac{1}{2\pi} \int_{-\infty}^{\infty} \|f(j\omega)\|^2\, d\omega \\
&= \|G\|_\infty^2 \|f\|_2^2.
\end{aligned}$$

To show that $\|G\|_\infty$ is the least upper bound, first choose a frequency ω_0 where $\bar{\sigma}[G(j\omega)]$ is maximum; that is,

$$\bar{\sigma}[G(j\omega_0)] = \|G\|_\infty,$$

and denote the singular value decomposition of $G(j\omega_0)$ by

$$G(j\omega_0) = \bar{\sigma} u_1(j\omega_0) v_1^*(j\omega_0) + \sum_{i=2}^{r} \sigma_i u_i(j\omega_0) v_i^*(j\omega_0)$$

where r is the rank of $G(j\omega_0)$ and u_i, v_i have unit length.

Next we assume that $G(s)$ has real coefficients and we shall construct a function $f(s) \in \mathcal{H}_2$ with real coefficients so that the norm is approximately achieved. [It will be clear in the following that the proof is much simpler if f is allowed to have complex coefficients, which is necessary when $G(s)$ has complex coefficients.]

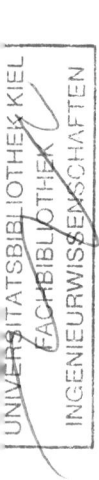

If $\omega_0 < \infty$, write $v_1(j\omega_0)$ as

$$v_1(j\omega_0) = \begin{bmatrix} \alpha_1 e^{j\theta_1} \\ \alpha_2 e^{j\theta_2} \\ \vdots \\ \alpha_q e^{j\theta_q} \end{bmatrix}$$

where $\alpha_i \in \mathbb{R}$ is such that $\theta_i \in (-\pi, 0]$ and q is the column dimension of G. Now let $0 \leq \beta_i \leq \infty$ be such that

$$\theta_i = \angle \left(\frac{\beta_i - j\omega_0}{\beta_i + j\omega_0} \right)$$

(with $\beta_i \to \infty$ if $\theta_i = 0$) and let f be given by

$$f(s) = \begin{bmatrix} \alpha_1 \frac{\beta_1 - s}{\beta_1 + s} \\ \alpha_2 \frac{\beta_2 - s}{\beta_2 + s} \\ \vdots \\ \alpha_q \frac{\beta_q - s}{\beta_q + s} \end{bmatrix} \hat{f}(s)$$

(with 1 replacing $\frac{\beta_i - s}{\beta_i + s}$ if $\theta_i = 0$), where a scalar function $\hat{f}$ is chosen so that

$$|\hat{f}(j\omega)| = \begin{cases} c & \text{if } |\omega - \omega_0| < \epsilon \text{ or } |\omega + \omega_0| < \epsilon \\ 0 & \text{otherwise} \end{cases}$$

where ϵ is a small positive number and c is chosen so that $\hat{f}$ has unit 2-norm (i.e., $c = \sqrt{\pi/2\epsilon}$). This, in turn, implies that f has unit 2-norm. Then

$$\|Gf\|_2^2 \approx \frac{1}{2\pi} \left[\overline{\sigma}\left[G(-j\omega_0)\right]^2 \pi + \overline{\sigma}\left[G(j\omega_0)\right]^2 \pi \right]$$
$$= \overline{\sigma}\left[G(j\omega_0)\right]^2 = \|G\|_\infty^2.$$

Similarly, if $\omega_0 = \infty$, the conclusion follows by letting $\omega_0 \to \infty$ in the foregoing. □

Illustrative MATLAB Commands:

≫ [sv, w]=sigma(A, B, C, D);   % frequency response of the singular values; or

≫ w=logspace(l, h, n); sv=sigma(A, B, C, D, w);   % n points between 10^l and 10^h.

Related MATLAB Commands: semilogx, semilogy, bode, freqs, nichols, frsp, vsvd, vplot, pkvnorm

4.3 Computing $\mathcal{L}_2$ and $\mathcal{H}_2$ Norms

Let $G(s) \in \mathcal{L}_2$ and recall that the $\mathcal{L}_2$ norm of G is defined as

$$\begin{aligned}
\|G\|_2 &:= \sqrt{\frac{1}{2\pi} \int_{-\infty}^{\infty} \text{trace}\{G^*(j\omega)G(j\omega)\}\, d\omega} \\
&= \|g\|_2 \\
&= \sqrt{\int_{-\infty}^{\infty} \text{trace}\{g^*(t)g(t)\}\, dt}
\end{aligned}$$

where $g(t)$ denotes the convolution kernel of G.

It is easy to see that the $\mathcal{L}_2$ norm defined previously is finite iff the transfer matrix G is strictly proper; that is, $G(\infty) = 0$. Hence, we will generally assume that the transfer matrix is strictly proper whenever we refer to the $\mathcal{L}_2$ norm of G (of course, this also applies to $\mathcal{H}_2$ norms). One straightforward way of computing the $\mathcal{L}_2$ norm is to use contour integral. Suppose G is strictly proper; then we have

$$\begin{aligned}
\|G\|_2^2 &= \frac{1}{2\pi} \int_{-\infty}^{\infty} \text{trace}\{G^*(j\omega)G(j\omega)\}\, d\omega \\
&= \frac{1}{2\pi j} \oint \text{trace}\{G^\sim(s)G(s)\}\, ds.
\end{aligned}$$

The last integral is a contour integral along the imaginary axis and around an infinite semicircle in the left-half plane; the contribution to the integral from this semicircle equals zero because G is strictly proper. By the residue theorem, $\|G\|_2^2$ equals the sum of the residues of $\text{trace}\{G^\sim(s)G(s)\}$ at its poles in the left-half plane.

Although $\|G\|_2$ can, in principle, be computed from its definition or from the method just suggested, it is useful in many applications to have alternative characterizations and to take advantage of the state-space representations of G. The computation of a $\mathcal{RH}_2$ transfer matrix norm is particularly simple.

Lemma 4.4 *Consider a transfer matrix*

$$G(s) = \left[\begin{array}{c|c} A & B \\ \hline C & 0 \end{array}\right]$$

with A stable. Then we have

$$\|G\|_2^2 = \text{trace}(B^*QB) = \text{trace}(CPC^*) \tag{4.1}$$

where Q and P are observability and controllability Gramians that can be obtained from the following Lyapunov equations:

$$AP + PA^* + BB^* = 0 \qquad A^*Q + QA + C^*C = 0.$$

Proof. Since G is stable, we have

$$g(t) = \mathcal{L}^{-1}(G) = \begin{cases} Ce^{At}B, & t \geq 0 \\ 0, & t < 0 \end{cases}$$

and

$$\begin{aligned} \|G\|_2^2 &= \int_0^\infty \text{trace}\{g^*(t)g(t)\}\, dt = \int_0^\infty \text{trace}\{g(t)g(t)^*\}\, dt \\ &= \int_0^\infty \text{trace}\{B^* e^{A^*t} C^* C e^{At} B\}\, dt = \int_0^\infty \text{trace}\{C e^{At} B B^* e^{A^*t} C^*\}\, dt. \end{aligned}$$

The lemma follows from the fact that the controllability Gramian of (A, B) and the observability Gramian of (C, A) can be represented as

$$Q = \int_0^\infty e^{A^*t} C^* C e^{At}\, dt, \quad P = \int_0^\infty e^{At} B B^* e^{A^*t}\, dt,$$

which can also be obtained from

$$AP + PA^* + BB^* = 0 \quad A^*Q + QA + C^*C = 0.$$

□

To compute the $\mathcal{L}_2$ norm of a rational transfer function, $G(s) \in \mathcal{RL}_2$, using the state-space approach, let $G(s) = [G(s)]_+ + [G(s)]_-$ with $G_+ \in \mathcal{RH}_2$ and $G_- \in \mathcal{RH}_2^\perp$; then

$$\|G\|_2^2 = \|[G(s)]_+\|_2^2 + \|[G(s)]_-\|_2^2$$

where $\|[G(s)]_+\|_2$ and $\|[G(s)]_-\|_2 = \|[G(-s)]_+\|_2 = \|([G(s)]_-)^\sim\|_2$ can be computed using the preceding lemma.

Still another useful characterization of the $\mathcal{H}_2$ norm of G is in terms of hypothetical input-output experiments. Let e_i denote the ith standard basis vector of $\mathbb{R}^m$, where m is the input dimension of the system. Apply the impulsive input $\delta(t)e_i$ [$\delta(t)$ is the unit impulse] and denote the output by $z_i(t) (= g(t)e_i)$. Assume $D = 0$; then $z_i \in \mathcal{L}_{2+}$ and

$$\|G\|_2^2 = \sum_{i=1}^m \|z_i\|_2^2.$$

Note that this characterization of the $\mathcal{H}_2$ norm can be appropriately generalized for nonlinear time-varying systems; see Chen and Francis [1992] for an application of this norm in sampled-data control.

Example 4.1 Consider a transfer matrix

$$G = \begin{bmatrix} \dfrac{3(s+3)}{(s-1)(s+2)} & \dfrac{2}{s-1} \\ \dfrac{s+1}{(s+2)(s+3)} & \dfrac{1}{s-4} \end{bmatrix} = G_s + G_u$$

4.4. Computing $\mathcal{L}_\infty$ and $\mathcal{H}_\infty$ Norms

with

$$G_s = \left[\begin{array}{cc|cc} -2 & 0 & -1 & 0 \\ 0 & -3 & 2 & 0 \\ \hline 1 & 0 & 0 & 0 \\ 1 & 1 & 0 & 0 \end{array}\right], \quad G_u = \left[\begin{array}{cc|cc} 1 & 0 & 4 & 2 \\ 0 & 4 & 0 & 1 \\ \hline 1 & 0 & 0 & 0 \\ 0 & 1 & 0 & 0 \end{array}\right].$$

Then the command **h2norm(G_s)** gives $\|G_s\|_2 = 0.6055$ and **h2norm(cjt(G_u))** gives $\|G_u\|_2 = 3.182$. Hence $\|G\|_2 = \sqrt{\|G_s\|_2^2 + \|G_u\|_2^2} = 3.2393$.

Illustrative MATLAB Commands:

≫ **P = gram(A, B); Q = gram(A′, C′);** or **P = lyap(A, B ∗ B′);**

≫ **[G_s, G_u] = sdecomp(G);** % decompose into stable and antistable parts.

4.4 Computing $\mathcal{L}_\infty$ and $\mathcal{H}_\infty$ Norms

We shall first consider, as in the $\mathcal{L}_2$ case, how to compute the ∞ norm of an $\mathcal{RL}_\infty$ transfer matrix. Let $G(s) \in \mathcal{RL}_\infty$ and recall that the $\mathcal{L}_\infty$ norm of a matrix rational transfer function G is defined as

$$\|G\|_\infty := \sup_\omega \overline{\sigma}\{G(j\omega)\}.$$

The computation of the $\mathcal{L}_\infty$ norm of G is complicated and requires a search. A control engineering interpretation of the infinity norm of a scalar transfer function G is the distance in the complex plane from the origin to the farthest point on the Nyquist plot of G, and it also appears as the peak value on the Bode magnitude plot of $|G(j\omega)|$. Hence the ∞ norm of a transfer function can, in principle, be obtained graphically.

To get an estimate, set up a fine grid of frequency points:

$$\{\omega_1, \cdots, \omega_N\}.$$

Then an estimate for $\|G\|_\infty$ is

$$\max_{1 \leq k \leq N} \overline{\sigma}\{G(j\omega_k)\}.$$

This value is usually read directly from a Bode singular value plot. The $\mathcal{RL}_\infty$ norm can also be computed in state-space.

Lemma 4.5 *Let $\gamma > 0$ and*

$$G(s) = \left[\begin{array}{c|c} A & B \\ \hline C & D \end{array}\right] \in \mathcal{RL}_\infty. \tag{4.2}$$

Then $\|G\|_\infty < \gamma$ if and only if $\bar{\sigma}(D) < \gamma$ and the Hamiltonian matrix H has no eigenvalues on the imaginary axis where

$$H := \begin{bmatrix} A + BR^{-1}D^*C & BR^{-1}B^* \\ -C^*(I + DR^{-1}D^*)C & -(A + BR^{-1}D^*C)^* \end{bmatrix} \quad (4.3)$$

and $R = \gamma^2 I - D^*D$.

Proof. Let $\Phi(s) = \gamma^2 I - G^\sim(s)G(s)$. Then it is clear that $\|G\|_\infty < \gamma$ if and only if $\Phi(j\omega) > 0$ for all $\omega \in \mathbb{R}$. Since $\Phi(\infty) = R > 0$ and since $\Phi(j\omega)$ is a continuous function of ω, $\Phi(j\omega) > 0$ for all $\omega \in \mathbb{R}$ if and only if $\Phi(j\omega)$ is nonsingular for all $\omega \in \mathbb{R} \cup \{\infty\}$; that is, $\Phi(s)$ has no imaginary axis zero. Equivalently, $\Phi^{-1}(s)$ has no imaginary axis pole. It is easy to compute by some simple algebra that

$$\Phi^{-1}(s) = \left[\begin{array}{c|c} H & \begin{bmatrix} BR^{-1} \\ -C^*DR^{-1} \end{bmatrix} \\ \hline \begin{bmatrix} R^{-1}D^*C & R^{-1}B^* \end{bmatrix} & R^{-1} \end{array} \right].$$

Thus the conclusion follows if the above realization has neither uncontrollable modes nor unobservable modes on the imaginary axis. Assume that $j\omega_0$ is an eigenvalue of H but not a pole of $\Phi^{-1}(s)$. Then $j\omega_0$ must be either an unobservable mode of $([\begin{array}{cc} R^{-1}D^*C & R^{-1}B^* \end{array}], H)$ or an uncontrollable mode of $(H, \begin{bmatrix} BR^{-1} \\ -C^*DR^{-1} \end{bmatrix})$. Now suppose $j\omega_0$ is an unobservable mode of $([\begin{array}{cc} R^{-1}D^*C & R^{-1}B^* \end{array}], H)$. Then there exists an $x_0 = \begin{bmatrix} x_1 \\ x_2 \end{bmatrix} \neq 0$ such that

$$Hx_0 = j\omega_0 x_0, \quad [\begin{array}{cc} R^{-1}D^*C & R^{-1}B^* \end{array}] x_0 = 0.$$

These equations can be simplified to

$$\begin{aligned} (j\omega_0 I - A)x_1 &= 0 \\ (j\omega_0 I + A^*)x_2 &= -C^*Cx_1 \\ D^*Cx_1 + B^*x_2 &= 0. \end{aligned}$$

Since A has no imaginary axis eigenvalues, we have $x_1 = 0$ and $x_2 = 0$. This contradicts our assumption, and hence the realization has no unobservable modes on the imaginary axis.

Similarly, a contradiction will also be arrived at if $j\omega_0$ is assumed to be an uncontrollable mode of $(H, \begin{bmatrix} BR^{-1} \\ -C^*DR^{-1} \end{bmatrix})$. □

4.4. Computing $\mathcal{L}_\infty$ and $\mathcal{H}_\infty$ Norms

Bisection Algorithm

Lemma 4.5 suggests the following bisection algorithm to compute $\mathcal{RL}_\infty$ norm:

(a) Select an upper bound γ_u and a lower bound γ_l such that $\gamma_l \leq \|G\|_\infty \leq \gamma_u$;

(b) If $(\gamma_u - \gamma_l)/\gamma_l \leq$ specified level, stop; $\|G\| \approx (\gamma_u + \gamma_l)/2$. Otherwise go to the next step;

(c) Set $\gamma = (\gamma_l + \gamma_u)/2$;

(d) Test if $\|G\|_\infty < \gamma$ by calculating the eigenvalues of H for the given γ;

(e) If H has an eigenvalue on $j\mathbb{R}$, set $\gamma_l = \gamma$; otherwise set $\gamma_u = \gamma$; go back to step (b).

Of course, the above algorithm applies to $\mathcal{H}_\infty$ norm computation as well. Thus $\mathcal{L}_\infty$ norm computation requires a search, over either γ or ω, in contrast to $\mathcal{L}_2$ ($\mathcal{H}_2$) norm computation, which does not. A somewhat analogous situation occurs for constant matrices with the norms $\|M\|_2^2 = \text{trace}(M^*M)$ and $\|M\|_\infty = \bar{\sigma}[M]$. In principle, $\|M\|_2^2$ can be computed exactly with a finite number of operations, as can the test for whether $\bar{\sigma}(M) < \gamma$ (e.g., $\gamma^2 I - M^*M > 0$), but the value of $\bar{\sigma}(M)$ cannot. To compute $\bar{\sigma}(M)$, we must use some type of iterative algorithm.

Remark 4.2 It is clear that $\|G\|_\infty < \gamma$ iff $\|\gamma^{-1}G\|_\infty < 1$. Hence, there is no loss of generality in assuming $\gamma = 1$. This assumption will often be made in the remainder of this book. It is also noted that there are other fast algorithms to carry out the preceding norm computation; nevertheless, this bisection algorithm is the simplest. ◇

Additional interpretations can be given for the $\mathcal{H}_\infty$ norm of a stable matrix transfer function. When $G(s)$ is a single-input and single-output system, the $\mathcal{H}_\infty$ norm of the $G(s)$ can be regarded as the largest possible amplification factor of the system's steady-state response to sinusoidal excitations. For example, the steady-state response of the system with respect to a sinusoidal input $u(t) = U\sin(\omega_0 t + \phi)$ is

$$y(t) = U|G(j\omega_0)|\sin(\omega_0 t + \phi + \angle G(j\omega_0))$$

and thus the maximum possible amplification factor is $\sup_{\omega_0} |G(j\omega_0)|$, which is precisely the $\mathcal{H}_\infty$ norm of the transfer function.

In the multiple-input and multiple-output case, the $\mathcal{H}_\infty$ norm of a transfer matrix $G \in \mathcal{RH}_\infty$ can also be regarded as the largest possible amplification factor of the system's steady-state response to sinusoidal excitations in the following sense: Let the sinusoidal inputs be

$$u(t) = \begin{bmatrix} u_1 \sin(\omega_0 t + \phi_1) \\ u_2 \sin(\omega_0 t + \phi_2) \\ \vdots \\ u_q \sin(\omega_0 t + \phi_q) \end{bmatrix}, \quad \hat{u} = \begin{bmatrix} u_1 \\ u_2 \\ \vdots \\ u_q \end{bmatrix}.$$

Then the steady-state response of the system can be written as

$$y(t) = \begin{bmatrix} y_1 \sin(\omega_0 t + \theta_1) \\ y_2 \sin(\omega_0 t + \theta_2) \\ \vdots \\ y_p \sin(\omega_0 t + \theta_p) \end{bmatrix}, \quad \hat{y} = \begin{bmatrix} y_1 \\ y_2 \\ \vdots \\ y_p \end{bmatrix}$$

for some y_i, θ_i, $i = 1, 2, \ldots, p$, and furthermore,

$$\|G\|_\infty = \sup_{\phi_i, \omega_o, \hat{u}} \frac{\|\hat{y}\|}{\|\hat{u}\|}$$

where $\|\cdot\|$ is the Euclidean norm. The details are left as an exercise.

Example 4.2 Consider a mass/spring/damper system as shown in Figure 4.2.

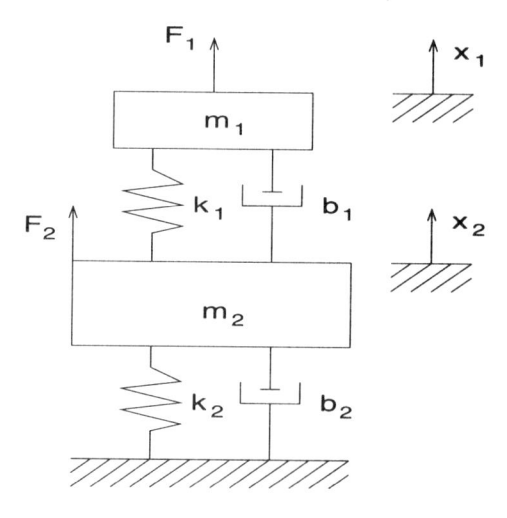

Figure 4.2: A two-mass/spring/damper system

The dynamical system can be described by the following differential equations:

$$\begin{bmatrix} \dot{x}_1 \\ \dot{x}_2 \\ \dot{x}_3 \\ \dot{x}_4 \end{bmatrix} = A \begin{bmatrix} x_1 \\ x_2 \\ x_3 \\ x_4 \end{bmatrix} + B \begin{bmatrix} F_1 \\ F_2 \end{bmatrix}$$

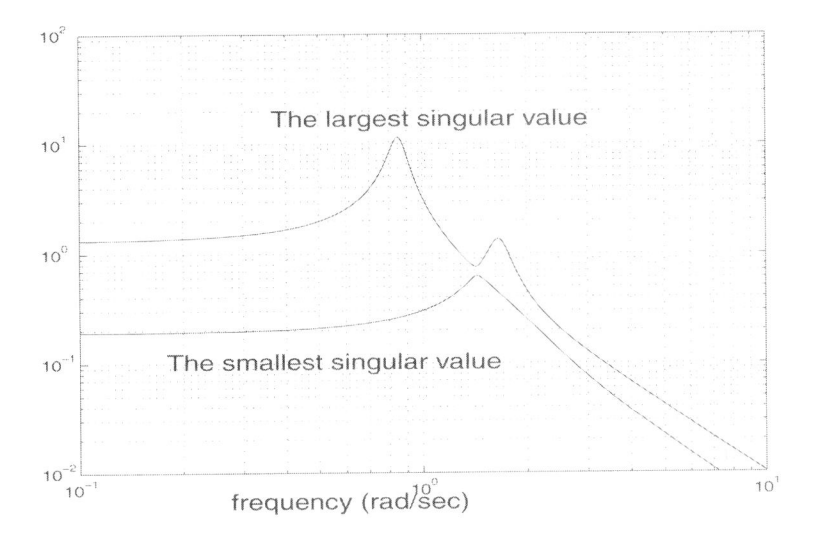

4.4. Computing $\mathcal{L}_\infty$ and $\mathcal{H}_\infty$ Norms

Figure 4.3: $\|G\|_\infty$ is the peak of the largest singular value of $G(j\omega)$

with

$$A = \begin{bmatrix} 0 & 0 & 1 & 0 \\ 0 & 0 & 0 & 1 \\ -\dfrac{k_1}{m_1} & \dfrac{k_1}{m_1} & -\dfrac{b_1}{m_1} & \dfrac{b_1}{m_1} \\ \dfrac{k_1}{m_2} & -\dfrac{k_1+k_2}{m_2} & \dfrac{b_1}{m_2} & -\dfrac{b_1+b_2}{m_2} \end{bmatrix}, \quad B = \begin{bmatrix} 0 & 0 \\ 0 & 0 \\ \dfrac{1}{m_1} & 0 \\ 0 & \dfrac{1}{m_2} \end{bmatrix}.$$

Suppose that $G(s)$ is the transfer matrix from (F_1, F_2) to (x_1, x_2); that is,

$$C = \begin{bmatrix} 1 & 0 & 0 & 0 \\ 0 & 1 & 0 & 0 \end{bmatrix}, \quad D = 0,$$

and suppose $k_1 = 1$, $k_2 = 4$, $b_1 = 0.2$, $b_2 = 0.1$, $m_1 = 1$, and $m_2 = 2$ with appropriate units. The following MATLAB commands generate the singular value Bode plot of the above system as shown in Figure 4.3.

≫ **G=pck(A,B,C,D);**

≫ **hinfnorm(G,0.0001)** or **linfnorm(G,0.0001)** % relative error ≤ 0.0001

≫ **w=logspace(-1,1,200);** % **200** points between $1 = 10^{-1}$ and $10 = 10^1$;

≫ **Gf=frsp(G,w);** % computing frequency response;

≫ **[u,s,v]=vsvd(Gf);** % SVD at each frequency;

≫ **vplot('liv, lm', s), grid** % plot both singular values and grid.

Then the $\mathcal{H}_\infty$ norm of this transfer matrix is $\|G(s)\|_\infty = 11.47$, which is shown as the peak of the largest singular value Bode plot in Figure 4.3. Since the peak is achieved at $\omega_{\max} = 0.8483$, exciting the system using the following sinusoidal input

$$\begin{bmatrix} F_1 \\ F_2 \end{bmatrix} = \begin{bmatrix} 0.9614 \sin(0.8483t) \\ 0.2753 \sin(0.8483t - 0.12) \end{bmatrix}$$

gives the steady-state response of the system as

$$\begin{bmatrix} x_1 \\ x_2 \end{bmatrix} = \begin{bmatrix} 11.47 \times 0.9614 \sin(0.8483t - 1.5483) \\ 11.47 \times 0.2753 \sin(0.8483t - 1.4283) \end{bmatrix}.$$

This shows that the system response will be amplified 11.47 times for an input signal at the frequency $\omega_{\max}$, which could be undesirable if F_1 and F_2 are disturbance force and x_1 and x_2 are the positions to be kept steady.

Example 4.3 Consider a two-by-two transfer matrix

$$G(s) = \begin{bmatrix} \dfrac{10(s+1)}{s^2 + 0.2s + 100} & \dfrac{1}{s+1} \\ \dfrac{s+2}{s^2 + 0.1s + 10} & \dfrac{5(s+1)}{(s+2)(s+3)} \end{bmatrix}.$$

A state-space realization of G can be obtained using the following MATLAB commands:

≫ **G11=nd2sys([10,10],[1,0.2,100]);**

≫ **G12=nd2sys(1,[1,1]);**

≫ **G21=nd2sys([1,2],[1,0.1,10]);**

≫ **G22=nd2sys([5,5],[1,5,6]);**

≫ **G=sbs(abv(G11,G21),abv(G12,G22));**

Next, we set up a frequency grid to compute the frequency response of G and the singular values of $G(j\omega)$ over a suitable range of frequency.

≫ **w=logspace(0,2,200);** % **200** points between $1 = 10^0$ and $100 = 10^2$;

≫ **Gf=frsp(G,w);** % computing frequency response;

≫ **[u,s,v]=vsvd(Gf);** % SVD at each frequency;

4.5. Notes and References

≫ **vplot('liv, lm', s), grid** % plot both singular values and grid;

≫ **pkvnorm(s)** % find the norm from the frequency response of the singular values.

The singular values of $G(j\omega)$ are plotted in Figure 4.4, which gives an estimate of $\|G\|_\infty \approx 32.861$. The state-space bisection algorithm described previously leads to $\|G\|_\infty = 50.25 \pm 0.01$ and the corresponding MATLAB command is

≫ **hinfnorm(G,0.0001)** or **linfnorm(G,0.0001)** % relative error ≤ 0.0001.

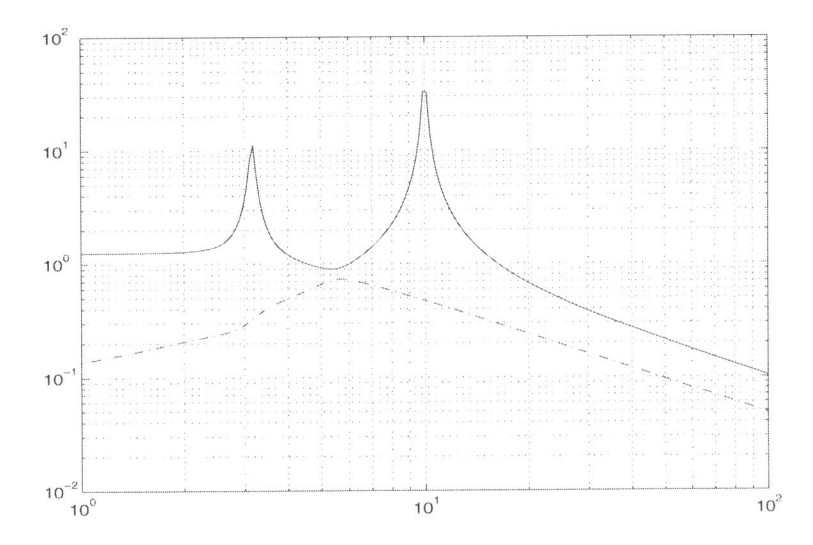

Figure 4.4: The largest and the smallest singular values of $G(j\omega)$

The preceding computational results show clearly that the graphical method can lead to a wrong answer for a lightly damped system if the frequency grid is not sufficiently dense. Indeed, we would get $\|G\|_\infty \approx 43.525, 48.286$ and 49.737 from the graphical method if $400, 800$, and 1600 frequency points are used, respectively.

Related MATLAB Commands: linfnorm, vnorm, getiv, scliv, var2con, xtract, xtracti

4.5 Notes and References

The basic concept of function spaces presented in this chapter can be found in any standard functional analysis textbook, for instance, Naylor and Sell [1982] and Gohberg and Goldberg [1981]. The system theoretical interpretations of the norms and function spaces can be found in Desoer and Vidyasagar [1975]. The bisection $\mathcal{L}_\infty$ norm computational algorithm was first developed in Boyd, Balakrishnan, and Kabamba [1989]. A more efficient $\mathcal{L}_\infty$ norm computational algorithm is presented in Bruinsma and Steinbuch [1990].

4.6 Problems

Problem 4.1 Let $G(s)$ be a matrix in $\mathcal{RH}_\infty$. Prove that

$$\left\| \begin{bmatrix} G \\ I \end{bmatrix} \right\|_\infty^2 = \|G\|_\infty^2 + 1.$$

Problem 4.2 (Parseval relation) Let $f(t), g(t) \in \mathcal{L}_2$, $F(j\omega) = \mathcal{F}\{f(t)\}$, and $G(j\omega) = \mathcal{F}\{g(t)\}$. Show that

$$\int_{-\infty}^{\infty} f(t)g(t)dt = \frac{1}{2\pi} \int_{-\infty}^{\infty} F(j\omega)G^*(j\omega)d\omega$$

and

$$\int_{-\infty}^{\infty} |f(t)|^2 dt = \frac{1}{2\pi} \int_{-\infty}^{\infty} |F(j\omega)|^2 d\omega.$$

Note that

$$F(j\omega) = \int_{-\infty}^{\infty} f(t) e^{-j\omega t} dt, \quad f(t) = \mathcal{F}^{-1}(F(j\omega)) = \frac{1}{2\pi} \int_{-\infty}^{\infty} F(j\omega) e^{j\omega t} d\omega.$$

where $\mathcal{F}^{-1}$ denotes the inverse Fourier transform.

Problem 4.3 Suppose A is stable. Show

$$\int_{-\infty}^{\infty} (j\omega I - A)^{-1} d\omega = \pi.$$

Suppose $G(s) = \left[\begin{array}{c|c} A & B \\ \hline C & 0 \end{array} \right] \in \mathcal{RH}_\infty$ and let $Q = Q^*$ be the observability Gramian. Use the above formula to show that

$$\frac{1}{2\pi} \int_{-\infty}^{\infty} G^\sim(j\omega) G(j\omega) d\omega = B^* Q B.$$

[Hint: Use the fact that $G^\sim(s)G(s) = F^\sim(s) + F(s)$ and $F(s) = B^*Q(sI - A)^{-1}B$.]

Problem 4.4 Compute the 2-norm and ∞-norm of the following systems:

$$G_1(s) = \begin{bmatrix} \dfrac{1}{s+1} & \dfrac{s+3}{(s+1)(s-2)} \\ \dfrac{10}{s-2} & \dfrac{5}{s+3} \end{bmatrix}, \quad G_2(s) = \left[\begin{array}{cc|c} 1 & 0 & 1 \\ 2 & 3 & 1 \\ \hline 1 & 2 & 0 \end{array} \right]$$

4.6. Problems

$$G_3(s) = \left[\begin{array}{cc|c} -1 & -2 & 1 \\ 1 & 0 & 0 \\ \hline 2 & 3 & 0 \end{array}\right], \quad G_4(s) = \left[\begin{array}{cccc|cc} -1 & -2 & -3 & 1 & 2 \\ 1 & 0 & 0 & 0 & 1 \\ 0 & 1 & 0 & 2 & 0 \\ \hline 1 & 0 & 0 & 1 & 0 \\ 0 & 1 & 1 & 0 & 2 \end{array}\right].$$

Problem 4.5 Let $r(t) = \sin\omega t$ be the input signal to a plant

$$G(s) = \frac{\omega_n^2}{s^2 + 2\xi\omega_n s + \omega_n^2}$$

with $0 < \xi < 1/\sqrt{2}$. Find the steady-state response of the system $y(t)$. Also find the frequency ω that gives the largest magnitude steady-state response of $y(t)$.

Problem 4.6 Let $G(s) \in \mathcal{RH}_\infty$ be a $p \times q$ transfer matrix and $y = G(s)u$. Suppose

$$u(t) = \begin{bmatrix} u_1 \sin(\omega_0 t + \phi_1) \\ u_2 \sin(\omega_0 t + \phi_2) \\ \vdots \\ u_q \sin(\omega_0 t + \phi_q) \end{bmatrix}, \quad \hat{u} = \begin{bmatrix} u_1 \\ u_2 \\ \vdots \\ u_q \end{bmatrix}.$$

Show that the steady-state response of the system is given by

$$y(t) = \begin{bmatrix} y_1 \sin(\omega_0 t + \theta_1) \\ y_2 \sin(\omega_0 t + \theta_2) \\ \vdots \\ y_p \sin(\omega_0 t + \theta_p) \end{bmatrix}, \quad \hat{y} = \begin{bmatrix} y_1 \\ y_2 \\ \vdots \\ y_p \end{bmatrix}$$

for some y_i and θ_i, $i = 1, 2, \ldots, p$. Show that $\sup_{\phi_i, \omega_o, \|\hat{u}\|_2 \leq 1} \|\hat{y}\|_2 = \|G\|_\infty$.

Problem 4.7 Write a MATLAB program to plot, versus γ, the distance from the imaginary axis to the nearest eigenvalue of the Hamiltonian matrix for a given state-space model with stable A. Try it on

$$\begin{bmatrix} \dfrac{s+1}{(s+2)(s+3)} & \dfrac{s}{s+1} \\ \dfrac{s^2-2}{(s+3)(s+4)} & \dfrac{s+4}{(s+1)(s+2)} \end{bmatrix}.$$

Read off the value of the $\mathcal{H}_\infty$-norm. Compare with the MATLAB function *hinfnorm*.

Problem 4.8 Let $G(s) = \dfrac{1}{(s^2 + 2\xi s + 1)(s + 1)}$. Compute $\|G(s)\|_\infty$ using the Bode plot and state-space algorithm, respectively for $\xi = 1, 0.1, 0.01, 0.001$ and compare the results.

Chapter 5

Internal Stability

This chapter introduces the feedback structure and discusses its stability and various stability tests. The arrangement of this chapter is as follows: Section 5.1 discusses the necessity for introducing feedback structure and describes the general feedback configuration. Section 5.2 defines the well-posedness of the feedback loop. Section 5.3 introduces the notion of internal stability and various stability tests. Section 5.4 introduces the stable coprime factorizations of rational matrices. The stability conditions in terms of various coprime factorizations are also considered in this section.

5.1 Feedback Structure

In designing control systems, there are several fundamental issues that transcend the boundaries of specific applications. Although they may differ for each application and may have different levels of importance, these issues are generic in their relationship to control design objectives and procedures. Central to these issues is the requirement to provide satisfactory performance in the face of modeling errors, system variations, and uncertainty. Indeed, this requirement was the original motivation for the development of feedback systems. Feedback is only required when system performance cannot be achieved because of uncertainty in system characteristics. A more detailed treatment of model uncertainties and their representations will be discussed in Chapter 8.

For the moment, assuming we are given a model including a representation of uncertainty that we believe adequately captures the essential features of the plant, the next step in the controller design process is to determine what structure is necessary to achieve the desired performance. Prefiltering input signals (or open-loop control) can change the dynamic response of the model set but cannot reduce the effect of uncertainty. If the uncertainty is too great to achieve the desired accuracy of response, then a feedback structure is required. The mere assumption of a feedback structure, however, does not guarantee a reduction of uncertainty, and there are many obstacles to achieving the uncertainty-reducing benefits of feedback. In particular, since for any reasonable model set representing a physical system uncertainty becomes large and the phase is completely unknown at sufficiently high frequencies, the loop gain must be small at those frequencies to avoid destabilizing the high-frequency system dynamics. Even worse is that the feedback system actually increases uncertainty and sensitivity in the frequency ranges where uncertainty is significantly large. In other words, because of the type of sets required to model physical systems reasonably and because of the restriction that our controllers be causal, we cannot use feedback (or any other control structure) to cause our closed-loop model set to be a proper subset of the open-loop model set. Often, what can be achieved with intelligent use of feedback is a significant reduction of uncertainty for certain signals of importance with a small increase spread over other signals. Thus, the feedback design problem centers around the tradeoff involved in reducing the overall impact of uncertainty. This tradeoff also occurs, for example, when using feedback to reduce command/disturbance error while minimizing response degradation due to measurement noise. To be of practical value, a design technique must provide means for performing these tradeoffs. We shall discuss these tradeoffs in more detail in the next chapter.

To focus our discussion, we shall consider the standard feedback configuration shown in Figure 5.1. It consists of the interconnected plant P and controller K forced by command r, sensor noise n, plant input disturbance d_i, and plant output disturbance d. In general, all signals are assumed to be multivariable, and all transfer matrices are assumed to have appropriate dimensions.

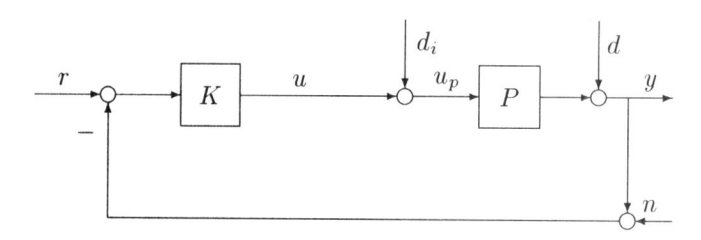

Figure 5.1: Standard feedback configuration

5.2 Well-Posedness of Feedback Loop

Assume that the plant P and the controller K in Figure 5.1 are fixed real rational proper transfer matrices. Then the first question one would ask is whether the feedback interconnection makes sense or is physically realizable. To be more specific, consider a simple example where

$$P = -\frac{s-1}{s+2}, \quad K = 1$$

5.2. Well-Posedness of Feedback Loop

are both proper transfer functions. However,

$$u = \frac{(s+2)}{3}(r - n - d) + \frac{s-1}{3}d_i.$$

That is, the transfer functions from the external signals $r - n - d$ and d_i to u are not proper. Hence, the feedback system is not physically realizable.

Definition 5.1 A feedback system is said to be *well-posed* if all closed-loop transfer matrices are well-defined and proper.

Now suppose that all the external signals r, n, d, and d_i are specified and that the closed-loop transfer matrices from them to u are, respectively, well-defined and proper. Then y and all other signals are also well-defined and the related transfer matrices are proper. Furthermore, since the transfer matrices from d and n to u are the same and differ from the transfer matrix from r to u by only a sign, the system is well-posed if and only if the transfer matrix from $\begin{bmatrix} d_i \\ d \end{bmatrix}$ to u exists and is proper.

To be consistent with the notation used in the rest of this book, we shall denote

$$\hat{K} := -K \tag{5.1}$$

and regroup the external input signals into the feedback loop as w_1 and w_2 and regroup the input signals of the plant and the controller as e_1 and e_2. Then the feedback loop with the plant and the controller can be simply represented as in Figure 5.2 and the system is well-posed if and only if the transfer matrix from $\begin{bmatrix} w_1 \\ w_2 \end{bmatrix}$ to e_1 exists and is proper.

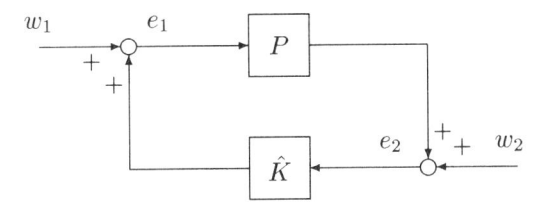

Figure 5.2: Internal stability analysis diagram

Lemma 5.1 *The feedback system in Figure 5.2 is well-posed if and only if*

$$I - \hat{K}(\infty)P(\infty) \tag{5.2}$$

is invertible.

Proof. The system in Figure 5.2 can be represented in equation form as

$$e_1 = w_1 + \hat{K}e_2$$
$$e_2 = w_2 + Pe_1.$$

Then an expression for e_1 can be obtained as

$$(I - \hat{K}P)e_1 = w_1 + \hat{K}w_2.$$

Thus well-posedness is equivalent to the condition that $(I - \hat{K}P)^{-1}$ exists and is proper. But this is equivalent to the condition that the constant term of the transfer function $I - \hat{K}P$ is invertible. □

It is straightforward to show that equation (5.2) is equivalent to either one of the following two conditions:

$$\begin{bmatrix} I & -\hat{K}(\infty) \\ -P(\infty) & I \end{bmatrix} \text{ is invertible;} \qquad (5.3)$$

$$I - P(\infty)\hat{K}(\infty) \text{ is invertible.}$$

The well-posedness condition is simple to state in terms of state-space realizations. Introduce realizations of P and $\hat{K}$:

$$P = \left[\begin{array}{c|c} A & B \\ \hline C & D \end{array}\right], \quad \hat{K} = \left[\begin{array}{c|c} \hat{A} & \hat{B} \\ \hline \hat{C} & \hat{D} \end{array}\right].$$

Then $P(\infty) = D$, $\hat{K}(\infty) = \hat{D}$ and the well-posedness condition in equation (5.3) is equivalent to the invertibility of $\begin{bmatrix} I & -\hat{D} \\ -D & I \end{bmatrix}$. Fortunately, in most practical cases we shall have $D = 0$, and hence well-posedness for most practical control systems is guaranteed.

5.3 Internal Stability

Consider a system described by the standard block diagram in Figure 5.2 and assume that the system is well-posed.

Definition 5.2 The system of Figure 5.2 is said to be *internally stable* if the transfer matrix

$$\begin{bmatrix} I & -\hat{K} \\ -P & I \end{bmatrix}^{-1} = \begin{bmatrix} (I - \hat{K}P)^{-1} & \hat{K}(I - P\hat{K})^{-1} \\ P(I - \hat{K}P)^{-1} & (I - P\hat{K})^{-1} \end{bmatrix} \qquad (5.4)$$

$$= \begin{bmatrix} I + \hat{K}(I - P\hat{K})^{-1}P & \hat{K}(I - P\hat{K})^{-1} \\ (I - P\hat{K})^{-1}P & (I - P\hat{K})^{-1} \end{bmatrix}$$

from (w_1, w_2) to (e_1, e_2) belongs to $\mathcal{RH}_\infty$.

5.3. Internal Stability

Note that to check internal stability, it is necessary (and sufficient) to test whether each of the four transfer matrices in equation (5.4) is in $\mathcal{H}_\infty$. Stability cannot be concluded even if three of the four transfer matrices in equation (5.4) are in $\mathcal{H}_\infty$. For example, let an interconnected system transfer function be given by

$$P = \frac{s-1}{s+1}, \qquad \hat{K} = -\frac{1}{s-1}.$$

Then it is easy to compute

$$\begin{bmatrix} e_1 \\ e_2 \end{bmatrix} = \begin{bmatrix} \dfrac{s+1}{s+2} & -\dfrac{s+1}{(s-1)(s+2)} \\ \dfrac{s-1}{s+2} & \dfrac{s+1}{s+2} \end{bmatrix} \begin{bmatrix} w_1 \\ w_2 \end{bmatrix},$$

which shows that the system is not internally stable although three of the four transfer functions are stable.

Remark 5.1 Internal stability is a basic requirement for a practical feedback system. This is because all interconnected systems may be unavoidably subject to some nonzero initial conditions and some (possibly small) errors, and it cannot be tolerated in practice that such errors at some locations will lead to unbounded signals at some other locations in the closed-loop system. Internal stability guarantees that all signals in a system are bounded provided that the injected signals (at any locations) are bounded. ◇

However, there are some special cases under which determining system stability is simple.

Corollary 5.2 *Suppose $\hat{K} \in \mathcal{RH}_\infty$. Then the system in Figure 5.2 is internally stable if and only if it is well-posed and $P(I - \hat{K}P)^{-1} \in \mathcal{RH}_\infty$.*

Proof. The necessity is obvious. To prove the sufficiency, it is sufficient to show that $(I - P\hat{K})^{-1} \in \mathcal{RH}_\infty$. But this follows from

$$(I - P\hat{K})^{-1} = I + (I - P\hat{K})^{-1} P\hat{K}$$

and $(I - P\hat{K})^{-1} P, \hat{K} \in \mathcal{RH}_\infty$. □

Also, we have the following:

Corollary 5.3 *Suppose $P \in \mathcal{RH}_\infty$. Then the system in Figure 5.2 is internally stable if and only if it is well-posed and $\hat{K}(I - P\hat{K})^{-1} \in \mathcal{RH}_\infty$.*

Corollary 5.4 *Suppose $P \in \mathcal{RH}_\infty$ and $\hat{K} \in \mathcal{RH}_\infty$. Then the system in Figure 5.2 is internally stable if and only if $(I - P\hat{K})^{-1} \in \mathcal{RH}_\infty$, or, equivalently, $\det(I - P(s)\hat{K}(s))$ has no zeros in the closed right-half plane.*

Note that all the previous discussions and conclusions apply equally to infinite dimensional plants and controllers. To study the more general case, we shall limit our discussions to finite dimensional systems and define

$$n_k := \text{ number of open right-half plane (rhp) poles of } \hat{K}(s)$$
$$n_p := \text{ number of open right-half plane (rhp) poles of } P(s).$$

Theorem 5.5 *The system is internally stable if and only if it is well-posed and*

(i) the number of open rhp poles of $P(s)\hat{K}(s) = n_k + n_p$;

(ii) $(I - P(s)\hat{K}(s))^{-1}$ *is stable.*

Proof. It is easy to show that $P\hat{K}$ and $(I - P\hat{K})^{-1}$ have the following realizations:

$$P\hat{K} = \left[\begin{array}{cc|c} A & B\hat{C} & B\hat{D} \\ 0 & \hat{A} & \hat{B} \\ \hline C & D\hat{C} & D\hat{D} \end{array}\right], \quad (I - P\hat{K})^{-1} = \left[\begin{array}{c|c} \bar{A} & \bar{B} \\ \hline \bar{C} & \bar{D} \end{array}\right]$$

where

$$\bar{A} = \left[\begin{array}{cc} A & B\hat{C} \\ 0 & \hat{A} \end{array}\right] + \left[\begin{array}{c} B\hat{D} \\ \hat{B} \end{array}\right](I - D\hat{D})^{-1}\left[\begin{array}{cc} C & D\hat{C} \end{array}\right]$$

$$\bar{B} = \left[\begin{array}{c} B\hat{D} \\ \hat{B} \end{array}\right](I - D\hat{D})^{-1}$$

$$\bar{C} = (I - D\hat{D})^{-1}\left[\begin{array}{cc} C & D\hat{C} \end{array}\right]$$

$$\bar{D} = (I - D\hat{D})^{-1}.$$

Hence, the system is internally stable iff $\bar{A}$ is stable. (see Problem 5.2.)

Now suppose that the system is internally stable; then $(I - P\hat{K})^{-1} \in \mathcal{RH}_\infty$. So we only need to show that given condition (ii), condition (i) is necessary and sufficient for the internal stability. This follows by noting that $(\bar{A}, \bar{B})$ is stabilizable iff

$$\left(\left[\begin{array}{cc} A & B\hat{C} \\ 0 & \hat{A} \end{array}\right], \left[\begin{array}{c} B\hat{D} \\ \hat{B} \end{array}\right]\right) \tag{5.5}$$

is stabilizable; and $(\bar{C}, \bar{A})$ is detectable iff

$$\left(\left[\begin{array}{cc} C & D\hat{C} \end{array}\right], \left[\begin{array}{cc} A & B\hat{C} \\ 0 & \hat{A} \end{array}\right]\right) \tag{5.6}$$

is detectable. But conditions (5.5) and (5.6) are equivalent to condition (i). $\square$

Condition (i) in the preceding theorem implies that there is no unstable pole/zero cancellation in forming the product $P\hat{K}$.

5.4. Coprime Factorization over $\mathcal{RH}_\infty$

The preceding theorem is, in fact, the basis for the classical control theory, where the stability is checked only for one closed-loop transfer function with the implicit assumption that the controller itself is stable (and most probably also minimum phase; or at least marginally stable and minimum phase with the condition that any imaginary axis pole of the controller is not in the same location as any zero of the plant).

Example 5.1 Let P and $\hat{K}$ be two-by-two transfer matrices

$$P = \begin{bmatrix} \dfrac{1}{s-1} & 0 \\ 0 & \dfrac{1}{s+1} \end{bmatrix}, \quad \hat{K} = \begin{bmatrix} \dfrac{1-s}{s+1} & -1 \\ 0 & -1 \end{bmatrix}.$$

Then

$$P\hat{K} = \begin{bmatrix} \dfrac{-1}{s+1} & \dfrac{-1}{s-1} \\ 0 & \dfrac{-1}{s+1} \end{bmatrix}, \quad (I - P\hat{K})^{-1} = \begin{bmatrix} \dfrac{s+1}{s+2} & -\dfrac{(s+1)^2}{(s+2)^2(s-1)} \\ 0 & \dfrac{s+1}{s+2} \end{bmatrix}.$$

So the closed-loop system is not stable even though

$$\det(I - P\hat{K}) = \frac{(s+2)^2}{(s+1)^2}$$

has no zero in the closed right-half plane and the number of unstable poles of $P\hat{K} = n_k + n_p = 1$. Hence, in general, $\det(I - P\hat{K})$ having no zeros in the closed right-half plane does not necessarily imply $(I - P\hat{K})^{-1} \in \mathcal{RH}_\infty$.

5.4 Coprime Factorization over $\mathcal{RH}_\infty$

Recall that two polynomials $m(s)$ and $n(s)$, with, for example, real coefficients, are said to be *coprime* if their greatest common divisor is 1 (equivalent, they have no common zeros). It follows from Euclid's algorithm[1] that two polynomials m and n are coprime iff there exist polynomials $x(s)$ and $y(s)$ such that $xm + yn = 1$; such an equation is called a Bezout identity. Similarly, two transfer functions $m(s)$ and $n(s)$ in $\mathcal{RH}_\infty$ are said to be *coprime over* $\mathcal{RH}_\infty$ if there exists $x, y \in \mathcal{RH}_\infty$ such that

$$xm + yn = 1.$$

---

[1] See, for example, Kailath [1980], pages 140–141.

The more primitive, but equivalent, definition is that m and n are coprime if every common divisor of m and n is invertible in $\mathcal{RH}_\infty$; that is,

$$h, mh^{-1}, nh^{-1} \in \mathcal{RH}_\infty \Longrightarrow h^{-1} \in \mathcal{RH}_\infty.$$

More generally, we have the following:

Definition 5.3 Two matrices M and N in $\mathcal{RH}_\infty$ are *right coprime over* $\mathcal{RH}_\infty$ if they have the same number of columns and if there exist matrices X_r and Y_r in $\mathcal{RH}_\infty$ such that

$$\begin{bmatrix} X_r & Y_r \end{bmatrix} \begin{bmatrix} M \\ N \end{bmatrix} = X_r M + Y_r N = I.$$

Similarly, two matrices $\tilde{M}$ and $\tilde{N}$ in $\mathcal{RH}_\infty$ are *left coprime over* $\mathcal{RH}_\infty$ if they have the same number of rows and if there exist matrices X_l and Y_l in $\mathcal{RH}_\infty$ such that

$$\begin{bmatrix} \tilde{M} & \tilde{N} \end{bmatrix} \begin{bmatrix} X_l \\ Y_l \end{bmatrix} = \tilde{M} X_l + \tilde{N} Y_l = I.$$

Note that these definitions are equivalent to saying that the matrix $\begin{bmatrix} M \\ N \end{bmatrix}$ is left invertible in $\mathcal{RH}_\infty$ and the matrix $\begin{bmatrix} \tilde{M} & \tilde{N} \end{bmatrix}$ is right invertible in $\mathcal{RH}_\infty$. These two equations are often called Bezout identities.

Now let P be a proper real rational matrix. A *right coprime factorization (rcf)* of P is a factorization $P = NM^{-1}$, where N and M are right coprime over $\mathcal{RH}_\infty$. Similarly, a *left coprime factorization (lcf)* has the form $P = \tilde{M}^{-1}\tilde{N}$, where $\tilde{N}$ and $\tilde{M}$ are left-coprime over $\mathcal{RH}_\infty$. A matrix $P(s) \in \mathcal{R}_p(s)$ is said to have *double coprime factorization* if there exist a right coprime factorization $P = NM^{-1}$, a left coprime factorization $P = \tilde{M}^{-1}\tilde{N}$, and $X_r, Y_r, X_l, Y_l \in \mathcal{RH}_\infty$ such that

$$\begin{bmatrix} X_r & Y_r \\ -\tilde{N} & \tilde{M} \end{bmatrix} \begin{bmatrix} M & -Y_l \\ N & X_l \end{bmatrix} = I. \tag{5.7}$$

Of course, implicit in these definitions is the requirement that both M and $\tilde{M}$ be square and nonsingular.

Theorem 5.6 *Suppose $P(s)$ is a proper real rational matrix and*

$$P = \left[\begin{array}{c|c} A & B \\ \hline C & D \end{array}\right]$$

is a stabilizable and detectable realization. Let F and L be such that $A+BF$ and $A+LC$ are both stable, and define

$$\begin{bmatrix} M & -Y_l \\ N & X_l \end{bmatrix} = \left[\begin{array}{c|cc} A+BF & B & -L \\ \hline F & I & 0 \\ C+DF & D & I \end{array}\right] \tag{5.8}$$

5.4. Coprime Factorization over $\mathcal{RH}_\infty$

$$\begin{bmatrix} X_r & Y_r \\ -\tilde{N} & \tilde{M} \end{bmatrix} = \left[\begin{array}{c|cc} A+LC & -(B+LD) & L \\ \hline F & I & 0 \\ C & -D & I \end{array}\right]. \tag{5.9}$$

Then $P = NM^{-1} = \tilde{M}^{-1}\tilde{N}$ are rcf and lcf, respectively, and, furthermore, equation (5.7) is satisfied.

Proof. The theorem follows by verifying equation (5.7). □

Remark 5.2 Note that if P is stable, then we can take $X_r = X_l = I$, $Y_r = Y_l = 0$, $N = \tilde{N} = P$, $M = \tilde{M} = I$. ◇

Remark 5.3 The coprime factorization of a transfer matrix can be given a feedback-control interpretation. For example, right coprime factorization comes out naturally from changing the control variable by a state feedback. Consider the state-space equations for a plant P:

$$\begin{aligned} \dot{x} &= Ax + Bu \\ y &= Cx + Du. \end{aligned}$$

Next, introduce a state feedback and change the variable

$$v := u - Fx$$

where F is such that $A + BF$ is stable. Then we get

$$\begin{aligned} \dot{x} &= (A+BF)x + Bv \\ u &= Fx + v \\ y &= (C+DF)x + Dv. \end{aligned}$$

Evidently, from these equations, the transfer matrix from v to u is

$$M(s) = \left[\begin{array}{c|c} A+BF & B \\ \hline F & I \end{array}\right]$$

and that from v to y is

$$N(s) = \left[\begin{array}{c|c} A+BF & B \\ \hline C+DF & D \end{array}\right]$$

Therefore,

$$u = Mv, \quad y = Nv$$

so that $y = NM^{-1}u$; that is, $P = NM^{-1}$. ◇

We shall now see how coprime factorizations can be used to obtain alternative characterizations of internal stability conditions. Consider again the standard stability analysis diagram in Figure 5.2. We begin with any rcf's and lcf's of P and $\hat{K}$:

$$P = NM^{-1} = \tilde{M}^{-1}\tilde{N} \tag{5.10}$$

$$\hat{K} = UV^{-1} = \tilde{V}^{-1}\tilde{U}. \tag{5.11}$$

Lemma 5.7 *Consider the system in Figure 5.2. The following conditions are equivalent:*

1. *The feedback system is internally stable.*

2. $\begin{bmatrix} M & U \\ N & V \end{bmatrix}$ *is invertible in $\mathcal{RH}_\infty$.*

3. $\begin{bmatrix} \tilde{V} & -\tilde{U} \\ -\tilde{N} & \tilde{M} \end{bmatrix}$ *is invertible in $\mathcal{RH}_\infty$.*

4. $\tilde{M}V - \tilde{N}U$ *is invertible in $\mathcal{RH}_\infty$.*

5. $\tilde{V}M - \tilde{U}N$ *is invertible in $\mathcal{RH}_\infty$.*

Proof. Note that the system is internally stable if

$$\begin{bmatrix} I & -\hat{K} \\ -P & I \end{bmatrix}^{-1} \in \mathcal{RH}_\infty$$

or, equivalently,

$$\begin{bmatrix} I & \hat{K} \\ P & I \end{bmatrix}^{-1} \in \mathcal{RH}_\infty \tag{5.12}$$

Now

$$\begin{bmatrix} I & \hat{K} \\ P & I \end{bmatrix} = \begin{bmatrix} I & UV^{-1} \\ NM^{-1} & I \end{bmatrix} = \begin{bmatrix} M & U \\ N & V \end{bmatrix} \begin{bmatrix} M^{-1} & 0 \\ 0 & V^{-1} \end{bmatrix}$$

so that

$$\begin{bmatrix} I & \hat{K} \\ P & I \end{bmatrix}^{-1} = \begin{bmatrix} M & 0 \\ 0 & V \end{bmatrix} \begin{bmatrix} M & U \\ N & V \end{bmatrix}^{-1}$$

Since the matrices

$$\begin{bmatrix} M & 0 \\ 0 & V \end{bmatrix}, \begin{bmatrix} M & U \\ N & V \end{bmatrix}$$

are right coprime (this fact is left as an exercise for the reader), equation (5.12) holds iff

$$\begin{bmatrix} M & U \\ N & V \end{bmatrix}^{-1} \in \mathcal{RH}_\infty$$

5.4. Coprime Factorization over $\mathcal{RH}_\infty$

This proves the equivalence of conditions 1 and 2. The equivalence of conditions 1 and 3 is proved similarly.

Conditions 4 and 5 are implied by conditions 2 and 3 from the following equation:

$$\begin{bmatrix} \tilde{V} & -\tilde{U} \\ -\tilde{N} & \tilde{M} \end{bmatrix} \begin{bmatrix} M & U \\ N & V \end{bmatrix} = \begin{bmatrix} \tilde{V}M - \tilde{U}N & 0 \\ 0 & \tilde{M}V - \tilde{N}U \end{bmatrix}$$

Since the left-hand side of the above equation is invertible in $\mathcal{RH}_\infty$, so is the right-hand side. Hence, conditions 4 and 5 are satisfied. We only need to show that either condition 4 or condition 5 implies condition 1. Let us show that condition 5 implies condition 1; this is obvious since

$$\begin{bmatrix} I & \hat{K} \\ P & I \end{bmatrix}^{-1} = \begin{bmatrix} I & \tilde{V}^{-1}\tilde{U} \\ NM^{-1} & I \end{bmatrix}^{-1}$$

$$= \begin{bmatrix} M & 0 \\ 0 & I \end{bmatrix} \begin{bmatrix} \tilde{V}M & \tilde{U} \\ N & I \end{bmatrix}^{-1} \begin{bmatrix} \tilde{V} & 0 \\ 0 & I \end{bmatrix} \in \mathcal{RH}_\infty$$

if $\begin{bmatrix} \tilde{V}M & \tilde{U} \\ N & I \end{bmatrix}^{-1} \in \mathcal{RH}_\infty$ or if condition 5 is satisfied. $\square$

Combining Lemma 5.7 and Theorem 5.6, we have the following corollary.

Corollary 5.8 *Let P be a proper real rational matrix and $P = NM^{-1} = \tilde{M}^{-1}\tilde{N}$ be the corresponding rcf and lcf over $\mathcal{RH}_\infty$. Then there exists a controller*

$$\hat{K}_0 = U_0 V_0^{-1} = \tilde{V}_0^{-1} \tilde{U}_0$$

with $U_0, V_0, \tilde{U}_0,$ and $\tilde{V}_0$ in $\mathcal{RH}_\infty$ such that

$$\begin{bmatrix} \tilde{V}_0 & -\tilde{U}_0 \\ -\tilde{N} & \tilde{M} \end{bmatrix} \begin{bmatrix} M & U_0 \\ N & V_0 \end{bmatrix} = \begin{bmatrix} I & 0 \\ 0 & I \end{bmatrix} \quad (5.13)$$

Furthermore, let F and L be such that $A+BF$ and $A+LC$ are stable. Then a particular set of state-space realizations for these matrices can be given by

$$\begin{bmatrix} M & U_0 \\ N & V_0 \end{bmatrix} = \left[\begin{array}{c|cc} A+BF & B & -L \\ \hline F & I & 0 \\ C+DF & D & I \end{array} \right] \quad (5.14)$$

$$\begin{bmatrix} \tilde{V}_0 & -\tilde{U}_0 \\ -\tilde{N} & \tilde{M} \end{bmatrix} = \left[\begin{array}{c|cc} A+LC & -(B+LD) & L \\ \hline F & I & 0 \\ C & -D & I \end{array} \right] \quad (5.15)$$

Proof. The idea behind the choice of these matrices is as follows. Using the observer theory, find a controller $\hat{K}_0$ achieving internal stability; for example

$$\hat{K}_0 := \left[\begin{array}{c|c} A + BF + LC + LDF & -L \\ \hline F & 0 \end{array} \right] \quad (5.16)$$

Perform factorizations

$$\hat{K}_0 = U_0 V_0^{-1} = \tilde{V}_0^{-1} \tilde{U}_0,$$

which are analogous to the ones performed on P. Then Lemma 5.7 implies that each of the two left-hand side block matrices of equation (5.13) must be invertible in $\mathcal{RH}_\infty$. In fact, equation (5.13) is satisfied by comparing it with equation (5.7). □

Finding a coprime factorization for a scalar transfer function is fairly easy. Let $P(s) = \text{num}(s)/\text{den}(s)$ where $\text{num}(s)$ and $\text{den}(s)$ are the numerator and the denominator polynomials of $P(s)$, and let $\alpha(s)$ be a stable polynomial of the same order as $\text{den}(s)$. Then $P(s) = n(s)/m(s)$ with $n(s) = \text{num}(s)/\alpha(s)$ and $m(s) = \text{den}(s)/\alpha(s)$ is a coprime factorization. However, finding an $x(s) \in \mathcal{H}_\infty$ and a $y(s) \in \mathcal{H}_\infty$ such that $x(s)n(s) + y(s)m(s) = 1$ needs much more work.

Example 5.2 Let $P(s) = \dfrac{s-2}{s(s+3)}$ and $\alpha = (s+1)(s+3)$. Then $P(s) = n(s)/m(s)$ with $n(s) = \dfrac{s-2}{(s+1)(s+3)}$ and $m(s) = \dfrac{s}{s+1}$ forms a coprime factorization. To find an $x(s) \in \mathcal{H}_\infty$ and a $y(s) \in \mathcal{H}_\infty$ such that $x(s)n(s) + y(s)m(s) = 1$, consider a stabilizing controller for P: $\hat{K} = -\dfrac{s-1}{s+10}$. Then $\hat{K} = u/v$ with $u = \hat{K}$ and $v = 1$ is a coprime factorization and

$$m(s)v(s) - n(s)u(s) = \frac{(s+11.7085)(s+2.214)(s+0.077)}{(s+1)(s+3)(s+10)} =: \beta(s)$$

Then we can take

$$x(s) = -u(s)/\beta(s) = \frac{(s-1)(s+1)(s+3)}{(s+11.7085)(s+2.214)(s+0.077)}$$

$$y(s) = v(s)/\beta(s) = \frac{(s+1)(s+3)(s+10)}{(s+11.7085)(s+2.214)(s+0.077)}$$

MATLAB programs can be used to find the appropriate F and L matrices in state-space so that the desired coprime factorization can be obtained. Let $A \in \mathbb{R}^{n \times n}$, $B \in \mathbb{R}^{n \times m}$ and $C \in \mathbb{R}^{p \times n}$. Then an F and an L can be obtained from

≫ **F=-lqr(A, B, eye(n), eye(m));** % or

```
>> F=-place(A, B, Pf);   % Pf= poles of A+BF
>> L = -lqr(A', C', eye(n), eye(p))';   % or
>> L = -place(A', C', Pl)';   % Pl=poles of A+LC.
```

5.5 Notes and References

The presentation of this chapter is based primarily on Doyle [1984]. The discussion of internal stability and coprime factorization can also be found in Francis [1987], Vidyasagar [1985], and Nett, Jacobson, and Balas [1984].

5.6 Problems

Problem 5.1 Recall that a feedback system is said to be internally stable if all closed-loop transfer functions are stable. Describe the conditions for internal stability of the following feedback system:

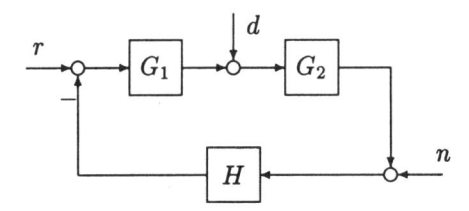

How can the stability conditions be simplified if $H(s)$ and $G_1(s)$ are both stable?

Problem 5.2 Show that $\begin{bmatrix} I & -\hat{K} \\ -P & I \end{bmatrix}^{-1} \in \mathcal{RH}_\infty$ if and only if

$$\bar{A} := \begin{bmatrix} A & B\hat{C} \\ 0 & \hat{A} \end{bmatrix} + \begin{bmatrix} B\hat{D} \\ \hat{B} \end{bmatrix} (I - D\hat{D})^{-1} \begin{bmatrix} C & D\hat{C} \end{bmatrix}$$

is stable.

Problem 5.3 Suppose $N, M, U, V \in \mathcal{RH}_\infty$ and NM^{-1} and UV^{-1} are right coprime factorizations, respectively. Show that

$$\begin{bmatrix} M & 0 \\ 0 & V \end{bmatrix} \begin{bmatrix} M & U \\ N & V \end{bmatrix}^{-1}$$

is also a right coprime factorization.

Problem 5.4 Let $G(s) = \dfrac{s-1}{(s+2)(s-3)}$. Find a stable coprime factorization $G = n(s)/m(s)$ and $x, y \in \mathcal{RH}_\infty$ such that $xn + ym = 1$.

Problem 5.5 Let $N(s) = \dfrac{(s-1)(s+\alpha)}{(s+2)(s+3)(s+\beta)}$ and $M(s) = \dfrac{(s-3)(s+\alpha)}{(s+3)(s+\beta)}$. Show that (N, M) is also a coprime factorization of the G in Problem 5.4 for any $\alpha > 0$ and $\beta > 0$.

Problem 5.6 Let $G = NM^{-1}$ be a right coprime factorization over $\mathcal{RH}_\infty$. It is called a normalized coprime factorization if $N^\sim N + M^\sim M = I$. Now consider scalar transfer function G. Then the following procedure can be used to find a normalized coprime factorization: (a) Let $G = n/m$ be any coprime factorization over $\mathcal{RH}_\infty$. (b) Find a stable and minimum phase spectral factor w such that $w^\sim w = n^\sim n + m^\sim m$. Let $N = n/w$ and $M = m/w$; then $G = N/M$ is a normalized coprime factorization. Find a normalized coprime factorization for Problem 5.4.

Problem 5.7 The following procedure constructs a normalized right coprime factorization when G is strictly proper:

1. Get a stabilizable, detectable realization A, B, C.

2. Do the MATLAB command $F = -\mathrm{lqr}(A, B, C'C, I)$.

3. Set
$$\begin{bmatrix} N \\ M \end{bmatrix}(s) = \left[\begin{array}{c|c} A+BF & B \\ \hline C & 0 \\ F & I \end{array}\right]$$

Verify that the procedure produces factors that satisfy $G = NM^{-1}$. Now try the procedure on
$$G(s) = \begin{bmatrix} \dfrac{1}{s-1} & \dfrac{1}{s-2} \\ \dfrac{2}{s} & \dfrac{1}{s+2} \end{bmatrix}$$

Verify numerically that
$$N(j\omega)^* N(j\omega) + M(j\omega)^* M(j\omega) = I, \quad \forall \omega. \tag{5.17}$$

Problem 5.8 Use the procedure in Problem 5.7 to find the normalized right coprime factorization for
$$G_1(s) = \begin{bmatrix} \dfrac{1}{s+1} & \dfrac{s+3}{(s+1)(s-2)} \\ \dfrac{10}{s-2} & \dfrac{5}{s+3} \end{bmatrix}$$

$$G_2(s) = \begin{bmatrix} \dfrac{2(s+1)(s+2)}{s(s+3)(s+4)} & \dfrac{s+2}{(s+1)(s+3)} \end{bmatrix}$$

5.6. Problems

$$G_3(s) = \left[\begin{array}{ccc|ccc} -1 & -2 & 1 & 1 & 2 & 3 \\ 0 & 2 & -1 & 3 & 2 & 1 \\ -4 & -3 & -2 & 1 & 1 & 1 \\ \hline 1 & 1 & 1 & 0 & 0 & 0 \\ 2 & 3 & 4 & 0 & 0 & 0 \end{array}\right]$$

$$G_4(s) = \left[\begin{array}{cc|cc} -1 & -2 & 1 & 2 \\ 0 & 1 & 2 & 1 \\ \hline 1 & 1 & 0 & 0 \\ 1 & 1 & 0 & 0 \end{array}\right]$$

Problem 5.9 Define the normalized left coprime factorization and describe a procedure to find such factorizations for strictly proper transfer matrices.

Chapter 6

Performance Specifications and Limitations

In this chapter, we consider further the feedback system properties and discuss how to achieve desired performance using feedback control. We also consider the mathematical formulations of optimal $\mathcal{H}_2$ and $\mathcal{H}_\infty$ control problems. A key step in the optimal control design is the selection of weighting functions. We shall give some guidelines to such selection process using some SISO examples. We shall also discuss in some detail the design limitations imposed by bandwidth constraints, the open-loop right-half plane zeros, and the open-loop right-half plane poles using Bode's gain and phase relation, Bode's sensitivity integral relation, and the Poisson integral formula.

6.1 Feedback Properties

In this section, we discuss the properties of a feedback system. In particular, we consider the benefit of the feedback structure and the concept of design tradeoffs for conflicting objectives — namely, how to achieve the benefits of feedback in the face of uncertainties.

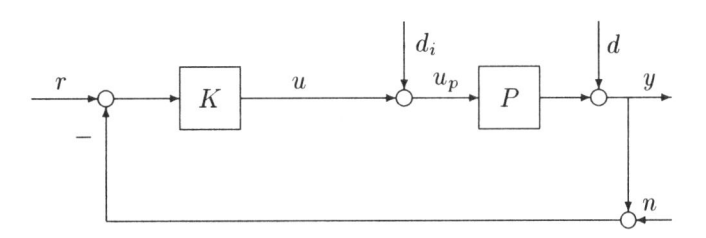

Figure 6.1: Standard feedback configuration

Consider again the feedback system shown in Figure 5.1. For convenience, the system diagram is shown again in Figure 6.1. For further discussion, it is convenient to define the *input loop transfer matrix*, L_i, and *output loop transfer matrix*, L_o, as

$$L_i = KP, \quad L_o = PK,$$

respectively, where L_i is obtained from breaking the loop at the input (u) of the plant while L_o is obtained from breaking the loop at the output (y) of the plant. The *input sensitivity* matrix is defined as the transfer matrix from d_i to u_p:

$$S_i = (I + L_i)^{-1}, \quad u_p = S_i d_i.$$

The *output sensitivity* matrix is defined as the transfer matrix from d to y:

$$S_o = (I + L_o)^{-1}, \quad y = S_o d.$$

The *input* and *output complementary sensitivity* matrices are defined as

$$T_i = I - S_i = L_i(I + L_i)^{-1}$$

$$T_o = I - S_o = L_o(I + L_o)^{-1},$$

respectively. (The word *complementary* is used to signify the fact that T is the complement of S, $T = I - S$.) The matrix $I + L_i$ is called the *input return difference matrix* and $I + L_o$ is called the *output return difference matrix*.

It is easy to see that the closed-loop system, if it is internally stable, satisfies the following equations:

$$y = T_o(r - n) + S_o P d_i + S_o d \quad (6.1)$$
$$r - y = S_o(r - d) + T_o n - S_o P d_i \quad (6.2)$$
$$u = K S_o(r - n) - K S_o d - T_i d_i \quad (6.3)$$
$$u_p = K S_o(r - n) - K S_o d + S_i d_i. \quad (6.4)$$

These four equations show the fundamental benefits and design objectives inherent in feedback loops. For example, equation (6.1) shows that the effects of disturbance d on the plant output can be made "small" by making the output sensitivity function S_o small. Similarly, equation (6.4) shows that the effects of disturbance d_i on the plant input can be made small by making the input sensitivity function S_i small. The notion of smallness for a transfer matrix in a certain range of frequencies can be made explicit using frequency-dependent singular values, for example, $\bar{\sigma}(S_o) < 1$ over a frequency range would mean that the effects of disturbance d at the plant output are effectively desensitized over that frequency range.

Hence, good disturbance rejection at the plant output (y) would require that

$$\bar{\sigma}(S_o) = \bar{\sigma}\left((I + PK)^{-1}\right) = \frac{1}{\underline{\sigma}(I + PK)} \quad \text{(for disturbance at plant output, } d\text{)},$$

$$\bar{\sigma}(S_o P) = \bar{\sigma}\left((I + PK)^{-1} P\right) = \bar{\sigma}(P S_i) \quad \text{(for disturbance at plant input, } d_i\text{)}$$

6.1. Feedback Properties

be made small and good disturbance rejection at the plant input (u_p) would require that

$$\overline{\sigma}(S_i) = \overline{\sigma}\left((I+KP)^{-1}\right) = \frac{1}{\underline{\sigma}(I+KP)} \quad \text{(for disturbance at plant input, } d_i\text{),}$$
$$\overline{\sigma}(S_i K) = \overline{\sigma}\left(K(I+PK)^{-1}\right) = \overline{\sigma}(KS_o) \quad \text{(for disturbance at plant output, } d\text{)}$$

be made small, particularly in the low-frequency range where d and d_i are usually significant.

Note that

$$\underline{\sigma}(PK) - 1 \leq \underline{\sigma}(I+PK) \leq \underline{\sigma}(PK) + 1$$
$$\underline{\sigma}(KP) - 1 \leq \underline{\sigma}(I+KP) \leq \underline{\sigma}(KP) + 1$$

then

$$\frac{1}{\underline{\sigma}(PK)+1} \leq \overline{\sigma}(S_o) \leq \frac{1}{\underline{\sigma}(PK)-1}, \quad \text{if } \underline{\sigma}(PK) > 1$$
$$\frac{1}{\underline{\sigma}(KP)+1} \leq \overline{\sigma}(S_i) \leq \frac{1}{\underline{\sigma}(KP)-1}, \quad \text{if } \underline{\sigma}(KP) > 1$$

These equations imply that

$$\overline{\sigma}(S_o) \ll 1 \iff \underline{\sigma}(PK) \gg 1$$
$$\overline{\sigma}(S_i) \ll 1 \iff \underline{\sigma}(KP) \gg 1.$$

Now suppose P and K are invertible; then

$$\underline{\sigma}(PK) \gg 1 \text{ or } \underline{\sigma}(KP) \gg 1 \iff \overline{\sigma}(S_o P) = \overline{\sigma}\left((I+PK)^{-1}P\right) \approx \overline{\sigma}(K^{-1}) = \frac{1}{\underline{\sigma}(K)}$$

$$\underline{\sigma}(PK) \gg 1 \text{ or } \underline{\sigma}(KP) \gg 1 \iff \overline{\sigma}(KS_o) = \overline{\sigma}\left(K(I+PK)^{-1}\right) \approx \overline{\sigma}(P^{-1}) = \frac{1}{\underline{\sigma}(P)}$$

Hence good performance at plant output (y) requires, in general, large output loop gain $\underline{\sigma}(L_o) = \underline{\sigma}(PK) \gg 1$ in the frequency range where d is significant for desensitizing d and large enough controller gain $\underline{\sigma}(K) \gg 1$ in the frequency range where d_i is significant for desensitizing d_i. Similarly, good performance at plant input (u_p) requires, in general, large input loop gain $\underline{\sigma}(L_i) = \underline{\sigma}(KP) \gg 1$ in the frequency range where d_i is significant for desensitizing d_i and large enough plant gain $\underline{\sigma}(P) \gg 1$ in the frequency range where d is significant, which cannot be changed by controller design, for desensitizing d. [In general, $S_o \neq S_i$ unless K and P are square and diagonal, which is true if P is a scalar system. Hence, small $\overline{\sigma}(S_o)$ does not necessarily imply small $\overline{\sigma}(S_i)$; in other words, good disturbance rejection at the output does not necessarily mean good disturbance rejection at the plant input.]

Hence, *good multivariable feedback loop design boils down to achieving high loop (and possibly controller) gains in the necessary frequency range.*

Despite the simplicity of this statement, feedback design is by no means trivial. This is true because loop gains cannot be made arbitrarily high over arbitrarily large frequency ranges. Rather, they must satisfy certain performance tradeoff and design limitations. A major performance tradeoff, for example, concerns commands and disturbance error reduction versus stability under the model uncertainty. Assume that the plant model is perturbed to $(I + \Delta)P$ with Δ stable, and assume that the system is nominally stable (i.e., the closed-loop system with $\Delta = 0$ is stable). Now the perturbed closed-loop system is stable if

$$\det(I + (I + \Delta)PK) = \det(I + PK)\det(I + \Delta T_o)$$

has no right-half plane zero. This would, in general, amount to requiring that $\|\Delta T_o\|$ be small or that $\bar{\sigma}(T_o)$ be small at those frequencies where Δ is significant, typically at high-frequency range, which, in turn, implies that the loop gain, $\bar{\sigma}(L_o)$, should be small at those frequencies.

Still another tradeoff is with the sensor noise error reduction. The conflict between the disturbance rejection and the sensor noise reduction is evident in equation (6.1). Large $\underline{\sigma}(L_o(j\omega))$ values over a large frequency range make errors due to d small. However, they also make errors due to n large because this noise is "passed through" over the same frequency range, that is,

$$y = T_o(r - n) + S_o P d_i + S_o d \approx (r - n)$$

Note that n is typically significant in the high-frequency range. Worst still, large loop gains outside of the bandwidth of P — that is, $\underline{\sigma}(L_o(j\omega)) \gg 1$ or $\underline{\sigma}(L_i(j\omega)) \gg 1$ while $\bar{\sigma}(P(j\omega)) \ll 1$ — can make the control activity (u) quite unacceptable, which may cause the saturation of actuators. This follows from

$$u = KS_o(r - n - d) - T_i d_i = S_i K(r - n - d) - T_i d_i \approx P^{-1}(r - n - d) - d_i$$

Here, we have assumed P to be square and invertible for convenience. The resulting equation shows that disturbances and sensor noise are actually amplified at u whenever the frequency range significantly exceeds the bandwidth of P, since for ω such that $\bar{\sigma}(P(j\omega)) \ll 1$ we have

$$\underline{\sigma}[P^{-1}(j\omega)] = \frac{1}{\bar{\sigma}[P(j\omega)]} \gg 1$$

Similarly, the controller gain, $\bar{\sigma}(K)$, should also be kept not too large in the frequency range where the loop gain is small in order not to saturate the actuators. This is because for small loop gain $\bar{\sigma}(L_o(j\omega)) \ll 1$ or $\bar{\sigma}(L_i(j\omega)) \ll 1$

$$u = KS_o(r - n - d) - T_i d_i \approx K(r - n - d)$$

Therefore, it is desirable to keep $\bar{\sigma}(K)$ not too large when the loop gain is small.

To summarize the above discussion, we note that good performance requires in some frequency range, typically some low-frequency range $(0, \omega_l)$,

$$\underline{\sigma}(PK) \gg 1, \quad \underline{\sigma}(KP) \gg 1, \quad \underline{\sigma}(K) \gg 1$$

and good robustness and good sensor noise rejection require in some frequency range, typically some high-frequency range (ω_h, ∞),

$$\bar{\sigma}(PK) \ll 1, \quad \bar{\sigma}(KP) \ll 1, \quad \bar{\sigma}(K) \leq M$$

where M is not too large. These design requirements are shown graphically in Figure 6.2. The specific frequencies ω_l and ω_h depend on the specific applications and the knowledge one has of the disturbance characteristics, the modeling uncertainties, and the sensor noise levels.

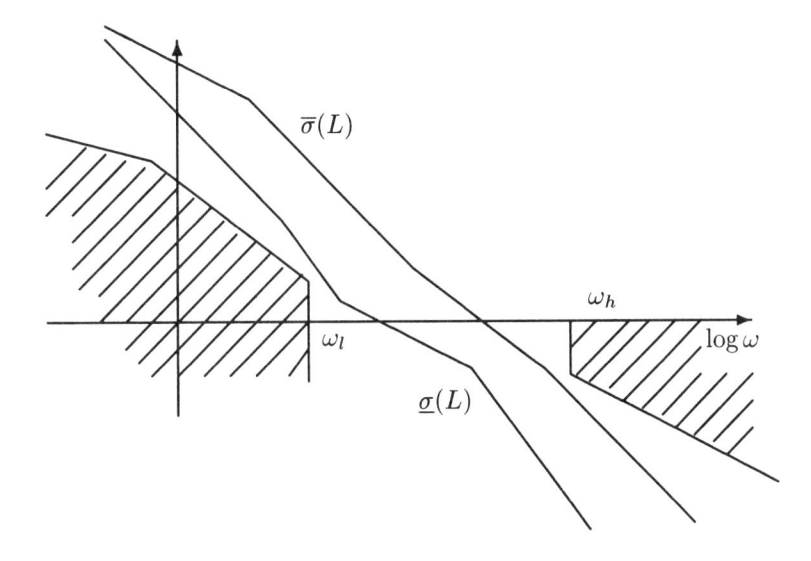

Figure 6.2: Desired loop gain

6.2 Weighted $\mathcal{H}_2$ and $\mathcal{H}_\infty$ Performance

In this section, we consider how to formulate some performance objectives into mathematically tractable problems. As shown in Section 6.1, the performance objectives of a feedback system can usually be specified in terms of requirements on the sensitivity functions and/or complementary sensitivity functions or in terms of some other closed-loop transfer functions. For instance, the performance criteria for a scalar system may be specified as requiring

$$\begin{cases} |S(j\omega)| \leq \varepsilon, & \forall \omega \leq \omega_0, \\ |S(j\omega)| \leq M, & \forall \omega > \omega_0 \end{cases}$$

where $S(j\omega) = 1/(1 + P(j\omega)K(j\omega))$. However, it is much more convenient to reflect the system performance objectives by choosing appropriate weighting functions. For example, the preceding performance objective can be written as

$$|W_e(j\omega)S(j\omega)| \leq 1, \quad \forall \omega$$

with

$$|W_e(j\omega)| = \begin{cases} 1/\varepsilon, & \forall \omega \leq \omega_0 \\ 1/M, & \forall \omega > \omega_0 \end{cases}$$

To use W_e in control design, a rational transfer function $W_e(s)$ is usually used to approximate the foregoing frequency response.

The advantage of using weighted performance specifications is obvious in multivariable system design. First, some components of a vector signal are usually more important than others. Second, each component of the signal may not be measured in the same units; for example, some components of the output error signal may be measured in terms of length, and others may be measured in terms of voltage. Therefore, weighting functions are essential to make these components comparable. Also, we might be primarily interested in rejecting errors in a certain frequency range (for example, low frequencies); hence some frequency-dependent weights must be chosen.

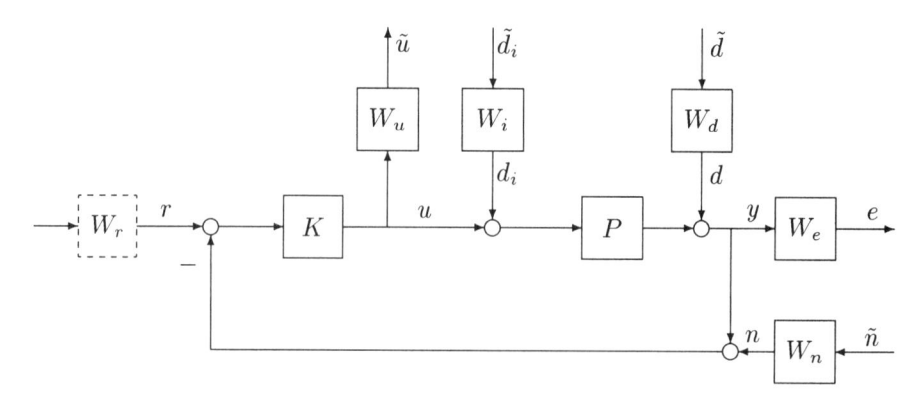

Figure 6.3: Standard feedback configuration with weights

In general, we shall modify the standard feedback diagram in Figure 6.1 into Figure 6.3. The weighting functions in Figure 6.3 are chosen to reflect the design objectives and knowledge of the disturbances and sensor noise. For example, W_d and W_i may be chosen to reflect the frequency contents of the disturbances d and d_i or they may be used to model the disturbance power spectrum depending on the nature of signals involved in the practical systems. The weighting matrix W_n is used to model the frequency contents of the sensor noise while W_e may be used to reflect the requirements on the shape of certain closed-loop transfer functions (for example, the shape of the output sensitivity function). Similarly, W_u may be used to reflect some restrictions on the control or actuator signals, and the dashed precompensator W_r is an optional element used to achieve deliberate command shaping or to represent a nonunity feedback system in equivalent unity feedback form.

6.2. Weighted $\mathcal{H}_2$ and $\mathcal{H}_\infty$ Performance

It is, in fact, essential that some appropriate weighting matrices be used in order to utilize the optimal control theory discussed in this book (i.e., $\mathcal{H}_2$ and $\mathcal{H}_\infty$ theory). So a very important step in the controller design process is to choose the appropriate weights, W_e, W_d, W_u, and possibly W_n, W_i, W_r. The appropriate choice of weights for a particular practical problem is not trivial. In many occasions, as in the scalar case, the weights are chosen purely as a design parameter without any physical bases, so these weights may be treated as tuning parameters that are chosen by the designer to achieve the best compromise between the conflicting objectives. The selection of the weighting matrices should be guided by the expected system inputs and the relative importance of the outputs.

Hence, control design may be regarded as a process of choosing a controller K such that certain weighted signals are made small in some sense. There are many different ways to define the smallness of a signal or transfer matrix, as we have discussed in the last chapter. Different definitions lead to different control synthesis methods, and some are much harder than others. A control engineer should make a judgment of the mathematical complexity versus engineering requirements.

Next, we introduce two classes of performance formulations: $\mathcal{H}_2$ and $\mathcal{H}_\infty$ criteria. For the simplicity of presentation, we shall assume that $d_i = 0$ and $n = 0$.

$\mathcal{H}_2$ Performance

Assume, for example, that the disturbance $\tilde{d}$ can be approximately modeled as an impulse with random input direction; that is,

$$\tilde{d}(t) = \eta \delta(t)$$

and

$$E(\eta \eta^*) = I$$

where E denotes the expectation. We may choose to minimize the expected energy of the error e due to the disturbance $\tilde{d}$:

$$E\left\{\|e\|_2^2\right\} = E\left\{\int_0^\infty \|e\|^2\, dt\right\} = \|W_e S_o W_d\|_2^2$$

In general, a controller minimizing only the above criterion can lead to a very large control signal u that could cause saturation of the actuators as well as many other undesirable problems. Hence, for a realistic controller design, it is necessary to include the control signal u in the cost function. Thus, our design criterion would usually be something like this:

$$E\left\{\|e\|_2^2 + \rho^2 \|\tilde{u}\|_2^2\right\} = \left\|\begin{bmatrix} W_e S_o W_d \\ \rho W_u K S_o W_d \end{bmatrix}\right\|_2^2$$

with some appropriate choice of weighting matrix W_u and scalar ρ. The parameter ρ clearly defines the tradeoff we discussed earlier between good disturbance rejection at the output and control effort (or disturbance and sensor noise rejection at the actuators). Note that ρ can be set to $\rho = 1$ by an appropriate choice of W_u. This problem can be viewed as minimizing the *energy* consumed by the system in order to reject the disturbance d.

This type of problem was the dominant paradigm in the 1960s and 1970s and is usually referred to as linear quadratic Gaussian control, or simply as LQG. (Such problems will also be referred to as $\mathcal{H}_2$ mixed-sensitivity problems for consistency with the $\mathcal{H}_\infty$ problems discussed next.) The development of this paradigm stimulated extensive research efforts and is responsible for important technological innovation, particularly in the area of estimation. The theoretical contributions include a deeper understanding of linear systems and improved computational methods for complex systems through state-space techniques. The major limitation of this theory is the lack of formal treatment of uncertainty in the plant itself. By allowing only additive noise for uncertainty, the stochastic theory ignored this important practical issue. Plant uncertainty is particularly critical in feedback systems. (See Paganini [1995,1996] for some recent results on robust $\mathcal{H}_2$ control theory.)

$\mathcal{H}_\infty$ Performance

Although the $\mathcal{H}_2$ norm (or $\mathcal{L}_2$ norm) may be a meaningful performance measure and although LQG theory can give efficient design compromises under certain disturbance and plant assumptions, the $\mathcal{H}_2$ norm suffers a major deficiency. This deficiency is due to the fact that the tradeoff between disturbance error reduction and sensor noise error reduction is not the only constraint on feedback design. The problem is that these performance tradeoffs are often overshadowed by a second limitation on high loop gains — namely, the requirement for tolerance to uncertainties. Though a controller may be designed using FDLTI models, the design must be implemented and operated with a real physical plant. The properties of physical systems (in particular, the ways in which they deviate from finite-dimensional linear models) put strict limitations on the frequency range over which the loop gains may be large.

A solution to this problem would be to put explicit constraints on the loop gain in the cost function. For instance, one may chose to minimize

$$\sup_{\|\tilde{d}\|_2 \leq 1} \|e\|_2 = \|W_e S_o W_d\|_\infty$$

subject to some restrictions on the control energy or control bandwidth:

$$\sup_{\|\tilde{d}\|_2 \leq 1} \|\tilde{u}\|_2 = \|W_u K S_o W_d\|_\infty$$

Or, more frequently, one may introduce a parameter ρ and a mixed criterion

$$\sup_{\|\tilde{d}\|_2 \leq 1} \left\{ \|e\|_2^2 + \rho^2 \|\tilde{u}\|_2^2 \right\} = \left\| \begin{bmatrix} W_e S_o W_d \\ \rho W_u K S_o W_d \end{bmatrix} \right\|_\infty^2$$

6.3. Selection of Weighting Functions

Alternatively, if the system robust stability margin is the major concern, the weighted complementary sensitivity has to be limited. Thus the whole cost function may be

$$\left\| \begin{bmatrix} W_e S_o W_d \\ \rho W_1 T_o W_2 \end{bmatrix} \right\|_\infty$$

where W_1 and W_2 are the frequency-dependent uncertainty scaling matrices. These design problems are usually called $\mathcal{H}_\infty$ mixed-sensitivity problems. For a scalar system, an $\mathcal{H}_\infty$ norm minimization problem can also be viewed as minimizing the maximum magnitude of the system's steady-state response with respect to the worst-case sinusoidal inputs.

6.3 Selection of Weighting Functions

The selection of weighting functions for a specific design problem often involves ad hoc fixing, many iterations, and fine tuning. It is very hard to give a general formula for the weighting functions that will work in every case. Nevertheless, we shall try to give some guidelines in this section by looking at a typical SISO problem.

Consider an SISO feedback system shown in Figure 6.1. Then the tracking error is $e = r - y = S(r - d) + Tn - SPd_i$. So, as we have discussed earlier, we must keep $|S|$ small over a range of frequencies, typically low frequencies where r and d are significant. To motivate the choice of our performance weighting function W_e, let $L = PK$ be a standard second-order system

$$L = \frac{\omega_n^2}{s(s + 2\xi\omega_n)}$$

It is well-known from the classical control theory that the quality of the (step) time response can be quantified by rise time t_r, settling time t_s, and percent overshoot $100M_p\%$. Furthermore, these performance indices can be approximately calculated as

$$t_r \approx \frac{0.6 + 2.16\xi}{\omega_n}, \ 0.3 \leq \xi \leq 0.8; \ t_s \approx \frac{4}{\xi\omega_n}; \ M_p = e^{-\frac{\pi\xi}{\sqrt{1-\xi^2}}}, \ 0 < \xi < 1$$

The key points to note are that (1) the speed of the system response is proportional to ω_n and (2) the overshoot of the system response is determined only by the damping ratio ξ. It is well known that the frequency ω_n and the damping ratio ξ can be essentially captured in the frequency domain by the open-loop crossover frequency and the phase margin or the bandwidth and the resonant peak of the closed-loop complementary sensitivity function T.

Since our performance objectives are closely related to the sensitivity function, we shall consider in some detail how these time domain indices or, equivalently, ω_n and ξ are related to the frequency response of the sensitivity function

$$S = \frac{1}{1+L} = \frac{s(s + 2\xi\omega_n)}{s^2 + 2\xi\omega_n s + \omega_n^2}$$

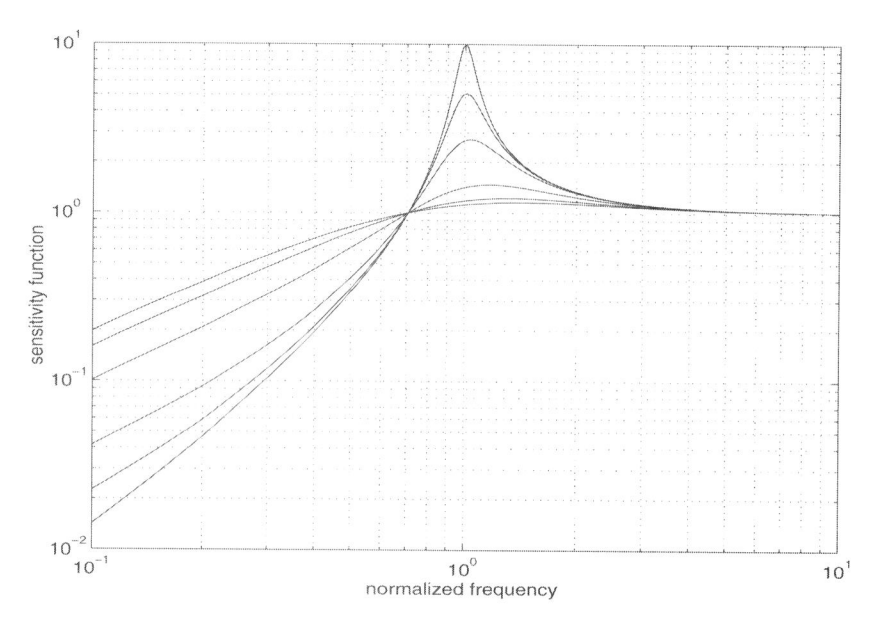

Figure 6.4: Sensitivity function S for $\xi = 0.05, 0.1, 0.2, 0.5, 0.8$, and 1 with normalized frequency (ω/ω_n)

The frequency response of the sensitivity function S is shown in Figure 6.4. Note that $|S(j\omega_n/\sqrt{2})| = 1$. We can regard the closed-loop bandwidth $\omega_b \approx \omega_n/\sqrt{2}$, since beyond this frequency the closed-loop system will not be able to track the reference and the disturbance will actually be amplified.

Next, note that

$$M_s := \|S\|_\infty = |S(j\omega_{\max})| = \frac{\alpha\sqrt{\alpha^2 + 4\xi^2}}{\sqrt{(1-\alpha^2)^2 + 4\xi^2\alpha^2}}$$

where $\alpha = \sqrt{0.5 + 0.5\sqrt{1 + 8\xi^2}}$ and $\omega_{\max} = \alpha\omega_n$. For example, $M_s = 5.123$ when $\xi = 0.1$. The relationship between ξ and M_s is shown in Figure 6.5. It is clear that the overshoot can be excessive if M_s is large. Hence a good control design should not have a very large M_s.

Now suppose we are given the time domain performance specifications then we can determine the corresponding requirements in frequency domain in terms of the bandwidth ω_b and the peak sensitivity M_s. Hence a good control design should result in a sensitivity function S satisfying both the bandwidth ω_b and the peak sensitivity M_s requirements, as shown in Figure 6.6. These requirements can be approximately represented as

$$|S(s)| \leq \left|\frac{s}{s/M_s + \omega_b}\right|, \quad s = j\omega, \ \forall \ \omega$$

6.3. Selection of Weighting Functions

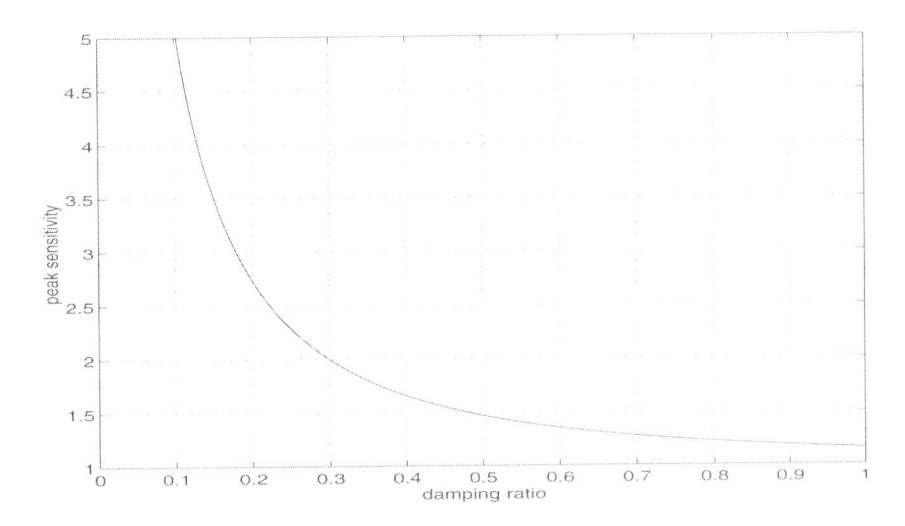

Figure 6.5: Peak sensitivity M_s versus damping ratio ξ

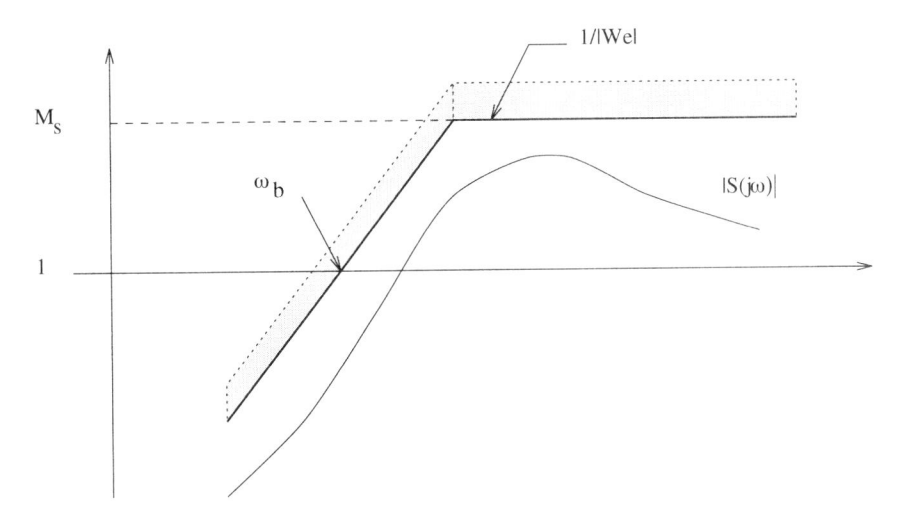

Figure 6.6: Performance weight W_e and desired S

Or, equivalently, $|W_eS| \leq 1$ with

$$W_e = \frac{s/M_s + \omega_b}{s} \qquad (6.5)$$

The preceding discussion applies in principle to most control design and hence the preceding weighting function can, in principle, be used as a candidate weighting function in an initial design. Since the steady-state error with respect to a step input is given by $|S(0)|$, it is clear that $|S(0)| = 0$ if the closed-loop system is stable and $\|W_eS\|_\infty < \infty$.

Unfortunately, the optimal control techniques described in this book cannot be used *directly* for problems with such weighting functions since these techniques assume that all unstable poles of the system (including plant and all performance and control weighting functions) are stabilizable by the control and detectable from the measurement outputs, which is clearly not satisfied if W_e has an imaginary axis pole since W_e is not detectable from the measurement. We shall discuss in Chapter 14 how such problems can be reformulated so that the techniques described in this book can be applied. A theory dealing directly with such problems is available but is much more complicated both theoretically and computationally and does not seem to offer much advantage.

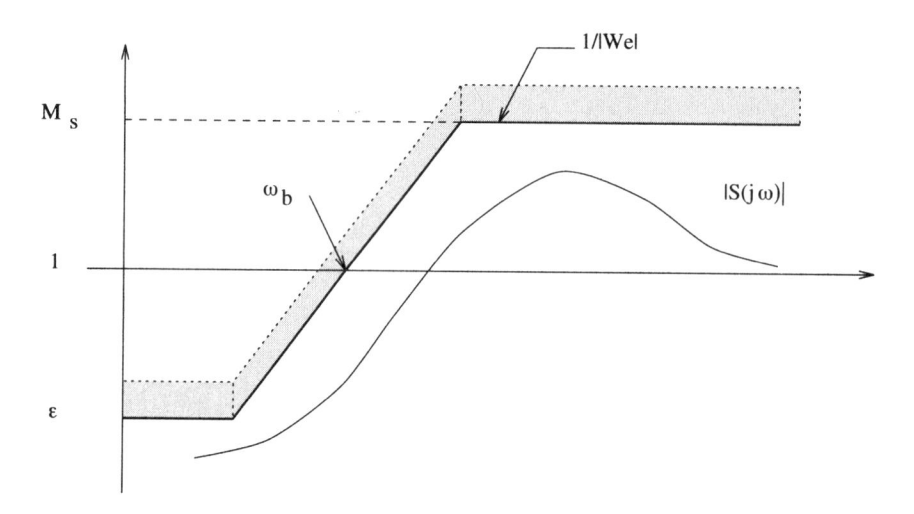

Figure 6.7: Practical performance weight W_e and desired S

Now instead of perfect tracking for step input, suppose we only need the steady-state error with respect to a step input to be no greater than ϵ (i.e., $|S(0)| \leq \epsilon$); then it is sufficient to choose a weighting function W_e satisfying $|W_e(0)| \geq 1/\epsilon$ so that $\|W_eS\|_\infty \leq 1$ can be achieved. A possible choice of W_e can be obtained by modifying the weighting function in equation (6.5):

$$W_e = \frac{s/M_s + \omega_b}{s + \omega_b\varepsilon} \qquad (6.6)$$

6.3. Selection of Weighting Functions

Hence, for practical purpose, one can usually choose a suitable ε, as shown in Figure 6.7, to satisfy the performance specifications. If a steeper transition between low-frequency and high-frequency is desired, the weight W_e can be modified as follows:

$$W_e = \left(\frac{s/\sqrt[k]{M_s} + \omega_b}{s + \omega_b \sqrt[k]{\varepsilon}} \right)^k \quad (6.7)$$

for some integer $k \geq 1$.

The selection of control weighting function W_u follows similarly from the preceding discussion by considering the control signal equation

$$u = KS(r - n - d) - Td_i$$

The magnitude of $|KS|$ in the low-frequency range is essentially limited by the allowable cost of control effort and saturation limit of the actuators; hence, in general, the maximum gain M_u of KS can be fairly large, while the high-frequency gain is essentially limited by the controller bandwidth (ω_{bc}) and the (sensor) noise frequencies. Ideally, one would like to roll off as fast as possible beyond the desired control bandwidth so that the high-frequency noises are attenuated as much as possible. Hence a candidate weight W_u would be

$$W_u = \frac{s + \omega_{bc}/M_u}{\omega_{bc}} \quad (6.8)$$

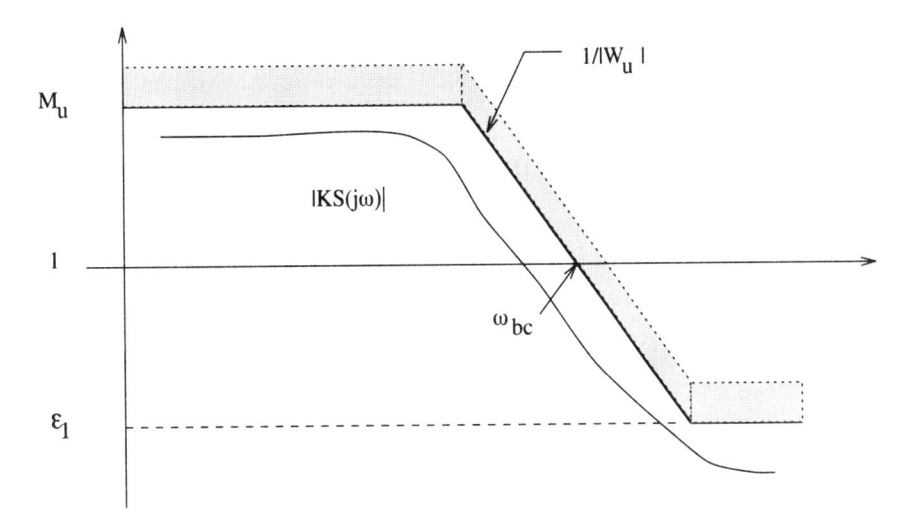

Figure 6.8: Control weight W_u and desired KS

However, again the optimal control design techniques developed in this book cannot be applied directly to a problem with an improper control weighting function. Hence we shall introduce a far away pole to make W_u proper:

$$W_u = \frac{s + \omega_{bc}/M_u}{\varepsilon_1 s + \omega_{bc}} \tag{6.9}$$

for a small $\varepsilon_1 > 0$, as shown in Figure 6.8. Similarly, if a faster rolloff is desired, one may choose

$$W_u = \left(\frac{s + \omega_{bc}/\sqrt[k]{M_u}}{\sqrt[k]{\varepsilon_1} s + \omega_{bc}}\right)^k \tag{6.10}$$

for some integer $k \geq 1$.

The weights for MIMO problems can be initially chosen as diagonal matrices with each diagonal term chosen in the foregoing form.

6.4 Bode's Gain and Phase Relation

One important problem that arises frequently is concerned with the level of performance that can be achieved in feedback design. It has been shown in Section 6.1 that the feedback design goals are inherently conflicting, and a tradeoff must be performed among different design objectives. It is also known that the fundamental requirements, such as stability and robustness, impose inherent limitations on the feedback properties irrespective of design methods, and the design limitations become more severe in the presence of right-half plane zeros and poles in the open-loop transfer function.

In the classical feedback theory, Bode's gain-phase integral relation (see Bode [1945]) has been used as an important tool to express design constraints in scalar systems. This integral relation says that the phase of a stable and minimum phase transfer function is determined uniquely by the magnitude of the transfer function. More precisely, let $L(s)$ be a stable and minimum phase transfer function: then

$$\angle L(j\omega_0) = \frac{1}{\pi} \int_{-\infty}^{\infty} \frac{d\ln|L|}{d\nu} \ln\coth\frac{|\nu|}{2} d\nu \tag{6.11}$$

where $\nu := \ln(\omega/\omega_0)$. The function $\ln\coth\frac{|\nu|}{2} = \ln\frac{e^{|\nu|/2} + e^{-|\nu|/2}}{e^{|\nu|/2} - e^{-|\nu|/2}}$ is plotted in Figure 6.9.

Note that $\ln\coth\frac{|\nu|}{2}$ decreases rapidly as ω deviates from ω_0 and hence the integral depends mostly on the behavior of $\frac{d\ln|L(j\omega)|}{d\nu}$ near the frequency ω_0. This is clear from the following integration:

$$\frac{1}{\pi} \int_{-\alpha}^{\alpha} \ln\coth\frac{|\nu|}{2} d\nu = \begin{cases} 1.1406 \text{ (rad)}, & \alpha = \ln 3 \\ 1.3146 \text{ (rad)}, & \alpha = \ln 5 \\ 1.443 \text{ (rad)}, & \alpha = \ln 10 \end{cases} = \begin{cases} 65.3°, & \alpha = \ln 3 \\ 75.3°, & \alpha = \ln 5 \\ 82.7°, & \alpha = \ln 10. \end{cases}$$

6.4. Bode's Gain and Phase Relation

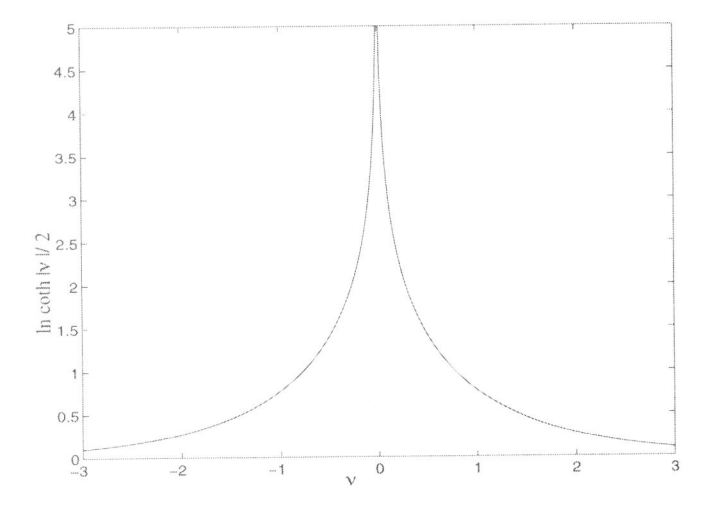

Figure 6.9: The function $\ln \coth \frac{|\nu|}{2}$ vs ν

Note that $\dfrac{d \ln |L(j\omega)|}{d\nu}$ is the slope of the Bode plot, which is generally negative for almost all frequencies. It follows that $\angle L(j\omega_0)$ will be large if the gain L attenuates slowly near ω_0 and small if it attenuates rapidly near ω_0. For example, suppose the slope $\dfrac{d \ln |L(j\omega)|}{d\nu} = -\ell$; that is, $(-20\ell$ dB per decade), in the neighborhood of ω_0; then it is reasonable to expect

$$\angle L(j\omega_0) < \begin{cases} -\ell \times 65.3^\circ, & \text{if the slope of } L = -\ell \text{ for } \tfrac{1}{3} \le \tfrac{\omega}{\omega_0} \le 3 \\ -\ell \times 75.3^\circ, & \text{if the slope of } L = -\ell \text{ for } \tfrac{1}{5} \le \tfrac{\omega}{\omega_0} \le 5 \\ -\ell \times 82.7^\circ, & \text{if the slope of } L = -\ell \text{ for } \tfrac{1}{10} \le \tfrac{\omega}{\omega_0} \le 10. \end{cases}$$

The behavior of $\angle L(j\omega)$ is particularly important near the crossover frequency ω_c, where $|L(j\omega_c)| = 1$ since $\pi + \angle L(j\omega_c)$ is the phase margin of the feedback system. Further, the return difference is given by

$$|1 + L(j\omega_c)| = |1 + L^{-1}(j\omega_c)| = 2 \left| \sin \frac{\pi + \angle L(j\omega_c)}{2} \right|,$$

which must not be too small for good stability robustness. If $\pi + \angle L(j\omega_c)$ is forced to be very small by rapid gain attenuation, the feedback system will amplify disturbances and exhibit little uncertainty tolerance at and near ω_c. Since it is generally required that the loop transfer function L roll off as fast as possible in the high-frequency range, it is reasonable to expect that $\angle L(j\omega_c)$ is at most $-\ell \times 90^\circ$ if the slope of $L(j\omega)$ is $-\ell$ near ω_c. Thus it is important to keep the slope of L near ω_c not much smaller than -1 for a reasonably wide range of frequencies in order to guarantee some reasonable performance. The conflict between attenuation rate and loop quality near crossover is thus clearly evident.

Bode's gain and phase relation can be extended to stable and nonminimum phase transfer functions easily. Let $z_1, z_2, \ldots, z_k$ be the right-half plane zeros of $L(s)$, then L can be factorized as

$$L(s) = \frac{-s+z_1}{s+z_1}\frac{-s+z_2}{s+z_2}\cdots\frac{-s+z_k}{s+z_k}L_{\mathrm{mp}}(s)$$

where L_{mp} is stable and minimum phase and $|L(j\omega)| = |L_{\mathrm{mp}}(j\omega)|$. Hence

$$\angle L(j\omega_0) = \angle L_{\mathrm{mp}}(j\omega_0) + \angle \prod_{i=1}^{k} \frac{-j\omega_0+z_i}{j\omega_0+z_i}$$

$$= \frac{1}{\pi}\int_{-\infty}^{\infty}\frac{d\ln|L_{\mathrm{mp}}|}{d\nu}\ln\coth\frac{|\nu|}{2}d\nu + \sum_{i=1}^{k}\angle\frac{-j\omega_0+z_i}{j\omega_0+z_i},$$

which gives

$$\angle L(j\omega_0) = \frac{1}{\pi}\int_{-\infty}^{\infty}\frac{d\ln|L|}{d\nu}\ln\coth\frac{|\nu|}{2}d\nu + \sum_{i=1}^{k}\angle\frac{-j\omega_0+z_i}{j\omega_0+z_i}. \qquad (6.12)$$

Since $\angle\dfrac{-j\omega_0+z_i}{j\omega_0+z_i} \leq 0$ for each i, a nonminimum phase zero contributes an additional phase lag and imposes limitations on the rolloff rate of the open-loop gain. For example, suppose L has a zero at $z > 0$; then

$$\phi_1(\omega_0/z) := \angle\frac{-j\omega_0+z}{j\omega_0+z}\bigg|_{\omega_0=z,z/2,z/4} = -90°, -53.13°, -28°,$$

as shown in Figure 6.10. Since the slope of $|L|$ near the crossover frequency is, in general, no greater than -1, which means that the phase due to the minimum phase part, L_{mp}, of L will, in general, be no greater than $-90°$, the crossover frequency (or the closed-loop bandwidth) must satisfy

$$\omega_c < z/2 \qquad (6.13)$$

in order to guarantee the closed-loop stability and some reasonable closed-loop performance.

Next suppose L has a pair of complex right-half zeros at $z = x \pm jy$ with $x > 0$; then

$$\phi_2(\omega_0/|z|) := \angle\frac{-j\omega_0+z}{j\omega_0+z}\frac{-j\omega_0+\bar{z}}{j\omega_0+\bar{z}}\bigg|_{\omega_0=|z|,|z|/2,|z|/3,|z|/4}$$

$$\approx \begin{cases} -180°, & -106.26°, & -73.7°, & -56°, & \mathrm{Re}(z) \gg \Im(z) \\ -180°, & -86.7°, & -55.9°, & -41.3°, & \mathrm{Re}(z) \approx \Im(z) \\ -360°, & 0°, & 0°, & 0°, & \mathrm{Re}(z) \ll \Im(z) \end{cases}$$

6.4. Bode's Gain and Phase Relation

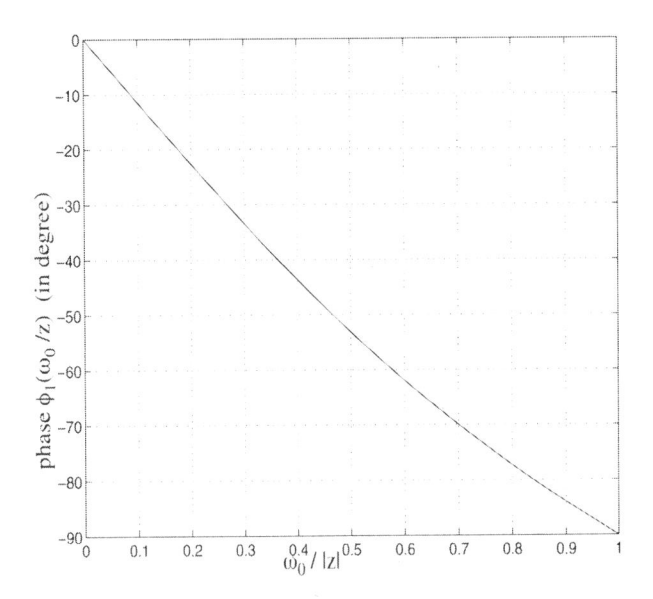

Figure 6.10: Phase $\phi_1(\omega_0/z)$ due to a real zero $z > 0$

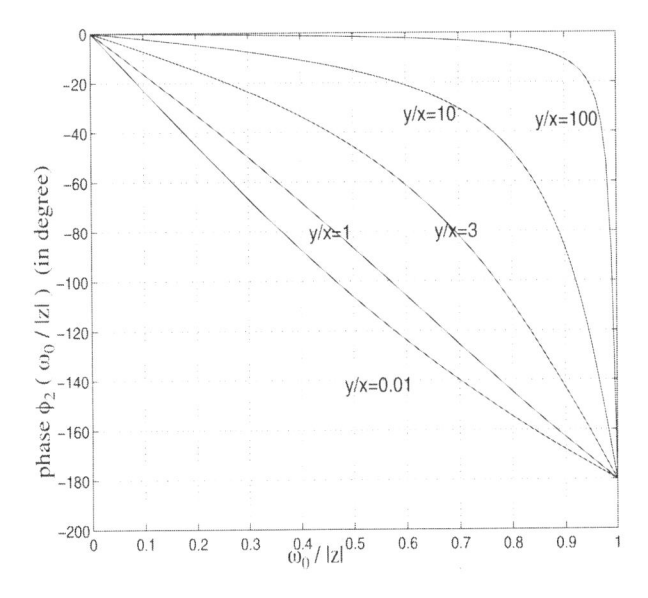

Figure 6.11: Phase $\phi_2(\omega_0/|z|)$ due to a pair of complex zeros: $z = x \pm jy$ and $x > 0$

as shown in Figure 6.11. In this case we conclude that the crossover frequency must satisfy

$$\omega_c < \begin{cases} |z|/4, & \operatorname{Re}(z) \gg \Im(z) \\ |z|/3, & \operatorname{Re}(z) \approx \Im(z) \\ |z|, & \operatorname{Re}(z) \ll \Im(z) \end{cases} \qquad (6.14)$$

in order to guarantee the closed-loop stability and some reasonable closed-loop performance.

6.5 Bode's Sensitivity Integral

In this section, we consider the design limitations imposed by the bandwidth constraints and the right-half plane poles and zeros using Bode's sensitivity integral and Poisson integral. Let L be the open-loop transfer function with at least two more poles than zeros and let $p_1, p_2, \ldots, p_m$ be the open right-half plane poles of L. Then the following Bode's sensitivity integral holds:

$$\int_0^\infty \ln|S(j\omega)|d\omega = \pi \sum_{i=1}^m \operatorname{Re}(p_i) \qquad (6.15)$$

In the case where L is stable, the integral simplifies to

$$\int_0^\infty \ln|S(j\omega)|d\omega = 0 \qquad (6.16)$$

These integrals show that there will exist a frequency range over which the magnitude of the sensitivity function exceeds one if it is to be kept below one at other frequencies, as illustrated in Figure 6.12. This is the so-called water bed effect.

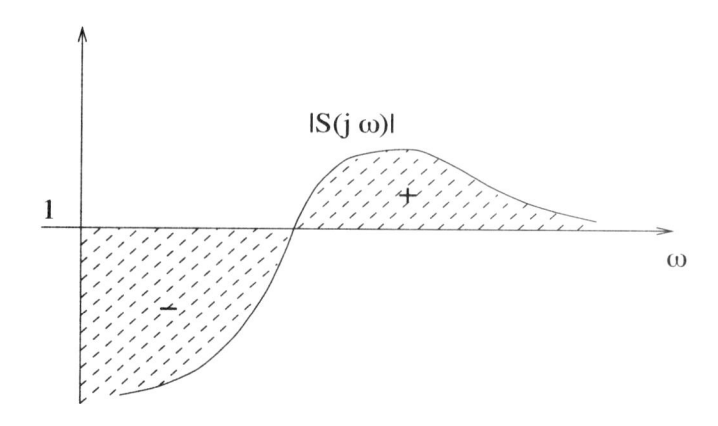

Figure 6.12: Water bed effect of sensitivity function

6.5. Bode's Sensitivity Integral

Suppose that the feedback system is designed such that the level of sensitivity reduction is given by
$$|S(j\omega)| \leq \epsilon < 1, \quad \forall \omega \in [0, \omega_l]$$
where $\epsilon > 0$ is a given constant.

Bandwidth constraints in feedback design typically require that the open-loop transfer function be small above a specified frequency, and that it roll off at a rate of more than one pole-zero excess above that frequency. These constraints are commonly needed to ensure stability robustness despite the presence of modeling uncertainty in the plant model, particularly at high frequencies. One way of quantifying such bandwidth constraints is by requiring the open-loop transfer function to satisfy
$$|L(j\omega)| \leq \frac{M_h}{\omega^{1+\beta}} \leq \tilde{\epsilon} < 1, \quad \forall \omega \in [\omega_h, \infty)$$
where $\omega_h > \omega_l$, and $M_h > 0$, $\beta > 0$ are some given constants.

Note that for $\omega \geq \omega_h$,
$$|S(j\omega)| \leq \frac{1}{1 - |L(j\omega)|} \leq \frac{1}{1 - \frac{M_h}{\omega^{1+\beta}}}$$

and

$$\begin{aligned}
-\int_{\omega_h}^{\infty} \ln\left(1 - \frac{M_h}{\omega^{1+\beta}}\right) d\omega &= \sum_{i=1}^{\infty} \int_{\omega_h}^{\infty} \frac{1}{i}\left(\frac{M_h}{\omega^{1+\beta}}\right)^i d\omega \\
&= \sum_{i=1}^{\infty} \frac{1}{i} \frac{\omega_h}{i(1+\beta) - 1}\left(\frac{M_h}{\omega_h^{1+\beta}}\right)^i \\
&\leq \frac{\omega_h}{\beta} \sum_{i=1}^{\infty} \frac{1}{i}\left(\frac{M_h}{\omega_h^{1+\beta}}\right)^i = -\frac{\omega_h}{\beta} \ln\left(1 - \frac{M_h}{\omega_h^{1+\beta}}\right) \\
&\leq -\frac{\omega_h}{\beta} \ln(1 - \tilde{\epsilon}).
\end{aligned}$$

Then
$$\pi \sum_{i=1}^{m} \text{Re}(p_i) = \int_0^{\infty} \ln|S(j\omega)| d\omega$$

$$\begin{aligned}
&= \int_0^{\omega_l} \ln|S(j\omega)|d\omega + \int_{\omega_l}^{\omega_h} \ln|S(j\omega)|d\omega + \int_{\omega_h}^{\infty} \ln|S(j\omega)|d\omega \\
&\leq \omega_l \ln \epsilon + (\omega_h - \omega_l) \max_{\omega \in [\omega_l, \omega_h]} \ln|S(j\omega)| - \int_{\omega_h}^{\infty} \ln\left(1 - \frac{M_h}{\omega^{1+\beta}}\right) d\omega \\
&\leq \omega_l \ln \epsilon + (\omega_h - \omega_l) \max_{\omega \in [\omega_l, \omega_h]} \ln|S(j\omega)| - \frac{\omega_h}{\beta} \ln(1 - \tilde{\epsilon}),
\end{aligned}$$

which gives
$$\max_{\omega \in [\omega_l, \omega_h]} |S(j\omega)| \geq e^{\alpha} \left(\frac{1}{\epsilon}\right)^{\frac{\omega_l}{\omega_h - \omega_l}} (1 - \tilde{\epsilon})^{\frac{\omega_h}{\beta(\omega_h - \omega_l)}}$$

where
$$\alpha = \frac{\pi \sum_{i=1}^{m} \text{Re}(p_i)}{\omega_h - \omega_l}.$$

The above lower bound shows that the sensitivity can be very significant in the transition band.

Next, using the Poisson integral relation, we investigate the design constraints on sensitivity properties imposed by open-loop nonminimum phase zeros. Suppose L has at least one more poles than zeros and suppose $z = x_0 + jy_0$ with $x_0 > 0$ is a right-half plane zero of L. Then

$$\int_{-\infty}^{\infty} \ln |S(j\omega)| \frac{x_0}{x_0^2 + (\omega - y_0)^2} d\omega = \pi \ln \prod_{i=1}^{m} \left|\frac{z + p_i}{z - p_i}\right| \qquad (6.17)$$

This integral implies that the sensitivity reduction ability of the system may be severely limited by the open-loop unstable poles and nonminimum phase zeros, especially when these poles and zeros are close to each other.

Define
$$\theta(z) := \int_{-\omega_l}^{\omega_l} \frac{x_0}{x_0^2 + (\omega - y_0)^2} d\omega$$

Then
$$\pi \ln \prod_{i=1}^{m} \left|\frac{z + p_i}{z - p_i}\right| = \int_{-\infty}^{\infty} \ln |S(j\omega)| \frac{x_0}{x_0^2 + (\omega - y_0)^2} d\omega$$
$$\leq (\pi - \theta(z)) \ln \|S(j\omega)\|_{\infty} + \theta(z) \ln(\epsilon),$$

which gives
$$\|S(s)\|_{\infty} \geq \left(\frac{1}{\epsilon}\right)^{\frac{\theta(z)}{\pi - \theta(z)}} \left(\prod_{i=1}^{m} \left|\frac{z + p_i}{z - p_i}\right|\right)^{\frac{\pi}{\pi - \theta(z)}}$$

This lower bound on the maximum sensitivity shows that for a nonminimum phase system, its sensitivity must increase significantly beyond one at certain frequencies if the sensitivity reduction is to be achieved at other frequencies.

6.6 Analyticity Constraints

Let $p_1, p_2, \ldots, p_m$ and $z_1, z_2, \ldots, z_k$ be the open right-half plane poles and zeros of L, respectively. Suppose that the closed-loop system is stable. Then

$$S(p_i) = 0, \quad T(p_i) = 1, \quad i = 1, 2, \ldots, m$$

6.6. Analyticity Constraints

and
$$S(z_j) = 1, \quad T(z_j) = 0, \quad j = 1, 2, \ldots, k$$

The internal stability of the feedback system is guaranteed by satisfying these analyticity (or interpolation) conditions. On the other hand, these conditions also impose severe limitations on the achievable performance of the feedback system.

Suppose $S = (I+L)^{-1}$ and $T = L(I+L)^{-1}$ are stable. Then $p_1, p_2, \ldots, p_m$ are the right-half plane zeros of S and $z_1, z_2, \ldots, z_k$ are the right-half plane zeros of T. Let

$$B_p(s) = \prod_{i=1}^{m} \frac{s - p_i}{s + p_i}, \quad B_z(s) = \prod_{j=1}^{k} \frac{s - z_j}{s + z_j}$$

Then $|B_p(j\omega)| = 1$ and $|B_z(j\omega)| = 1$ for all frequencies and, moreover,

$$B_p^{-1}(s)S(s) \in \mathcal{H}_\infty, \quad B_z^{-1}(s)T(s) \in \mathcal{H}_\infty.$$

Hence, by the maximum modulus theorem, we have

$$\|S(s)\|_\infty = \|B_p^{-1}(s)S(s)\|_\infty \geq |B_p^{-1}(z)S(z)|$$

for any z with $\text{Re}(z) > 0$. Let z be a right-half plane zero of L; then

$$\|S(s)\|_\infty \geq |B_p^{-1}(z)| = \prod_{i=1}^{m} \left| \frac{z + p_i}{z - p_i} \right|$$

Similarly, one can obtain

$$\|T(s)\|_\infty \geq |B_z^{-1}(p)| = \prod_{j=1}^{k} \left| \frac{p + z_j}{p - z_j} \right|$$

where p is a right-half plane pole of L.

The weighted problem can be considered in the same fashion. Let W_e be a weight such that $W_e S$ is stable. Then

$$\|W_e(s)S(s)\|_\infty \geq |W_e(z)| \prod_{i=1}^{m} \left| \frac{z + p_i}{z - p_i} \right|$$

Now suppose $W_e(s) = \dfrac{s/M_s + \omega_b}{s + \omega_b \epsilon}$, $\|W_e S\|_\infty \leq 1$, and z is a real right-half plane zero. Then

$$\frac{z/M_s + \omega_b}{z + \omega_b \epsilon} \leq \prod_{i=1}^{m} \left| \frac{z - p_i}{z + p_i} \right| =: \alpha,$$

which gives

$$\omega_b \leq \frac{z}{1 - \alpha\epsilon}\left(\alpha - \frac{1}{M_s}\right) \approx z\left(\alpha - \frac{1}{M_s}\right)$$

where $\alpha = 1$ if L has no right-half plane poles. This shows that the bandwidth of the closed-loop must be much smaller than the right-half plane zero. Similar conclusions can be arrived at for complex right-half plane zeros.

6.7 Notes and References

The loop-shaping design is well-known for SISO systems in the classical control theory. The idea was extended to MIMO systems by Doyle and Stein [1981] using the LQG design technique. The limitations of the loop-shaping design are discussed in detail in Stein and Doyle [1991]. Chapter 16 presents another loop-shaping method using $\mathcal{H}_\infty$ control theory, which has the potential to overcome the limitations of the LQG/LTR method. Some additional discussions on the choice of weighting functions can be found in Skogestad and Postlethwaite [1996]. The design tradeoffs and limitations for SISO systems are discussed in detail in Bode [1945], Horowitz [1963], and Doyle, Francis, and Tannenbaum [1992]. The monograph by Freudenberg and Looze [1988] contains many multivariable generalizations. The multivariable generalization of Bode's integral relation can be found in Chen [1995], on which Section 6.5 is based. Some related results can be found in Boyd and Desoer [1985]. Additional related results can be found in a recent book by Seron, Braslavsky, and Goodwin [1997].

6.8 Problems

Problem 6.1 Let P be an open-loop plant. It is desired to design a controller so that the overshoot $\leq 10\%$ and settling time ≤ 10 sec. Estimate the allowable peak sensitivity M_s and the closed-loop bandwidth.

Problem 6.2 Let $L_1 = \dfrac{1}{s(s+1)^2}$ be an open-loop transfer function of a unity feedback system. Find the phase margin, overshoot, settling time, and the corresponding M_s.

Problem 6.3 Repeat Problem 6.2 with

$$L_2 = \frac{100(s+10)}{(s+1)(s+2)(s+20)}.$$

Problem 6.4 Let $P = \dfrac{10(1-s)}{s(s+10)}$. Use classical loop-shaping method to design a lead or lag controller so that the system has at least $30°$ phase margin and as large a crossover frequency as possible.

Problem 6.5 Use the root locus method to show that a nonminimum phase system cannot be stabilized by a very high-gain controller.

Problem 6.6 Let $P = \dfrac{5}{(1-s)(s+2)}$. Design a lead or lag controller so that the system has at least $30°$ phase margin with loop gain ≥ 2 for any frequency $\omega \leq 0.1$ and the smallest possible bandwidth (or crossover frequency).

Problem 6.7 Use the root locus method to show that an unstable system cannot be stabilized by a very low gain controller.

6.8. Problems

Problem 6.8 Consider the unity-feedback loop with proper controller $K(s)$ and strictly proper plant $P(s)$, both assumed square. Assume internal stability.

1. Let $w(s)$ be a scalar weighting function, assumed in $\mathcal{RH}_\infty$. Define

$$\epsilon = \|w(I+PK)^{-1}\|_\infty, \quad \delta = \|K(I+PK)^{-1}\|_\infty$$

so ϵ measures, say, disturbance attenuation and δ measures, say, control effort. Derive the following inequality, which shows that ϵ and δ cannot both be small simultaneously in general. For every Re $s_0 \geq 0$

$$|w(s_0)| \leq \epsilon + |w(s_0)|\sigma_{\min}[P(s_0)]\delta.$$

2. If we want very good disturbance attenuation at a particular frequency, you might guess that we need high controller gain at that frequency. Fix ω with $j\omega$ not a pole of $P(s)$, and suppose

$$\epsilon := \sigma_{\max}[(I+PK)^{-1}(j\omega)] < 1.$$

Derive a lower bound for $\sigma_{\min}[K(j\omega)]$. This lower bound should blow up as $\epsilon \to 0$.

Problem 6.9 Suppose that P is proper and has one right half plane zero at $s = z > 0$. Suppose that $y = \frac{w}{1+PK}$, where w is a unit step at time $t = 0$, and that our performance specification is

$$|y(t)| \leq \begin{cases} \alpha, & \text{if } 0 \leq t \leq T; \\ \beta, & \text{if } T < t \end{cases}$$

for some $\alpha > 1 > \beta > 0$. Show that for a proper, internally stabilizing, LTI controller K to exist that meets the specification, we must have that

$$\ln\left(\frac{\alpha - \beta}{\alpha - 1}\right) \leq zT.$$

What tradeoffs does this imply?

Problem 6.10 Let K be a stabilizing controller for the plant

$$P = \frac{s - \alpha}{(s - \beta)(s + \gamma)}$$

$\alpha > 0, \beta > 0, \gamma \geq 0$. Suppose $|S(j\omega)| \leq \delta < 1$, $\forall \omega \in [-\omega_0, \omega_0]$ where

$$S(s) = \frac{1}{1 + PK}.$$

Find a lower bound for $\|S\|_\infty$ and calculate the lower bound for $\alpha = 1$, $\beta = 2$, $\gamma = 10$, $\delta = 0.2$, and $\omega_0 = 1$.

Chapter 7

Balanced Model Reduction

Simple linear models/controllers are normally preferred over complex ones in control system design for some obvious reasons: They are much easier to do analysis and synthesis with. Furthermore, simple controllers are easier to implement and are more reliable because there are fewer things to go wrong in the hardware or bugs to fix in the software. In the case when the system is infinite dimensional, the model/controller approximation becomes essential. In this chapter we consider the problem of reducing the order of a linear multivariable dynamical system. There are many ways to reduce the order of a dynamical system. However, we shall study only one of them: the balanced truncation method. The main advantage of this method is that it is simple and performs fairly well.

A model order-reduction problem can, in general, be stated as follows: Given a full-order model $G(s)$, find a lower-order model (say, an rth order model G_r), such that G and G_r are close in some sense. Of course, there are many ways to define the closeness of an approximation. For example, one may desire that the reduced model be such that

$$G = G_r + \Delta_a$$

and Δ_a is small in some norm. This model reduction is usually called an *additive* model reduction problem. We shall be only interested in $\mathcal{L}_\infty$ norm approximation in this book. Once the norm is chosen, the additive model reduction problem can be formulated as

$$\inf_{\deg(G_r) \leq r} \|G - G_r\|_\infty.$$

In general, a practical model reduction problem is inherently frequency-weighted (i.e., the requirement on the approximation accuracy at one frequency range can be drastically different from the requirement at another frequency range). These problems can, in general, be formulated as frequency-weighted model reduction problems:

$$\inf_{\deg(G_r) \leq r} \|W_o(G - G_r)W_i\|_\infty$$

with an appropriate choice of W_i and W_o. We shall see in this chapter how the balanced realization can give an effective approach to the aforementioned model reduction problems.

7.1 Lyapunov Equations

Testing stability, controllability, and observability of a system is very important in linear system analysis and synthesis. However, these tests often have to be done indirectly. In that respect, the Lyapunov theory is sometimes useful. Consider the following Lyapunov equation:

$$A^*Q + QA + H = 0 \qquad (7.1)$$

with given real matrices A and H. It is well known that this equation has a unique solution iff $\lambda_i(A) + \bar{\lambda}_j(A) \neq 0, \forall i, j$. In this section, we shall study the relationships between the stability of A and the solution of Q. The following results are standard.

Lemma 7.1 *Assume that A is stable, then the following statements hold:*

(i) $Q = \int_0^\infty e^{A^*t} H e^{At} dt$.

(ii) $Q > 0$ if $H > 0$ and $Q \geq 0$ if $H \geq 0$.

(iii) *If $H \geq 0$, then (H, A) is observable iff $Q > 0$.*

An immediate consequence of part (iii) is that, given a stable matrix A, a pair (C, A) is observable if and only if the solution to the following Lyapunov equation

$$A^*Q + QA + C^*C = 0$$

is positive definite, where Q is the *observability Gramian*. Similarly, a pair (A, B) is controllable if and only if the solution to

$$AP + PA^* + BB^* = 0$$

is positive definite, where P is the *controllability Gramian*.

In many applications, we are given the solution of the Lyapunov equation and need to conclude the stability of the matrix A.

Lemma 7.2 *Suppose Q is the solution of the Lyapunov equation (7.1), then*

(i) $\mathrm{Re}\lambda_i(A) \leq 0$ *if $Q > 0$ and $H \geq 0$.*

(ii) *A is stable if $Q > 0$ and $H > 0$.*

(iii) *A is stable if $Q \geq 0$, $H \geq 0$, and (H, A) is detectable.*

7.2. Balanced Realizations

Proof. Let λ be an eigenvalue of A and $v \neq 0$ be a corresponding eigenvector, then $Av = \lambda v$. Premultiply equation (7.1) by v^* and postmultiply equation (7.1) by v to get

$$2\text{Re}\,\lambda(v^*Qv) + v^*Hv = 0.$$

Now if $Q > 0$, then $v^*Qv > 0$, and it is clear that $\text{Re}\,\lambda \leq 0$ if $H \geq 0$ and $\text{Re}\,\lambda < 0$ if $H > 0$. Hence (i) and (ii) hold. To see (iii), we assume $\text{Re}\,\lambda \geq 0$. Then we must have $v^*Hv = 0$ (i.e., $Hv = 0$). This implies that λ is an unstable and unobservable mode, which contradicts the assumption that (H, A) is detectable. □

7.2 Balanced Realizations

Although there are infinitely many different state-space realizations for a given transfer matrix, some particular realizations have proven to be very useful in control engineering and signal processing. Here we will only introduce one class of realizations for stable transfer matrices that are most useful in control applications. To motivate the class of realizations, we first consider some simple facts.

Lemma 7.3 *Let* $\left[\begin{array}{c|c} A & B \\ \hline C & D \end{array}\right]$ *be a state-space realization of a (not necessarily stable) transfer matrix $G(s)$. Suppose that there exists a symmetric matrix*

$$P = P^* = \begin{bmatrix} P_1 & 0 \\ 0 & 0 \end{bmatrix}$$

with P_1 nonsingular such that

$$AP + PA^* + BB^* = 0.$$

Now partition the realization (A, B, C, D) compatibly with P as

$$\left[\begin{array}{cc|c} A_{11} & A_{12} & B_1 \\ A_{21} & A_{22} & B_2 \\ \hline C_1 & C_2 & D \end{array}\right].$$

Then $\left[\begin{array}{c|c} A_{11} & B_1 \\ \hline C_1 & D \end{array}\right]$ *is also a realization of G. Moreover, (A_{11}, B_1) is controllable if A_{11} is stable.*

Proof. Use the partitioned P and (A, B, C) to get

$$0 = AP + PA^* + BB^* = \begin{bmatrix} A_{11}P_1 + P_1A_{11}^* + B_1B_1^* & P_1A_{21}^* + B_1B_2^* \\ A_{21}P_1 + B_2B_1^* & B_2B_2^* \end{bmatrix},$$

which gives $B_2 = 0$ and $A_{21} = 0$ since P_1 is nonsingular. Hence, part of the realization is not controllable:

$$\left[\begin{array}{cc|c} A_{11} & A_{12} & B_1 \\ A_{21} & A_{22} & B_2 \\ \hline C_1 & C_2 & D \end{array}\right] = \left[\begin{array}{cc|c} A_{11} & A_{12} & B_1 \\ 0 & A_{22} & 0 \\ \hline C_1 & C_2 & D \end{array}\right] = \left[\begin{array}{c|c} A_{11} & B_1 \\ \hline C_1 & D \end{array}\right].$$

Finally, it follows from Lemma 7.1 that (A_{11}, B_1) is controllable if A_{11} is stable. □

We also have the following:

Lemma 7.4 *Let* $\left[\begin{array}{c|c} A & B \\ \hline C & D \end{array}\right]$ *be a state-space realization of a (not necessarily stable) transfer matrix $G(s)$. Suppose that there exists a symmetric matrix*

$$Q = Q^* = \left[\begin{array}{cc} Q_1 & 0 \\ 0 & 0 \end{array}\right]$$

with Q_1 nonsingular such that

$$QA + A^*Q + C^*C = 0.$$

Now partition the realization (A, B, C, D) compatibly with Q as

$$\left[\begin{array}{cc|c} A_{11} & A_{12} & B_1 \\ A_{21} & A_{22} & B_2 \\ \hline C_1 & C_2 & D \end{array}\right].$$

Then $\left[\begin{array}{c|c} A_{11} & B_1 \\ \hline C_1 & D \end{array}\right]$ *is also a realization of G. Moreover, (C_1, A_{11}) is observable if A_{11} is stable.*

The preceding two lemmas suggest that to obtain a minimal realization from a stable nonminimal realization, one only needs to eliminate all states corresponding to the zero block diagonal term of the controllability Gramian P and the observability Gramian Q. In the case where P is not block diagonal, the following procedure can be used to eliminate noncontrollable subsystems:

1. Let $G(s) = \left[\begin{array}{c|c} A & B \\ \hline C & D \end{array}\right]$ be a stable realization.

2. Compute the controllability Gramian $P \geq 0$ from

$$AP + PA^* + BB^* = 0.$$

3. Diagonalize P to get $P = \left[\begin{array}{cc} U_1 & U_2 \end{array}\right] \left[\begin{array}{cc} \Lambda_1 & 0 \\ 0 & 0 \end{array}\right] \left[\begin{array}{cc} U_1 & U_2 \end{array}\right]^*$ with $\Lambda_1 > 0$ and $\left[\begin{array}{cc} U_1 & U_2 \end{array}\right]$ unitary.

7.2. Balanced Realizations

4. Then $G(s) = \left[\begin{array}{c|c} U_1^* A U_1 & U_1^* B \\ \hline C U_1 & D \end{array}\right]$ is a controllable realization.

A dual procedure can also be applied to eliminate nonobservable subsystems.

Now assume that $\Lambda_1 > 0$ is diagonal and is partitioned as $\Lambda_1 = \text{diag}(\Lambda_{11}, \Lambda_{12})$ such that $\lambda_{\max}(\Lambda_{12}) \ll \lambda_{\min}(\Lambda_{11})$; then it is tempting to conclude that one can also discard those states corresponding to Λ_{12} without causing much error. However, this is not necessarily true, as shown in the following example.

Example 7.1 Consider a stable transfer function

$$G(s) = \frac{3s + 18}{s^2 + 3s + 18}.$$

Then $G(s)$ has a state-space realization given by

$$G(s) = \left[\begin{array}{cc|c} -1 & -4/\alpha & 1 \\ 4\alpha & -2 & 2\alpha \\ \hline -1 & 2/\alpha & 0 \end{array}\right]$$

where α is any nonzero number. It is easy to check that the controllability Gramian of the realization is given by

$$P = \begin{bmatrix} 0.5 & \\ & \alpha^2 \end{bmatrix}.$$

Since the last diagonal term of P can be made arbitrarily small by making α small, the controllability of the corresponding state can be made arbitrarily weak. If the state corresponding to the last diagonal term of P is removed, we get a transfer function

$$\hat{G} = \left[\begin{array}{c|c} -1 & 1 \\ \hline -1 & 0 \end{array}\right] = \frac{-1}{s+1},$$

which is not close to the original transfer function in any sense. The problem may be easily detected if one checks the observability Gramian Q, which is

$$Q = \begin{bmatrix} 0.5 & \\ & 1/\alpha^2 \end{bmatrix}.$$

Since $1/\alpha^2$ is very large if α is small, this shows that the state corresponding to the last diagonal term is strongly observable.

This example shows that the controllability (or observability) Gramian alone cannot give an accurate indication of the dominance of the system states in the input/output behavior. This motivates the introduction of a balanced realization that gives balanced Gramians for controllability and observability.

Suppose $G = \left[\begin{array}{c|c} A & B \\ \hline C & D \end{array}\right]$ is stable (i.e., A is stable). Let P and Q denote the controllability Gramian and observability Gramian, respectively. Then by Lemma 7.1, P and Q satisfy the following Lyapunov equations:

$$AP + PA^* + BB^* = 0 \tag{7.2}$$

$$A^*Q + QA + C^*C = 0, \tag{7.3}$$

and $P \geq 0$, $Q \geq 0$. Furthermore, the pair (A, B) is controllable iff $P > 0$, and (C, A) is observable iff $Q > 0$.

Suppose the state is transformed by a nonsingular T to $\hat{x} = Tx$ to yield the realization

$$G = \left[\begin{array}{c|c} \hat{A} & \hat{B} \\ \hline \hat{C} & \hat{D} \end{array}\right] = \left[\begin{array}{c|c} TAT^{-1} & TB \\ \hline CT^{-1} & D \end{array}\right].$$

Then the Gramians are transformed to $\hat{P} = TPT^*$ and $\hat{Q} = (T^{-1})^*QT^{-1}$. Note that $\hat{P}\hat{Q} = TPQT^{-1}$, and therefore the eigenvalues of the product of the Gramians are invariant under state transformation.

Consider the similarity transformation T, which gives the eigenvector decomposition

$$PQ = T^{-1}\Lambda T, \quad \Lambda = \text{diag}(\lambda_1 I_{s_1}, \ldots, \lambda_N I_{s_N}).$$

Then the columns of T^{-1} are eigenvectors of PQ corresponding to the eigenvalues $\{\lambda_i\}$. Later, it will be shown that PQ has a real diagonal Jordan form and that $\Lambda \geq 0$, which are consequences of $P \geq 0$ and $Q \geq 0$.

Although the eigenvectors are not unique, in the case of a minimal realization they can always be chosen such that

$$\hat{P} = TPT^* = \Sigma,$$

$$\hat{Q} = (T^{-1})^*QT^{-1} = \Sigma,$$

where $\Sigma = \text{diag}(\sigma_1 I_{s_1}, \sigma_2 I_{s_2}, \ldots, \sigma_N I_{s_N})$ and $\Sigma^2 = \Lambda$. This new realization with controllability and observability Gramians $\hat{P} = \hat{Q} = \Sigma$ will be referred to as a *balanced realization* (also called internally balanced realization). The decreasingly ordered numbers, $\sigma_1 > \sigma_2 > \ldots > \sigma_N \geq 0$, are called the *Hankel singular values* of the system.

More generally, if a realization of a stable system is not minimal, then there is a transformation such that the controllability and observability Gramians for the transformed realization are diagonal and the controllable and observable subsystem is balanced. This is a consequence of the following matrix fact.

7.2. Balanced Realizations

Theorem 7.5 *Let P and Q be two positive semidefinite matrices. Then there exists a nonsingular matrix T such that*

$$TPT^* = \begin{bmatrix} \Sigma_1 & & \\ & \Sigma_2 & \\ & & 0 \\ & & & 0 \end{bmatrix}, \quad (T^{-1})^*QT^{-1} = \begin{bmatrix} \Sigma_1 & & & 0 \\ & 0 & & \\ & & \Sigma_3 & \\ & & & 0 \end{bmatrix}$$

respectively, with Σ_1, Σ_2, Σ_3 diagonal and positive definite.

Proof. Since P is a positive semidefinite matrix, there exists a transformation T_1 such that

$$T_1 P T_1^* = \begin{bmatrix} I & 0 \\ 0 & 0 \end{bmatrix}$$

Now let

$$(T_1^*)^{-1} Q T_1^{-1} = \begin{bmatrix} Q_{11} & Q_{12} \\ Q_{12}^* & Q_{22} \end{bmatrix}$$

and there exists a unitary matrix U_1 such that

$$U_1 Q_{11} U_1^* = \begin{bmatrix} \Sigma_1^2 & 0 \\ 0 & 0 \end{bmatrix}, \quad \Sigma_1 > 0$$

Let

$$(T_2^*)^{-1} = \begin{bmatrix} U_1 & 0 \\ 0 & I \end{bmatrix}$$

and then

$$(T_2^*)^{-1}(T_1^*)^{-1} Q T_1^{-1}(T_2)^{-1} = \begin{bmatrix} \Sigma_1^2 & 0 & \hat{Q}_{121} \\ 0 & 0 & \hat{Q}_{122} \\ \hat{Q}_{121}^* & \hat{Q}_{122}^* & Q_{22} \end{bmatrix}$$

But $Q \geq 0$ implies $\hat{Q}_{122} = 0$. So now let

$$(T_3^*)^{-1} = \begin{bmatrix} I & 0 & 0 \\ 0 & I & 0 \\ -\hat{Q}_{121}^* \Sigma_1^{-2} & 0 & I \end{bmatrix}$$

giving

$$(T_3^*)^{-1}(T_2^*)^{-1}(T_1^*)^{-1} Q T_1^{-1}(T_2)^{-1}(T_3)^{-1} = \begin{bmatrix} \Sigma_1^2 & 0 & 0 \\ 0 & 0 & 0 \\ 0 & 0 & Q_{22} - \hat{Q}_{121}^* \Sigma_1^{-2} \hat{Q}_{121} \end{bmatrix}$$

Next find a unitary matrix U_2 such that

$$U_2 (Q_{22} - \hat{Q}_{121}^* \Sigma_1^{-2} \hat{Q}_{121}) U_2^* = \begin{bmatrix} \Sigma_3 & 0 \\ 0 & 0 \end{bmatrix}, \quad \Sigma_3 > 0$$

Define
$$(T_4^*)^{-1} = \begin{bmatrix} \Sigma_1^{-1/2} & 0 & 0 \\ 0 & I & 0 \\ 0 & 0 & U_2 \end{bmatrix}$$

and let
$$T = T_4 T_3 T_2 T_1$$

Then
$$TPT^* = \begin{bmatrix} \Sigma_1 & & & \\ & \Sigma_2 & & \\ & & 0 & \\ & & & 0 \end{bmatrix}, \quad (T^*)^{-1}QT^{-1} = \begin{bmatrix} \Sigma_1 & & & \\ & 0 & & \\ & & \Sigma_3 & \\ & & & 0 \end{bmatrix}$$

with $\Sigma_2 = I$. □

Corollary 7.6 *The product of two positive semidefinite matrices is similar to a positive semidefinite matrix.*

Proof. Let P and Q be any positive semidefinite matrices. Then it is easy to see that with the transformation given previously
$$TPQT^{-1} = \begin{bmatrix} \Sigma_1^2 & 0 \\ 0 & 0 \end{bmatrix}$$
□

Corollary 7.7 *For any stable system* $G = \left[\begin{array}{c|c} A & B \\ \hline C & D \end{array}\right]$, *there exists a nonsingular transformation* T *such that* $G = \left[\begin{array}{c|c} TAT^{-1} & TB \\ \hline CT^{-1} & D \end{array}\right]$ *has controllability Gramian* P *and observability Gramian* Q *given by*

$$P = \begin{bmatrix} \Sigma_1 & & & \\ & \Sigma_2 & & \\ & & 0 & \\ & & & 0 \end{bmatrix}, \quad Q = \begin{bmatrix} \Sigma_1 & & & \\ & 0 & & \\ & & \Sigma_3 & \\ & & & 0 \end{bmatrix}$$

respectively, with $\Sigma_1, \Sigma_2, \Sigma_3$ *diagonal and positive definite.*

7.2. Balanced Realizations

In the special case where $\left[\begin{array}{c|c} A & B \\ \hline C & D \end{array}\right]$ is a minimal realization, a balanced realization can be obtained through the following simplified procedure:

1. Compute the controllability and observability Gramians $P > 0, Q > 0$.

2. Find a matrix R such that $P = R^*R$.

3. Diagonalize RQR^* to get $RQR^* = U\Sigma^2 U^*$.

4. Let $T^{-1} = R^*U\Sigma^{-1/2}$. Then $TPT^* = (T^*)^{-1}QT^{-1} = \Sigma$ and $\left[\begin{array}{c|c} TAT^{-1} & TB \\ \hline CT^{-1} & D \end{array}\right]$ is balanced.

Assume that the Hankel singular values of the system are decreasingly ordered so that $\Sigma = \text{diag}(\sigma_1 I_{s_1}, \sigma_2 I_{s_2}, \ldots, \sigma_N I_{s_N})$ with $\sigma_1 > \sigma_2 > \ldots > \sigma_N$ and suppose $\sigma_r \gg \sigma_{r+1}$ for some r. Then the balanced realization implies that those states corresponding to the singular values of $\sigma_{r+1}, \ldots, \sigma_N$ are less controllable and less observable than those states corresponding to $\sigma_1, \ldots, \sigma_r$. Therefore, truncating those less controllable and less observable states will not lose much information about the system.

Two other closely related realizations are called *input normal realization* with $P = I$ and $Q = \Sigma^2$, and *output normal realization* with $P = \Sigma^2$ and $Q = I$. Both realizations can be obtained easily from the balanced realization by a suitable scaling on the states.

Next we shall derive some simple and useful bounds for the $\mathcal{H}_\infty$ norm and the $\mathcal{L}_1$ norm of a stable system.

Theorem 7.8 *Suppose*

$$G(s) = \left[\begin{array}{c|c} A & B \\ \hline C & 0 \end{array}\right] \in \mathcal{RH}_\infty$$

is a balanced realization; that is, there exists

$$\Sigma = \text{diag}(\sigma_1 I_{s_1}, \sigma_2 I_{s_2}, \ldots, \sigma_N I_{s_N}) \geq 0$$

with $\sigma_1 > \sigma_2 > \ldots > \sigma_N \geq 0$, such that

$$A\Sigma + \Sigma A^* + BB^* = 0 \qquad A^*\Sigma + \Sigma A + C^*C = 0$$

Then

$$\sigma_1 \leq \|G\|_\infty \leq \int_0^\infty \|g(t)\| \, dt \leq 2\sum_{i=1}^N \sigma_i$$

where $g(t) = Ce^{At}B$.

Remark 7.1 It should be clear that the inequalities stated in the theorem do not depend on a particular state-space realization of $G(s)$. However, use of the balanced realization does make the proof simple. ◇

Proof. Let $G(s)$ have the following state-space realization:

$$\begin{aligned} \dot{x} &= Ax + Bw \\ z &= Cx. \end{aligned} \quad (7.4)$$

Assume without loss of generality that (A, B) is controllable and (C, A) is observable. Then Σ is nonsingular. Next, differentiate $x(t)^*\Sigma^{-1}x(t)$ along the solution of equation (7.4) for any given input w as follows:

$$\frac{d}{dt}(x^*\Sigma^{-1}x) = \dot{x}^*\Sigma^{-1}x + x^*\Sigma^{-1}\dot{x} = x^*(A^*\Sigma^{-1} + \Sigma^{-1}A)x + 2\langle w, B^*\Sigma^{-1}x\rangle$$

Using the equation involving controllability Gramian to substitute for $A^*\Sigma^{-1} + \Sigma^{-1}A$ and completion of the squares gives

$$\frac{d}{dt}(x^*\Sigma^{-1}x) = \|w\|^2 - \|w - B^*\Sigma^{-1}x\|^2$$

Integration from $t = -\infty$ to $t = 0$ with $x(-\infty) = 0$ and $x(0) = x_0$ gives

$$x_0^*\Sigma^{-1}x_0 = \|w\|_2^2 - \|w - B^*\Sigma^{-1}x\|_2^2 \leq \|w\|_2^2$$

Let $w = B^*\Sigma^{-1}x$; then $\dot{x} = (A + BB^*\Sigma^{-1})x = -\Sigma A^*\Sigma^{-1}x \implies x \in \mathcal{L}_2[-\infty, 0)$
$\implies w \in \mathcal{L}_2[-\infty, 0)$ and

$$\inf_{w \in \mathcal{L}_2[-\infty,0)} \left\{ \|w\|_2^2 \mid x(0) = x_0 \right\} = x_0^*\Sigma^{-1}x_0.$$

Given $x(0) = x_0$ and $w = 0$ for $t \geq 0$, the norm of $z(t) = Ce^{At}x_0$ can be found from

$$\int_0^\infty \|z(t)\|^2 dt = \int_0^\infty x_0^* e^{A^*t} C^* C e^{At} x_0 dt = x_0^* \Sigma x_0$$

To show $\sigma_1 \leq \|G\|_\infty$, note that

$$\|G\|_\infty = \sup_{w \in \mathcal{L}_2(-\infty,\infty)} \frac{\|g * w\|_2}{\|w\|_2} = \sup_{w \in \mathcal{L}_2(-\infty,\infty)} \frac{\sqrt{\int_{-\infty}^\infty \|z(t)\|^2 dt}}{\sqrt{\int_{-\infty}^\infty \|w(t)\|^2 dt}}$$

$$\geq \sup_{w \in \mathcal{L}_2(-\infty,0]} \frac{\sqrt{\int_0^\infty \|z(t)\|^2 dt}}{\sqrt{\int_{-\infty}^0 \|w(t)\|^2 dt}} = \sup_{x_0 \neq 0} \sqrt{\frac{x_0^*\Sigma x_0}{x_0^*\Sigma^{-1}x_0}} = \sigma_1$$

We shall now show the other inequalities. Since

$$G(s) := \int_0^\infty g(t)e^{-st} dt, \ \text{Re}(s) > 0,$$

7.2. Balanced Realizations

by the definition of $\mathcal{H}_\infty$ norm, we have

$$\begin{aligned}
\|G\|_\infty &= \sup_{\mathrm{Re}(s)>0} \left\| \int_0^\infty g(t) e^{-st} dt \right\| \\
&\leq \sup_{\mathrm{Re}(s)>0} \int_0^\infty \|g(t) e^{-st}\| dt \\
&\leq \int_0^\infty \|g(t)\| dt.
\end{aligned}$$

To prove the last inequality, let e_i be the ith unit vector and define

$$E_1 = \begin{bmatrix} e_1 & \cdots & e_{s_1} \end{bmatrix}, \quad E_2 = \begin{bmatrix} e_{s_1+1} & \cdots & e_{s_1+s_2} \end{bmatrix}, \quad \ldots,$$

$$E_N = \begin{bmatrix} e_{s_1+\cdots+s_{N-1}+1} & \cdots & e_{s_1+\cdots+s_N} \end{bmatrix}.$$

Then $\sum_{i=1}^N E_i E_i^* = I$ and

$$\begin{aligned}
\int_0^\infty \|g(t)\| dt &= \int_0^\infty \left\| C e^{At/2} \sum_{i=1}^N E_i E_i^* e^{At/2} B \right\| dt \\
&\leq \sum_{i=1}^N \int_0^\infty \left\| C e^{At/2} E_i E_i^* e^{At/2} B \right\| dt \\
&\leq \sum_{i=1}^N \int_0^\infty \left\| C e^{At/2} E_i \right\| \left\| E_i^* e^{At/2} B \right\| dt \\
&\leq \sum_{i=1}^N \sqrt{\int_0^\infty \left\| C e^{At/2} E_i \right\|^2 dt} \sqrt{\int_0^\infty \left\| E_i^* e^{At/2} B \right\|^2 dt} \leq 2 \sum_{i=1}^N \sigma_i
\end{aligned}$$

where we have used Cauchy-Schwarz inequality and the following relations:

$$\int_0^\infty \left\| C e^{At/2} E_i \right\|^2 dt = \int_0^\infty \lambda_{\max} \left(E_i^* e^{A^*t/2} C^* C e^{At/2} E_i \right) dt = 2\lambda_{\max} \left(E_i^* \Sigma E_i \right) = 2\sigma_i$$

$$\int_0^\infty \left\| E_i^* e^{At/2} B \right\|^2 dt = \int_0^\infty \lambda_{\max} \left(E_i^* e^{At/2} BB^* e^{A^*t/2} E_i \right) dt = 2\lambda_{\max} \left(E_i^* \Sigma E_i \right) = 2\sigma_i$$

$\square$

Example 7.2 Consider a system

$$G(s) = \left[\begin{array}{cc|c} -1 & -2 & 1 \\ 1 & 0 & 0 \\ \hline 2 & 3 & 0 \end{array} \right]$$

It is easy to show that the Hankel singular values of G are $\sigma_1 = 1.6061$ and $\sigma_2 = 0.8561$. The $\mathcal{H}_\infty$ norm of G is $\|G\|_\infty = 2.972$ and the $\mathcal{L}_1$ norm of $g(t)$ can be computed as

$$\int_0^\infty |g(t)|dt = h_1 + h_2 + h_3 + h_4 + \ldots$$

where $h_i, i = 1, 2, \ldots$ are the variations of the step response of G shown in Figure 7.1, which gives $\int_0^\infty |g(t)|dt \approx 3.5$. (See Problem 7.2.)

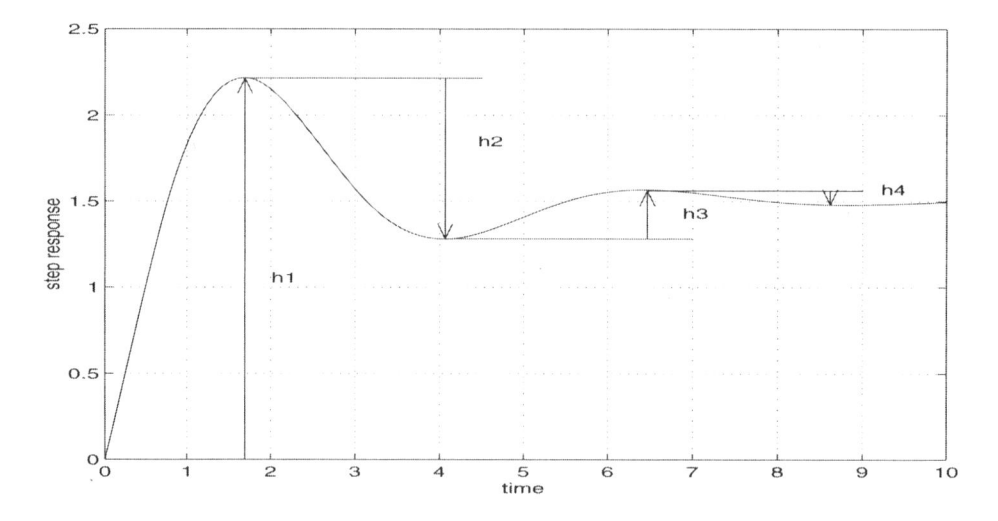

Figure 7.1: Estimating the $\mathcal{L}_1$ norm of $g(t)$

So we have

$$1.6061 = \sigma_1 \leq \|G\|_\infty = 2.972 \leq \int_0^\infty |g(t)|dt = 3.5 \leq 2(\sigma_1 + \sigma_2) = 4.9244.$$

Illustrative MATLAB Commands:

≫ [Ab, Bb, Cb, sig, Tinv]=balreal(A, B, C);   % sig is a vector of Hankel singular values and Tinv = T^{-1};

≫ [G$_\mathbf{b}$, sig] = sysbal(G);

Related MATLAB Commands: ssdelete, ssselect, modred, strunc

7.3 Model Reduction by Balanced Truncation

Consider a stable system $G \in \mathcal{RH}_\infty$ and suppose $G = \left[\begin{array}{c|c} A & B \\ \hline C & D \end{array}\right]$ is a balanced realization (i.e., its controllability and observability Gramians are equal and diagonal). Denote the balanced Gramians by Σ; then

$$A\Sigma + \Sigma A^* + BB^* = 0 \tag{7.5}$$

$$A^*\Sigma + \Sigma A + C^*C = 0. \tag{7.6}$$

Now partition the balanced Gramian as $\Sigma = \begin{bmatrix} \Sigma_1 & 0 \\ 0 & \Sigma_2 \end{bmatrix}$ and partition the system accordingly as

$$G = \left[\begin{array}{cc|c} A_{11} & A_{12} & B_1 \\ A_{21} & A_{22} & B_2 \\ \hline C_1 & C_2 & D \end{array}\right].$$

The following theorem characterizes the properties of these subsystems.

Theorem 7.9 *Assume that Σ_1 and Σ_2 have no diagonal entries in common. Then both subsystems (A_{ii}, B_i, C_i), $i = 1, 2$ are asymptotically stable.*

Proof. It is clearly sufficient to show that A_{11} is asymptotically stable. The proof for the stability of A_{22} is similar. Note that equations (7.5) and (7.6) can be written in terms of their partitioned matrices as

$$A_{11}\Sigma_1 + \Sigma_1 A_{11}^* + B_1 B_1^* = 0 \tag{7.7}$$

$$\Sigma_1 A_{11} + A_{11}^* \Sigma_1 + C_1^* C_1 = 0 \tag{7.8}$$

$$A_{21}\Sigma_1 + \Sigma_2 A_{12}^* + B_2 B_1^* = 0 \tag{7.9}$$

$$\Sigma_2 A_{21} + A_{12}^* \Sigma_1 + C_2^* C_1 = 0 \tag{7.10}$$

$$A_{22}\Sigma_2 + \Sigma_2 A_{22}^* + B_2 B_2^* = 0 \tag{7.11}$$

$$\Sigma_2 A_{22} + A_{22}^* \Sigma_2 + C_2^* C_2 = 0. \tag{7.12}$$

By Lemma 7.3 or Lemma 7.4, Σ_1 can be assumed to be positive definite without loss of generality. Then it is obvious that $\lambda_i(A_{11}) \leq 0$ by Lemma 7.2. Assume that A_{11} is not asymptotically stable; then there exists an eigenvalue at $j\omega$ for some ω. Let V be a basis matrix for $\text{Ker}(A_{11} - j\omega I)$. Then we have

$$(A_{11} - j\omega I)V = 0, \tag{7.13}$$

which gives

$$V^*(A_{11}^* + j\omega I) = 0.$$

Equations (7.7) and (7.8) can be rewritten as

$$(A_{11} - j\omega I)\Sigma_1 + \Sigma_1(A_{11}^* + j\omega I) + B_1 B_1^* = 0 \qquad (7.14)$$
$$\Sigma_1(A_{11} - j\omega I) + (A_{11}^* + j\omega I)\Sigma_1 + C_1^* C_1 = 0. \qquad (7.15)$$

Multiplication of equation (7.15) from the right by V and from the left by V^* gives $V^* C_1^* C_1 V = 0$, which is equivalent to

$$C_1 V = 0.$$

Multiplication of equation (7.15) from the right by V now gives

$$(A_{11}^* + j\omega I)\Sigma_1 V = 0.$$

Analogously, first multiply equation (7.14) from the right by $\Sigma_1 V$ and from the left by $V^* \Sigma_1$ to obtain

$$B_1^* \Sigma_1 V = 0.$$

Then multiply equation (7.14) from the right by $\Sigma_1 V$ to get

$$(A_{11} - j\omega I)\Sigma_1^2 V = 0.$$

It follows that the columns of $\Sigma_1^2 V$ are in $\text{Ker}(A_{11} - j\omega I)$. Therefore, there exists a matrix $\bar{\Sigma}_1$ such that

$$\Sigma_1^2 V = V \bar{\Sigma}_1^2.$$

Since $\bar{\Sigma}_1^2$ is the restriction of Σ_1^2 to the space spanned by V, it follows that it is possible to choose V such that $\bar{\Sigma}_1^2$ is diagonal. It is then also possible to choose $\bar{\Sigma}_1$ diagonal and such that the diagonal entries of $\bar{\Sigma}_1$ are a subset of the diagonal entries of Σ_1.

Multiply equation (7.9) from the right by $\Sigma_1 V$ and equation (7.10) by V to get

$$A_{21}\Sigma_1^2 V + \Sigma_2 A_{12}^* \Sigma_1 V = 0$$
$$\Sigma_2 A_{21} V + A_{12}^* \Sigma_1 V = 0,$$

which gives

$$(A_{21} V)\bar{\Sigma}_1^2 = \Sigma_2^2 (A_{21} V).$$

This is a Sylvester equation in $(A_{21} V)$. Because $\bar{\Sigma}_1^2$ and Σ_2^2 have no diagonal entries in common, it follows that

$$A_{21} V = 0 \qquad (7.16)$$

is the unique solution. Now equations (7.16) and (7.13) imply that

$$\begin{bmatrix} A_{11} & A_{12} \\ A_{21} & A_{22} \end{bmatrix} \begin{bmatrix} V \\ 0 \end{bmatrix} = j\omega \begin{bmatrix} V \\ 0 \end{bmatrix},$$

which means that the A-matrix of the original system has an eigenvalue at $j\omega$. This contradicts the fact that the original system is asymptotically stable. Therefore, A_{11} must be asymptotically stable. □

7.3. Model Reduction by Balanced Truncation

Corollary 7.10 *If Σ has distinct singular values, then every subsystem is asymptotically stable.*

The stability condition in Theorem 7.9 is only sufficient as shown in the following example.

Example 7.3 Note that

$$\frac{(s-1)(s-2)}{(s+1)(s+2)} = \left[\begin{array}{cc|c} -2 & -2.8284 & -2 \\ 0 & -1 & -1.4142 \\ \hline 2 & 1.4142 & 1 \end{array}\right]$$

is a balanced realization with $\Sigma = I$, and every subsystem of the realization is stable. On the other hand,

$$\frac{s^2 - s + 2}{s^2 + s + 2} = \left[\begin{array}{cc|c} -1 & 1.4142 & 1.4142 \\ -1.4142 & 0 & 0 \\ \hline -1.4142 & 0 & 1 \end{array}\right]$$

is also a balanced realization with $\Sigma = I$, but one of the subsystems is not stable.

Theorem 7.11 *Suppose $G(s) \in \mathcal{RH}_\infty$ and*

$$G(s) = \left[\begin{array}{cc|c} A_{11} & A_{12} & B_1 \\ A_{21} & A_{22} & B_2 \\ \hline C_1 & C_2 & D \end{array}\right]$$

is a balanced realization with Gramian $\Sigma = \mathrm{diag}(\Sigma_1, \Sigma_2)$

$$\begin{aligned} \Sigma_1 &= \mathrm{diag}(\sigma_1 I_{s_1}, \sigma_2 I_{s_2}, \ldots, \sigma_r I_{s_r}) \\ \Sigma_2 &= \mathrm{diag}(\sigma_{r+1} I_{s_{r+1}}, \sigma_{r+2} I_{s_{r+2}}, \ldots, \sigma_N I_{s_N}) \end{aligned}$$

and

$$\sigma_1 > \sigma_2 > \cdots > \sigma_r > \sigma_{r+1} > \sigma_{r+2} > \cdots > \sigma_N$$

where σ_i has multiplicity s_i, $i = 1, 2, \ldots, N$ and $s_1 + s_2 + \cdots + s_N = n$. Then the truncated system

$$G_r(s) = \left[\begin{array}{c|c} A_{11} & B_1 \\ \hline C_1 & D \end{array}\right]$$

is balanced and asymptotically stable. Furthermore,

$$\|G(s) - G_r(s)\|_\infty \leq 2(\sigma_{r+1} + \sigma_{r+2} + \cdots + \sigma_N)$$

Proof. The stability of G_r follows from Theorem 7.9. We shall first show the one step model reduction. Hence we shall assume $\Sigma_2 = \sigma_N I_{s_N}$. Define the approximation error

$$E_{11} := \left[\begin{array}{cc|c} A_{11} & A_{12} & B_1 \\ A_{21} & A_{22} & B_2 \\ \hline C_1 & C_2 & D \end{array}\right] - \left[\begin{array}{c|c} A_{11} & B_1 \\ \hline C_1 & D \end{array}\right]$$

$$= \left[\begin{array}{ccc|c} A_{11} & 0 & 0 & B_1 \\ 0 & A_{11} & A_{12} & B_1 \\ 0 & A_{21} & A_{22} & B_2 \\ \hline -C_1 & C_1 & C_2 & 0 \end{array}\right]$$

Apply a similarity transformation T to the preceding state-space realization with

$$T = \left[\begin{array}{ccc} I/2 & I/2 & 0 \\ I/2 & -I/2 & 0 \\ 0 & 0 & I \end{array}\right], \quad T^{-1} = \left[\begin{array}{ccc} I & I & 0 \\ I & -I & 0 \\ 0 & 0 & I \end{array}\right]$$

to get

$$E_{11} = \left[\begin{array}{ccc|c} A_{11} & 0 & A_{12}/2 & B_1 \\ 0 & A_{11} & -A_{12}/2 & 0 \\ A_{21} & -A_{21} & A_{22} & B_2 \\ \hline 0 & -2C_1 & C_2 & 0 \end{array}\right]$$

Consider a dilation of $E_{11}(s)$:

$$E(s) = \left[\begin{array}{cc} E_{11}(s) & E_{12}(s) \\ E_{21}(s) & E_{22}(s) \end{array}\right]$$

$$= \left[\begin{array}{cccc|cc} A_{11} & 0 & A_{12}/2 & B_1 & 0 \\ 0 & A_{11} & -A_{12}/2 & 0 & \sigma_N \Sigma_1^{-1} C_1^* \\ A_{21} & -A_{21} & A_{22} & B_2 & -C_2^* \\ \hline 0 & -2C_1 & C_2 & 0 & 2\sigma_N I \\ -2\sigma_N B_1^* \Sigma_1^{-1} & 0 & -B_2^* & 2\sigma_N I & 0 \end{array}\right]$$

$$=: \left[\begin{array}{c|c} \tilde{A} & \tilde{B} \\ \hline \tilde{C} & \tilde{D} \end{array}\right]$$

Then it is easy to verify that

$$\tilde{P} = \left[\begin{array}{ccc} \Sigma_1 & 0 & 0 \\ 0 & \sigma_N^2 \Sigma_1^{-1} & 0 \\ 0 & 0 & 2\sigma_N I_{s_N} \end{array}\right]$$

satisfies

$$\tilde{A}\tilde{P} + \tilde{P}\tilde{A}^* + \tilde{B}\tilde{B}^* = 0$$
$$\tilde{P}\tilde{C}^* + \tilde{B}\tilde{D}^* = 0$$

7.3. Model Reduction by Balanced Truncation

Using these two equations, we have

$$E(s)E^\sim(s) = \left[\begin{array}{cc|c} \tilde{A} & -\tilde{B}\tilde{B}^* & \tilde{B}\tilde{D}^* \\ 0 & -\tilde{A}^* & \tilde{C}^* \\ \hline \tilde{C} & -\tilde{D}\tilde{B}^* & \tilde{D}\tilde{D}^* \end{array}\right]$$

$$= \left[\begin{array}{cc|c} \tilde{A} & -\tilde{A}\tilde{P}-\tilde{P}\tilde{A}^*-\tilde{B}\tilde{B}^* & \tilde{P}\tilde{C}^*+\tilde{B}\tilde{D}^* \\ 0 & -\tilde{A}^* & \tilde{C}^* \\ \hline \tilde{C} & -\tilde{C}\tilde{P}-\tilde{D}\tilde{B}^* & \tilde{D}\tilde{D}^* \end{array}\right]$$

$$= \left[\begin{array}{cc|c} \tilde{A} & 0 & 0 \\ 0 & -\tilde{A}^* & \tilde{C}^* \\ \hline \tilde{C} & 0 & \tilde{D}\tilde{D}^* \end{array}\right]$$

$$= \tilde{D}\tilde{D}^* = 4\sigma_N^2 I$$

where the second equality is obtained by applying a similarity transformation

$$T = \left[\begin{array}{cc} I & \tilde{P} \\ 0 & I \end{array}\right]$$

Hence $\|E_{11}\|_\infty \leq \|E\|_\infty = 2\sigma_N$, which is the desired result.

The remainder of the proof is achieved by using the order reduction by one-step results and by noting that $G_k(s) = \left[\begin{array}{c|c} A_{11} & B_1 \\ \hline C_1 & D \end{array}\right]$ obtained by the "kth" order partitioning is internally balanced with balanced Gramian given by

$$\Sigma_1 = \text{diag}(\sigma_1 I_{s_1}, \sigma_2 I_{s_2}, \ldots, \sigma_k I_{s_k})$$

Let $E_k(s) = G_{k+1}(s) - G_k(s)$ for $k = 1, 2, \ldots, N-1$ and let $G_N(s) = G(s)$. Then

$$\bar{\sigma}[E_k(j\omega)] \leq 2\sigma_{k+1}$$

since $G_k(s)$ is a reduced-order model obtained from the internally balanced realization of $G_{k+1}(s)$ and the bound for one-step order reduction holds.

Noting that

$$G(s) - G_r(s) = \sum_{k=r}^{N-1} E_k(s)$$

by the definition of $E_k(s)$, we have

$$\bar{\sigma}[G(j\omega) - G_r(j\omega)] \leq \sum_{k=r}^{N-1} \bar{\sigma}[E_k(j\omega)] \leq 2\sum_{k=r}^{N-1} \sigma_{k+1}$$

This is the desired upper bound. □

A useful consequence of the preceding theorem is the following corollary.

Corollary 7.12 Let $\sigma_i, i = 1, \ldots, N$ be the Hankel singular values of $G(s) \in \mathcal{RH}_\infty$. Then
$$\|G(s) - G(\infty)\|_\infty \leq 2(\sigma_1 + \ldots + \sigma_N)$$

The above bound can be tight for some systems.

Example 7.4 Consider an nth-order transfer function
$$G(s) = \sum_{i=1}^{n} \frac{b_i}{s + a_i},$$
with $a_i > 0$ and $b_i > 0$. Then $\|G(s)\|_\infty = G(0) = \sum_{i=1}^{n} b_i/a_i$ and $G(s)$ has the following state-space realization:

$$G = \left[\begin{array}{cccc|c} -a_1 & & & & \sqrt{b_1} \\ & -a_2 & & & \sqrt{b_2} \\ & & \ddots & & \vdots \\ & & & -a_n & \sqrt{b_n} \\ \hline \sqrt{b_1} & \sqrt{b_2} & \cdots & \sqrt{b_n} & 0 \end{array}\right]$$

and the controllability and observability Gramians of the realization are given by
$$P = Q = \left[\frac{\sqrt{b_i b_j}}{a_i + a_j}\right]$$

It is easy to see that $\sigma_i = \lambda_i(P) = \lambda_i(Q)$ and
$$\sum_{i=1}^{n} \sigma_i = \sum_{i=1}^{n} \lambda_i(P) = \text{trace}(P) = \sum_{i=1}^{n} \frac{b_i}{2a_i} = \frac{1}{2} G(0) = \frac{1}{2} \|G\|_\infty$$

In particular, let $a_i = b_i = \alpha^{2i}$; then $P = Q \to \frac{1}{2} I_n$ (i.e., $\sigma_j \to \frac{1}{2}$ as $\alpha \to \infty$). This example also shows that even when the Hankel singular values are extremely close, they may not be regarded as repeated singular values.

The model reduction bound can also be loose for systems with Hankel singular values close to each other.

7.3. Model Reduction by Balanced Truncation

Example 7.5 Consider the balanced realization of a fourth-order system:

$$G(s) = \frac{(s-0.99)(s-2)(s-3)(s-4)}{(s+1)(s+2)(s+3)(s+4)}$$

$$= \left[\begin{array}{cccc|c} -9.2e+00 & -5.7e+00 & -2.7e+00 & 1.3e+00 & -4.3e+00 \\ 5.7e+00 & -8.1e-07 & -6.4e-01 & 1.5e-06 & 1.3e-03 \\ -2.7e+00 & 6.4e-01 & -7.9e-01 & 7.1e-01 & -1.3e+00 \\ -1.3e+00 & 1.5e-06 & -7.1e-01 & -2.7e-06 & -2.3e-03 \\ \hline 4.3e+00 & 1.3e-03 & 1.3e+00 & -2.3e-03 & 1.0e+00 \end{array}\right]$$

with Hankel singular values given by

$$\sigma_1 = 0.9998, \quad \sigma_2 = 0.9988, \quad \sigma_3 = 0.9963, \sigma_4 = 0.9923.$$

The approximation errors and the estimated bounds are listed in the following table. The table shows that the actual error for an rth-order approximation is almost the same as $2\sigma_{r+1}$, which would be the estimated bound if we regard $\sigma_{r+1} = \sigma_{r+2} = \cdots = \sigma_4$. In general, it is not hard to construct an nth-order system so that the rth-order balanced model reduction error is approximately $2\sigma_{r+1}$ but the error bound is arbitrarily close to $2(n-r)\sigma_{r+1}$. One method to construct such a system is as follows: Let $G(s)$ be a stable all-pass function, that is, $G^\sim(s)G(s) = I$. Then there is a balanced realization for G so that the controllability and observability Gramians are $P = Q = I$. Next, make a very small perturbation to the balanced realization, then the perturbed system has a balanced realization with distinct singular values and $P = Q \approx I$. This perturbed system will have the desired properties.

r	0	1	2	3
$\|G - G_r\|_\infty$	1.9997	1.9983	1.9933	1.9845
Bounds: $2\sum_{i=r+1}^{4} \sigma_i$	7.9744	5.9748	3.9772	1.9845
$2\sigma_{r+1}$	1.9996	1.9976	1.9926	1.9845

The balanced realization and truncation can be done using the following MATLAB commands:

> ≫ **[G$_b$, sig]** = **sysbal(G)**; % find a balanced realization G_b and the Hankel singular values sig.

> ≫ **G$_r$** = **strunc(G$_b$, 2)**; % truncate to the second-order.

Related MATLAB Commands: reordsys, resid, Hankmr

7.4 Frequency-Weighted Balanced Model Reduction

This section considers the extension of the balanced truncation method to the frequency-weighted case. Given the original full-order model $G \in \mathcal{RH}_\infty$, the input weighting matrix $W_i \in \mathcal{RH}_\infty$, and the output weighting matrix $W_o \in \mathcal{RH}_\infty$, our objective is to find a lower-order model G_r such that

$$\|W_o(G - G_r)W_i\|_\infty$$

is made as small as possible. Assume that $G, W_i,$ and W_o have the following state-space realizations:

$$G = \left[\begin{array}{c|c} A & B \\ \hline C & 0 \end{array}\right], \quad W_i = \left[\begin{array}{c|c} A_i & B_i \\ \hline C_i & D_i \end{array}\right], \quad W_o = \left[\begin{array}{c|c} A_o & B_o \\ \hline C_o & D_o \end{array}\right]$$

with $A \in \mathbb{R}^{n \times n}$. Note that there is no loss of generality in assuming $D = G(\infty) = 0$ since otherwise it can be eliminated by replacing G_r with $D + G_r$.

Now the state-space realization for the weighted transfer matrix is given by

$$W_o G W_i = \left[\begin{array}{ccc|c} A & 0 & BC_i & BD_i \\ B_oC & A_o & 0 & 0 \\ 0 & 0 & A_i & B_i \\ \hline D_oC & C_o & 0 & 0 \end{array}\right] =: \left[\begin{array}{c|c} \bar{A} & \bar{B} \\ \hline \bar{C} & 0 \end{array}\right]$$

Let $\bar{P}$ and $\bar{Q}$ be the solutions to the following Lyapunov equations:

$$\bar{A}\bar{P} + \bar{P}\bar{A}^* + \bar{B}\bar{B}^* = 0 \qquad (7.17)$$
$$\bar{Q}\bar{A} + \bar{A}^*\bar{Q} + \bar{C}^*\bar{C} = 0 \qquad (7.18)$$

Then the input weighted Gramian P and the output weighted Gramian Q are defined by

$$P := \begin{bmatrix} I_n & 0 \end{bmatrix} \bar{P} \begin{bmatrix} I_n \\ 0 \end{bmatrix}, \quad Q := \begin{bmatrix} I_n & 0 \end{bmatrix} \bar{Q} \begin{bmatrix} I_n \\ 0 \end{bmatrix}$$

It can be shown easily that P and Q satisfy the following lower-order equations:

$$\begin{bmatrix} A & BC_i \\ 0 & A_i \end{bmatrix} \begin{bmatrix} P & P_{12} \\ P_{12}^* & P_{22} \end{bmatrix} + \begin{bmatrix} P & P_{12} \\ P_{12}^* & P_{22} \end{bmatrix} \begin{bmatrix} A & BC_i \\ 0 & A_i \end{bmatrix}^* + \begin{bmatrix} BD_i \\ B_i \end{bmatrix} \begin{bmatrix} BD_i \\ B_i \end{bmatrix}^* = 0$$
$$(7.19)$$

$$\begin{bmatrix} Q & Q_{12} \\ Q_{12}^* & Q_{22} \end{bmatrix} \begin{bmatrix} A & 0 \\ B_oC & A_o \end{bmatrix} + \begin{bmatrix} A & 0 \\ B_oC & A_o \end{bmatrix}^* \begin{bmatrix} Q & Q_{12} \\ Q_{12}^* & Q_{22} \end{bmatrix} + \begin{bmatrix} C^*D_o^* \\ C_o^* \end{bmatrix} \begin{bmatrix} C^*D_o^* \\ C_o^* \end{bmatrix}^* = 0$$
$$(7.20)$$

The computation can be further reduced if $W_i = I$ or $W_o = I$. In the case of $W_i = I$, P can be obtained from

$$PA^* + AP + BB^* = 0 \qquad (7.21)$$

7.4. Frequency-Weighted Balanced Model Reduction

while in the case of $W_o = I$, Q can be obtained from

$$QA + A^*Q + C^*C = 0 \qquad (7.22)$$

Now let T be a nonsingular matrix such that

$$TPT^* = (T^{-1})^*QT^{-1} = \begin{bmatrix} \Sigma_1 & \\ & \Sigma_2 \end{bmatrix}$$

(i.e., balanced) with $\Sigma_1 = \text{diag}(\sigma_1 I_{s_1}, \ldots, \sigma_r I_{s_r})$ and $\Sigma_2 = \text{diag}(\sigma_{r+1} I_{s_{r+1}}, \ldots, \sigma_N I_{s_N})$ and partition the system accordingly as

$$\left[\begin{array}{c|c} TAT^{-1} & TB \\ \hline CT^{-1} & 0 \end{array}\right] = \left[\begin{array}{cc|c} A_{11} & A_{12} & B_1 \\ A_{21} & A_{22} & B_2 \\ \hline C_1 & C_2 & 0 \end{array}\right]$$

Then a reduced-order model G_r is obtained as

$$G_r = \left[\begin{array}{c|c} A_{11} & B_1 \\ \hline C_1 & 0 \end{array}\right]$$

Unfortunately, there is generally no known a priori error bound for the approximation error and the reduced-order model G_r is not guaranteed to be stable either.

A very special frequency-weighted model reduction problem is the relative error model reduction problem where the objective is to find a reduced-order model G_r so that

$$G_r = G(I + \Delta_{\text{rel}})$$

and $\|\Delta_{\text{rel}}\|_\infty$ is made as small as possible. Δ_{rel} is usually called the *relative error*. In the case where G is square and invertible, this problem can be simply formulated as

$$\min_{\deg G_r \leq r} \left\| G^{-1}(G - G_r) \right\|_\infty.$$

Of course, the dual approximation problem

$$G_r = (I + \Delta_{\text{rel}})G$$

can be obtained by taking the transpose of G. It turns out that the approximation G_r obtained below also serves as a multiplicative approximation:

$$G = G_r(I + \Delta_{\text{mul}})$$

where Δ_{mul} is usually called the *multiplicative error*.

Error bounds can be derived if the frequency-weighted balanced truncation method is applied to the relative and multiplicative approximations.

Theorem 7.13 Let $G, G^{-1} \in \mathcal{RH}_\infty$ be an nth-order square transfer matrix with a state-space realization

$$G(s) = \left[\begin{array}{c|c} A & B \\ \hline C & D \end{array}\right]$$

Let P and Q be the solutions to

$$PA^* + AP + BB^* = 0 \quad (7.23)$$

$$Q(A - BD^{-1}C) + (A - BD^{-1}C)^*Q + C^*(D^{-1})^*D^{-1}C = 0 \quad (7.24)$$

Suppose

$$P = Q = \text{diag}(\sigma_1 I_{s_1}, \ldots, \sigma_r I_{s_r}, \sigma_{r+1} I_{s_{r+1}}, \ldots, \sigma_N I_{s_N}) = \text{diag}(\Sigma_1, \Sigma_2)$$

with $\sigma_1 > \sigma_2 > \ldots > \sigma_N \geq 0$, and let the realization of G be partitioned compatibly with Σ_1 and Σ_2 as

$$G(s) = \left[\begin{array}{cc|c} A_{11} & A_{12} & B_1 \\ A_{21} & A_{22} & B_2 \\ \hline C_1 & C_2 & D \end{array}\right]$$

Then

$$G_r(s) = \left[\begin{array}{c|c} A_{11} & B_1 \\ \hline C_1 & D \end{array}\right]$$

is stable and minimum phase. Furthermore,

$$\|\Delta_{\text{rel}}\|_\infty \leq \prod_{i=r+1}^N \left(1 + 2\sigma_i(\sqrt{1+\sigma_i^2} + \sigma_i)\right) - 1$$

$$\|\Delta_{\text{mul}}\|_\infty \leq \prod_{i=r+1}^N \left(1 + 2\sigma_i(\sqrt{1+\sigma_i^2} + \sigma_i)\right) - 1$$

Related MATLAB Commands: srelbal, sfrwtbal

7.5 Notes and References

Balanced realization was first introduced by Mullis and Roberts [1976] to study roundoff noise in digital filters. Moore [1981] proposed the balanced truncation method for model-reduction. The stability properties of the reduced-order model were shown by Pernebo and Silverman [1982]. The error bound for the balanced model reduction was shown by Enns [1984a, 1984b], and Glover [1984] subsequently gave an independent proof. The frequency-weighted balanced model-reduction method was also introduced by Enns [1984a, 1984b]. The error bounds for the relative error are derived in Zhou [1995]. Other related results are shown in Green [1988]. Other weighted model-reduction methods can be found in Al-Saggaf and Franklin [1988], Glover [1986b], Glover, Limebeer and Hung [1992], Green [1988], Hung and Glover [1986], Zhou [1995], and references therein. Discrete-time balance model-reduction results can be found in Al-Saggaf and Franklin [1987], Hinrichsen and Pritchard [1990], and references therein.

7.6 Problems

Problem 7.1 Use the following relation

$$\frac{d}{dt}\left(e^{At}Qe^{A^*t}\right) = Ae^{At}Qe^{A^*t} + e^{At}Qe^{A^*t}A$$

to show that $P = \int_0^\infty e^{At}Qe^{A^*t}dt$ solves

$$AP + PA^* + Q = 0$$

if A is stable.

Problem 7.2 Let $G(s) \in \mathcal{H}_\infty$ and let $g(t)$ be the inverse Laplace transform of $G(s)$. Let $h_i, i = 1, 2, \ldots$ be the variations of the step response of G. Show that

$$\int_0^\infty |g(t)|dt = h_1 + h_2 + h_3 + h_4 + \ldots$$

Problem 7.3 Let $Q \geq 0$ be the solution to

$$QA + A^*Q + C^*C = 0$$

Suppose Q has m zero eigenvalues. Show that there is a nonsingular matrix T such that

$$\left[\begin{array}{c|c} TAT^{-1} & TB \\ \hline CT^{-1} & D \end{array}\right] = \left[\begin{array}{cc|c} A_{11} & 0 & B_1 \\ A_{21} & A_{22} & B_2 \\ \hline C_1 & 0 & D \end{array}\right], \quad A_{22} \in \mathbb{R}^{m \times m}.$$

Apply the above result to the following state-space model:

$$A = \begin{bmatrix} -4 & -7 & -2 \\ 1 & 0 & 0 \\ -1 & 1 & 0 \end{bmatrix}, \quad B = \begin{bmatrix} 1 & 2 \\ 0 & -1 \\ 0 & 2 \end{bmatrix}, \quad C = \begin{bmatrix} 0 & 2 & 1 \\ 1 & 1 & 0 \end{bmatrix}, \quad D = 0$$

Problem 7.4 Let

$$G(s) = \sum_{i=1}^{5} \frac{\alpha^{2i}}{s + \alpha^{2i}}$$

Find a balanced realization for each of the following α:

$$\alpha = 2, 4, 20, 100.$$

Discuss the behavior of the Hankel singular values as $\alpha \to \infty$.

Problem 7.5 Find a transformation so that $TPT^* = \Sigma^2, (T^*)^{-1}QT^{-1} = I$. (This realization is called output normalized realization.)

Problem 7.6 Consider the model reduction error:

$$E_{11} = \left[\begin{array}{ccc|c} A_{11} & 0 & A_{12}/2 & B_1 \\ 0 & A_{11} & -A_{12}/2 & 0 \\ A_{21} & -A_{21} & A_{22} & B_2 \\ \hline 0 & -2C_1 & C_2 & 0 \end{array}\right] =: \left[\begin{array}{c|c} A_e & B_e \\ \hline C_e & 0 \end{array}\right].$$

Show that

$$\tilde{P} = \begin{bmatrix} \Sigma_1 & 0 & \\ 0 & \sigma_N^2 \Sigma_1^{-1} & 0 \\ 0 & 0 & 2\sigma_N I_{s_N} \end{bmatrix}$$

satisfies

$$A_e \tilde{P} + \tilde{P} A_e^* + B_e B_e^* + \frac{1}{4\sigma_N^2} \tilde{P} C_e^* C_e \tilde{P} = 0.$$

Problem 7.7 Suppose P and Q are the controllability and observability Gramians of $G(s) = C(sI - A)^{-1}B \in \mathcal{RH}_\infty$. Let $G_d(z) = G(s)|_{s=\frac{z+1}{z-1}} = C_d(zI - A_d)^{-1}B_d + D_d$. Compute the controllability and observability Gramians P_d and Q_d and compare PQ and $P_d Q_d$.

Problem 7.8 Note that a delay can be approximated as

$$e^{-\tau s} \approx \left(\frac{1 - \frac{\tau}{2n}s}{1 + \frac{\tau}{2n}s}\right)^n$$

for a sufficiently large n. Let a process model $\dfrac{e^{-s}}{1+Ts}$ be approximated by

$$G(s) = \left(\frac{1 - 0.05s}{1 + 0.05s}\right)^{10} \frac{1}{1 + sT}$$

For each $T = 0, 0.01, 0.1, 1, 10$, find a reduced-order model, if possible, using balanced truncation such that the approximation error is no greater than 0.1.

Chapter 8

Uncertainty and Robustness

In this chapter we briefly describe various types of uncertainties that can arise in physical systems, and we single out "unstructured uncertainties" as generic errors that are associated with all design models. We obtain robust stability tests for systems under various model uncertainty assumptions through the use of the small gain theorem. We also obtain some sufficient conditions for robust performance under unstructured uncertainties. The difficulty associated with MIMO robust performance design and the role of plant condition numbers for systems with skewed performance and uncertainty specifications are revealed. A simple example is also used to indicate the fundamental difference between the robustness of an SISO system and that of a MIMO system. In particular, we show that applying the SISO analysis/design method to a MIMO system may lead to erroneous results.

8.1 Model Uncertainty

Most control designs are based on the use of a design model. The relationship between models and the reality they represent is subtle and complex. A mathematical model provides a map from inputs to responses. The quality of a model depends on how closely its responses match those of the true plant. Since no single fixed model can respond exactly like the true plant, we need, at the very least, a set of maps. However, the modeling problem is much deeper — the universe of mathematical models from which a model set is chosen is distinct from the universe of physical systems. Therefore, a model set that includes the true physical plant can never be constructed. It is necessary for the engineer to make a leap of faith regarding the applicability of a particular design based on a mathematical model. To be practical, a design technique must help make this leap small by accounting for the inevitable inadequacy of models. A good model should be simple enough to facilitate design, yet complex enough to give the engineer confidence that designs based on the model will work on the true plant.

The term *uncertainty* refers to the differences or errors between models and reality, and whatever mechanism is used to express these errors will be called a *representation of uncertainty*. Representations of uncertainty vary primarily in terms of the amount of structure they contain. This reflects both our knowledge of the physical mechanisms that cause differences between the model and the plant and our ability to represent these mechanisms in a way that facilitates convenient manipulation. For example, consider the problem of bounding the magnitude of the effect of some uncertainty on the output of a nominally fixed linear system. A useful measure of uncertainty in this context is to provide a bound on the power spectrum of the output's deviation from its nominal response. In the simplest case, this power spectrum is assumed to be independent of the input. This is equivalent to assuming that the uncertainty is generated by an additive noise signal with a bounded power spectrum; the uncertainty is represented as additive noise. Of course, no physical system is linear with additive noise, but some aspects of physical behavior are approximated quite well using this model. This type of uncertainty received a great deal of attention in the literature during the 1960s and 1970s, and elegant solutions are obtained for many interesting problems (e.g., white noise propagation in linear systems, Wiener and Kalman filtering, and LQG optimal control). Unfortunately, LQG optimal control did not address uncertainty adequately and hence had less practical impact than might have been hoped.

Generally, the deviation's power spectrum of the true output from the nominal will depend significantly on the input. For example, an additive noise model is entirely inappropriate for capturing uncertainty arising from variations in the material properties of physical plants. The actual construction of model sets for more general uncertainty can be quite difficult. For example, a set membership statement for the parameters of an otherwise known FDLTI model is a highly structured representation of uncertainty. It typically arises from the use of linear incremental models at various operating points (e.g., aerodynamic coefficients in flight control vary with flight environment and aircraft configurations, and equation coefficients in power plant control vary with aging, slag buildup, coal composition, etc.). In each case, the amounts of variation and any known relationships between parameters can be expressed by confining the parameters to appropriately defined subsets of parameter space. However, for certain classes of signals (e.g., high-frequency), the parameterized FDLTI model fails to describe the plant because the plant will always have dynamics that are not represented in the fixed order model.

In general, we are forced to use not just a single parameterized model but model sets that allow for plant dynamics that are not explicitly represented in the model structure. A simple example of this involves using frequency domain bounds on transfer functions to describe a model set. To use such sets to describe physical systems, the bounds must roughly grow with frequency. In particular, at sufficiently high frequencies, phase is completely unknown (i.e., $\pm 180°$ uncertainties). This is a consequence of dynamic properties that inevitably occur in physical systems. This gives a less structured representation of uncertainty.

8.1. Model Uncertainty

Examples of less structured representations of uncertainty are direct set membership statements for the transfer function matrix of the model. For instance, the statement

$$P_\Delta(s) = P(s) + W_1(s)\Delta(s)W_2(s), \quad \bar{\sigma}[\Delta(j\omega)] < 1, \ \forall \omega \geq 0, \tag{8.1}$$

where W_1 and W_2 are stable transfer matrices that characterize the spatial and frequency structure of the uncertainty, confines the matrix P_Δ to a neighborhood of the nominal model P. In particular, if $W_1 = I$ and $W_2 = w(s)I$, where $w(s)$ is a scalar function, then P_Δ describes a disk centered at P with radius $w(j\omega)$ at each frequency, as shown in Figure 8.1. The statement does not imply a mechanism or structure that gives rise to Δ. The uncertainty may be caused by parameter changes, as mentioned previously or by neglected dynamics, or by a host of other unspecified effects. An alternative statement to equation (8.1) is the so-called multiplicative form:

$$P_\Delta(s) = (I + W_1(s)\Delta(s)W_2(s))P(s). \tag{8.2}$$

This statement confines P_Δ to a normalized neighborhood of the nominal model P. An advantage of equation (8.2) over (8.1) is that in equation (8.2) compensated transfer functions have the same uncertainty representation as the raw model (i.e., the weighting functions apply to PK as well as P). Some other alternative set membership statements will be discussed later.

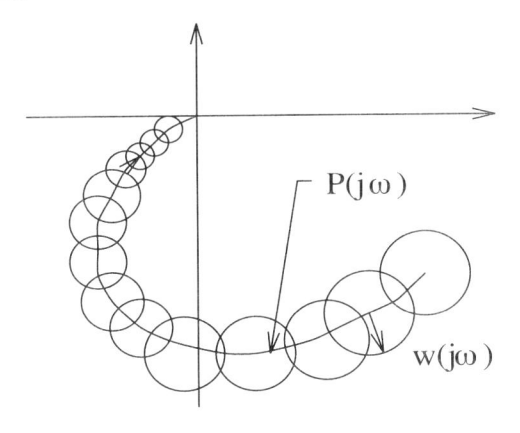

Figure 8.1: Nyquist diagram of an uncertain model

The best choice of uncertainty representation for a specific FDLTI model depends, of course, on the errors the model makes. In practice, it is generally possible to represent some of these errors in a highly structured parameterized form. These are usually the low-frequency error components. There are always remaining higher-frequency errors, however, which cannot be covered this way. These are caused by such effects as infinite-dimensional electromechanical resonance, time delays, diffusion processes, etc. Fortunately, the less structured representations, such as equations (8.1) and (8.2), are well suited to represent this latter class of errors. Consequently, equations (8.1) and

(8.2) have become widely used "generic" uncertainty representations for FDLTI models. An important point is that the construction of the weighting matrices W_1 and W_2 for multivariable systems is not trivial.

Motivated from these observations, we will focus for the moment on the multiplicative description of uncertainty. We will assume that P_Δ in equation (8.2) remains a strictly proper FDLTI system for all Δ. More general perturbations (e.g., time varying, infinite dimensional, nonlinear) can also be covered by this set provided they are given appropriate "conic sector" interpretations via Parseval's theorem. This connection is developed in [Safonov, 1980] and [Zames, 1966] and will not be pursued here.

When used to represent the various high-frequency mechanisms mentioned previously, the weighting functions in equation (8.2) commonly have the properties illustrated in Figure 8.2. They are small ($\ll 1$) at low frequencies and increase to unity and above at higher frequencies. The growth with frequency inevitably occurs because phase uncertainties eventually exceed ± 180 degrees and magnitude deviations eventually exceed the nominal transfer function magnitudes. Readers who are skeptical about this reality are encouraged to try a few experiments with physical devices.

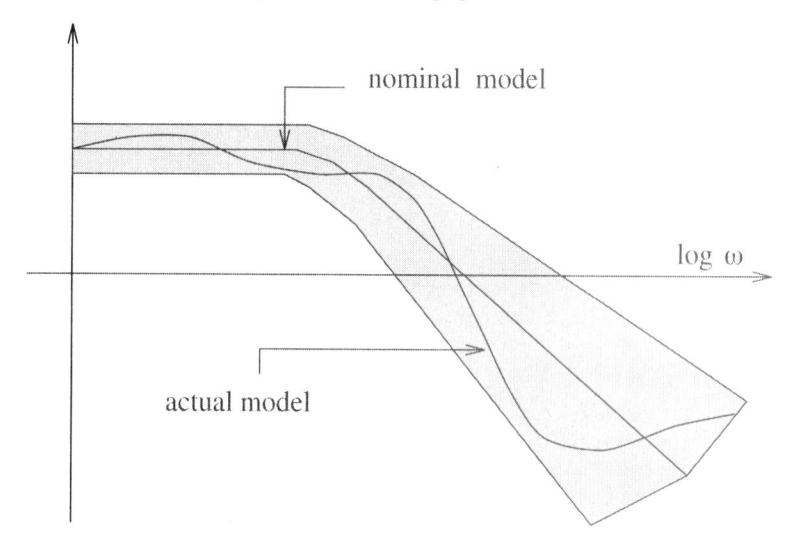

Figure 8.2: Typical behavior of multiplicative uncertainty: $p_\delta(s) = [1 + w(s)\delta(s)]p(s)$

Also note that the representation of uncertainty in equation (8.2) can be used to include perturbation effects that are in fact certain. A nonlinear element may be quite accurately modeled, but because our design techniques cannot effectively deal with the nonlinearity, it is treated as a conic sector nonlinearity.[1] As another example, we may deliberately choose to ignore various known dynamic characteristics in order to achieve a simple nominal design model. One such instance is the model reduction process discussed in the last chapter.

---

[1] See, for example, Safonov [1980] and Zames [1966].

8.1. Model Uncertainty

Example 8.1 Let a dynamical system be described by

$$P(s,\alpha,\beta) = \frac{10\left((2+0.2\alpha)s^2 + (2+0.3\alpha+0.4\beta)s + (1+0.2\beta)\right)}{(s^2+0.5s+1)(s^2+2s+3)(s^2+3s+6)}, \quad \alpha,\ \beta \in [-1,1]$$

Then for each frequency, all possible frequency responses with varying parameters α and β are in a box, as shown in Figure 8.3. We can also obtain an unstructured uncertainty bound for this system. In fact, we have

$$P(s,\alpha,\beta) \in \{P_0 + W\Delta \mid \|\Delta\| \leq 1\}$$

with $P_0 := P(s,0,0)$ and

$$W(s) = P(s,1,1) - P(s,0,0) = \frac{10\left(0.2s^2 + 0.7s + 0.2\right)}{(s^2+0.5s+1)(s^2+2s+3)(s^2+3s+6)}$$

The frequency response $P_0 + W\Delta$ is shown in Figure 8.3 as circles.

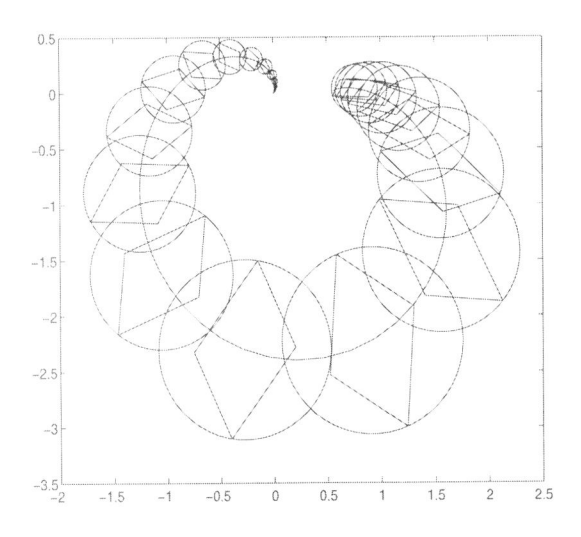

Figure 8.3: Nyquist diagram of uncertain system and disk covering

Another way to bound the frequency response is to treat α and β as norm bounded uncertainties; that is,

$$P(s,\alpha,\beta) \in \{P_0 + W_1\Delta_1 + W_2\Delta_2 \mid \|\Delta_i\|_\infty \leq 1\}$$

with $P_0 = P(s,0,0)$ and

$$W_1 = \frac{10(0.2s^2 + 0.3s)}{(s^2+0.5s+1)(s^2+2s+3)(s^2+3s+6)},$$

$$W_2 = \frac{10(0.4s + 0.2)}{(s^2 + 0.5s + 1)(s^2 + 2s + 3)(s^2 + 3s + 6)}$$

It is in fact easy to show that

$$\{P_0 + W_1\Delta_1 + W_2\Delta_2 \mid \|\Delta_i\|_\infty \leq 1\} = \{P_0 + W\Delta \mid \|\Delta\|_\infty \leq 1\}$$

with $|W| = |W_1| + |W_2|$. The frequency response $P_0 + W\Delta$ is shown in Figure 8.4. This bounding is clearly more conservative.

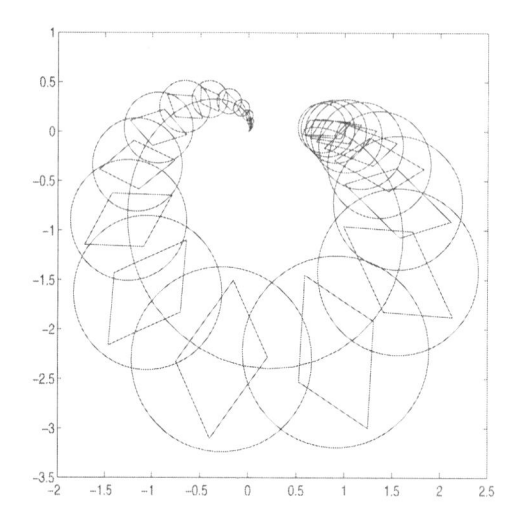

Figure 8.4: A conservative covering

Example 8.2 Consider a process control model

$$G(s) = \frac{ke^{-\tau s}}{Ts + 1}, \quad 4 \leq k \leq 9,\ 2 \leq T \leq 3,\ 1 \leq \tau \leq 2.$$

Take the nominal model as

$$G_0(s) = \frac{6.5}{(2.5s + 1)(1.5s + 1)}$$

Then for each frequency, all possible frequency responses are in a box, as shown in Figure 8.5. To obtain an unstructured uncertainty bound for this system, plot the error

$$\Delta_a(j\omega) = G(j\omega) - G_0(j\omega)$$

for a set of parameters, as shown in Figure 8.6, and then use the MATLAB command **ginput** to pick a set of upper-bound frequency responses and use **fitmag** to fit a stable and minimum phase transfer function to the upper-bound frequency responses.

8.1. Model Uncertainty

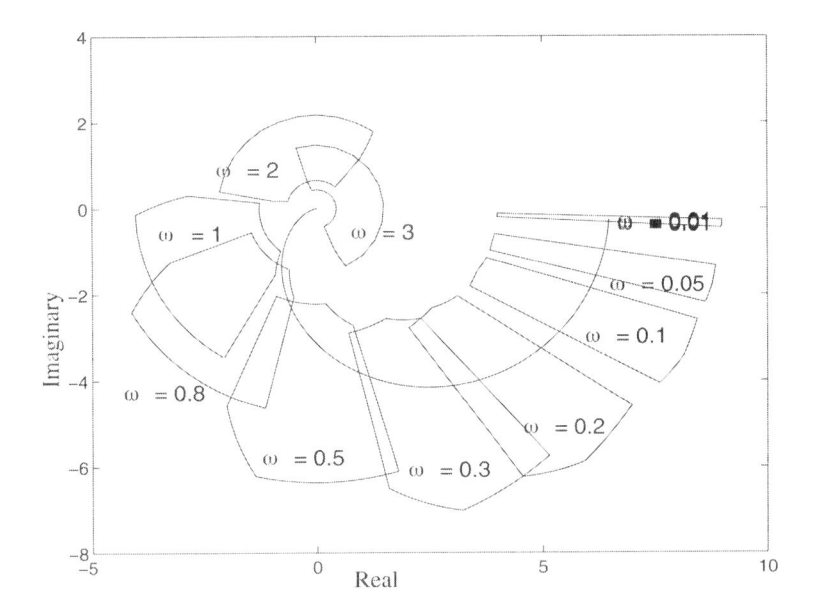

Figure 8.5: Uncertain delay system and G_0

≫ **mf= ginput(50)**   % pick 50 points: the first column of mf is the frequency points and the second column of mf is the corresponding magnitude responses.

≫ **magg=vpck(mf(:,2),mf(:,1));**   % pack them as a varying matrix.

≫ **W$_a$ =fitmag(magg);**   % choose the order of W_a online. A third-order W_a is sufficient for this example.

≫ **[A,B,C,D]=unpck(W$_a$)**   % converting into state-space.

≫ **[Z, P, K]=ss2zp(A,B,C,D)**   % converting into zero/pole/gain form.

We get
$$W_a(s) = \frac{0.0376(s+116.4808)(s+7.4514)(s+0.2674)}{(s+1.2436)(s+0.5575)(s+4.9508)}$$
and the frequency response of W_a is also plotted in Figure 8.6. Similarly, define the multiplicative uncertainty
$$\Delta_m(s) := \frac{G(s) - G_0(s)}{G_0(s)}$$
and a W_m can be found such that $|\Delta_m(j\omega)| \leq |W_m(j\omega)|$, as shown in Figure 8.7. A W_m is given by
$$W_m = \frac{2.8169(s+0.212)(s^2+2.6128s+1.732)}{s^2+2.2425s+2.6319}$$

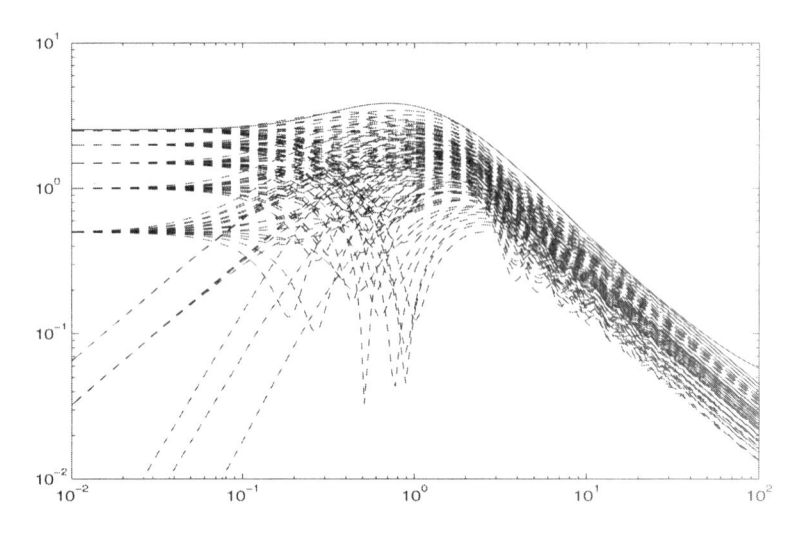

Figure 8.6: Δ_a (dashed line) and a bound W_a (solid line)

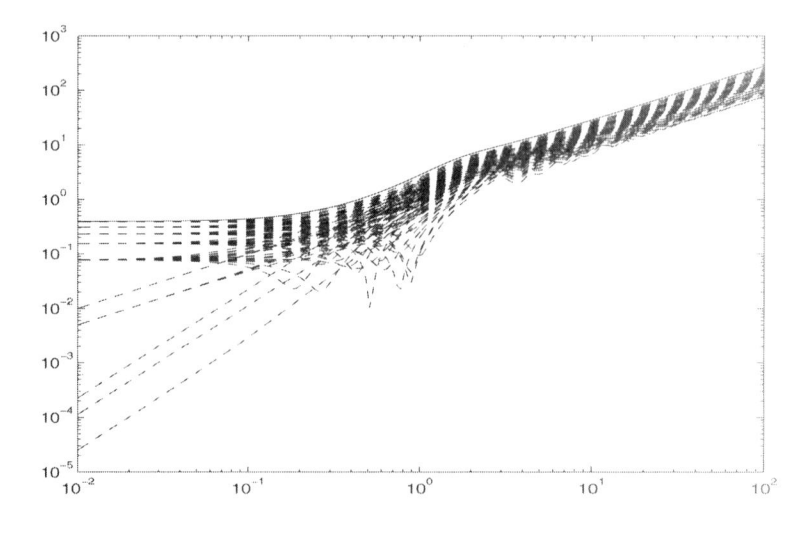

Figure 8.7: Δ_m (dashed line) and a bound W_m (solid line)

Note that this W_m is not proper since G_0 and G do not have the same relative degrees. To get a proper W_m, we need to choose a nominal model G_0 having the same the relative order as that of G.

The following terminologies are used in this book:

Definition 8.1 Given the description of an uncertainty model set Π and a set of performance objectives, suppose $P \in \Pi$ is the nominal design model and K is the resulting controller. Then the closed-loop feedback system is said to have

Nominal Stability (NS): if K internally stabilizes the nominal model P.

Robust Stability (RS): if K internally stabilizes every plant belonging to Π.

Nominal Performance (NP): if the performance objectives are satisfied for the nominal plant P.

Robust Performance (RP): if the performance objectives are satisfied for every plant belonging to Π.

The nominal stability and performance can be easily checked using various standard techniques. The conditions for which the robust stability and robust performance are satisfied under various assumptions on the uncertainty set Π will be considered in the following sections.

Related MATLAB Commands: magfit, drawmag, fitsys, genphase, vunpck, vabs, vinv, vimag, vreal, vcjt, vebe

8.2 Small Gain Theorem

This section and the next section consider the stability test of a nominally stable system under unstructured perturbations. The basis for the robust stability criteria derived in the sequel is the so-called *small gain theorem*.

Consider the interconnected system shown in Figure 8.8 with $M(s)$ a stable $p \times q$ transfer matrix.

Theorem 8.1 (Small Gain Theorem) *Suppose $M \in \mathcal{RH}_\infty$ and let $\gamma > 0$. Then the interconnected system shown in Figure 8.8 is well-posed and internally stable for all $\Delta(s) \in \mathcal{RH}_\infty$ with*

(a) $\|\Delta\|_\infty \leq 1/\gamma$ if and only if $\|M(s)\|_\infty < \gamma$

(b) $\|\Delta\|_\infty < 1/\gamma$ if and only if $\|M(s)\|_\infty \leq \gamma$

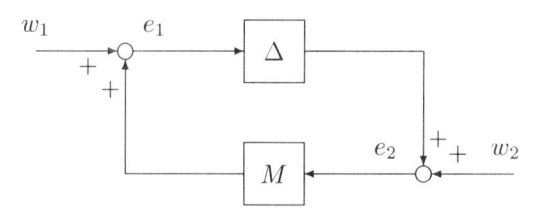

Figure 8.8: $M - \Delta$ loop for stability analysis

Proof. We shall only prove part (a). The proof for part (b) is similar. Without loss of generality, assume $\gamma = 1$.

(Sufficiency) It is clear that $M(s)\Delta(s)$ is stable since both $M(s)$ and $\Delta(s)$ are stable. Thus by Theorem 5.5 (or Corollary 5.4) the closed-loop system is stable if $\det(I - M\Delta)$ has no zero in the closed right-half plane for all $\Delta \in \mathcal{RH}_\infty$ and $\|\Delta\|_\infty \leq 1$. Equivalently, the closed-loop system is stable if

$$\inf_{s \in \overline{\mathbb{C}}_+} \underline{\sigma}(I - M(s)\Delta(s)) \neq 0$$

for all $\Delta \in \mathcal{RH}_\infty$ and $\|\Delta\|_\infty \leq 1$. But this follows from

$$\inf_{s \in \overline{\mathbb{C}}_+} \underline{\sigma}(I - M(s)\Delta(s)) \geq 1 - \sup_{s \in \overline{\mathbb{C}}_+} \bar{\sigma}(M(s)\Delta(s)) = 1 - \|M(s)\Delta(s)\|_\infty \geq 1 - \|M(s)\|_\infty > 0.$$

(Necessity) This will be shown by contradiction. Suppose $\|M\|_\infty \geq 1$. We will show that there exists a $\Delta \in \mathcal{RH}_\infty$ with $\|\Delta\|_\infty \leq 1$ such that $\det(I - M(s)\Delta(s))$ has a zero on the imaginary axis, so the system is unstable. Suppose $\omega_0 \in \mathbb{R}_+ \cup \{\infty\}$ is such that $\bar{\sigma}(M(j\omega_0)) \geq 1$. Let $M(j\omega_0) = U(j\omega)\Sigma(j\omega_0)V^*(j\omega_0)$ be a singular value decomposition with

$$U(j\omega_0) = \begin{bmatrix} u_1 & u_2 & \cdots & u_p \end{bmatrix}$$

$$V(j\omega_0) = \begin{bmatrix} v_1 & v_2 & \cdots & v_q \end{bmatrix}$$

$$\Sigma(j\omega_0) = \begin{bmatrix} \sigma_1 & & \\ & \sigma_2 & \\ & & \ddots \end{bmatrix}$$

To obtain a contradiction, it now suffices to construct a $\Delta \in \mathcal{RH}_\infty$ such that $\Delta(j\omega_0) = \frac{1}{\sigma_1} v_1 u_1^*$ and $\|\Delta\|_\infty \leq 1$. Indeed, for such $\Delta(s)$,

$$\det(I - M(j\omega_0)\Delta(j\omega_0)) = \det(I - U\Sigma V^* v_1 u_1^*/\sigma_1) = 1 - u_1^* U\Sigma V^* v_1/\sigma_1 = 0$$

and thus the closed-loop system is either not well-posed (if $\omega_0 = \infty$) or unstable (if $\omega \in \mathbb{R}$). There are two different cases:

8.2. Small Gain Theorem

(1) $\omega_0 = 0$ or ∞: then U and V are real matrices. In this case, $\Delta(s)$ can be chosen as

$$\Delta = \frac{1}{\sigma_1} v_1 u_1^* \in \mathbb{R}^{q \times p}$$

(2) $0 < \omega_0 < \infty$: write u_1 and v_1 in the following form:

$$u_1^* = \begin{bmatrix} u_{11} e^{j\theta_1} & u_{12} e^{j\theta_2} & \cdots & u_{1p} e^{j\theta_p} \end{bmatrix}, \quad v_1 = \begin{bmatrix} v_{11} e^{j\phi_1} \\ v_{12} e^{j\phi_2} \\ \vdots \\ v_{1q} e^{j\phi_q} \end{bmatrix}$$

where $u_{1i} \in \mathbb{R}$ and $v_{1j} \in \mathbb{R}$ are chosen so that $\theta_i, \phi_j \in [-\pi, 0)$ for all i, j.

Choose $\beta_i \geq 0$ and $\alpha_j \geq 0$ so that

$$\angle \left(\frac{\beta_i - j\omega_0}{\beta_i + j\omega_0} \right) = \theta_i, \quad \angle \left(\frac{\alpha_j - j\omega_0}{\alpha_j + j\omega_0} \right) = \phi_j$$

for $i = 1, 2, \ldots, p$ and $j = 1, 2, \ldots, q$. Let

$$\Delta(s) = \frac{1}{\sigma_1} \begin{bmatrix} v_{11} \frac{\alpha_1 - s}{\alpha_1 + s} \\ \vdots \\ v_{1q} \frac{\alpha_q - s}{\alpha_q + s} \end{bmatrix} \begin{bmatrix} u_{11} \frac{\beta_1 - s}{\beta_1 + s} & \cdots & u_{1p} \frac{\beta_p - s}{\beta_p + s} \end{bmatrix} \in \mathcal{RH}_\infty$$

Then $\|\Delta\|_\infty = 1/\sigma_1 \leq 1$ and $\Delta(j\omega_0) = \frac{1}{\sigma_1} v_1 u_1^*$.

□

The theorem still holds even if Δ and M are infinite dimensional. This is summarized as the following corollary.

Corollary 8.2 *The following statements are equivalent:*

(i) The system is well-posed and internally stable for all $\Delta \in \mathcal{H}_\infty$ with $\|\Delta\|_\infty < 1/\gamma$;

(ii) The system is well-posed and internally stable for all $\Delta \in \mathcal{RH}_\infty$ with $\|\Delta\|_\infty < 1/\gamma$;

(iii) The system is well-posed and internally stable for all $\Delta \in \mathbb{C}^{q \times p}$ with $\|\Delta\| < 1/\gamma$;

(iv) $\|M\|_\infty \leq \gamma$.

Remark 8.1 It can be shown that the small gain condition is sufficient to guarantee internal stability even if Δ is a nonlinear and time-varying "stable" operator with an appropriately defined stability notion, see Desoer and Vidyasagar [1975]. ◇

The following lemma shows that if $\|M\|_\infty > \gamma$, there exists a destabilizing Δ with $\|\Delta\|_\infty < 1/\gamma$ such that the closed-loop system has poles in the open right-half plane. (This is stronger than what is given in the proof of Theorem 8.1.)

Lemma 8.3 *Suppose $M \in \mathcal{RH}_\infty$ and $\|M\|_\infty > \gamma$. Then there exists a $\sigma_0 > 0$ such that for any given $\sigma \in [0, \sigma_0]$ there exists a $\Delta \in \mathcal{RH}_\infty$ with $\|\Delta\|_\infty < 1/\gamma$ such that $\det(I - M(s)\Delta(s))$ has a zero on the axis $\mathrm{Re}(s) = \sigma$.*

Proof. Without loss of generality, assume $\gamma = 1$. Since $M \in \mathcal{RH}_\infty$ and $\|M\|_\infty > 1$, there exists a $0 < \omega_0 < \infty$ such that $\|M(j\omega_0)\| > 1$. Given any γ such that $1 < \gamma < \|M(j\omega_0)\|$, there is a sufficiently small $\sigma_0 > 0$ such that

$$\min_{\sigma \in [0,\sigma_0]} \|M(\sigma + j\omega_0)\| \geq \gamma$$

and

$$\sqrt{\frac{\omega_0^2 + (\sigma_0 + \alpha)^2}{\omega_0^2 + (\sigma_0 - \alpha)^2}} \sqrt{\frac{\omega_0^2 + (\sigma_0 + \beta)^2}{\omega_0^2 + (\sigma_0 - \beta)^2}} < \gamma$$

for any $\alpha \geq 0$ and $\beta \geq 0$.

Now let $\sigma \in [0, \sigma_0]$ and let $M(\sigma + j\omega_0) = U\Sigma V^*$ be a singular value decomposition with

$$U = \begin{bmatrix} u_1 & u_2 & \cdots & u_p \end{bmatrix}$$
$$V = \begin{bmatrix} v_1 & v_2 & \cdots & v_q \end{bmatrix}$$
$$\Sigma = \begin{bmatrix} \sigma_1 & & & \\ & \sigma_2 & & \\ & & \ddots & \end{bmatrix}.$$

Write u_1 and v_1 in the following form:

$$u_1^* = \begin{bmatrix} u_{11}e^{j\theta_1} & u_{12}e^{j\theta_2} & \cdots & u_{1p}e^{j\theta_p} \end{bmatrix}, \quad v_1 = \begin{bmatrix} v_{11}e^{j\phi_1} \\ v_{12}e^{j\phi_2} \\ \vdots \\ v_{1q}e^{j\phi_q} \end{bmatrix}$$

where $u_{1i} \in \mathbb{R}$ and $v_{1j} \in \mathbb{R}$ are chosen so that $\theta_i, \phi_j \in [-\pi, 0)$ for all i, j. Choose $\beta_i \geq 0$ and $\alpha_j \geq 0$ so that

$$\angle\left(\frac{\beta_i - \sigma - j\omega_0}{\beta_i + \sigma + j\omega_0}\right) = \theta_i, \quad \angle\left(\frac{\alpha_j - \sigma - j\omega_0}{\alpha_j + \sigma + j\omega_0}\right) = \phi_j$$

for $i = 1, 2, \ldots, p$ and $j = 1, 2, \ldots, q$. Let

$$\Delta(s) = \frac{1}{\sigma_1} \begin{bmatrix} \tilde{\alpha}_1 v_{11} \frac{\alpha_1 - s}{\alpha_1 + s} \\ \vdots \\ \tilde{\alpha}_q v_{1q} \frac{\alpha_q - s}{\alpha_q + s} \end{bmatrix} \begin{bmatrix} \tilde{\beta}_1 u_{11} \frac{\beta_1 - s}{\beta_1 + s} & \cdots & \tilde{\beta}_p u_{1p} \frac{\beta_p - s}{\beta_p + s} \end{bmatrix} \in \mathcal{RH}_\infty$$

where
$$\tilde{\alpha}_i := \sqrt{\frac{\omega_0^2 + (\sigma + \alpha_i)^2}{\omega_0^2 + (\sigma - \alpha_i)^2}}, \quad \tilde{\beta}_j := \sqrt{\frac{\omega_0^2 + (\sigma + \beta_j)^2}{\omega_0^2 + (\sigma - \beta_j)^2}}$$

Then
$$\|\Delta\|_\infty \le \frac{\max_i \{\tilde{\alpha}_i\} \max_j \{\tilde{\beta}_j\}}{\sigma_1} \le \frac{\max_i \{\tilde{\alpha}_i\} \max_j \{\tilde{\beta}_j\}}{\gamma} < 1$$

and
$$\Delta(\sigma + j\omega_0) = \frac{1}{\sigma_1} v_1 u_1^*$$

$$\det(I - M(\sigma + j\omega_0)\Delta(\sigma + j\omega_0)) = 0$$

Hence $s = \sigma + j\omega_0$ is a zero for the transfer function $\det(I - M(s)\Delta(s))$. $\square$

The preceding lemma plays a key role in the necessity proofs of many robust stability tests in the sequel.

8.3 Stability under Unstructured Uncertainties

The small gain theorem in the last section will be used here to derive robust stability tests under various assumptions of model uncertainties. The modeling error Δ will again be assumed to be stable. (Most of the robust stability tests discussed in the sequel can be generalized easily to the unstable Δ case with some mild assumptions on the number of unstable poles of the uncertain model; we encourage readers to fill in the details.) In addition, we assume that the modeling error Δ is suitably scaled with weighting functions W_1 and W_2 (i.e., the uncertainty can be represented as $W_1 \Delta W_2$).

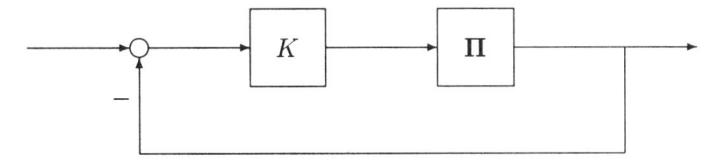

Figure 8.9: Unstructured robust stability analysis

We shall consider the standard setup shown in Figure 8.9, where Π is the set of uncertain plants with $P \in \Pi$ as the nominal plant and with K as the internally stabilizing controller for P. The sensitivity and complementary sensitivity matrix functions are defined, as usual, as
$$S_o = (I + PK)^{-1}, \quad T_o = I - S_o$$
and
$$S_i = (I + KP)^{-1}, \quad T_i = I - S_i.$$

Recall that the closed-loop system is well-posed and internally stable if and only if

$$\begin{bmatrix} I & K \\ -\Pi & I \end{bmatrix}^{-1} = \begin{bmatrix} (I+K\Pi)^{-1} & -K(I+\Pi K)^{-1} \\ (I+\Pi K)^{-1}\Pi & (I+\Pi K)^{-1} \end{bmatrix} \in \mathcal{RH}_\infty$$

for all $\Pi \in \mathbf{\Pi}$.

8.3.1 Additive Uncertainty

We assume that the model uncertainty can be represented by an additive perturbation:

$$\Pi = P + W_1 \Delta W_2.$$

Theorem 8.4 *Let* $\mathbf{\Pi} = \{P + W_1 \Delta W_2 : \Delta \in \mathcal{RH}_\infty\}$ *and let* K *be a stabilizing controller for the nominal plant* P. *Then the closed-loop system is well-posed and internally stable for all* $\|\Delta\|_\infty < 1$ *if and only if* $\|W_2 K S_o W_1\|_\infty \leq 1$.

Proof. Let $\Pi = P + W_1 \Delta W_2 \in \mathbf{\Pi}$. Then

$$\begin{bmatrix} I & K \\ -\Pi & I \end{bmatrix}^{-1}$$

$$= \begin{bmatrix} (I + KS_o W_1 \Delta W_2)^{-1} S_i & -KS_o(I + W_1 \Delta W_2 K S_o)^{-1} \\ (I + S_o W_1 \Delta W_2 K)^{-1} S_o(P + W_1 \Delta W_2) & S_o(I + W_1 \Delta W_2 K S_o)^{-1} \end{bmatrix}$$

is well-posed and internally stable if $(I + \Delta W_2 K S_o W_1)^{-1} \in \mathcal{RH}_\infty$ since

$$\det(I + KS_o W_1 \Delta W_2) = \det(I + W_1 \Delta W_2 K S_o) = \det(I + S_o W_1 \Delta W_2 K)$$
$$= \det(I + \Delta W_2 K S_o W_1).$$

But $(I + \Delta W_2 K S_o W_1)^{-1} \in \mathcal{RH}_\infty$ is guaranteed if $\|\Delta W_2 K S_o W_1\|_\infty < 1$ (small gain theorem). Hence $\|W_2 K S_o W_1\|_\infty \leq 1$ is sufficient for robust stability.

To show the necessity, note that robust stability implies that

$$K(I + \Pi K)^{-1} = K S_o(I + W_1 \Delta W_2 K S_o)^{-1} \in \mathcal{RH}_\infty$$

for all admissible Δ. This, in turn, implies that

$$\Delta W_2 K(I + \Pi K)^{-1} W_1 = I - (I + \Delta W_2 K S_o W_1)^{-1} \in \mathcal{RH}_\infty$$

for all admissible Δ. By the small gain theorem, this is true for all $\Delta \in \mathcal{RH}_\infty$ with $\|\Delta\|_\infty < 1$ only if $\|W_2 K S_o W_1\|_\infty \leq 1$. $\square$

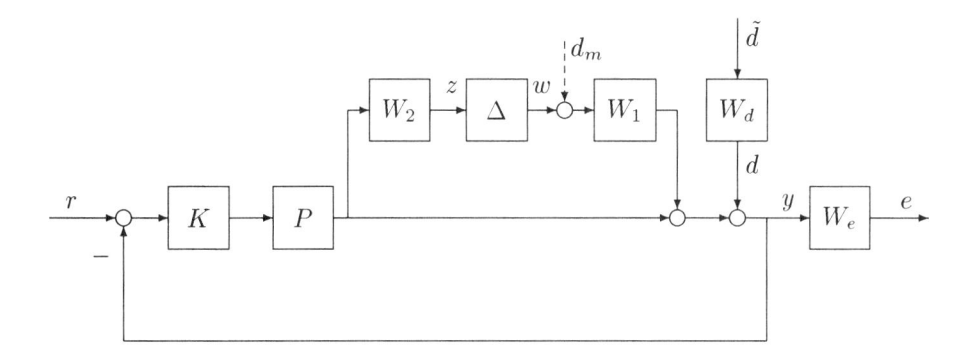

8.3. Stability under Unstructured Uncertainties

Figure 8.10: Output multiplicative perturbed systems

8.3.2 Multiplicative Uncertainty

In this section, we assume that the system model is described by the following set of multiplicative perturbations:

$$\Pi = (I + W_1 \Delta W_2)P$$

with $W_1, W_2, \Delta \in \mathcal{RH}_\infty$. Consider the feedback system shown in Figure 8.10.

Theorem 8.5 *Let $\Pi = \{(I + W_1\Delta W_2)P : \Delta \in \mathcal{RH}_\infty\}$ and let K be a stabilizing controller for the nominal plant P. Then the closed-loop system is well-posed and internally stable for all $\Delta \in \mathcal{RH}_\infty$ with $\|\Delta\|_\infty < 1$ if and only if $\|W_2 T_o W_1\|_\infty \leq 1$.*

Proof. We shall first prove that the condition is necessary for robust stability. Suppose $\|W_2 T_o W_1\|_\infty > 1$. Then by Lemma 8.3, for any given sufficiently small $\sigma > 0$, there is a $\Delta \in \mathcal{RH}_\infty$ with $\|\Delta\|_\infty < 1$ such that $(I + \Delta W_2 T_o W_1)^{-1}$ has poles on the axis $\text{Re}(s) = \sigma$. This implies that

$$(I + \Pi K)^{-1} = S_o(I + W_1 \Delta W_2 T_o)^{-1}$$

has poles on the axis $\text{Re}(s) = \sigma$ since σ can always be chosen so that the unstable poles are not cancelled by the zeros of S_o. Hence $\|W_2 T_o W_1\|_\infty \leq 1$ is necessary for robust stability. The sufficiency follows from the small gain theorem. □

8.3.3 Coprime Factor Uncertainty

As another example, consider a left coprime factor perturbed plant described in Figure 8.11.

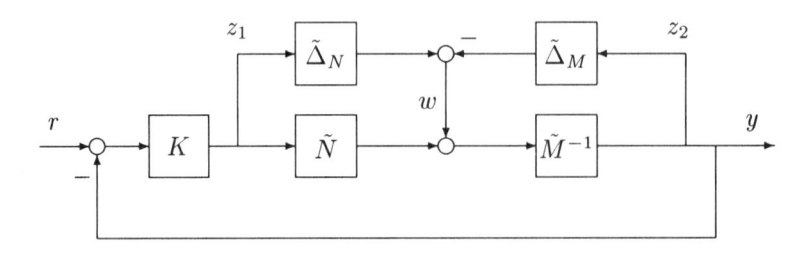

Figure 8.11: Left coprime factor perturbed systems

Theorem 8.6 *Let*
$$\Pi = (\tilde{M} + \tilde{\Delta}_M)^{-1}(\tilde{N} + \tilde{\Delta}_N)$$
with $\tilde{M}, \tilde{N}, \tilde{\Delta}_M, \tilde{\Delta}_N \in \mathcal{RH}_\infty$. The transfer matrices $(\tilde{M}, \tilde{N})$ are assumed to be a stable left coprime factorization of P (i.e., $P = \tilde{M}^{-1}\tilde{N}$), and K internally stabilizes the nominal system P. Define $\Delta := \begin{bmatrix} \tilde{\Delta}_N & \tilde{\Delta}_M \end{bmatrix}$. Then the closed-loop system is well-posed and internally stable for all $\|\Delta\|_\infty < 1$ if and only if

$$\left\| \begin{bmatrix} K \\ I \end{bmatrix} (I + PK)^{-1} \tilde{M}^{-1} \right\|_\infty \leq 1$$

Proof. Let $K = UV^{-1}$ be a right coprime factorization over $\mathcal{RH}_\infty$. By Lemma 5.7, the closed-loop system is internally stable if and only if

$$\left((\tilde{N} + \tilde{\Delta}_N)U + (\tilde{M} + \tilde{\Delta}_M)V\right)^{-1} \in \mathcal{RH}_\infty \tag{8.3}$$

Since K stabilizes P, $(\tilde{N}U + \tilde{M}V)^{-1} \in \mathcal{RH}_\infty$. Hence equation (8.3) holds if and only if

$$\left(I + (\tilde{\Delta}_N U + \tilde{\Delta}_M V)(\tilde{N}U + \tilde{M}V)^{-1}\right)^{-1} \in \mathcal{RH}_\infty$$

By the small gain theorem, the above is true for all $\|\Delta\|_\infty < 1$ if and only if

$$\left\| \begin{bmatrix} U \\ V \end{bmatrix} (\tilde{N}U + \tilde{M}V)^{-1} \right\|_\infty = \left\| \begin{bmatrix} K \\ I \end{bmatrix} (I + PK)^{-1} \tilde{M}^{-1} \right\|_\infty \leq 1$$

$\square$

8.3. Stability under Unstructured Uncertainties

8.3.4 Unstructured Robust Stability Tests

Table 8.1 summaries robust stability tests on the plant uncertainties under various assumptions. All of the tests pertain to the standard setup shown in Figure 8.9, where $\mathbf{\Pi}$ is the set of uncertain plants with $P \in \mathbf{\Pi}$ as the nominal plant and with K as the internally stabilizing controller of P.

\multicolumn{3}{c}{$W_1 \in \mathcal{RH}_\infty \quad W_2 \in \mathcal{RH}_\infty \quad \Delta \in \mathcal{RH}_\infty \quad \|\Delta\|_\infty < 1$}		
Perturbed Model Sets $\mathbf{\Pi}$	Representative Types of Uncertainty Characterized	Robust Stability Tests
$(I + W_1 \Delta W_2) P$	output (sensor) errors neglected HF dynamics uncertain rhp zeros	$\|W_2 T_o W_1\|_\infty \leq 1$
$P(I + W_1 \Delta W_2)$	input (actuators) errors neglected HF dynamics uncertain rhp zeros	$\|W_2 T_i W_1\|_\infty \leq 1$
$(I + W_1 \Delta W_2)^{-1} P$	LF parameter errors uncertain rhp poles	$\|W_2 S_o W_1\|_\infty \leq 1$
$P(I + W_1 \Delta W_2)^{-1}$	LF parameter errors uncertain rhp poles	$\|W_2 S_i W_1\|_\infty \leq 1$
$P + W_1 \Delta W_2$	additive plant errors neglected HF dynamics uncertain rhp zeros	$\|W_2 K S_o W_1\|_\infty \leq 1$
$P(I + W_1 \Delta W_2 P)^{-1}$	LF parameter errors uncertain rhp poles	$\|W_2 S_o P W_1\|_\infty \leq 1$
$(\tilde{M} + \tilde{\Delta}_M)^{-1}(\tilde{N} + \tilde{\Delta}_N)$ $P = \tilde{M}^{-1}\tilde{N}$ $\Delta = \begin{bmatrix} \tilde{\Delta}_N & \tilde{\Delta}_M \end{bmatrix}$	LF parameter errors neglected HF dynamics uncertain rhp poles & zeros	$\left\| \begin{bmatrix} K \\ I \end{bmatrix} S_o \tilde{M}^{-1} \right\|_\infty \leq 1$
$(N + \Delta_N)(M + \Delta_M)^{-1}$ $P = NM^{-1}$ $\Delta = \begin{bmatrix} \Delta_N \\ \Delta_M \end{bmatrix}$	LF parameter errors neglected HF dynamics uncertain rhp poles & zeros	$\|M^{-1} S_i [K\ I]\|_\infty \leq 1$

Table 8.1: Unstructured robust stability tests (HF: high frequency, LF: low frequency)

Table 8.1 should be interpreted as follows:

UNSTRUCTURED ANALYSIS THEOREM

Given **NS & Perturbed Model Sets**

Then **Closed-Loop Robust Stability**

if and only if **Robust Stability Tests**

The table also indicates representative types of physical uncertainties that can be usefully represented by cone-bounded perturbations inserted at appropriate locations. For example, the representation $P_\Delta = (I + W_1 \Delta W_2)P$ in the first row is useful for output errors at high frequencies (HF), covering such things as unmodeled high-frequency dynamics of sensors or plants, including diffusion processes, transport lags, electromechanical resonances, etc. The representation $P_\Delta = P(I + W_1 \Delta W_2)$ in the second row covers similar types of errors at the inputs. Both cases should be contrasted with the third and the fourth rows, which treat $P(I + W_1 \Delta W_2)^{-1}$ and $(I + W_1 \Delta W_2)^{-1} P$. These representations are more useful for variations in modeled dynamics, such as low-frequency (LF) errors produced by parameter variations with operating conditions, with aging, or across production copies of the same plant.

Note from the table that the stability requirements on Δ do not limit our ability to represent variations in either the number or locations of rhp singularities, as can be seen from some simple examples.

Example 8.3 Suppose an uncertain *system with changing numbers of right-half plane poles* is described by

$$P_\Delta = \left\{ \frac{1}{s - \delta} : \delta \in \mathbb{R}, \ |\delta| \leq 1 \right\}.$$

Then $P_1 = \dfrac{1}{s-1} \in P_\Delta$ has one right-half plane pole and $P_2 = \dfrac{1}{s+1} \in P_\Delta$ has no right-half plane pole. Nevertheless, the set of P_Δ can be covered by a set of feedback uncertain plants:

$$P_\Delta \subset \mathbf{\Pi} := \left\{ P(1 + \delta P)^{-1} : \delta \in \mathcal{RH}_\infty, \ \|\delta\|_\infty \leq 1 \right\}$$

with $P = \dfrac{1}{s}$.

Example 8.4 As another example, consider the following set of plants:

$$P_\Delta = \frac{s+1+\alpha}{(s+1)(s+2)}, \ |\alpha| \leq 2.$$

This set of plants has *changing numbers of right-half plane zeros* since the plant has no right-half plane zero when $\alpha = 0$ and has one right-half plane zero when $\alpha = -2$. The uncertain plant can be covered by a set of multiplicative perturbed plants:

$$P_\Delta \subset \Pi := \left\{ \frac{1}{s+2}(1 + \frac{2\delta}{s+1}), \ \delta \in \mathcal{RH}_\infty, \ \|\delta\|_\infty \leq 1 \right\}.$$

It should be noted that this covering can be quite conservative.

8.4 Robust Performance

Consider the perturbed system shown in Figure 8.12 with the set of perturbed models described by a set Π. Suppose the weighting matrices $W_d, W_e \in \mathcal{RH}_\infty$ and the perfor-

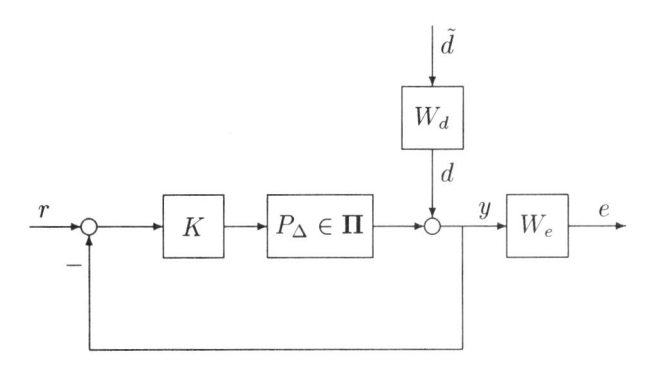

Figure 8.12: Diagram for robust performance analysis

mance criterion is to keep the error e as small as possible in some sense for all possible models belonging to the set Π. In general, the set Π can be either a parameterized set or an unstructured set such as those described in Table 8.1. The performance specifications are usually specified in terms of the magnitude of each component e in the time domain with respect to bounded disturbances, or, alternatively and more conveniently, some requirements on the closed-loop frequency response of the transfer matrix between $\tilde{d}$ and e (say, integral of square error or the magnitude of the steady-state error with respect to sinusoidal disturbances). The former design criterion leads to the so-called

$\mathcal{L}_1$-optimal control framework and the latter leads to $\mathcal{H}_2$ and $\mathcal{H}_\infty$ design frameworks, respectively. In this section, we will focus primarily on the $\mathcal{H}_\infty$ performance objectives with unstructured model uncertainty descriptions. The performance under structured uncertainty will be considered in Chapter 10.

Suppose the performance criterion is to keep the worst-case energy of the error e as small as possible over all $\tilde{d}$ of unit energy, for example,

$$\sup_{\|\tilde{d}\|_2 \leq 1} \|e\|_2 \leq \epsilon$$

for some small ϵ. By scaling the error e (i.e., by properly selecting W_e) we can assume without loss of generality that $\epsilon = 1$.

Let $T_{e\tilde{d}}$ denote the transfer matrix between $\tilde{d}$ and e, then

$$T_{e\tilde{d}} = W_e(I + P_\Delta K)^{-1}W_d, \quad P_\Delta \in \Pi. \tag{8.4}$$

Then the robust performance criterion in this case can be described as requiring that the closed-loop system be robustly stable and that

$$\|T_{e\tilde{d}}\|_\infty \leq 1, \quad \forall P_\Delta \in \Pi. \tag{8.5}$$

More specifically, an output multiplicatively perturbed system will be analyzed first. The analysis for other classes of models can be done analogously. The perturbed model can be described as

$$\Pi := \{(I + W_1 \Delta W_2)P : \Delta \in \mathcal{RH}_\infty, \|\Delta\|_\infty < 1\} \tag{8.6}$$

with $W_1, W_2 \in \mathcal{RH}_\infty$. The explicit system diagram is as shown in Figure 8.10. For this class of models, we have

$$T_{e\tilde{d}} = W_e S_o (I + W_1 \Delta W_2 T_o)^{-1} W_d,$$

and the robust performance is satisfied iff

$$\|W_2 T_o W_1\|_\infty \leq 1$$

and

$$\|T_{e\tilde{d}}\|_\infty \leq 1, \ \forall \Delta \in \mathcal{RH}_\infty, \ \|\Delta\|_\infty < 1.$$

The exact analysis for this robust performance problem is not trivial and will be given in Chapter 10. However, some sufficient conditions are relatively easy to obtain by bounding these two inequalities, and they may shed some light on the nature of these problems. It will be assumed throughout that the controller K internally stabilizes the nominal plant P.

Theorem 8.7 *Suppose $P_\Delta \in \{(I + W_1 \Delta W_2)P : \Delta \in \mathcal{RH}_\infty, \|\Delta\|_\infty < 1\}$ and K internally stabilizes P. Then the system robust performance is guaranteed if either one of the following conditions is satisfied:*

8.4. Robust Performance

(i) for each frequency ω

$$\overline{\sigma}(W_d)\overline{\sigma}(W_eS_o) + \overline{\sigma}(W_1)\overline{\sigma}(W_2T_o) \leq 1; \qquad (8.7)$$

(ii) for each frequency ω

$$\kappa(W_1^{-1}W_d)\overline{\sigma}(W_eS_oW_d) + \overline{\sigma}(W_2T_oW_1) \leq 1 \qquad (8.8)$$

where W_1 and W_d are assumed to be invertible and $\kappa(W_1^{-1}W_d)$ is the condition number.

Proof. It is obvious that both condition (8.7) and condition (8.8) guarantee that $\|W_2T_oW_1\|_\infty \leq 1$. So it is sufficient to show that $\|T_{e\tilde{d}}\|_\infty \leq 1, \forall \Delta \in \mathcal{RH}_\infty, \|\Delta\|_\infty < 1$. Now for any frequency ω, it is easy to see that

$$\begin{aligned}
\overline{\sigma}(T_{e\tilde{d}}) &\leq \overline{\sigma}(W_eS_o)\overline{\sigma}[(I + W_1\Delta W_2T_o)^{-1}]\overline{\sigma}(W_d) \\
&= \frac{\overline{\sigma}(W_eS_o)\overline{\sigma}(W_d)}{\underline{\sigma}(I + W_1\Delta W_2T_o)} \leq \frac{\overline{\sigma}(W_eS_o)\overline{\sigma}(W_d)}{1 - \overline{\sigma}(W_1\Delta W_2T_o)} \\
&\leq \frac{\overline{\sigma}(W_eS_o)\overline{\sigma}(W_d)}{1 - \overline{\sigma}(W_1)\overline{\sigma}(W_2T_o)\overline{\sigma}(\Delta)}.
\end{aligned}$$

Hence condition (8.7) guarantees $\overline{\sigma}(T_{e\tilde{d}}) \leq 1$ for all $\Delta \in \mathcal{RH}_\infty$ with $\|\Delta\|_\infty < 1$ at all frequencies.

Similarly, suppose W_1 and W_d are invertible; write

$$T_{e\tilde{d}} = W_eS_oW_d(W_1^{-1}W_d)^{-1}(I + \Delta W_2T_oW_1)^{-1}(W_1^{-1}W_d),$$

and then

$$\overline{\sigma}(T_{e\tilde{d}}) \leq \frac{\overline{\sigma}(W_eS_oW_d)\kappa(W_1^{-1}W_d)}{1 - \overline{\sigma}(W_2T_oW_1)\overline{\sigma}(\Delta)}.$$

Hence by condition (8.8), $\overline{\sigma}(T_{e\tilde{d}}) \leq 1$ is guaranteed for all $\Delta \in \mathcal{RH}_\infty$ with $\|\Delta\|_\infty < 1$ at all frequencies. $\square$

Remark 8.2 It is not hard to show that either one of the conditions in the theorem is also necessary for scalar valued systems.

Remark 8.3 Suppose $\kappa(W_1^{-1}W_d) \approx 1$ (weighting matrices satisfying this condition are usually called round weights). This is particularly the case if $W_1 = w_1(s)I$ and $W_d = w_d(s)I$. Recall that $\overline{\sigma}(W_eS_oW_d) \leq 1$ is the necessary and sufficient condition for nominal performance and that $\overline{\sigma}(W_2T_oW_1) \leq 1$ is the necessary and sufficient condition for robust stability. Hence the condition (ii) in Theorem 8.7 is almost guaranteed by NP + RS (i.e., RP is almost guaranteed by NP + RS). Since RP implies NP + RS, we have NP + RS $\approx$ RP. (In contrast, such a conclusion cannot be drawn in the skewed case, which will be considered in the next section.) Since condition (ii) implies NP+RS, we can also conclude that condition (ii) is almost equivalent to RP (i.e., beside being sufficient, it is almost necessary). $\diamond$

Remark 8.4 Note that in light of the equivalence relation between the robust stability and nominal performance, it is reasonable to conjecture that the preceding robust performance problem is equivalent to the robust stability problem in Figure 8.9 with the uncertainty model set given by

$$\Pi := (I + W_d \Delta_e W_e)^{-1}(I + W_1 \Delta W_2)P$$

and $\|\Delta_e\|_\infty < 1$, $\|\Delta\|_\infty < 1$, as shown in Figure 8.13. This conjecture is indeed true; however, the equivalent model uncertainty is structured, and the exact stability analysis for such systems is not trivial and will be studied in Chapter 10. ◇

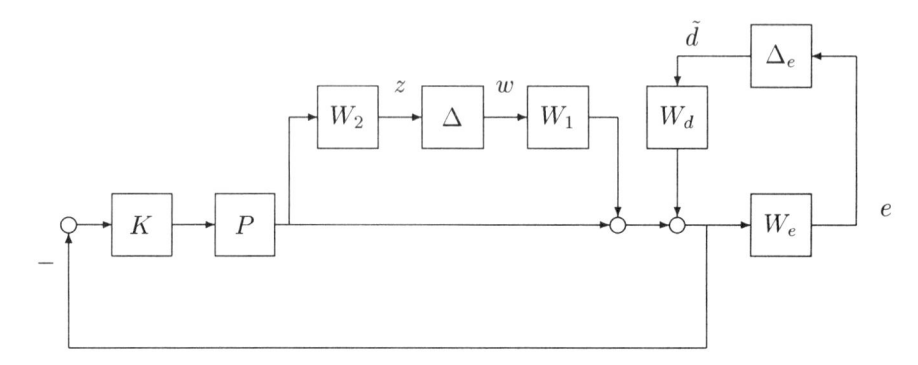

Figure 8.13: Robust performance with unstructured uncertainty vs. robust stability with structured uncertainty

Remark 8.5 Note that if W_1 and W_d are invertible, then $T_{e\tilde{d}}$ can also be written as

$$T_{e\tilde{d}} = W_e S_o W_d \left[I + (W_1^{-1} W_d)^{-1} \Delta W_2 T_o W_1 (W_1^{-1} W_d)\right]^{-1}.$$

So another alternative sufficient condition for robust performance can be obtained as

$$\overline{\sigma}(W_e S_o W_d) + \kappa(W_1^{-1} W_d)\overline{\sigma}(W_2 T_o W_1) \leq 1.$$

A similar situation also occurs in the skewed case below. We will not repeat all these variations. ◇

8.5 Skewed Specifications

We now consider the system with skewed specifications (i.e., the uncertainty and performance are not measured at the same location). For instance, the system performance is still measured in terms of output sensitivity, but the uncertainty model is in input multiplicative form:

$$\Pi := \{P(I + W_1 \Delta W_2) : \Delta \in \mathcal{RH}_\infty, \|\Delta\|_\infty < 1\}.$$

8.5. Skewed Specifications

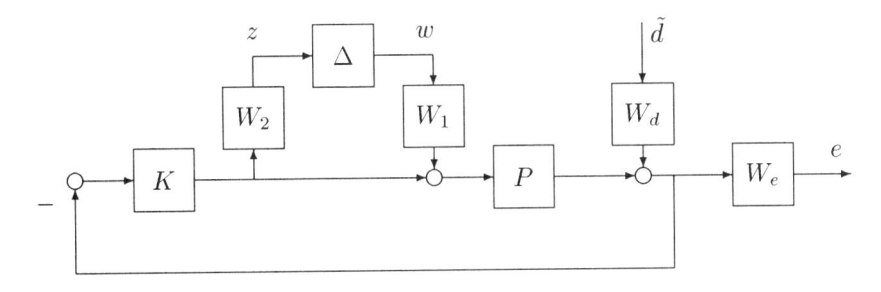

Figure 8.14: Skewed problems

The system block diagram is shown in Figure 8.14.

For systems described by this class of models, the robust stability condition becomes

$$\|W_2 T_i W_1\|_\infty \leq 1,$$

and the nominal performance condition becomes

$$\|W_e S_o W_d\|_\infty \leq 1.$$

To consider the robust performance, let $\tilde{T}_{e\tilde{d}}$ denote the transfer matrix from $\tilde{d}$ to e. Then

$$\begin{aligned}
\tilde{T}_{e\tilde{d}} &= W_e S_o (I + P W_1 \Delta W_2 K S_o)^{-1} W_d \\
&= W_e S_o W_d \left[I + (W_d^{-1} P W_1) \Delta (W_2 T_i W_1)(W_d^{-1} P W_1)^{-1} \right]^{-1}.
\end{aligned}$$

The last equality follows if W_1, W_d, and P are invertible and, if W_2 is invertible, can also be written as

$$\tilde{T}_{e\tilde{d}} = W_e S_o W_d (W_1^{-1} W_d)^{-1} \left[I + (W_1^{-1} P W_1) \Delta (W_2 P^{-1} W_2^{-1})(W_2 T_o W_1) \right]^{-1} (W_1^{-1} W_d).$$

Then the following results follow easily.

Theorem 8.8 *Suppose $P_\Delta \in \Pi = \{P(I + W_1 \Delta W_2) : \Delta \in \mathcal{RH}_\infty, \|\Delta\|_\infty < 1\}$ and K internally stabilizes P. Assume that $P, W_1, W_2,$ and W_d are square and invertible. Then the system robust performance is guaranteed if either one of the following conditions is satisfied:*

(i) for each frequency ω

$$\overline{\sigma}(W_e S_o W_d) + \kappa(W_d^{-1} P W_1) \overline{\sigma}(W_2 T_i W_1) \leq 1; \tag{8.9}$$

(ii) for each frequency ω

$$\kappa(W_1^{-1} W_d) \overline{\sigma}(W_e S_o W_d) + \overline{\sigma}(W_1^{-1} P W_1) \overline{\sigma}(W_2 P^{-1} W_2^{-1}) \overline{\sigma}(W_2 T_o W_1) \leq 1. \tag{8.10}$$

Remark 8.6 If the appropriate invertibility conditions are not satisfied, then an alternative sufficient condition for robust performance can be given by

$$\bar{\sigma}(W_d)\bar{\sigma}(W_e S_o) + \bar{\sigma}(PW_1)\bar{\sigma}(W_2 K S_o) \leq 1.$$

Similar to the previous case, there are many different variations of sufficient conditions although equation (8.10) may be the most useful one. ◇

Remark 8.7 It is important to note that in this case, the robust stability condition is given in terms of $L_i = KP$ while the nominal performance condition is given in terms of $L_o = PK$. These classes of problems are called *skewed problems* or problems with skewed specifications.[2] Since, in general, $PK \neq KP$, the robust stability margin or tolerances for uncertainties at the plant input and output are generally not the same. ◇

Remark 8.8 It is also noted that the robust performance condition is related to the condition number of the weighted nominal model. So, in general, if the weighted nominal model is ill-conditioned at the range of critical frequencies, then the robust performance condition may be far more restrictive than the robust stability condition and the nominal performance condition together. For simplicity, assume $W_1 = I$, $W_d = I$ and $W_2 = w_t I$, where $w_t \in \mathcal{RH}_\infty$ is a scalar function. Further, P is assumed to be invertible. Then the robust performance condition (8.10) can be written as

$$\bar{\sigma}(W_e S_o) + \kappa(P)\bar{\sigma}(w_t T_o) \leq 1, \forall \omega.$$

Comparing these conditions with those obtained for nonskewed problems shows that the condition related to robust stability is scaled by the condition number of the plant.[3] Since $\kappa(P) \geq 1$, it is clear that the skewed specifications are much harder to satisfy if the plant is not well conditioned. This problem will be discussed in more detail in Section 10.3.3 of Chapter 10. ◇

Remark 8.9 Suppose K is invertible, then $\tilde{T}_{e\tilde{d}}$ can be written as

$$\tilde{T}_{e\tilde{d}} = W_e K^{-1}(I + T_i W_1 \Delta W_2)^{-1} S_i K W_d.$$

Assume further that $W_e = I$, $W_d = w_s I$, $W_2 = I$, where $w_s \in \mathcal{RH}_\infty$ is a scalar function. Then a sufficient condition for robust performance is given by

$$\kappa(K)\bar{\sigma}(S_i w_s) + \bar{\sigma}(T_i W_1) \leq 1, \forall \omega,$$

with $\kappa(K) := \bar{\sigma}(K)\bar{\sigma}(K^{-1})$. This is equivalent to treating the input multiplicative plant uncertainty as the output multiplicative controller uncertainty. ◇

---

[2] See Stein and Doyle [1991].
[3] Alternative condition can be derived so that the condition related to nominal performance is scaled by the condition number.

8.5. Skewed Specifications

The fact that the condition number appeared in the robust performance test for skewed problems can be given another interpretation by considering two sets of plants $\mathbf{\Pi}_1$ and $\mathbf{\Pi}_2$, as shown in Figure 8.15 and below.

$$\mathbf{\Pi}_1 := \{P(I + w_t\Delta) : \Delta \in \mathcal{RH}_\infty, \|\Delta\|_\infty < 1\}$$
$$\mathbf{\Pi}_2 := \{(I + \tilde{w}_t\Delta)P : \Delta \in \mathcal{RH}_\infty, \|\Delta\|_\infty < 1\}.$$

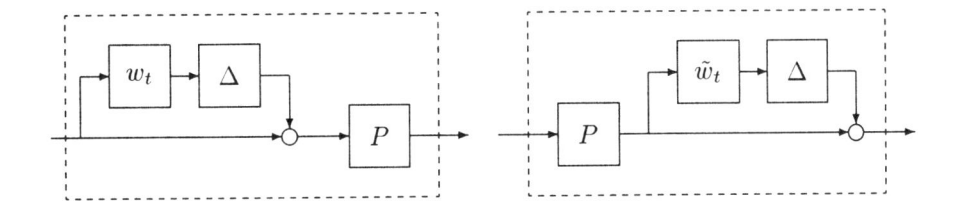

Figure 8.15: Converting input uncertainty to output uncertainty

Assume that P is invertible; then

$$\mathbf{\Pi}_2 \supseteq \mathbf{\Pi}_1 \quad \text{if} \quad |\tilde{w}_t| \geq |w_t|\kappa(P) \quad \forall \omega$$

since $P(I + w_t\Delta) = (I + w_t P\Delta P^{-1})P$.

The condition number of a transfer matrix can be very high at high frequency, which may significantly limit the achievable performance. The example below, taken from the textbook by Franklin, Powell, and Workman [1990, page 788], shows that the condition number shown in Figure 8.16 may increase with the frequency:

$$P(s) = \left[\begin{array}{ccc|cc} -0.2 & 0.1 & 1 & 0 & 1 \\ -0.05 & 0 & 0 & 0 & 0.7 \\ 0 & 0 & -1 & 1 & 0 \\ \hline 1 & 0 & 0 & 0 & 0 \\ 0 & 1 & 0 & 0 & 0 \end{array}\right] = \frac{1}{a(s)}\left[\begin{array}{cc} s & (s+1)(s+0.07) \\ -0.05 & 0.7(s+1)(s+0.13) \end{array}\right]$$

where $a(s) = (s+1)(s+0.1707)(s+0.02929)$.

It is appropriate to point out that the skewed problem setup, although more complicated than that of the nonskewed problem, is particularly suitable for control system design. To be more specific, consider the transfer function from w and $\tilde{d}$ to z and e:

$$\begin{bmatrix} z \\ e \end{bmatrix} = G(s) \begin{bmatrix} w \\ \tilde{d} \end{bmatrix}$$

where

$$G(s) := \begin{bmatrix} -W_2 T_i W_1 & -W_2 K S_o W_d \\ W_e S_o P W_1 & W_e S_o W_d \end{bmatrix}$$
$$= \begin{bmatrix} -W_2 & 0 \\ 0 & W_e \end{bmatrix} \begin{bmatrix} K \\ I \end{bmatrix} (I + PK)^{-1} \begin{bmatrix} P & I \end{bmatrix} \begin{bmatrix} W_1 & 0 \\ 0 & W_d \end{bmatrix}$$

154   UNCERTAINTY AND ROBUSTNESS

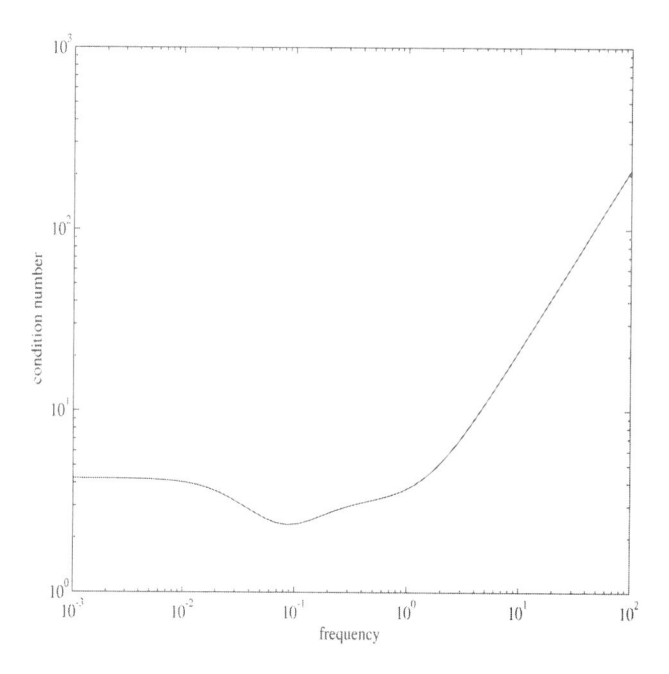

Figure 8.16: Condition number $\kappa(\omega) = \bar{\sigma}(P(j\omega))/\underline{\sigma}(P(j\omega))$

Then a suitable performance criterion is to make $\|G(s)\|_\infty$ small. Indeed, small $\|G(s)\|_\infty$ implies that T_i, KS_o, S_oP and S_o are small in some suitable frequency ranges, which are the desired design specifications discussed in Section 6.1 of Chapter 6. It will be clear in Chapter 16 and Chapter 17 that the $\|G\|_\infty$ is related to the robust stability margin in the gap metric, ν-gap metric, and normalized coprime factor perturbations. Therefore, making $\|G\|_\infty$ small is a suitable design approach.

8.6 Classical Control for MIMO Systems

In this section, we show through an example that the classical control theory may not be reliable when it is applied to MIMO system design.

Consider a symmetric spinning body with torque inputs, T_1 and T_2, along two orthogonal transverse axes, x and y, as shown in Figure 8.17. Assume that the angular velocity of the spinning body with respect to the z axis is constant, Ω. Assume further that the inertias of the spinning body with respect to the x, y, and z axes are I_1, $I_2 = I_1$, and I_3, respectively. Denote by ω_1 and ω_2 the angular velocities of the body with respect to the x and y axes, respectively. Then the Euler's equation of the spinning

8.6. Classical Control for MIMO Systems

body is given by

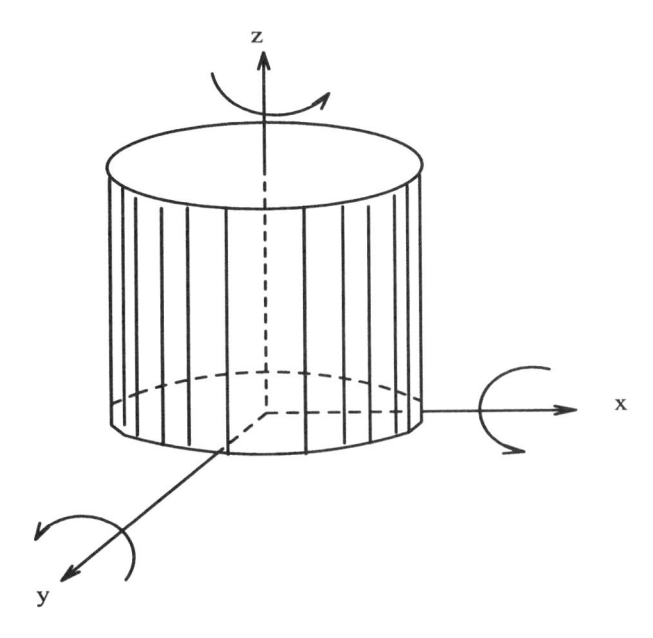

$$I_1\dot{\omega}_1 - \omega_2\Omega(I_1 - I_3) = T_1$$
$$I_1\dot{\omega}_2 - \omega_1\Omega(I_3 - I_1) = T_2$$

Figure 8.17: Spinning body

Define
$$\begin{bmatrix} u_1 \\ u_2 \end{bmatrix} := \begin{bmatrix} T_1/I_1 \\ T_2/I_1 \end{bmatrix}, \ a := (1 - I_3/I_1)\Omega.$$

Then the system dynamical equations can be written as

$$\begin{bmatrix} \dot{\omega}_1 \\ \dot{\omega}_2 \end{bmatrix} = \begin{bmatrix} 0 & a \\ -a & 0 \end{bmatrix} \begin{bmatrix} \omega_1 \\ \omega_2 \end{bmatrix} + \begin{bmatrix} u_1 \\ u_2 \end{bmatrix}$$

Now suppose that the angular rates ω_1 and ω_2 are measured in scaled and rotated coordinates:

$$\begin{bmatrix} y_1 \\ y_2 \end{bmatrix} = \frac{1}{\cos\theta}\begin{bmatrix} \cos\theta & \sin\theta \\ -\sin\theta & \cos\theta \end{bmatrix}\begin{bmatrix} \omega_1 \\ \omega_2 \end{bmatrix} = \begin{bmatrix} 1 & a \\ -a & 1 \end{bmatrix}\begin{bmatrix} \omega_1 \\ \omega_2 \end{bmatrix}$$

where $\tan\theta := a$. (There is no specific physical meaning for the measurements of y_1 and y_2 but they are assumed here only for the convenience of discussion.) Then the transfer matrix for the spinning body can be computed as

$$Y(s) = P(s)U(s)$$

with
$$P(s) = \frac{1}{s^2 + a^2} \begin{bmatrix} s - a^2 & a(s+1) \\ -a(s+1) & s - a^2 \end{bmatrix}$$

Suppose the control law is chosen to be a unit feedback $u = -y$. Then the sensitivity function and the complementary sensitivity function are given by

$$S = (I + P)^{-1} = \frac{1}{s+1} \begin{bmatrix} s & -a \\ a & s \end{bmatrix}, \quad T = P(I+P)^{-1} = \frac{1}{s+1} \begin{bmatrix} 1 & a \\ -a & 1 \end{bmatrix}$$

Note that each single loop has the open-loop transfer function as $\frac{1}{s}$, so each loop has 90° phase margin and ∞ gain margin.

Suppose one loop transfer function is perturbed, as shown in Figure 8.18.

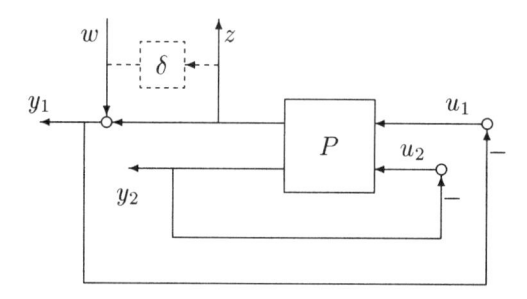

Figure 8.18: One-loop-at-a-time analysis

Denote
$$\frac{z(s)}{w(s)} = -T_{11} = -\frac{1}{s+1}$$

Then the maximum allowable perturbation is given by
$$\|\delta\|_\infty < \frac{1}{\|T_{11}\|_\infty} = 1,$$

which is independent of a. Similarly the maximum allowable perturbation on the other loop is also 1 by symmetry. However, if both loops are perturbed at the same time, then the maximum allowable perturbation is much smaller, as shown next.

Consider a multivariable perturbation, as shown in Figure 8.19; that is, $P_\Delta = (I + \Delta)P$, with
$$\Delta = \begin{bmatrix} \delta_{11} & \delta_{12} \\ \delta_{21} & \delta_{22} \end{bmatrix} \in \mathcal{RH}_\infty$$

a 2×2 transfer matrix such that $\|\Delta\|_\infty < \gamma$. Then by the small gain theorem, the system is robustly stable for every such Δ iff

$$\gamma \le \frac{1}{\|T\|_\infty} = \frac{1}{\sqrt{1+a^2}} \quad (\ll 1 \text{ if } a \gg 1).$$

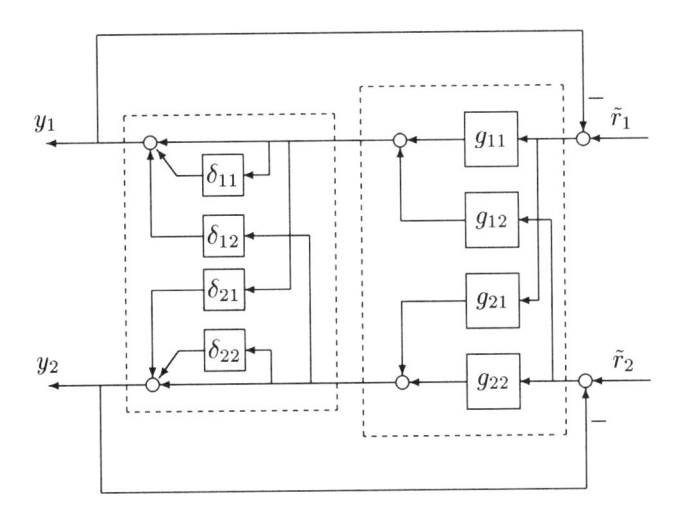

Figure 8.19: Simultaneous perturbations

In particular, consider
$$\Delta = \Delta_d = \begin{bmatrix} \delta_{11} & \\ & \delta_{22} \end{bmatrix} \in \mathbb{R}^{2 \times 2}.$$
Then the closed-loop system is stable for every such Δ iff
$$\det(I + T\Delta_d) = \frac{1}{(s+1)^2}\left(s^2 + (2 + \delta_{11} + \delta_{22})s + 1 + \delta_{11} + \delta_{22} + (1+a^2)\delta_{11}\delta_{22}\right)$$
has no zero in the closed right-half plane. Hence the stability region is given by
$$2 + \delta_{11} + \delta_{22} > 0$$
$$1 + \delta_{11} + \delta_{22} + (1+a^2)\delta_{11}\delta_{22} > 0.$$
It is easy to see that the system is unstable with
$$\delta_{11} = -\delta_{22} = \frac{1}{\sqrt{1+a^2}}.$$
The stability region for $a = 5$ is drawn in Figure 8.20, which shows how checking the axis misses nearby regions of instability, and that for $a \gg 5$, things just get that much worse. The hyperbola portion of the picture gets arbitrarily close to (0,0). This clearly shows that the analysis of a MIMO system using SISO methods can be misleading and can even give erroneous results. Hence an MIMO method has to be used.

8.7 Notes and References

The small gain theorem was first presented by Zames [1966]. The book by Desoer and Vidyasagar [1975] contains an extensive treatment and applications of this theorem in

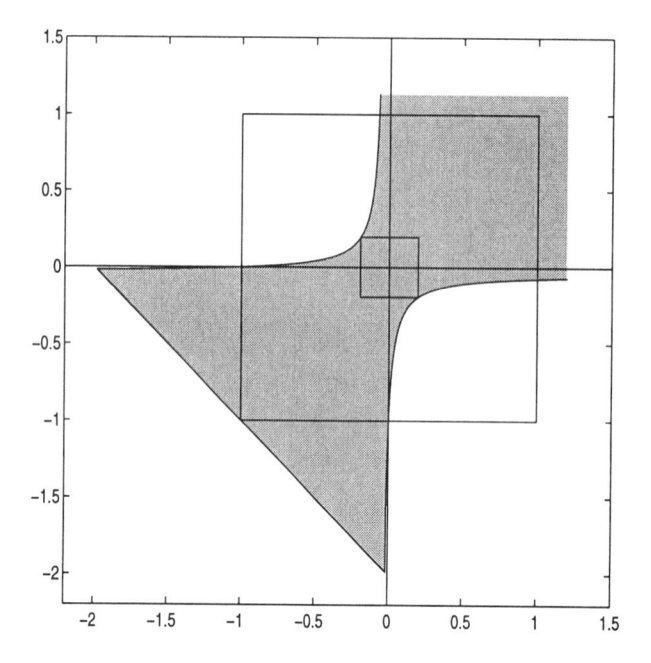

Figure 8.20: Stability region for $a = 5$

various forms. Robust stability conditions under various uncertainty assumptions are discussed in Doyle, Wall, and Stein [1982].

8.8 Problems

Problem 8.1 This problem shows that the stability margin is critically dependent on the type of perturbation. The setup is a unity-feedback loop with controller $K(s) = 1$ and plant $P_{\text{nom}}(s) + \Delta(s)$, where

$$P_{\text{nom}}(s) = \frac{10}{s^2 + 0.2s + 1}.$$

1. Assume $\Delta(s) \in \mathcal{RH}_\infty$. Compute the largest β such that the feedback system is internally stable for all $\|\Delta\|_\infty < \beta$.

2. Repeat but with $\Delta \in \mathbb{R}$.

Problem 8.2 Let $M \in \mathbb{C}^{p \times q}$ be a given complex matrix. Then it is shown in Qiu et al [1995] that $I - \Delta M$ is invertible for all $\Delta \in \mathbb{R}^{q \times p}$ such that $\overline{\sigma}(\Delta) \leq \gamma$ if and only if $\mu_\Delta(M) < 1/\gamma$, where

$$\mu_\Delta(M) = \inf_{\alpha \in (0,1]} \sigma_2 \left(\begin{bmatrix} \operatorname{Re} M & -\alpha \Im M \\ \alpha^{-1} \Im M & \operatorname{Re} M \end{bmatrix} \right).$$

8.8. Problems

It follows that $(I - \Delta M(s))^{-1} \in \mathcal{RH}_\infty$ for a given $M(s) \in \mathcal{RH}_\infty$ and all $\Delta \in \mathbb{R}^{q \times p}$ with $\bar{\sigma}(\Delta) \leq \gamma$ if and only if $\sup_\omega \mu_\Delta(M(j\omega)) < 1/\gamma$. Write a MATLAB program to compute $\mu_\Delta(M)$ and apply it to the preceding problem.

Problem 8.3 Let $G(s) = \dfrac{Ke^{-\tau s}}{Ts+1}$ and $K \in [10, 12], \tau \in [0, 0.5], T = 1$. Find a nominal model $G_o(s) \in \mathcal{RH}_\infty$ and a weighting function $W(s) \in \mathcal{RH}_\infty$ such that

$$G(s) \in \{G_o(s)(1 + W(s)\Delta(s)) : \quad \Delta \in \mathcal{H}_\infty, \quad \|\Delta\| \leq 1\}.$$

Problem 8.4 Think of an example of a physical system with the property that the number of unstable poles changes when the system undergoes a small change, for example, when a mass is perturbed slightly or the geometry is deformed slightly.

Problem 8.5 Let $\mathcal{X}$ be a space of scalar-valued transfer functions. A function $f(s)$ in $\mathcal{X}$ is a *unit* if $1/f(s)$ is in $\mathcal{X}$.

1. Prove that the set of units in $\mathcal{RH}_\infty$ is an open set, that is, if f is a unit, then

$$(\exists \epsilon > 0) \ (\forall g \in \mathcal{RH}_\infty) \ \|g\|_\infty < \epsilon \implies f+g \text{ is a unit}. \tag{8.11}$$

2. Here is an application of the preceding fact. Consider the unity feedback system with controller $k(s)$ and plant $p(s)$, both SISO, with $k(s)$ proper and $p(s)$ strictly proper. Do coprime factorizations over $\mathcal{RH}_\infty$:

$$p = \frac{n_p}{m_p}, \quad k = \frac{n_k}{m_k}.$$

Then the feedback system is internally stable iff $n_p n_k + m_p m_k$ is a unit in $\mathcal{RH}_\infty$. Assume it is a unit. Perturb $p(s)$ to

$$p = \frac{n_p + \Delta_n}{m_p + \Delta_m}, \quad \Delta_n, \Delta_m \in \mathcal{RH}_\infty.$$

Show that internal stability is preserved if $\|\Delta_n\|_\infty$ and $\|\Delta_m\|_\infty$ are small enough. The conclusion is that internal stability is preserved if the perturbations are small enough in the $\mathcal{H}_\infty$ norm.

3. Give an example of a unit $f(s)$ in $\mathcal{RH}_\infty$ such that equation (8.11) fails for the $\mathcal{H}_2$ norm, that is, such that

$$(\forall \epsilon > 0) \ (\exists g \in \mathcal{RH}_2) \ \|g\|_2 < \epsilon \text{ and } f+g \text{ is not a unit}.$$

What is the significance of this fact concerning robust stability?

Problem 8.6 Let Δ and M be square constant matrices. Prove that the following three conditions are equivalent:

1. $\begin{bmatrix} I & -\Delta \\ -M & I \end{bmatrix}$ is invertible;

2. $I - M\Delta$ is invertible;

3. $I - \Delta M$ is invertible.

Problem 8.7 Consider the unity feedback system

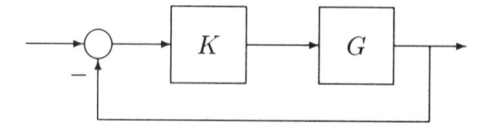

$$G(s) = \begin{bmatrix} \dfrac{1}{s} & \dfrac{1}{s} \\ \dfrac{1}{s} & \dfrac{1}{s} \end{bmatrix}$$

design a proper controller $K(s)$ to stabilize the feedback system internally. Now perturb $G(s)$ to

$$\begin{bmatrix} \dfrac{1+\epsilon}{s} & \dfrac{1}{s} \\ \dfrac{1}{s} & \dfrac{1}{s} \end{bmatrix}, \quad \epsilon \in \mathbb{R}.$$

Is the feedback system internally stable for all sufficiently small ϵ?

Problem 8.8 Consider the unity feedback system with $K(s) = 3$, $G(s) = \dfrac{1}{s-2}$. Compute by hand (i.e., without MATLAB) a normalized coprime factorization of $G(s)$. Considering perturbations Δ_N and Δ_M of the factors of $G(s)$, compute by hand the stability radius ϵ, that is, the least upper bound on $\left\| \begin{bmatrix} \Delta_N & \Delta_M \end{bmatrix} \right\|_\infty$ such that feedback stability is preserved.

Problem 8.9 Let a unit feedback system with a controller $K(s) = \dfrac{1}{s}$ and a nominal plant model $P_o(s) = \dfrac{s+1}{s^2 + 0.2s + 5}$. Construct a smallest destabilizing $\Delta \in \mathcal{RH}_\infty$ in the sense of $\|\Delta\|_\infty$ for each of the following cases:

(a) $P = P_o + \Delta$;

(b) $P = P_o(1 + W\Delta)$ with $W(s) = \dfrac{0.2(s+10)}{s+50}$;

8.8. Problems

(c) $P = \dfrac{N + \Delta_n}{M + \Delta_m}$, $N = \dfrac{2(s+1)}{(s+2)^2}$, $M = \dfrac{s^2 + 0.2s + 5}{(s+2)^2}$, and $\Delta = \begin{bmatrix} \Delta_n & \Delta_m \end{bmatrix}$.

Problem 8.10 This problem concerns the unity feedback system with controller $K(s)$ and plant

$$G(s) = \frac{1}{s+1} \begin{bmatrix} 1 & 2 \\ 3 & 4 \end{bmatrix}.$$

1. Take $K(s) = kI_2$ (k a real scalar) and find the range of k for internal stability.

2. Take

$$K(s) = \begin{bmatrix} k_1 & 0 \\ 0 & k_2 \end{bmatrix}$$

(k_1, k_2 real scalars) and find the region of (k_1, k_2) in $\mathbb{R}^2$ for internal stability.

Problem 8.11 (Kharitonov's Theorem) Let $a(s)$ be an interval polynomial

$$a(s) = [a_0^-, a_0^+] + [a_1^-, a_1^+]s + [a_2^-, a_2^+]s^2 + \cdots.$$

Kharitonov's theorem shows that $a(s)$ is stable if and only if the following four Kharitonov polynomials are stable:

$$K_1(s) = a_0^- + a_1^- s + a_2^+ s^2 + a_3^+ s^3 + a_4^- s^4 + a_5^- s^5 + a_6^+ s^6 + \cdots$$

$$K_2(s) = a_0^- + a_1^+ s + a_2^+ s^2 + a_3^- s^3 + a_4^- s^4 + a_5^+ s^5 + a_6^+ s^6 + \cdots$$

$$K_3(s) = a_0^+ + a_1^+ s + a_2^- s^2 + a_3^- s^3 + a_4^+ s^4 + a_5^+ s^5 + a_6^- s^6 + \cdots$$

$$K_4(s) = a_0^+ + a_1^- s + a_2^- s^2 + a_3^+ s^3 + a_4^+ s^4 + a_5^- s^5 + a_6^- s^6 + \cdots$$

Let $a_i := (a_i^- + a_i^+)/2$ and let

$$a_{\text{nom}}(s) = a_0 + a_1 s + a_2 s^2 + \cdots.$$

Find a least conservative $W(s)$ such that

$$\frac{a(s)}{a_{\text{nom}}(s)} \in \{1 + W(s)\Delta(s) \mid \|\Delta\|_\infty \le 1\}.$$

Problem 8.12 One of the main tools in this chapter was the small-gain theorem. One way to state it is as follows: Define a transfer matrix $F(s)$ in $\mathcal{RH}_\infty$ to be *contractive* if $\|F\|_\infty \le 1$ and *strictly contractive* if $\|F\|_\infty < 1$. Then for the unity feedback system the small gain theorem is this: If K is contractive and G is strictly contractive, then the feedback system is stable.

This problem concerns passivity and the passivity theorem. This is an important tool in the study of the stability of feedback systems, especially robotics, that is complementary to the small gain theorem.

Consider a system with a square transfer matrix $F(s)$ in $\mathcal{RH}_\infty$. This is said to be *passive* if
$$F(j\omega) + F(j\omega)^* \geq 0, \quad \forall \omega.$$
Here, the symbol ≥ 0 means that the matrix is positive semidefinite. If the system is SISO, the condition is equivalent to
$$\text{Re } F(j\omega) \geq 0, \quad \forall \omega;$$
that is, the Nyquist plot of F lies in the right-half plane. The system is *strictly passive* if $F - \epsilon I$ is passive for some $\epsilon > 0$.

1. Consider a mechanical system with input vector $u(t)$ (forces and torques) and output vector $y(t)$ (velocities) modeled by the equation
$$M\dot{y} + Ky = u$$
where M and K are symmetric, positive definite matrices. Show that this system is passive.

2. If F is passive, then $(I+F)^{-1} \in \mathcal{RH}_\infty$ and $(I+F)^{-1}(I-F)$ is contractive; if F is strictly passive, then $(I+F)^{-1}(I-F)$ is strictly contractive. Prove these statements for the case that F is SISO.

3. Using the results so far, show (in the MIMO case) that the unity feedback system is stable if K is passive and G is strictly passive.

Problem 8.13 Consider a SISO feedback system shown below with $P = P_o + W_2 \Delta_2$.

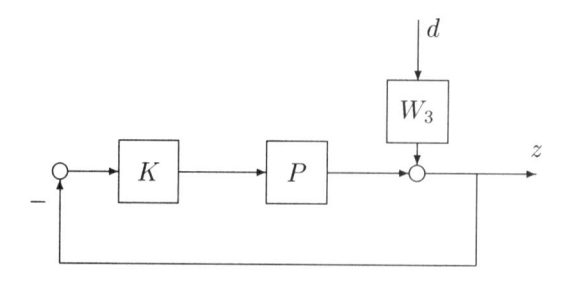

Assume that P_0 and P have the same number of right-half plane poles, W_2 is stable, and
$$|\text{Re}\{\Delta_2\}| \leq \alpha, \quad |\Im\{\Delta_2\}| \leq \beta.$$
Derive the necessary and sufficient conditions for the feedback system to be robustly stable.

8.8. Problems

Problem 8.14 Let $P = \begin{bmatrix} P_{11} & P_{12} \\ P_{21} & P_{22} \end{bmatrix} \in \mathcal{RH}_\infty$ be a two-by-two transfer matrix. Find sufficient (and necessary, if possible) conditions in each case so that $\mathcal{F}_u(P, \Delta)$ is stable for all possible stable Δ that satisfies the following conditions, respectively:

1. at each frequency
$$\text{Re}\,\Delta(j\omega) \geq 0, \quad |\Delta(j\omega)| < \alpha$$

2. at each frequency
$$\text{Re}\,\Delta(j\omega)e^{\pm j\theta} \geq 0, \quad |\Delta(j\omega)| < \alpha$$
where $\theta \geq 0$.

3. at each frequency
$$\text{Re}\,\Delta(j\omega) \geq 0, \Im\,\Delta(j\omega) \geq 0, \text{Re}\,\Delta(j\omega) + \Im\,\Delta(j\omega) < \alpha$$

Problem 8.15 Let $P = (I + \Delta W)P_0$ such that P and P_0 have the same number of unstable poles for all admissible Δ, $\|\Delta\|_\infty < \gamma$. Show that K robustly stabilizes P if and only if K stabilizes P_0 and

$$\left\|WP_0K(I + P_0K)^{-1}\right\|_\infty \leq 1.$$

Problem 8.16 Give appropriate generalizations of the preceding problem to other types of uncertainties.

Problem 8.17 Let $K = I$ and

$$P_0 = \begin{bmatrix} \dfrac{1}{s+1} & \dfrac{2}{s+3} \\ \dfrac{1}{s+1} & \dfrac{1}{s+1} \end{bmatrix}$$

1. Let $P = P_0 + \Delta$ with $\|\Delta\|_\infty \leq \gamma$. Determine the largest γ for robust stability.

2. Let $\Delta = \begin{bmatrix} k_1 & \\ & k_2 \end{bmatrix} \in \mathbb{R}^{2\times 2}$. Determine the stability region.

Problem 8.18 Repeat the preceding problem with

$$P_0 = \begin{bmatrix} \dfrac{s-1}{(s+1)^2} & \dfrac{5s+1}{(s+1)^2} \\ \dfrac{-1}{(s+1)^2} & \dfrac{s-1}{(s+1)^2} \end{bmatrix}.$$

Chapter 9

Linear Fractional Transformation

This chapter introduces a new matrix function: linear fractional transformation (LFT). We show that many interesting control problems can be formulated in an LFT framework and thus can be treated using the same technique.

9.1 Linear Fractional Transformations

This section introduces the matrix linear fractional transformations. It is well known from the one-complex-variable function theory that a mapping $F : \mathbb{C} \mapsto \mathbb{C}$ of the form

$$F(s) = \frac{a + bs}{c + ds}$$

with $a, b, c,$ and $d \in \mathbb{C}$ is called a *linear fractional transformation*. In particular, if $c \neq 0$ then $F(s)$ can also be written as

$$F(s) = \alpha + \beta s(1 - \gamma s)^{-1}$$

for some α, β and $\gamma \in \mathbb{C}$. The linear fractional transformation described above for scalars can be generalized to the matrix case.

Definition 9.1 Let M be a complex matrix partitioned as

$$M = \begin{bmatrix} M_{11} & M_{12} \\ M_{21} & M_{22} \end{bmatrix} \in \mathbb{C}^{(p_1 + p_2) \times (q_1 + q_2)},$$

and let $\Delta_\ell \in \mathbb{C}^{q_2 \times p_2}$ and $\Delta_u \in \mathbb{C}^{q_1 \times p_1}$ be two other complex matrices. Then we can formally define a *lower LFT* with respect to Δ_ℓ as the map

$$\mathcal{F}_\ell(M, \bullet) : \mathbb{C}^{q_2 \times p_2} \mapsto \mathbb{C}^{p_1 \times q_1}$$

165

with
$$\mathcal{F}_\ell(M, \Delta_\ell) := M_{11} + M_{12}\Delta_\ell(I - M_{22}\Delta_\ell)^{-1}M_{21}$$
provided that the inverse $(I - M_{22}\Delta_\ell)^{-1}$ exists. We can also define an *upper LFT* with respect to Δ_u as
$$\mathcal{F}_u(M, \bullet) : \mathbb{C}^{q_1 \times p_1} \mapsto \mathbb{C}^{p_2 \times q_2}$$
with
$$\mathcal{F}_u(M, \Delta_u) = M_{22} + M_{21}\Delta_u(I - M_{11}\Delta_u)^{-1}M_{12}$$
provided that the inverse $(I - M_{11}\Delta_u)^{-1}$ exists.

The matrix M in the preceding LFTs is called the *coefficient matrix*. The motivation for the terminologies of *lower* and *upper* LFTs should be clear from the following diagram representations of $\mathcal{F}_\ell(M, \Delta_\ell)$ and $\mathcal{F}_u(M, \Delta_u)$:

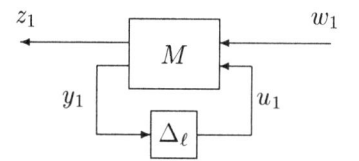

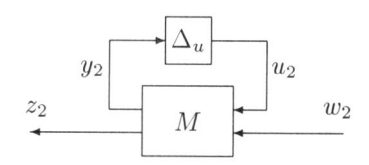

The diagram on the left represents the following set of equations:
$$\begin{bmatrix} z_1 \\ y_1 \end{bmatrix} = M \begin{bmatrix} w_1 \\ u_1 \end{bmatrix} = \begin{bmatrix} M_{11} & M_{12} \\ M_{21} & M_{22} \end{bmatrix} \begin{bmatrix} w_1 \\ u_1 \end{bmatrix},$$
$$u_1 = \Delta_\ell y_1$$

while the diagram on the right represents
$$\begin{bmatrix} y_2 \\ z_2 \end{bmatrix} = M \begin{bmatrix} u_2 \\ w_2 \end{bmatrix} = \begin{bmatrix} M_{11} & M_{12} \\ M_{21} & M_{22} \end{bmatrix} \begin{bmatrix} u_2 \\ w_2 \end{bmatrix},$$
$$u_2 = \Delta_u y_2.$$

It is easy to verify that the mapping defined on the left diagram is equal to $\mathcal{F}_\ell(M, \Delta_\ell)$ and the mapping defined on the right diagram is equal to $\mathcal{F}_u(M, \Delta_u)$. So from the above diagrams, $\mathcal{F}_\ell(M, \Delta_\ell)$ is a transformation obtained from closing the *lower* loop on the left diagram; similarly, $\mathcal{F}_u(M, \Delta_u)$ is a transformation obtained from closing the *upper* loop on the right diagram. In most cases, we shall use the general term *LFT* in referring to both upper and lower LFTs and assume that the context will distinguish the situations since one can use either of these notations to express a given object. Indeed, it is clear that $\mathcal{F}_u(N, \Delta) = \mathcal{F}_\ell(M, \Delta)$ with $N = \begin{bmatrix} M_{22} & M_{21} \\ M_{12} & M_{11} \end{bmatrix}$. It is usually not crucial which expression is used; however, it is often the case that one expression is more convenient than the other for a given problem. It should also be clear to the reader that in writing $\mathcal{F}_\ell(M, \Delta)$ [or $\mathcal{F}_u(M, \Delta)$] it is implied that Δ has compatible dimensions.

9.1. Linear Fractional Transformations

A useful interpretation of an LFT [e.g., $\mathcal{F}_\ell(M,\Delta)$] is that $\mathcal{F}_\ell(M,\Delta)$ has a nominal mapping, M_{11}, and is perturbed by Δ, while M_{12}, M_{21}, and M_{22} reflect a prior knowledge as to how the perturbation affects the nominal map, M_{11}. A similar interpretation can be applied to $\mathcal{F}_u(M,\Delta)$. This is why LFT is particularly useful in the study of perturbations, which is the focus of the next chapter.

The physical meaning of an LFT in control science is obvious if we take M as a proper transfer matrix. In that case, the LFTs defined previously are simply the closed-loop transfer matrices from $w_1 \mapsto z_1$ and $w_2 \mapsto z_2$, respectively; that is,

$$T_{zw1} = \mathcal{F}_\ell(M,\Delta_\ell), \qquad T_{zw2} = \mathcal{F}_u(M,\Delta_u)$$

where M may be the controlled plant and Δ may be either the system model uncertainties or the controllers.

Definition 9.2 An LFT, $\mathcal{F}_\ell(M,\Delta)$, is said to be *well-defined (or well-posed)* if $(I - M_{22}\Delta)$ is invertible.

Note that this definition is consistent with the well-posedness definition of the feedback system, which requires that the corresponding transfer matrix be invertible in $\mathcal{R}_p(s)$. It is clear that the study of an LFT that is not well-defined is meaningless; hence throughout this book, whenever an LFT is invoked, it will be assumed implicitly that it is well-defined. It is also clear from the definition that, for any M, $\mathcal{F}_\ell(M,0)$ is well-defined; hence any function that is not well-defined at the origin cannot be expressed as an LFT in terms of its variables. For example, $f(\delta) = 1/\delta$ is not an LFT of δ.

In some literature, LFT is used to refer to the following matrix functions:

$$(A + BQ)(C + DQ)^{-1} \qquad \text{or} \qquad (C + QD)^{-1}(A + QB)$$

where C is usually assumed to be invertible due to practical consideration. The following results follow from some simple algebra.

Lemma 9.1 *Suppose C is invertible. Then*

$$(A + BQ)(C + DQ)^{-1} = \mathcal{F}_\ell(M,Q)$$
$$(C + QD)^{-1}(A + QB) = \mathcal{F}_\ell(N,Q)$$

with

$$M = \begin{bmatrix} AC^{-1} & B - AC^{-1}D \\ C^{-1} & -C^{-1}D \end{bmatrix}, \quad N = \begin{bmatrix} C^{-1}A & C^{-1} \\ B - DC^{-1}A & -DC^{-1} \end{bmatrix}.$$

The converse also holds if M satisfies certain conditions.

Lemma 9.2 Let $\mathcal{F}_\ell(M, Q)$ be a given LFT with $M = \begin{bmatrix} M_{11} & M_{12} \\ M_{21} & M_{22} \end{bmatrix}$.

(a) If M_{12} is invertible, then

$$\mathcal{F}_\ell(M, Q) = (C + QD)^{-1}(A + QB)$$

with $A = M_{12}^{-1} M_{11}$, $B = M_{21} - M_{22} M_{12}^{-1} M_{11}$, $C = M_{12}^{-1}$, and $D = -M_{22} M_{12}^{-1}$; that is,

$$\begin{bmatrix} A & C \\ B & D \end{bmatrix} = \mathcal{F}_\ell \left(\begin{bmatrix} 0 & 0 & -I \\ M_{21} & 0 & M_{22} \\ \hline M_{11} & I & 0 \end{bmatrix}, -M_{12}^{-1} \right)$$

$$= \mathcal{F}_\ell \left(\begin{bmatrix} 0 & 0 & -I \\ M_{21} & 0 & M_{22} \\ \hline M_{11} & I & M_{12} + E \end{bmatrix}, E^{-1} \right)$$

for any nonsingular matrix E.

(b) If M_{21} is invertible, then

$$\mathcal{F}_\ell(M, Q) = (A + BQ)(C + DQ)^{-1}$$

with $A = M_{11} M_{21}^{-1}$, $B = M_{12} - M_{11} M_{21}^{-1} M_{22}$, $C = M_{21}^{-1}$, and $D = -M_{21}^{-1} M_{22}$; that is,

$$\begin{bmatrix} A & B \\ C & D \end{bmatrix} = \mathcal{F}_\ell \left(\begin{bmatrix} 0 & M_{12} & M_{11} \\ 0 & 0 & I \\ \hline -I & M_{22} & 0 \end{bmatrix}, -M_{21}^{-1} \right)$$

$$= \mathcal{F}_\ell \left(\begin{bmatrix} 0 & M_{12} & M_{11} \\ 0 & 0 & I \\ \hline -I & M_{22} & M_{21} + E \end{bmatrix}, E^{-1} \right)$$

for any nonsingular matrix E.

However, for an arbitrary LFT $\mathcal{F}_\ell(M, Q)$, neither M_{21} nor M_{12} is necessarily square and invertible; therefore, the alternative fractional formula is more restrictive.

It should be pointed out that some seemingly simple functions do not have simple LFT representations. For example,

$$(A + QB)(I + QD)^{-1}$$

cannot always be written in the form of $\mathcal{F}_\ell(M, Q)$ for some M; however, it can be written as

$$(A + QB)(I + QD)^{-1} = \mathcal{F}_\ell(N, \Delta)$$

9.1. Linear Fractional Transformations

with
$$N = \left[\begin{array}{c|cc} A & I & A \\ \hline -B & 0 & -B \\ D & 0 & D \end{array}\right], \quad \Delta = \left[\begin{array}{cc} Q & \\ & Q \end{array}\right].$$

Note that the dimension of Δ is twice of Q.

The following lemma shows that the inverse of an *LFT* is still an LFT.

Lemma 9.3 *Let* $M = \begin{bmatrix} M_{11} & M_{12} \\ M_{21} & M_{22} \end{bmatrix}$ *and M_{22} is nonsingular. Then*

$$(\mathcal{F}_u(M, \Delta))^{-1} = \mathcal{F}_u(N, \Delta)$$

with N, is given by

$$N = \begin{bmatrix} M_{11} - M_{12} M_{22}^{-1} M_{21} & -M_{12} M_{22}^{-1} \\ M_{22}^{-1} M_{21} & M_{22}^{-1} \end{bmatrix}.$$

LFT is a very convenient tool to formulate many mathematical objects. We shall illustrate this by the following two examples.

Simple Block Diagrams

A feedback system with the following block diagram

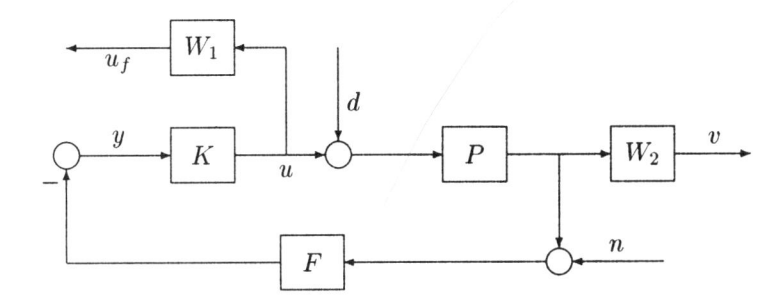

can be rearranged as an LFT:

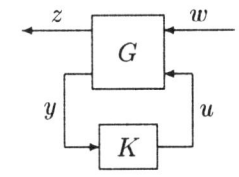

with
$$w = \begin{pmatrix} d \\ n \end{pmatrix}, \quad z = \begin{pmatrix} v \\ u_f \end{pmatrix}, \quad G = \left[\begin{array}{cc|c} W_2 P & 0 & W_2 P \\ 0 & 0 & W_1 \\ \hline -FP & -F & -FP \end{array}\right].$$

A state-space realization for the generalized plant G can be obtained by directly realizing the transfer matrix G using any standard multivariable realization techniques (e.g., Gilbert realization). However, the direct realization approach is usually complicated. Here we shall show another way to obtain the realization for G based on the realizations of each component. To simplify the expression, we shall assume that the plant P is strictly proper and P, F, W_1, and W_2 have, respectively, the following state-space realizations:

$$P = \left[\begin{array}{c|c} A_p & B_p \\ \hline C_p & 0 \end{array}\right], \quad F = \left[\begin{array}{c|c} A_f & B_f \\ \hline C_f & D_f \end{array}\right], \quad W_1 = \left[\begin{array}{c|c} A_u & B_u \\ \hline C_u & D_u \end{array}\right], \quad W_2 = \left[\begin{array}{c|c} A_v & B_v \\ \hline C_v & D_v \end{array}\right].$$

That is,
$$\dot{x}_p = A_p x_p + B_p(d + u), \quad y_p = C_p x_p,$$
$$\dot{x}_f = A_f x_f + B_f(y_p + n), \quad -y = C_f x_f + D_f(y_p + n),$$
$$\dot{x}_u = A_u x_u + B_u u, \quad u_f = C_u x_u + D_u u,$$
$$\dot{x}_v = A_v x_v + B_v y_p, \quad v = C_v x_v + D_v y_p.$$

Now define a new state vector
$$x = \begin{bmatrix} x_p \\ x_f \\ x_u \\ x_v \end{bmatrix}$$

and eliminate the variable y_p to get a realization of G as
$$\dot{x} = Ax + B_1 w + B_2 u$$
$$z = C_1 x + D_{11} w + D_{12} u$$
$$y = C_2 x + D_{21} w + D_{22} u$$

with

$$A = \begin{bmatrix} A_p & 0 & 0 & 0 \\ B_f C_p & A_f & 0 & 0 \\ 0 & 0 & A_u & 0 \\ B_v C_p & 0 & 0 & A_v \end{bmatrix}, \quad B_1 = \begin{bmatrix} B_p & 0 \\ 0 & B_f \\ 0 & 0 \\ 0 & 0 \end{bmatrix}, \quad B_2 = \begin{bmatrix} B_p \\ 0 \\ B_u \\ 0 \end{bmatrix}$$

$$C_1 = \begin{bmatrix} D_v C_p & 0 & 0 & C_v \\ 0 & 0 & C_u & 0 \end{bmatrix}, \quad D_{11} = 0, \quad D_{12} = \begin{bmatrix} 0 \\ D_u \end{bmatrix}$$

$$C_2 = \begin{bmatrix} -D_f C_p & -C_f & 0 & 0 \end{bmatrix}, \quad D_{21} = \begin{bmatrix} 0 & -D_f \end{bmatrix}, \quad D_{22} = 0.$$

9.1. Linear Fractional Transformations

Parametric Uncertainty: A Mass/Spring/Damper System

One natural type of uncertainty is unknown coefficients in a state-space model. To motivate this type of uncertainty description, we shall begin with a familiar mechanical system, shown in Figure 9.1.

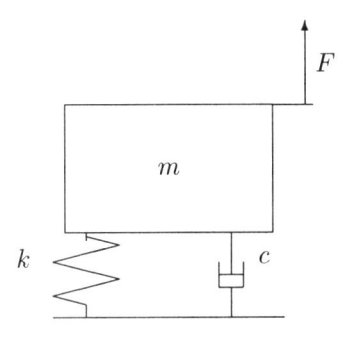

Figure 9.1: A mass/spring/damper system

The dynamical equation of the system motion can be described by

$$\ddot{x} + \frac{c}{m}\dot{x} + \frac{k}{m}x = \frac{F}{m}.$$

Suppose that the three physical parameters $m, c,$ and k are not known exactly, but are believed to lie in known intervals. In particular, the actual mass m is within 10% of a nominal mass, $\bar{m}$, the actual damping value c is within 20% of a nominal value of $\bar{c}$, and the spring stiffness is within 30% of its nominal value of $\bar{k}$. Now introducing perturbations $\delta_m, \delta_c,$ and δ_k, which are assumed to be unknown but lie in the interval $[-1, 1]$, the block diagram for the dynamical system is as shown in Figure 9.2.

It is easy to check that $\frac{1}{m}$ can be represented as an LFT in δ_m:

$$\frac{1}{m} = \frac{1}{\bar{m}(1+0.1\delta_m)} = \frac{1}{\bar{m}} - \frac{0.1}{\bar{m}}\delta_m(1+0.1\delta_m)^{-1} = \mathcal{F}_\ell(M_1, \delta_m)$$

with $M_1 = \begin{bmatrix} \frac{1}{\bar{m}} & -\frac{0.1}{\bar{m}} \\ 1 & -0.1 \end{bmatrix}$. Suppose that the input signals of the dynamical system are selected as $x_1 = x, x_2 = \dot{x}, F$, and the output signals are selected as $\dot{x}_1$ and $\dot{x}_2$. To represent the system model as an LFT of the natural uncertainty parameters $\delta_m, \delta_c,$ and δ_k, we shall first isolate the uncertainty parameters and denote the inputs and outputs of $\delta_k, \delta_c,$ and δ_m as y_k, y_c, y_m and u_k, u_c, u_m, respectively, as shown in Figure 9.3.

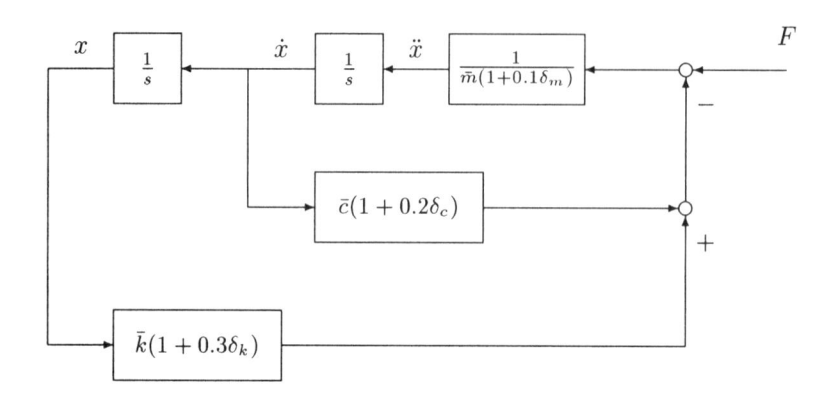

Figure 9.2: Block diagram of mass/spring/damper equation

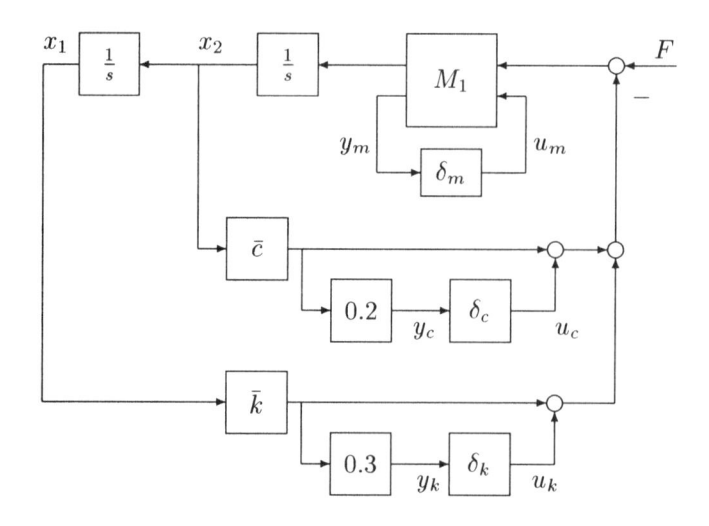

Figure 9.3: A block diagram for the mass/spring/damper system with uncertain parameters

9.2. Basic Principle

Then

$$\begin{bmatrix} \dot{x}_1 \\ \dot{x}_2 \\ y_k \\ y_c \\ y_m \end{bmatrix} = \begin{bmatrix} 0 & 1 & 0 & 0 & 0 & 0 \\ -\frac{\bar{k}}{\bar{m}} & -\frac{\bar{c}}{\bar{m}} & \frac{1}{\bar{m}} & -\frac{1}{\bar{m}} & -\frac{1}{\bar{m}} & -\frac{0.1}{\bar{m}} \\ 0.3\bar{k} & 0 & 0 & 0 & 0 & 0 \\ 0 & 0.2\bar{c} & 0 & 0 & 0 & 0 \\ -\bar{k} & -\bar{c} & 1 & -1 & -1 & -0.1 \end{bmatrix} \begin{bmatrix} x_1 \\ x_2 \\ F \\ u_k \\ u_c \\ u_m \end{bmatrix}, \quad \begin{bmatrix} u_k \\ u_c \\ u_m \end{bmatrix} = \Delta \begin{bmatrix} y_k \\ y_c \\ y_m \end{bmatrix}.$$

That is,

$$\begin{bmatrix} \dot{x}_1 \\ \dot{x}_2 \end{bmatrix} = \mathcal{F}_\ell(M, \Delta) \begin{bmatrix} x_1 \\ x_2 \\ F \end{bmatrix}$$

where

$$M = \begin{bmatrix} 0 & 1 & 0 & 0 & 0 & 0 \\ -\frac{\bar{k}}{\bar{m}} & -\frac{\bar{c}}{\bar{m}} & \frac{1}{\bar{m}} & -\frac{1}{\bar{m}} & -\frac{1}{\bar{m}} & -\frac{0.1}{\bar{m}} \\ 0.3\bar{k} & 0 & 0 & 0 & 0 & 0 \\ 0 & 0.2\bar{c} & 0 & 0 & 0 & 0 \\ -\bar{k} & -\bar{c} & 1 & -1 & -1 & -0.1 \end{bmatrix}, \quad \Delta = \begin{bmatrix} \delta_k & 0 & 0 \\ 0 & \delta_c & 0 \\ 0 & 0 & \delta_m \end{bmatrix}.$$

9.2 Basic Principle

We have studied two simple examples of the use of LFTs and, in particular, their role in modeling uncertainty. The basic principle at work here in writing a matrix LFT is often referred to as *"pulling out the Δ's"*. We will try to illustrate this with another picture. Consider a structure with four substructures interconnected in some known way, as shown in Figure 9.4. This diagram can be redrawn as a standard one via "pulling out the Δ's" in Figure 9.5.

Now the matrix M of the LFT can be obtained by computing the corresponding transfer matrix in the shadowed box.

We shall illustrate the preceding principle with an example. Consider an input/output relation

$$z = \frac{a + b\delta_2 + c\delta_1\delta_2^2}{1 + d\delta_1\delta_2 + e\delta_1^2} w =: Gw$$

where a, b, c, d, and e are given constants or transfer functions. We would like to write G as an LFT in terms of δ_1 and δ_2. We shall do this in three steps:

1. Draw a block diagram for the input/output relation with each δ separated as shown in Figure 9.6.

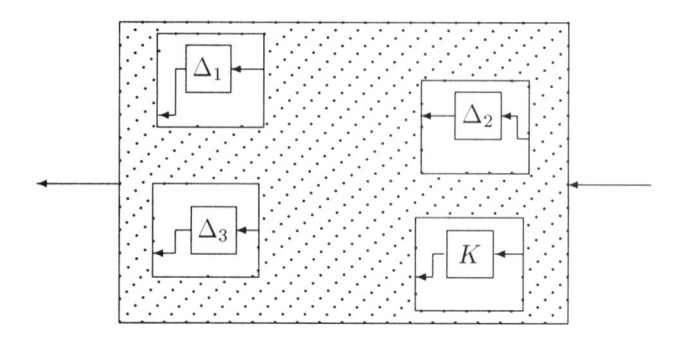

Figure 9.4: Multiple source of uncertain structure

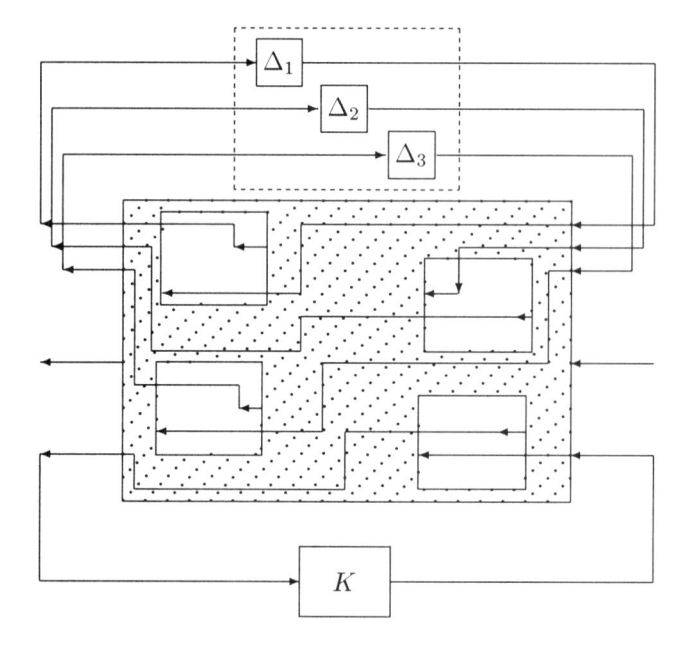

Figure 9.5: Pulling out the Δ's

9.2. Basic Principle

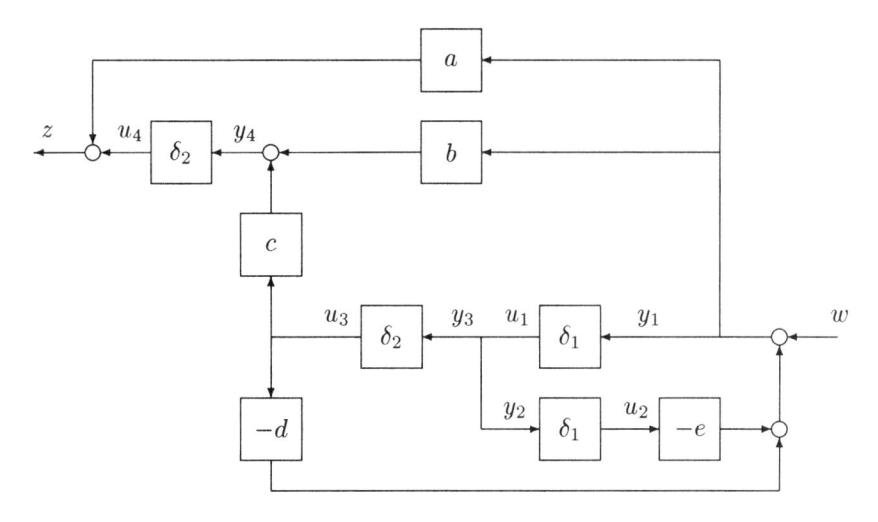

Figure 9.6: Block diagram for G

2. Mark the inputs and outputs of the δ's as y's and u's, respectively. (This is essentially *pulling out the Δ's*.)

3. Write z and y's in terms of w and u's with all δ's taken out. (This step is equivalent to computing the transformation in the shadowed box in Figure 9.5.)

$$\begin{bmatrix} y_1 \\ y_2 \\ y_3 \\ y_4 \\ z \end{bmatrix} = M \begin{bmatrix} u_1 \\ u_2 \\ u_3 \\ u_4 \\ w \end{bmatrix}$$

where

$$M = \begin{bmatrix} 0 & -e & -d & 0 & 1 \\ 1 & 0 & 0 & 0 & 0 \\ 1 & 0 & 0 & 0 & 0 \\ 0 & -be & -bd+c & 0 & b \\ \hline 0 & -ae & -ad & 1 & a \end{bmatrix}.$$

Then

$$z = \mathcal{F}_u(M, \Delta)w, \quad \Delta = \begin{bmatrix} \delta_1 I_2 & 0 \\ 0 & \delta_2 I_2 \end{bmatrix}.$$

All LFT examples in Section 9.1 can be obtained following the preceding steps.

For SIMULINK users, it is much easier to do all the computations using SIMULINK block diagrams, as shown in the following example.

Example 9.1 Consider the HIMAT (highly maneuverable aircraft) control problem from the μ Analysis and Synthesis Toolbox (Balas et al. [1994]). The system diagram is shown in Figure 9.7 where

$$W_{\text{del}} = \begin{bmatrix} \dfrac{50(s+100)}{s+10000} & 0 \\ 0 & \dfrac{50(s+100)}{s+10000} \end{bmatrix}, \quad W_p = \begin{bmatrix} \dfrac{0.5(s+3)}{s+0.03} & 0 \\ 0 & \dfrac{0.5(s+3)}{s+0.03} \end{bmatrix},$$

$$W_n = \begin{bmatrix} \dfrac{2(s+1.28)}{s+320} & 0 \\ 0 & \dfrac{2(s+1.28)}{s+320} \end{bmatrix},$$

$$P_0 = \left[\begin{array}{cccc|cc} -0.0226 & -36.6 & -18.9 & -32.1 & 0 & 0 \\ 0 & -1.9 & 0.983 & 0 & -0.414 & 0 \\ 0.0123 & -11.7 & -2.63 & 0 & -77.8 & 22.4 \\ 0 & 0 & 1 & 0 & 0 & 0 \\ \hline 0 & 57.3 & 0 & 0 & 0 & 0 \\ 0 & 0 & 0 & 57.3 & 0 & 0 \end{array}\right]$$

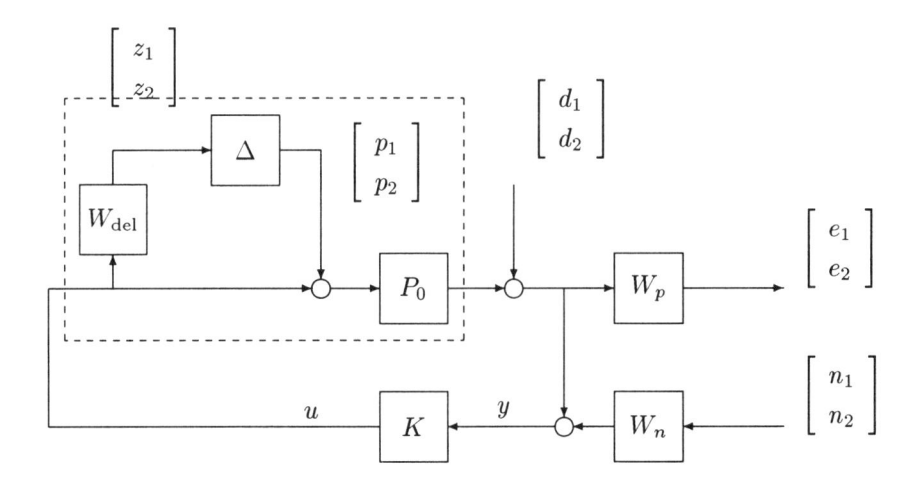

Figure 9.7: HIMAT closed-loop interconnection

9.2. Basic Principle

The open-loop interconnection is

$$\begin{bmatrix} z_1 \\ z_2 \\ e_1 \\ e_2 \\ y_1 \\ y_2 \end{bmatrix} = \hat{G}(s) \begin{bmatrix} p_1 \\ p_2 \\ d_1 \\ d_2 \\ n_1 \\ n_2 \\ u_1 \\ u_2 \end{bmatrix}.$$

The SIMULINK block diagram of this open-loop interconnection is shown in Figure 9.8.

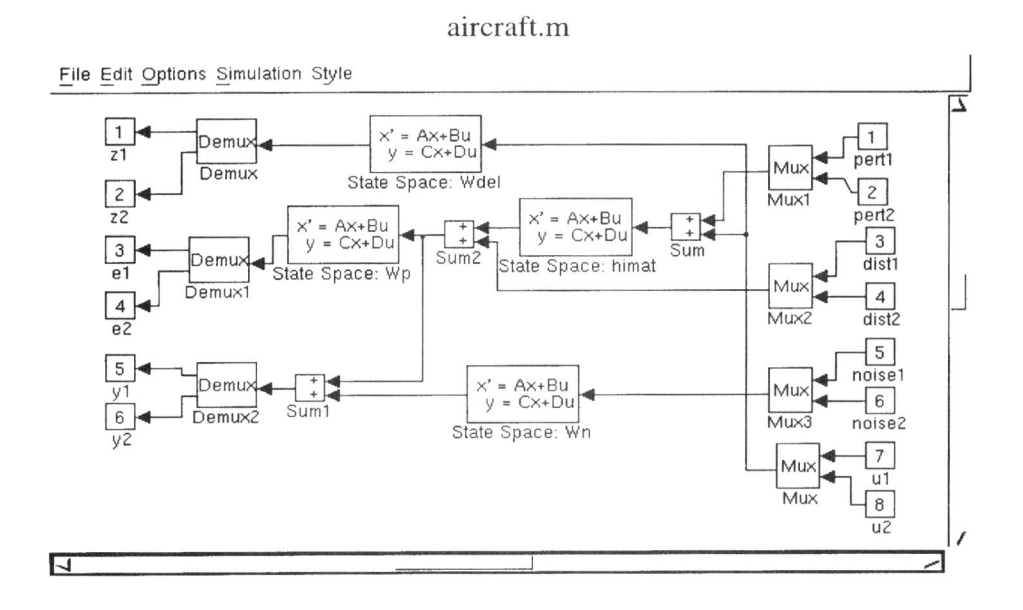

Figure 9.8: SIMULINK block diagram for HIMAT (aircraft.m)

The $\hat{G}(s) = \left[\begin{array}{c|c} A & B \\ \hline C & D \end{array} \right]$ can be computed by

$$\gg [\mathbf{A}, \mathbf{B}, \mathbf{C}, \mathbf{D}] = \text{linmod}('\text{aircraft}')$$

which gives

$$A = \begin{bmatrix} -10000I_2 & 0 & 0 & 0 & 0 & 0 & 0 & 0 \\ 0 & -0.0226 & -36.6 & -18.9 & -32.1 & 0 & 0 & 0 \\ 0 & 0 & -1.9 & 0.983 & 0 & 0 & 0 & 0 \\ 0 & 0.0123 & -11.7 & -2.63 & 0 & 0 & 0 & 0 \\ 0 & 0 & 0 & 1 & 0 & 0 & 0 & 0 \\ 0 & 0 & -54.087 & 0 & 0 & -0.018 & 0 & 0 \\ 0 & 0 & 0 & 0 & -54.087 & 0 & -0.018 & 0 \\ 0 & 0 & 0 & 0 & 0 & 0 & 0 & -320I_2 \end{bmatrix}$$

$$B = \begin{bmatrix} 0 & 0 & 0 & 0 & -703.5624 & 0 \\ 0 & 0 & 0 & 0 & 0 & -703.5624 \\ 0 & 0 & 0 & 0 & 0 & 0 \\ -0.4140 & 0 & 0 & 0 & -0.4140 & 0 \\ -77.8 & 22.4 & 0 & 0 & -77.8 & 22.4 \\ 0 & 0 & 0 & 0 & 0 & 0 \\ 0 & 0 & -0.9439I_2 & 0 & 0 & 0 \\ 0 & 0 & 0 & -25.2476I_2 & 0 & 0 \end{bmatrix}$$

$$C = \begin{bmatrix} 703.5624I_2 & 0 & 0 & 0 & 0 & 0 & 0 & 0 & 0 \\ 0 & 0 & 28.65 & 0 & 0 & -0.9439 & 0 & 0 & 0 \\ 0 & 0 & 0 & 0 & 28.65 & 0 & -0.9439 & 0 & 0 \\ 0 & 0 & 57.3 & 0 & 0 & 0 & 0 & 25.2476 & 0 \\ 0 & 0 & 0 & 0 & 57.3 & 0 & 0 & 0 & 25.2476 \end{bmatrix}$$

$$D = \begin{bmatrix} 0 & 0 & 0 & 0 & 0 & 0 & 50 & 0 \\ 0 & 0 & 0 & 0 & 0 & 0 & 0 & 50 \\ 0 & 0 & 0.5 & 0 & 0 & 0 & 0 & 0 \\ 0 & 0 & 0 & 0.5 & 0 & 0 & 0 & 0 \\ 0 & 0 & 1 & 0 & 2 & 0 & 0 & 0 \\ 0 & 0 & 0 & 1 & 0 & 2 & 0 & 0 \end{bmatrix}.$$

9.3 Redheffer Star Products

The most important property of LFTs is that any interconnection of LFTs is again an LFT. This property is by far the most often used and is the heart of LFT machinery. Indeed, it is not hard to see that most of the interconnection structures discussed earlier (e.g., feedback and cascade) can be viewed as special cases of the so-called *star product*.

9.3. Redheffer Star Products

Suppose that P and K are compatibly partitioned matrices

$$P = \begin{bmatrix} P_{11} & P_{12} \\ P_{21} & P_{22} \end{bmatrix}, \quad K = \begin{bmatrix} K_{11} & K_{12} \\ K_{21} & K_{22} \end{bmatrix}$$

such that the matrix product $P_{22}K_{11}$ is well-defined and square, and assume further that $I - P_{22}K_{11}$ is invertible. Then the *star product of P and K with respect to this partition* is defined as

$$P \star K := \begin{bmatrix} F_l(P, K_{11}) & P_{12}(I - K_{11}P_{22})^{-1} K_{12} \\ K_{21}(I - P_{22}K_{11})^{-1} P_{21} & F_u(K, P_{22}) \end{bmatrix}. \quad (9.1)$$

Note that this definition is dependent on the partitioning of the matrices P and K. In fact, this star product may be well-defined for one partition and not well-defined for another; however, we will not explicitly show this dependence because it is always clear from the context. In a block diagram, this dependence appears, as shown in Figure 9.9.

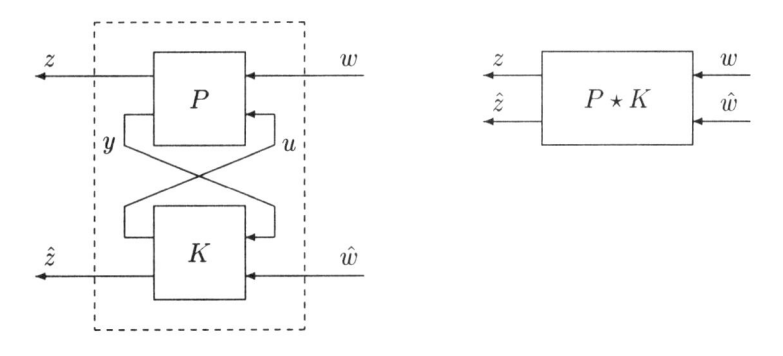

Figure 9.9: Interconnection of LFTs

Now suppose that P and K are transfer matrices with state-space representations:

$$P = \left[\begin{array}{c|cc} A & B_1 & B_2 \\ \hline C_1 & D_{11} & D_{12} \\ C_2 & D_{21} & D_{22} \end{array}\right] \quad K = \left[\begin{array}{c|cc} A_K & B_{K1} & B_{K2} \\ \hline C_{K1} & D_{K11} & D_{K12} \\ C_{K2} & D_{K21} & D_{K22} \end{array}\right].$$

Then the transfer matrix

$$P \star K : \begin{bmatrix} w \\ \hat{w} \end{bmatrix} \mapsto \begin{bmatrix} z \\ \hat{z} \end{bmatrix}$$

has a representation

$$P \star K = \left[\begin{array}{c|cc} \bar{A} & \bar{B}_1 & \bar{B}_2 \\ \hline \bar{C}_1 & \bar{D}_{11} & \bar{D}_{12} \\ \bar{C}_2 & \bar{D}_{21} & \bar{D}_{22} \end{array}\right] = \left[\begin{array}{c|c} \bar{A} & \bar{B} \\ \hline \bar{C} & \bar{D} \end{array}\right]$$

where

$$\bar{A} = \begin{bmatrix} A + B_2\tilde{R}^{-1}D_{K11}C_2 & B_2\tilde{R}^{-1}C_{K1} \\ B_{K1}R^{-1}C_2 & A_K + B_{K1}R^{-1}D_{22}C_{K1} \end{bmatrix}$$

$$\bar{B} = \begin{bmatrix} B_1 + B_2\tilde{R}^{-1}D_{K11}D_{21} & B_2\tilde{R}^{-1}D_{K12} \\ B_{K1}R^{-1}D_{21} & B_{K2} + B_{K1}R^{-1}D_{22}D_{K12} \end{bmatrix}$$

$$\bar{C} = \begin{bmatrix} C_1 + D_{12}D_{K11}R^{-1}C_2 & D_{12}\tilde{R}^{-1}C_{K1} \\ D_{K21}R^{-1}C_2 & C_{K2} + D_{K21}R^{-1}D_{22}C_{K1} \end{bmatrix}$$

$$\bar{D} = \begin{bmatrix} D_{11} + D_{12}D_{K11}R^{-1}D_{21} & D_{12}\tilde{R}^{-1}D_{K12} \\ D_{K21}R^{-1}D_{21} & D_{K22} + D_{K21}R^{-1}D_{22}D_{K12} \end{bmatrix}$$

$$R = I - D_{22}D_{K11}, \quad \tilde{R} = I - D_{K11}D_{22}.$$

In fact, it is easy to show that

$$\bar{A} = \begin{bmatrix} A & B_2 \\ C_2 & D_{22} \end{bmatrix} \star \begin{bmatrix} D_{K11} & C_{K1} \\ B_{K1} & A_K \end{bmatrix},$$

$$\bar{B} = \begin{bmatrix} B_1 & B_2 \\ D_{21} & D_{22} \end{bmatrix} \star \begin{bmatrix} D_{K11} & D_{K12} \\ B_{K1} & B_{K2} \end{bmatrix},$$

$$\bar{C} = \begin{bmatrix} C_1 & D_{12} \\ C_2 & D_{22} \end{bmatrix} \star \begin{bmatrix} D_{K11} & C_{K1} \\ D_{K21} & C_{K2} \end{bmatrix},$$

$$\bar{D} = \begin{bmatrix} D_{11} & D_{12} \\ D_{21} & D_{22} \end{bmatrix} \star \begin{bmatrix} D_{K11} & D_{K12} \\ D_{K21} & D_{K22} \end{bmatrix}.$$

The MATLAB command **starp** can be used to compute the star product:

$$\gg \mathbf{P} \star \mathbf{K} = \mathbf{starp}(\mathbf{P}, \mathbf{K}, \mathbf{dimy}, \mathbf{dimu})$$

where dimy and dimu are the dimensions of y and u, respectively. In the particular case when $\dim(\hat{z}) = 0$ and $\dim(\hat{w}) = 0$, we have

$$\gg \mathcal{F}_\ell(\mathbf{P}, \mathbf{K}) = \mathbf{starp}(\mathbf{P}, \mathbf{K})$$

9.4 Notes and References

This chapter is based on the lecture notes by Packard [1991] and the paper by Doyle, Packard, and Zhou [1991].

9.5 Problems

Problem 9.1 Find M and N matrices such that $\tilde{\Delta} = M\Delta N$, where Δ is block diagonal.

1. $\tilde{\Delta} = \begin{bmatrix} \Delta_1 & \Delta_2 \end{bmatrix}$

2. $\tilde{\Delta} = \begin{pmatrix} \Delta_1 & 0 & 0 \\ 0 & 0 & \Delta_2 \end{pmatrix}$

3. $\tilde{\Delta} = \begin{pmatrix} \Delta_1 & \begin{bmatrix} 0 & 0 \end{bmatrix} \\ \begin{bmatrix} \Delta_2 \\ 0 \end{bmatrix} & \Delta_3 \end{pmatrix}$

4. $\tilde{\Delta} = \begin{pmatrix} \Delta_1 & 0 & 0 \\ \Delta_2 & \Delta_3 & 0 \\ 0 & 0 & \Delta_4 \end{pmatrix}$

5. $\tilde{\Delta} = \begin{pmatrix} \Delta_1 & \Delta_3 & 0 \\ \Delta_2 & \Delta_4 & \Delta_6 \\ 0 & \Delta_5 & 0 \end{pmatrix}$

Problem 9.2 Let $G = (I - P(s)\Delta)^{-1}$. Find a matrix $M(s)$ such that $G = \mathcal{F}_u(M, \Delta)$.

Problem 9.3 Consider the unity feedback system with $G(s)$ of size 2×2. Suppose $G(s)$ has an uncertainty model of the form

$$G(s) = \begin{bmatrix} [1 + \Delta_{11}(s)]g_{11}(s) & [1 + \Delta_{12}(s)]g_{12}(s) \\ [1 + \Delta_{21}(s)]g_{21}(s) & [1 + \Delta_{22}(s)]g_{22}(s) \end{bmatrix}.$$

Suppose also that we wish to study robust stability of the feedback system. Pull out the Δ's and draw the appropriate block diagram in the form of a structured perturbation of a nominal system.

Problem 9.4 Let

$$P(s) = \left[\begin{array}{c|cc} A & B_1 & B_2 \\ \hline C_1 & D_{11} & D_{12} \\ C_2 & D_{21} & D_{22} \end{array}\right], \quad K(s) = \left[\begin{array}{c|c} \hat{A} & \hat{B} \\ \hline \hat{C} & \hat{D} \end{array}\right].$$

Find state-space realizations for $\mathcal{F}_\ell(P, K)$ and $\mathcal{F}_\ell(P, \hat{D})$.

Problem 9.5 Suppose D_{21} is nonsingular and

$$M(s) = \left[\begin{array}{c|cc} A & B_1 & B_2 \\ \hline C_1 & D_{11} & D_{12} \\ C_2 & D_{21} & D_{22} \end{array} \right].$$

Find a state-space realization for

$$\hat{M}(s) = \left[\begin{array}{cc} \left[\begin{array}{cc} 0 & M_{12} \\ 0 & 0 \end{array} \right] & \left[\begin{array}{c} M_{11} \\ I \end{array} \right] \\ \left[\begin{array}{cc} -I & M_{22} \end{array} \right] & M_{21} + E \end{array} \right]$$

where E is a constant matrix. Find the state-space realization for $\mathcal{F}_\ell(\hat{M}, E^{-1})$ when $E = I$.

Chapter 10

μ and μ Synthesis

It is noted that the robust stability and robust performance criteria derived in Chapter 8 vary with the assumptions about the uncertainty descriptions and performance requirements. We shall show in this chapter that they can all be treated in a unified framework using the LFT machinery introduced in the last chapter and the structured singular value to be introduced in this chapter. This, of course, does not mean that those special problems and their corresponding results are not important; on the contrary, they are sometimes very enlightening to our understanding of complex problems, such as those in which complex problems are formed from simple problems. On the other hand, a unified approach may relieve the mathematical burden of dealing with specific problems repeatedly. Furthermore, the unified framework introduced here will enable us to treat exactly the robust stability and robust performance problems for systems with multiple sources of uncertainties, which is a formidable problem from the standpoint of Chapter 8, in the same fashion as single unstructured uncertainty. Indeed, if a system is subject to multiple sources of uncertainties, in order to use the results in Chapter 8 for unstructured cases, it is necessary to reflect all sources of uncertainties from their known point of occurrence to a single reference location in the loop. Such reflected uncertainties invariably have a great deal of structure, which must then be "covered up" with a large, arbitrarily more conservative perturbation in order to maintain a simple cone-bounded representation at the reference location. Readers might have some idea about the conservativeness in such reflection based on the skewed specification problem, where an input multiplicative uncertainty of the plant is reflected at the output and the size of the reflected uncertainty is proportional to the condition number of the plant. In general, the reflected uncertainty may be proportional to the condition number of the transfer matrix between its original location and the reflected location. Thus it is highly desirable to treat the uncertainties as they are and where they are. The structured singular value is defined exactly for that purpose.

10.1 General Framework for System Robustness

As we illustrated in Chapter 9, any interconnected system may be rearranged to fit the general framework in Figure 10.1. Although the interconnection structure can become quite complicated for complex systems, many software packages, such as SIMULINK and μ Analysis and Synthesis Toolbox, are available that could be used to generate the interconnection structure from system components. Various modeling assumptions will be considered, and the impact of these assumptions on analysis and synthesis methods will be explored in this general framework.

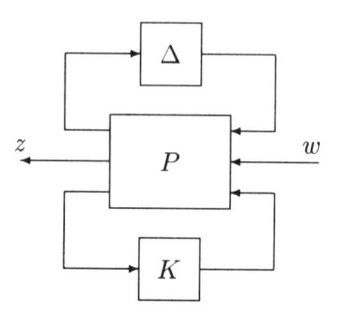

Figure 10.1: General framework

Note that uncertainty may be modeled in two ways, either as external inputs or as perturbations to the nominal model. The performance of a system is measured in terms of the behavior of the outputs or errors. The assumptions that characterize the uncertainty, performance, and nominal models determine the analysis techniques that must be used. The models are assumed to be FDLTI systems. The uncertain inputs are assumed to be either filtered white noise or weighted power or weighted $\mathcal{L}_p$ signals. Performance is measured as weighted output variances, or as power, or as weighted output $\mathcal{L}_p$ norms. The perturbations are assumed to be themselves FDLTI systems that are norm-bounded as input-output operators. Various combinations of these assumptions form the basis for all the standard linear system analysis tools.

Given that the nominal model is an FDLTI system, the interconnection system has the form

$$P(s) = \begin{bmatrix} P_{11}(s) & P_{12}(s) & P_{13}(s) \\ P_{21}(s) & P_{22}(s) & P_{23}(s) \\ P_{31}(s) & P_{32}(s) & P_{33}(s) \end{bmatrix}$$

and the closed-loop system is an LFT on the perturbation and the controller given by

$$\begin{aligned} z &= \mathcal{F}_u\left(\mathcal{F}_\ell(P,K),\Delta\right)w \\ &= \mathcal{F}_\ell\left(\mathcal{F}_u(P,\Delta),K\right)w. \end{aligned}$$

We shall focus our discussion in this section on analysis methods; therefore, the controller may be viewed as just another system component and absorbed into the

10.1. General Framework for System Robustness

interconnection structure. Denote

$$M(s) = \mathcal{F}_\ell(P(s), K(s)) = \begin{bmatrix} M_{11}(s) & M_{12}(s) \\ M_{21}(s) & M_{22}(s) \end{bmatrix}.$$

Then the general framework reduces to Figure 10.2, where

$$z = \mathcal{F}_u(M, \Delta)w = \left[M_{22} + M_{21}\Delta(I - M_{11}\Delta)^{-1} M_{12} \right] w.$$

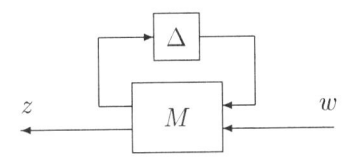

Figure 10.2: Analysis framework

Suppose $K(s)$ is a stabilizing controller for the nominal plant P. Then $M(s) \in \mathcal{RH}_\infty$. In general, the stability of $\mathcal{F}_u(M, \Delta)$ does not necessarily imply the internal stability of the closed-loop feedback system. However, they can be made equivalent with suitably chosen w and z. For example, consider again the multiplicatively perturbed system shown in Figure 10.3.

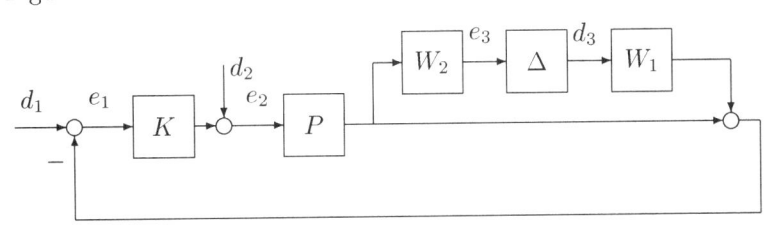

Figure 10.3: Multiplicatively perturbed systems

Now let

$$w := \begin{bmatrix} d_1 \\ d_2 \end{bmatrix}, \quad z := \begin{bmatrix} e_1 \\ e_2 \end{bmatrix}.$$

Then the system is robustly stable for all $\Delta(s) \in \mathcal{RH}_\infty$ with $\|\Delta\|_\infty < 1$ if and only if $\mathcal{F}_u(M, \Delta) \in \mathcal{RH}_\infty$ for all admissible Δ with $M_{11} = -W_2PK(I + PK)^{-1}W_1$, which is guaranteed by $\|M_{11}\|_\infty \leq 1$.

The analysis results presented in the previous chapters together with the associated synthesis tools are summarized in Table 10.1 with various uncertainty modeling assumptions.

However, the analysis is not so simple for systems with multiple sources of model uncertainties, including the robust performance problem for systems with unstructured

Input Assumptions	Performance Specifications	Perturbation Assumptions	Analysis Tests	Synthesis Methods
$E(w(t)w(\tau)^*)$ $= \delta(t-\tau)I$	$E(z(t)^*z(t)) \leq 1$	$\Delta = 0$	$\|M_{22}\|_2 \leq 1$	LQG
$w = U_0\delta(t)$ $E(U_0 U_0^*) = I$	$E(\|z\|_2^2) \leq 1$			Wiener-Hopf $\mathcal{H}_2$
$\|w\|_2 \leq 1$	$\|z\|_2 \leq 1$	$\Delta = 0$	$\|M_{22}\|_\infty \leq 1$	Singular Value Loop Shaping
$\|w\|_2 \leq 1$	Internal Stability	$\|\Delta\|_\infty < 1$	$\|M_{11}\|_\infty \leq 1$	$\mathcal{H}_\infty$

Table 10.1: General analysis for single source of uncertainty

uncertainty. As we shown in Chapter 9, if a system is built from components that are themselves uncertain, then, in general, the uncertainty in the system level is structured, involving typically a large number of real parameters. The stability analysis involving real parameters is much more difficult and will be discussed in Chapter 18. Here we shall simply cover the real parametric uncertainty with norm-bounded dynamical uncertainty. Moreover, the interconnection model M can always be chosen so that $\Delta(s)$ is block diagonal, and, by absorbing any weights, $\|\Delta\|_\infty < 1$. Thus we shall assume that $\Delta(s)$ takes the form of

$$\Delta(s) = \{\text{diag } [\delta_1 I_{r_1}, \ldots, \delta_s I_{r_S}, \Delta_1, \ldots, \Delta_F] : \delta_i(s) \in \mathcal{RH}_\infty, \Delta_j \in \mathcal{RH}_\infty\}$$

with $\|\delta_i\|_\infty < 1$ and $\|\Delta_j\|_\infty < 1$. Then the system is robustly stable iff the interconnected system in Figure 10.4 is stable.

The results of Table 10.1 can be applied to analysis of the system's robust stability in two ways:

(1) $\|M_{11}\|_\infty \leq 1$ implies stability, but not conversely, because this test ignores the known block diagonal structure of the uncertainties and is equivalent to regarding Δ as unstructured. This can be arbitrarily conservative in that stable systems can have arbitrarily large $\|M_{11}\|_\infty$.

10.2. Structured Singular Value

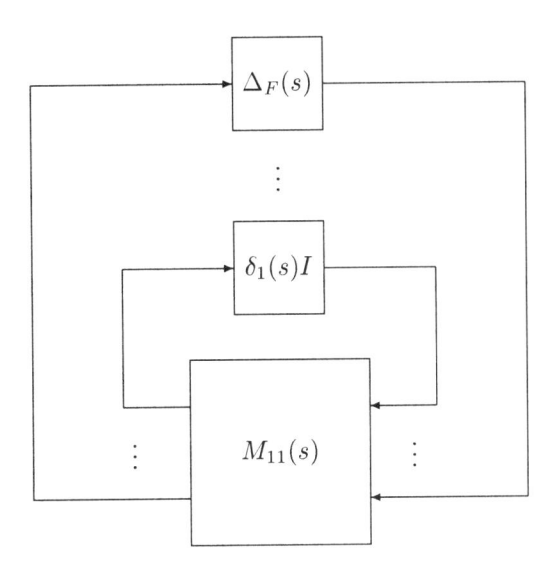

Figure 10.4: Robust stability analysis framework

(2) Test for each δ_i (Δ_j) individually (assuming no uncertainty in other channels). This test can be arbitrarily optimistic because it ignores interaction between the δ_i (Δ_j). This optimism is also clearly shown in the spinning body example in Section 8.6.

The difference between the stability margins (or bounds on Δ) obtained in (1) and (2) can be arbitrarily far apart. Only when the margins are close can conclusions be made about the general case with structured uncertainty.

The exact stability and performance analysis for systems with structured uncertainty requires a new matrix function called the structured singular value (SSV), which is denoted by μ.

10.2 Structured Singular Value

10.2.1 Definitions of μ

We shall motivate the definition of the structured singular value by asking the following question: Given a matrix $M \in \mathbb{C}^{p \times q}$, what is the smallest perturbation matrix $\Delta \in \mathbb{C}^{q \times p}$ in the sense of $\overline{\sigma}(\Delta)$ such that

$$\det(I - M\Delta) = 0?$$

That is, we are interested in finding

$$\alpha_{\min} := \inf \left\{ \overline{\sigma}(\Delta) : \det(I - M\Delta) = 0, \ \Delta \in \mathbb{C}^{q \times p} \right\}.$$

It is easy to see that

$$\alpha_{\min} = \inf\left\{\alpha : \det(I - \alpha M\Delta) = 0,\ \overline{\sigma}(\Delta) \leq 1,\ \Delta \in \mathbb{C}^{q\times p}\right\} = \frac{1}{\max\limits_{\overline{\sigma}(\Delta)\leq 1} \rho(M\Delta)}$$

and

$$\max_{\overline{\sigma}(\Delta)\leq 1} \rho(M\Delta) = \overline{\sigma}(M).$$

Hence the smallest norm of a "destabilizing" perturbation matrix is $1/\overline{\sigma}(M)$ with a smallest "destabilizing" Δ:

$$\Delta_{\text{des}} = \frac{1}{\overline{\sigma}(M)} v_1 u_1^*, \quad \det(I - M\Delta_{\text{des}}) = 0$$

where $M = \overline{\sigma}(M) u_1 v_1^* + \sigma_2 u_2 v_2^* + \cdots$ is a singular value decomposition.

So the reciprocal of the largest singular value of a matrix is a measure of the smallest "destabilizing" perturbation matrix. Hence it is instructive to introduce the following alternative definition for the largest singular value:

$$\overline{\sigma}(M) := \frac{1}{\inf\{\overline{\sigma}(\Delta) : \det(I - M\Delta) = 0,\ \Delta \in \mathbb{C}^{q\times p}\}}.$$

Next we consider a similar problem but with Δ structurally restricted. In particular, we consider the block diagonal matrix Δ. We shall consider two types of blocks: *repeated scalar* and *full* blocks. Let S and F represent the number of *repeated scalar* blocks and the number of *full* blocks, respectively. To bookkeep their dimensions, we introduce positive integers $r_1, \ldots, r_S;\ m_1, \ldots, m_F$. The ith repeated scalar block is $r_i \times r_i$, while the jth full block is $m_j \times m_j$. With those integers given, we define $\mathbf{\Delta} \subset \mathbb{C}^{n\times n}$ as

$$\mathbf{\Delta} = \left\{\text{diag}\ [\delta_1 I_{r_1}, \ldots, \delta_s I_{r_S}, \Delta_1, \ldots, \Delta_F] : \delta_i \in \mathbb{C}, \Delta_j \in \mathbb{C}^{m_j \times m_j}\right\}. \tag{10.1}$$

For consistency among all the dimensions, we must have

$$\sum_{i=1}^{S} r_i + \sum_{j=1}^{F} m_j = n.$$

Often, we will need norm-bounded subsets of $\mathbf{\Delta}$, and we introduce the following notation:

$$\mathbf{B\Delta} = \{\Delta \in \mathbf{\Delta} : \overline{\sigma}(\Delta) \leq 1\} \tag{10.2}$$

$$\mathbf{B}^{\circ}\mathbf{\Delta} = \{\Delta \in \mathbf{\Delta} : \overline{\sigma}(\Delta) < 1\} \tag{10.3}$$

where the superscript "o" symbolizes the open ball. To keep the notation as simple as possible in equation (10.1), we place all of the repeated scalar blocks first; in actuality, they can come in any order. Also, the full blocks do not have to be square, but restricting them as such saves a great deal in terms of notation.

10.2. Structured Singular Value

Now we ask a similar question: Given a matrix $M \in \mathbb{C}^{p \times q}$, what is the smallest perturbation matrix $\Delta \in \boldsymbol{\Delta}$ in the sense of $\bar{\sigma}(\Delta)$ such that

$$\det(I - M\Delta) = 0?$$

That is, we are interested in finding

$$\alpha_{\min} := \inf \left\{ \bar{\sigma}(\Delta) : \det(I - M\Delta) = 0, \ \Delta \in \boldsymbol{\Delta} \right\}.$$

Again we have

$$\alpha_{\min} = \inf \left\{ \alpha : \det(I - \alpha M\Delta) = 0, \ \Delta \in \mathbf{B}\boldsymbol{\Delta} \right\} = \frac{1}{\max_{\Delta \in \mathbf{B}\boldsymbol{\Delta}} \rho(M\Delta)}.$$

Similar to the unstructured case, we shall call $1/\alpha_{\min}$ the structured singular value and denote it by $\mu_{\boldsymbol{\Delta}}(M)$.

Definition 10.1 *For $M \in \mathbb{C}^{n \times n}$, $\mu_{\boldsymbol{\Delta}}(M)$ is defined as*

$$\mu_{\boldsymbol{\Delta}}(M) := \frac{1}{\min \left\{ \bar{\sigma}(\Delta) : \Delta \in \boldsymbol{\Delta}, \det(I - M\Delta) = 0 \right\}} \tag{10.4}$$

unless no $\Delta \in \boldsymbol{\Delta}$ makes $I - M\Delta$ singular, in which case $\mu_{\boldsymbol{\Delta}}(M) := 0$.

Remark 10.1 Without a loss in generality, the full blocks in the minimal norm Δ can each be chosen to be dyads (rank = 1). To see this, assume $S = 0$ (i.e., all blocks are full blocks). Suppose that $I - M\Delta$ is singular for some $\Delta \in \boldsymbol{\Delta}$. Then there is an $x \in \mathbb{C}^n$ such that $M\Delta x = x$. Now partition x conformably with Δ:

$$x = \begin{bmatrix} x_1 \\ x_2 \\ \vdots \\ x_F \end{bmatrix}, \quad x_i \in \mathbb{C}^{m_i}, i = 1, \ldots, F$$

and let

$$\tilde{\Delta}_i = \begin{cases} \dfrac{\Delta_i x_i x_i^*}{\|x_i\|^2}, & x_i \neq 0; \\ 0, & x_i = 0 \end{cases} \quad \text{for } i = 1, 2, \ldots, F.$$

Define

$$\tilde{\Delta} = \text{diag}\{\tilde{\Delta}_1, \tilde{\Delta}_2, \ldots, \tilde{\Delta}_F\}.$$

Then $\bar{\sigma}(\tilde{\Delta}) \leq \bar{\sigma}(\Delta)$, $\tilde{\Delta} x = \Delta x$, and thus $(I - M\tilde{\Delta})x = (I - M\Delta)x = 0$ (i.e., $I - M\tilde{\Delta}$ is also singular). Hence we have replaced a general perturbation Δ that satisfies the singularity condition with a perturbation $\tilde{\Delta}$ that is no larger [in the $\bar{\sigma}(\cdot)$ sense] and has rank 1 for each block but still satisfies the singularity condition. ◇

Lemma 10.1 $\mu_{\mathbf{\Delta}}(M) = \max_{\Delta \in \mathbf{B\Delta}} \rho(M\Delta)$

In view of this lemma, continuity of the function $\mu:\mathbb{C}^{n \times n} \to \mathbb{R}$ is apparent. In general, though, the function $\mu:\mathbb{C}^{n \times n} \to \mathbb{R}$ is not a norm, since it does not satisfy the triangle inequality; however, for any $\alpha \in \mathbb{C}$, $\mu(\alpha M) = |\alpha|\mu(M)$, so in some sense, it is related to how "big" the matrix is.

We can relate $\mu_{\mathbf{\Delta}}(M)$ to familiar linear algebra quantities when $\mathbf{\Delta}$ is one of two extreme sets.

- If $\mathbf{\Delta} = \{\delta I : \delta \in \mathbb{C}\}$ ($S=1, F=0, r_1=n$), then $\mu_{\mathbf{\Delta}}(M) = \rho(M)$, the spectral radius of M.

    Proof. The only Δ's in $\mathbf{\Delta}$ that satisfy the $\det(I - M\Delta) = 0$ constraint are reciprocals of nonzero eigenvalues of M. The smallest one of these is associated with the largest (magnitude) eigenvalue, so, $\mu_{\mathbf{\Delta}}(M) = \rho(M)$. □

- If $\mathbf{\Delta} = \mathbb{C}^{n \times n}$ ($S=0, F=1, m_1=n$), then $\mu_{\mathbf{\Delta}}(M) = \bar{\sigma}(M)$.

Obviously, for a general $\mathbf{\Delta}$, as in equation (10.1), we must have

$$\{\delta I_n : \delta \in \mathbb{C}\} \subset \mathbf{\Delta} \subset \mathbb{C}^{n \times n}. \tag{10.5}$$

Hence directly from the definition of μ and from the two preceding special cases, we conclude that

$$\rho(M) \leq \mu_{\mathbf{\Delta}}(M) \leq \bar{\sigma}(M). \tag{10.6}$$

These bounds alone are not sufficient for our purposes because the gap between ρ and $\bar{\sigma}$ can be arbitrarily large.

Example 10.1 Suppose

$$\Delta = \begin{bmatrix} \delta_1 & 0 \\ 0 & \delta_2 \end{bmatrix}$$

and consider

(1) $M = \begin{bmatrix} 0 & \beta \\ 0 & 0 \end{bmatrix}$ for any $\beta > 0$. Then $\rho(M) = 0$ and $\bar{\sigma}(M) = \beta$. But $\mu(M) = 0$ since $\det(I - M\Delta) = 1$ for all admissible Δ.

(2) $M = \begin{bmatrix} -1/2 & 1/2 \\ -1/2 & 1/2 \end{bmatrix}$. Then $\rho(M) = 0$ and $\bar{\sigma}(M) = 1$. Since

$$\det(I - M\Delta) = 1 + \frac{\delta_1 - \delta_2}{2},$$

it is easy to see that $\min\left\{\max_i |\delta_i| : 1 + \frac{\delta_1-\delta_2}{2} = 0\right\} = 1$, so $\mu(M) = 1$.

10.2. Structured Singular Value

Thus neither ρ nor $\bar{\sigma}$ provide useful bounds even in simple cases. The only time they do provide reliable bounds is when $\rho \approx \bar{\sigma}$.

However, the bounds can be refined by considering transformations on M that *do not affect* $\mu_\Delta(M)$, but *do affect* ρ and $\bar{\sigma}$. To do this, define the following two subsets of $\mathbb{C}^{n \times n}$:

$$\mathcal{U} = \{U \in \Delta : UU^* = I_n\} \tag{10.7}$$

$$\mathcal{D} = \left\{ \begin{array}{c} \mathrm{diag}\,[D_1, \ldots, D_S, d_1 I_{m_1}, \ldots, d_{F-1} I_{m_{F-1}}, I_{m_F}] : \\ D_i \in \mathbb{C}^{r_i \times r_i}, D_i = D_i^* > 0, d_j \in \mathbb{R}, d_j > 0 \end{array} \right\}. \tag{10.8}$$

Note that for any $\Delta \in \Delta, U \in \mathcal{U}$, and $D \in \mathcal{D}$,

$$U^* \in \mathcal{U} \quad U\Delta \in \Delta \quad \Delta U \in \Delta \quad \bar{\sigma}(U\Delta) = \bar{\sigma}(\Delta U) = \bar{\sigma}(\Delta) \tag{10.9}$$

$$D\Delta = \Delta D. \tag{10.10}$$

Consequently, we have the following:

Theorem 10.2 *For all $U \in \mathcal{U}$ and $D \in \mathcal{D}$*

$$\mu_\Delta(MU) = \mu_\Delta(UM) = \mu_\Delta(M) = \mu_\Delta(DMD^{-1}). \tag{10.11}$$

Proof. For all $D \in \mathcal{D}$ and $\Delta \in \Delta$,

$$\det(I - M\Delta) = \det(I - MD^{-1}\Delta D) = \det(I - DMD^{-1}\Delta)$$

since D commutes with Δ. Therefore $\mu_\Delta(M) = \mu_\Delta(DMD^{-1})$. Also, for each $U \in \mathcal{U}$, $\det(I - M\Delta) = 0$ if and only if $\det(I - MUU^*\Delta) = 0$. Since $U^*\Delta \in \Delta$ and $\bar{\sigma}(U^*\Delta) = \bar{\sigma}(\Delta)$, we get $\mu_\Delta(MU) = \mu_\Delta(M)$ as desired. The argument for UM is the same. □

Therefore, the bounds in equation (10.6) can be tightened to

$$\max_{U \in \mathcal{U}} \rho(UM) \leq \max_{\Delta \in \mathbf{B}\Delta} \rho(\Delta M) = \mu_\Delta(M) \leq \inf_{D \in \mathcal{D}} \bar{\sigma}(DMD^{-1}) \tag{10.12}$$

where the equality comes from Lemma 10.1. Note that the last element in the D matrix is normalized to 1 since for any nonzero scalar γ, $DMD^{-1} = (\gamma D) M (\gamma D)^{-1}$.

Remark 10.2 Note that the scaling set $\mathcal{D}$ in Theorem 10.2 and in inequality (10.12) is not necessarily restricted to being Hermitian. In fact, it can be replaced by any set of nonsingular matrices that satisfy equation (10.10). However, enlarging the set of scaling matrices does not improve the upper-bound in inequality (10.12). This can be shown as follows: Let D be any nonsingular matrix such that $D\Delta = \Delta D$. Then there exist a Hermitian matrix $0 < R = R^* \in \mathcal{D}$ and a unitary matrix U such that $D = UR$ and

$$\inf_D \bar{\sigma}\left(DMD^{-1}\right) = \inf_D \bar{\sigma}\left(URMR^{-1}U^*\right) = \inf_{R \in \mathcal{D}} \bar{\sigma}\left(RMR^{-1}\right).$$

Therefore, there is no loss of generality in assuming $\mathcal{D}$ to be Hermitian. ◇

10.2.2 Bounds

In this section we will concentrate on the bounds

$$\max_{U \in \mathcal{U}} \rho(UM) \le \mu_\Delta(M) \le \inf_{D \in \mathcal{D}} \bar{\sigma}\left(DMD^{-1}\right).$$

The lower bound is always an equality (Doyle [1982]).

Theorem 10.3 $\max_{U \in \mathcal{U}} \rho(MU) = \mu_\Delta(M)$

Unfortunately, the quantity $\rho(UM)$ can have multiple local maxima that are not global. Thus local search cannot be guaranteed to obtain μ, but can only yield a lower bound. For computation purposes one can derive a slightly different formulation of the lower bound as a power algorithm that is reminiscent of power algorithms for eigenvalues and singular values (Packard and Doyle [1988a, 1988b]). While there are open questions about convergence, the algorithm usually works quite well and has proven to be an effective method to compute μ.

The upper-bound can be reformulated as a convex optimization problem, so the global minimum can, in principle, be found. Unfortunately, the upper-bound is not always equal to μ. For block structures Δ satisfying $2S + F \le 3$, the upper-bound is always equal to $\mu_\Delta(M)$, and for block structures with $2S + F > 3$, there exist matrices for which μ is less than the infimum. This can be summarized in the following diagram, which shows for which cases the upper-bound is guaranteed to be equal to μ. See Packard and Doyle [1993] for details.

Theorem 10.4 $\mu_\Delta(M) = \inf_{D \in \mathcal{D}} \bar{\sigma}(DMD^{-1})$ if $2S + F \le 3$

S= \ F=	0	1	2	3	4
0		yes	yes	yes	no
1	yes	yes	no	no	no
2	no	no	no	no	no

Several of the boxes have connections with standard results.

- $S = 0, F = 1$: $\mu_\Delta(M) = \bar{\sigma}(M)$.

10.2. Structured Singular Value

- $S = 1$, $F = 0$: $\mu_\Delta(M) = \rho(M) = \inf_{D \in \mathcal{D}} \overline{\sigma}(DMD^{-1})$. This is a standard result in linear algebra. In fact, without a loss in generality, the matrix M can be assumed in Jordan canonical form. Now let

$$J_1 = \begin{bmatrix} \lambda & 1 & & & \\ & \lambda & 1 & & \\ & & \ddots & \ddots & \\ & & & \lambda & 1 \\ & & & & \lambda \end{bmatrix}, \quad D_1 = \begin{bmatrix} 1 & & & & \\ & k & & & \\ & & \ddots & & \\ & & & k^{n_1-2} & \\ & & & & k^{n_1-1} \end{bmatrix} \in \mathbb{C}^{n_1 \times n_1}.$$

Then $\inf_{D_1 \in \mathbb{C}^{n_1 \times n_1}} \overline{\sigma}(D_1 J_1 D_1^{-1}) = \lim_{k \to \infty} \overline{\sigma}(D_1 J_1 D_1^{-1}) = |\lambda|$. (Note that by Remark 10.2, the scaling matrix does not need to be Hermitian.) The conclusion follows by applying this result to each Jordan block.

That μ equals to the preceding upper-bound in this case is also equivalent to the fact that Lyapunov asymptotic stability and exponential stability are equivalent for discrete time systems. This is because $\rho(M) < 1$ (exponential stability of a discrete time system matrix M) implies for some nonsingular $D \in \mathbb{C}^{n \times n}$

$$\overline{\sigma}(DMD^{-1}) < 1 \quad \text{or} \quad (D^{-1})^* M^* D^* DMD^{-1} - I < 0,$$

which, in turn, is equivalent to the existence of a $P = D^*D > 0$ such that

$$M^* P M - P < 0$$

(Lyapunov asymptotic stability).

- $S = 0$, $F = 2$: This case was studied by Redheffer [1959].

- $S = 1$, $F = 1$: This is equivalent to a state-space characterization of the $\mathcal{H}_\infty$ norm of a discrete time transfer function.

- $S = 2$, $F = 0$: This is equivalent to the fact that for multidimensional systems (two dimensional, in fact), exponential stability is not equivalent to Lyapunov stability.

- $S = 0$, $F \geq 4$: For this case, the upper-bound is not always equal to μ. This is important, as these are the cases that arise most frequently in applications. Fortunately, the bound seems to be close to μ. The worst known example has a ratio of μ over the bound of about .85, and most systems are close to 1.

The preceding bounds are much more than just computational schemes. They are also theoretically rich and can unify a number of apparently different results in linear systems theory. There are several connections with Lyapunov asymptotic stability, two of which were hinted at previously, but there are further connections between the upper-bound scalings and solutions to Lyapunov and Riccati equations. Indeed, many major theorems in linear systems theory follow from the upper-bounds and from some results of linear fractional transformations. The lower bound can be viewed as a natural generalization of the maximum modulus theorem.

Of course, one of the most important uses of the upper-bound is as a computational scheme when combined with the lower bound. For reliable use of the μ theory, it is essential to have upper and lower bounds. Another important feature of the upper-bound is that it can be combined with H_∞ controller synthesis methods to yield an ad hoc μ-synthesis method. Note that the upper-bound when applied to transfer functions is simply a scaled H_∞ norm. This is exploited in the $D-K$ iteration procedure to perform approximate μ synthesis (Doyle [1982]), which will be briefly introduced in Section 10.4.

The upper and lower bounds of the structured singular value and the scaling matrix D can be computed using the MATLAB command

≫ **[bounds,rowd] = mu(M,blk)**

where the structure of the Δ is specified by a two-column matrix **blk**. for example, a

$$\Delta = \begin{bmatrix} \delta_1 I_2 & 0 & 0 & 0 & 0 & 0 \\ 0 & \delta_2 & 0 & 0 & 0 & 0 \\ 0 & 0 & \Delta_3 & 0 & 0 & 0 \\ 0 & 0 & 0 & \Delta_4 & 0 & 0 \\ 0 & 0 & 0 & 0 & \delta_5 I_3 & 0 \\ 0 & 0 & 0 & 0 & 0 & \Delta_6 \end{bmatrix}$$

$$\delta_1, \delta_2, \delta_5, \in \mathbb{C}, \quad \Delta_3 \in \mathbb{C}^{2\times 3}, \Delta_4 \in \mathbb{C}^{3\times 3}, \Delta_6 \in \mathbb{C}^{2\times 1}$$

can be specified by

$$\mathbf{blk} = \begin{bmatrix} 2 & 0 \\ 1 & 1 \\ 2 & 3 \\ 3 & 3 \\ 3 & 0 \\ 2 & 1 \end{bmatrix}.$$

Note that Δ_j is not required to be square. The outputs of the program include a 2×1 vector **bounds** containing the upper and lower bounds of $\mu_\Delta(M)$ and the row vector **rowd** containing the scaling D. The D matrix can be recovered by

≫ **[D_ℓ, D_r] = unwrapd(rowd, blk)**

where D_ℓ and D_r denote the left and right scaling matrices used in computing the upper-bound $\inf \overline{\sigma} \left(D_\ell M D_r^{-1} \right)$ when some full blocks are not necessarily square and they are equal if all full blocks are square.

10.2. Structured Singular Value

Example 10.2 Let

$$M = \begin{bmatrix} j & 2 & 2j & 0 & -1 & -1+3j & 2+3j \\ 3+j & 2-j & -1+j & 2+j & -1+j & 1 & -1+j \\ 3+j & j & 2+2j & -1+2j & 3-j & 3j & -1+j \\ -1+j & -1-j & j & 0 & 1-j & 2-j & 2+2j \\ 3 & j & 1+j & 3j & 1+j & 3j & -j \\ 1 & 3+2j & 2+2j & 3j & 1+2j & 2+j & -1+2j \\ 2+j & -1-j & -1 & 3+3j & 2+3j & 2j & 1-j \end{bmatrix}$$

and

$$\boldsymbol{\Delta} = \left\{ \Delta = \begin{bmatrix} \delta_1 I_2 & & & \\ & \delta_2 & & \\ & & \Delta_3 & \\ & & & \Delta_4 \end{bmatrix} : \delta_1, \delta_2 \in \mathbb{C}, \Delta_3 \in \mathbb{C}^{2\times 3}, \Delta_4 \in \mathbb{C}^{2\times 1} \right\}.$$

Then $\text{blk} = \begin{bmatrix} 2 & 0 \\ 1 & 1 \\ 2 & 3 \\ 2 & 1 \end{bmatrix}$ and the MATLAB program gives $\text{bounds} = \begin{bmatrix} 10.5955 & 10.5518 \end{bmatrix}$

and

$$D_\ell = \begin{bmatrix} D_1 & & & \\ & 0.7638 & & \\ & & 0.8809 I_3 & \\ & & & 1.0293 \end{bmatrix}$$

$$D_r = \begin{bmatrix} D_1 & & & \\ & 0.7638 & & \\ & & 0.8809 I_2 & \\ & & & 1.0293 I_2 \end{bmatrix}$$

where

$$D_1 = \begin{bmatrix} 1.0260 - 0.0657j & 0.2174 - 0.3471j \\ -0.0701 + 0.3871j & -0.4487 - 0.6953j \end{bmatrix}.$$

In fact, D_ℓ and D_r can be replaced by Hermitian matrices without changing the upper-bound by replacing D_1 with

$$\hat{D}_1 = \begin{bmatrix} 1.0992 & 0.0041 - 0.0591j \\ 0.0041 + 0.0591j & 0.9215 \end{bmatrix}$$

since $D_1 = U_1 \hat{D}_1$ and

$$U_1 = \begin{bmatrix} 0.9155 - 0.0713j & 0.2365 - 0.3177j \\ -0.1029 + 0.3824j & -0.5111 - 0.7629j \end{bmatrix}$$

is a unitary matrix.

10.2.3 Well-Posedness and Performance for Constant LFTs

Let M be a complex matrix partitioned as

$$M = \begin{bmatrix} M_{11} & M_{12} \\ M_{21} & M_{22} \end{bmatrix} \qquad (10.13)$$

and suppose there are two defined block structures, $\mathbf{\Delta}_1$ and $\mathbf{\Delta}_2$, which are compatible in size with M_{11} and M_{22}, respectively. Define a third structure $\mathbf{\Delta}$ as

$$\mathbf{\Delta} = \left\{ \begin{bmatrix} \Delta_1 & 0 \\ 0 & \Delta_2 \end{bmatrix} : \Delta_1 \in \mathbf{\Delta}_1, \Delta_2 \in \mathbf{\Delta}_2 \right\}. \qquad (10.14)$$

Now we may compute μ with respect to three structures. The notations we use to keep track of these computations are as follows: $\mu_1(\cdot)$ is with respect to $\mathbf{\Delta}_1$, $\mu_2(\cdot)$ is with respect to $\mathbf{\Delta}_2$, and $\mu_{\mathbf{\Delta}}(\cdot)$ is with respect to $\mathbf{\Delta}$. In view of these notations, $\mu_1(M_{11})$, $\mu_2(M_{22})$, and $\mu_{\mathbf{\Delta}}(M)$ all make sense, though, for instance, $\mu_1(M)$ does not.

This section is interested in following constant matrix problems:

- Determine whether the LFT $\mathcal{F}_\ell(M, \Delta_2)$ is well-defined for all $\Delta_2 \in \mathbf{\Delta}_2$ with $\overline{\sigma}(\Delta_2) \leq \beta \ (<\beta)$.

- If so, determine how "large" $\mathcal{F}_\ell(M, \Delta_2)$ can get for this norm-bounded set of perturbations.

Let $\Delta_2 \in \mathbf{\Delta}_2$. Recall that $\mathcal{F}_\ell(M, \Delta_2)$ is *well-defined* if $I - M_{22}\Delta_2$ is invertible. The first theorem is nothing more than a restatement of the definition of μ.

Theorem 10.5 *The linear fractional transformation $\mathcal{F}_\ell(M, \Delta_2)$ is well-defined*

(a) *for all $\Delta_2 \in \mathbf{B\Delta}_2$ if and only if $\mu_2(M_{22}) < 1$.*

(b) *for all $\Delta_2 \in \mathbf{B°\Delta}_2$ if and only if $\mu_2(M_{22}) \leq 1$.*

As the "perturbation" Δ_2 deviates from zero, the matrix $\mathcal{F}_\ell(M, \Delta_2)$ deviates from M_{11}. The range of values that $\mu_1(\mathcal{F}_\ell(M, \Delta_2))$ takes on is intimately related to $\mu_{\mathbf{\Delta}}(M)$, as shown in the following theorem:

10.2. Structured Singular Value

Theorem 10.6 *(main loop theorem)* *The following are equivalent:*

$$\mu_{\mathbf{\Delta}}(M) < 1 \quad \Longleftrightarrow \quad \begin{cases} \mu_2(M_{22}) < 1, \quad \text{and} \\ \\ \max_{\Delta_2 \in \mathbf{B}\mathbf{\Delta}_2} \mu_1(\mathcal{F}_\ell(M, \Delta_2)) < 1 \end{cases}$$

$$\mu_{\mathbf{\Delta}}(M) \leq 1 \quad \Longleftrightarrow \quad \begin{cases} \mu_2(M_{22}) \leq 1, \quad \text{and} \\ \\ \sup_{\Delta_2 \in \mathbf{B}^{\mathbf{O}}\mathbf{\Delta}_2} \mu_1(\mathcal{F}_\ell(M, \Delta_2)) \leq 1. \end{cases}$$

Proof. We shall only prove the first part of the equivalence. The proof for the second part is similar.

$\Leftarrow$ Let $\Delta_i \in \mathbf{\Delta}_i$ be given, with $\bar{\sigma}(\Delta_i) \leq 1$, and define $\Delta = \text{diag}\,[\Delta_1, \Delta_2]$. Obviously $\Delta \in \mathbf{\Delta}$. Now

$$\det(I - M\Delta) = \det\begin{bmatrix} I - M_{11}\Delta_1 & -M_{12}\Delta_2 \\ -M_{21}\Delta_1 & I - M_{22}\Delta_2 \end{bmatrix}. \tag{10.15}$$

By hypothesis $I - M_{22}\Delta_2$ is invertible, and hence $\det(I - M\Delta)$ becomes

$$\det(I - M_{22}\Delta_2)\det\left(I - M_{11}\Delta_1 - M_{12}\Delta_2(I - M_{22}\Delta_2)^{-1}M_{21}\Delta_1\right).$$

Collecting the Δ_1 terms leaves

$$\det(I - M\Delta) = \det(I - M_{22}\Delta_2)\det(I - \mathcal{F}_\ell(M, \Delta_2)\Delta_1). \tag{10.16}$$

But $\mu_1(\mathcal{F}_\ell(M, \Delta_2)) < 1$ and $\Delta_1 \in \mathbf{B}\mathbf{\Delta}_1$, so $I - \mathcal{F}_\ell(M, \Delta_2)\Delta_1$ must be nonsingular. Therefore, $I - M\Delta$ is nonsingular and, by definition, $\mu_{\mathbf{\Delta}}(M) < 1$.

$\Rightarrow$ Basically, the argument above is reversed. Again let $\Delta_1 \in \mathbf{B}\mathbf{\Delta}_1$ and $\Delta_2 \in \mathbf{B}\mathbf{\Delta}_2$ be given, and define $\Delta = \text{diag}\,[\Delta_1, \Delta_2]$. Then $\Delta \in \mathbf{B}\mathbf{\Delta}$ and, by hypothesis, $\det(I - M\Delta) \neq 0$. It is easy to verify from the definition of μ that (always)

$$\mu(M) \geq \max\{\mu_1(M_{11}), \mu_2(M_{22})\}.$$

We can see that $\mu_2(M_{22}) < 1$, which gives that $I - M_{22}\Delta_2$ is also nonsingular. Therefore, the expression in equation (10.16) is valid, giving

$$\det(I - M_{22}\Delta_2)\det(I - \mathcal{F}_\ell(M, \Delta_2)\Delta_1) = \det(I - M\Delta) \neq 0.$$

Obviously, $I - \mathcal{F}_\ell(M, \Delta_2)\Delta_1$ is nonsingular for all $\Delta_i \in \mathbf{B}\mathbf{\Delta}_i$, which indicates that the claim is true. □

Remark 10.3 This theorem forms the basis for all uses of μ in linear system robustness analysis, whether from a state-space, frequency domain, or Lyapunov approach. ◇

The role of the block structure $\boldsymbol{\Delta}_2$ in the main loop theorem is clear — it is the structure that the perturbations come from; however, the role of the perturbation structure $\boldsymbol{\Delta}_1$ is often misunderstood. Note that $\mu_1(\cdot)$ appears on the right-hand side of the theorem, so that the set $\boldsymbol{\Delta}_1$ defines what particular property of $\mathcal{F}_\ell(M, \Delta_2)$ is considered. As an example, consider the theorem applied with the two simple block structures considered right after Lemma 10.1. Define $\boldsymbol{\Delta}_1 := \{\delta_1 I_n : \delta_1 \in \mathbb{C}\}$. Hence, for $A \in \mathbb{C}^{n \times n}$, $\mu_1(A) = \rho(A)$. Likewise, define $\boldsymbol{\Delta}_2 = \mathbb{C}^{m \times m}$; then for $D \in \mathbb{C}^{m \times m}$, $\mu_2(D) = \bar{\sigma}(D)$. Now, let $\boldsymbol{\Delta}$ be the diagonal augmentation of these two sets, namely

$$\boldsymbol{\Delta} := \left\{ \begin{bmatrix} \delta_1 I_n & 0_{n \times m} \\ 0_{m \times n} & \Delta_2 \end{bmatrix} : \delta_1 \in \mathbb{C}, \Delta_2 \in \mathbb{C}^{m \times m} \right\} \subset \mathbb{C}^{(n+m) \times (n+m)}.$$

Let $A \in \mathbb{C}^{n \times n}, B \in \mathbb{C}^{n \times m}, C \in \mathbb{C}^{m \times n}$, and $D \in \mathbb{C}^{m \times m}$ be given, and interpret them as the state-space model of a discrete time system

$$\begin{aligned} x_{k+1} &= Ax_k + Bu_k \\ y_k &= Cx_k + Du_k. \end{aligned}$$

Let $M \in \mathbb{C}^{(n+m) \times (n+m)}$ be the block state-space matrix of the system

$$M = \begin{bmatrix} A & B \\ C & D \end{bmatrix}.$$

Applying the theorem with these data gives that the following are equivalent:

- The spectral radius of A satisfies $\rho(A) < 1$, and

$$\max_{\substack{\delta_1 \in \mathbb{C} \\ |\delta_1| \leq 1}} \bar{\sigma}\left(D + C\delta_1 (I - A\delta_1)^{-1} B\right) < 1. \tag{10.17}$$

- The maximum singular value of D satisfies $\bar{\sigma}(D) < 1$, and

$$\max_{\substack{\Delta_2 \in \mathbb{C}^{m \times m} \\ \bar{\sigma}(\Delta_2) \leq 1}} \rho\left(A + B\Delta_2 (I - D\Delta_2)^{-1} C\right) < 1. \tag{10.18}$$

- The structured singular value of M satisfies

$$\mu_{\boldsymbol{\Delta}}(M) < 1. \tag{10.19}$$

The first condition is recognized by two things: The system is stable, and the $\|\cdot\|_\infty$ norm on the transfer function from u to y is less than 1 (by replacing δ_1 with $\frac{1}{z}$):

$$\|G\|_\infty := \max_{\substack{z \in \mathbb{C} \\ |z| \geq 1}} \bar{\sigma}\left(D + C(zI - A)^{-1} B\right) = \max_{\substack{\delta_1 \in \mathbb{C} \\ |\delta_1| \leq 1}} \bar{\sigma}\left(D + C\delta_1(I - A\delta_1)^{-1} B\right).$$

10.2. Structured Singular Value

The second condition implies that $(I - D\Delta_2)^{-1}$ is well defined for all $\bar{\sigma}(\Delta_2) \leq 1$ and that a robust stability result holds for the uncertain difference equation

$$x_{k+1} = \left(A + B\Delta_2\left(I - D\Delta_2\right)^{-1} C\right) x_k$$

where Δ_2 is any element in $\mathbb{C}^{m \times m}$ with $\bar{\sigma}(\Delta_2) \leq 1$, but otherwise unknown.

This equivalence between the small gain condition, $\|G\|_\infty < 1$, and the stability robustness of the uncertain difference equation is well-known. This is the small gain theorem, in its necessary and sufficient form for linear, time invariant systems with one of the components norm bounded, but otherwise unknown. What is important to note is that both of these conditions are equivalent to a condition involving the structured singular value of the state-space matrix. Already we have seen that special cases of μ are the spectral radius and the maximum singular value. Here we see that other important linear system properties — namely, robust stability and input-output gain — are also related to a particular case of the structured singular value.

Example 10.3 Let M, Δ_1, and Δ_2 be defined as in the beginning of this section. Now suppose $\mu_2(M_{22}) < 1$. Find

$$\max_{\Delta_2 \in \mathbf{B\Delta_2}} \mu_1 \left(\mathcal{F}_\ell (M, \Delta_2)\right).$$

This can be done iteratively as follows:

$$\max_{\Delta_2 \in \mathbf{B\Delta_2}} \mu_1 \left(\mathcal{F}_\ell (M, \Delta_2)\right) = \alpha$$

$$\iff \max_{\Delta_2 \in \mathbf{B\Delta_2}} \mu_1 \left(\mathcal{F}_\ell \left(\begin{bmatrix} M_{11}/\alpha & M_{12}/\alpha \\ M_{21} & M_{22} \end{bmatrix}, \Delta_2\right)\right) = 1$$

$$\iff \mu_\Delta \left(\begin{bmatrix} M_{11}/\alpha & M_{12}/\alpha \\ M_{21} & M_{22} \end{bmatrix}\right) = 1.$$

Hence

$$\max_{\Delta_2 \in \mathbf{B\Delta_2}} \mu_1 \left(\mathcal{F}_\ell (M, \Delta_2)\right) = \left\{\alpha : \mu_\Delta \left(\begin{bmatrix} M_{11}/\alpha & M_{12}/\alpha \\ M_{21} & M_{22} \end{bmatrix}\right) = 1\right\}.$$

For example, let $\Delta_1 = \delta I_2$, $\Delta_2 \in \mathbb{C}^{2 \times 2}$:

$$A = \begin{bmatrix} 0.1 & 0.2 \\ 1 & 0 \end{bmatrix}, \quad B = \begin{bmatrix} 1 & 0 \\ 1 & 1 \end{bmatrix}, \quad C = \begin{bmatrix} 1 & 2 \\ 1 & 3 \end{bmatrix}, \quad D = \begin{bmatrix} 0.5 & 0 \\ 0 & 0.8 \end{bmatrix}.$$

Find

$$\alpha_{\max} = \sup_{\bar{\sigma}(\Delta_2) \leq 1} \rho(A + B\Delta_2(I - D\Delta_2)^{-1}C).$$

Define $\Delta = \begin{bmatrix} \delta I_2 & \\ & \Delta_2 \end{bmatrix}$. Then a bisection search can be done to find

$$\alpha_{\max} = \left\{\alpha : \mu_\Delta\left(\begin{bmatrix} A/\alpha & B/\alpha \\ C & D \end{bmatrix}\right) = 1\right\} = 21.77.$$

Related MATLAB Commands: unwrapp, muunwrap, dypert, sisorat

10.3 Structured Robust Stability and Performance

10.3.1 Robust Stability

The most well-known use of μ as a robustness analysis tool is in the frequency domain. Suppose $G(s)$ is a stable, real rational, multi-input, multioutput transfer function of a linear system. For clarity, assume G has q_1 inputs and p_1 outputs. Let Δ be a block structure, as in equation (10.1), and assume that the dimensions are such that $\Delta \subset \mathbb{C}^{q_1 \times p_1}$. We want to consider feedback perturbations to G that are themselves dynamical systems with the block diagonal structure of the set Δ.

Let $\mathcal{M}(\Delta)$ denote the set of all block diagonal and stable rational transfer functions that have block structures such as Δ.

$$\mathcal{M}(\Delta) := \{\Delta(\cdot) \in \mathcal{RH}_\infty : \Delta(s_o) \in \Delta \text{ for all } s_o \in \overline{\mathbb{C}}_+\}$$

Theorem 10.7 *Let $\beta > 0$. The loop shown below is well-posed and internally stable for all $\Delta(\cdot) \in \mathcal{M}(\Delta)$ with $\|\Delta\|_\infty < \frac{1}{\beta}$ if and only if*

$$\sup_{\omega \in \mathbb{R}} \mu_\Delta(G(j\omega)) \leq \beta.$$

Proof. ($\Longleftarrow$) Suppose $\sup_{s \in \overline{\mathbb{C}}_+} \mu_\Delta(G(s)) \leq \beta$. Then $\det(I - G(s)\Delta(s)) \neq 0$ for all $s \in \overline{\mathbb{C}}_+ \cup \{\infty\}$ whenever $\|\Delta\|_\infty < 1/\beta$ (i.e., the system is robustly stable). Now it is sufficient to show that

$$\sup_{s \in \overline{\mathbb{C}}_+} \mu_\Delta(G(s)) = \sup_{\omega \in \mathbb{R}} \mu_\Delta(G(j\omega)).$$

10.3. Structured Robust Stability and Performance

It is clear that

$$\sup_{s \in \overline{\mathbb{C}}_+} \mu_\Delta(G(s)) = \sup_{s \in \mathbb{C}_+} \mu_\Delta(G(s)) \geq \sup_{\omega} \mu_\Delta(G(j\omega)).$$

Now suppose $\sup_{s \in \mathbb{C}_+} \mu_\Delta(G(s)) > \beta$; then by the definition of μ, there is an $s_o \in \overline{\mathbb{C}}_+ \cup \{\infty\}$ and a complex structured Δ such that $\overline{\sigma}(\Delta) < 1/\beta$ and $\det(I - G(s_o)\Delta) = 0$. This implies that there is a $0 \leq \hat{\omega} \leq \infty$ and $0 < \alpha \leq 1$ such that $\det(I - G(j\hat{\omega})\alpha\Delta) = 0$. This, in turn, implies that $\mu_\Delta(G(j\hat{\omega})) > \beta$ since $\overline{\sigma}(\alpha\Delta) < 1/\beta$. In other words, $\sup_{s \in \mathbb{C}_+} \mu_\Delta(G(s)) \leq \sup_{\omega} \mu_\Delta(G(j\omega))$. The proof is complete.

($\Longrightarrow$) Suppose $\sup_{\omega \in \mathbb{R}} \mu_\Delta(G(j\omega)) > \beta$. Then there is a $0 < \omega_o < \infty$ such that $\mu_\Delta(G(j\omega_o)) > \beta$. By Remark 10.1, there is a complex $\Delta_c \in \mathbf{\Delta}$ that each full block has rank 1 and $\overline{\sigma}(\Delta_c) < 1/\beta$ such that $I - G(j\omega_o)\Delta_c$ is singular. Next, using the same construction used in the proof of the small gain theorem (Theorem 8.1), one can find a rational $\Delta(s)$ such that $\|\Delta(s)\|_\infty = \overline{\sigma}(\Delta_c) < 1/\beta$, $\Delta(j\omega_o) = \Delta_c$, and $\Delta(s)$ destabilizes the system. $\square$

Hence, the peak value on the μ plot of the frequency response determines the size of perturbations that the loop is robustly stable against.

Remark 10.4 The internal stability with a closed ball of uncertainties is more complicated. The following example is shown in Tits and Fan [1995]. Consider

$$G(s) = \frac{1}{s+1} \begin{bmatrix} 0 & -1 \\ 1 & 0 \end{bmatrix}$$

and $\Delta = \delta(s)I_2$. Then

$$\sup_{\omega \in \mathbb{R}} \mu_\Delta(G(j\omega)) = \sup_{\omega \in \mathbb{R}} \frac{1}{|j\omega + 1|} = \mu_\Delta(G(j0)) = 1.$$

On the other hand, $\mu_\Delta(G(s)) < 1$ for all $s \neq 0, s \in \overline{\mathbb{C}}_+$, and the only matrices in the form of $\Gamma = \gamma I_2$ with $|\gamma| \leq 1$ for which

$$\det(I - G(0)\Gamma) = 0$$

are the *complex* matrices $\pm jI_2$. Thus, clearly, $(I - G(s)\Delta(s))^{-1} \in \mathcal{RH}_\infty$ for all real rational $\Delta(s) = \delta(s)I_2$ with $\|\delta\|_\infty \leq 1$ since $\Delta(0)$ must be real. This shows that $\sup_{\omega \in \mathbb{R}} \mu_\Delta(G(j\omega)) < 1$ is not necessary for $(I - G(s)\Delta(s))^{-1} \in \mathcal{RH}_\infty$ with the closed ball of structured uncertainty $\|\Delta\|_\infty \leq 1$. Similar examples with no repeated blocks are generated by setting $G(s) = \frac{1}{s+1} M$, where M is any real matrix with $\mu_\Delta(M) = 1$ for which there is no real $\Delta \in \mathbf{\Delta}$ with $\overline{\sigma}(\Delta) = 1$ such that $\det(I - M\Delta) = 0$. For example, let

$$M = \begin{bmatrix} 0 & \beta \\ \gamma & \alpha \\ \gamma & -\alpha \end{bmatrix} \begin{bmatrix} -\beta & \alpha & \alpha \\ 0 & -\gamma & \gamma \end{bmatrix}, \quad \Delta = \left\{ \begin{bmatrix} \delta_1 & & \\ & \delta_2 & \\ & & \delta_3 \end{bmatrix}, \delta_i \in \mathbb{C} \right\}$$

with $\gamma^2 = \frac{1}{2}$ and $\beta^2 + 2\alpha^2 = 1$. Then it is shown in Packard and Doyle [1993] that $\mu_{\boldsymbol{\Delta}}(M) = 1$ and all $\Delta \in \boldsymbol{\Delta}$ with $\overline{\sigma}(\Delta) = 1$ that satisfy $\det(I - M\Delta) = 0$ must be complex. ◇

Remark 10.5 Let $\Delta \in \mathcal{RH}_\infty$ be a structured uncertainty and

$$G(s) = \begin{bmatrix} G_{11}(s) & G_{12}(s) \\ G_{21}(s) & G_{22}(s) \end{bmatrix} \in \mathcal{RH}_\infty.$$

Then $\mathcal{F}_u(G, \Delta) \in \mathcal{RH}_\infty$ does not necessarily imply $(I - G_{11}\Delta)^{-1} \in \mathcal{RH}_\infty$ whether Δ is in an open ball or is in a closed ball. For example, consider

$$G(s) = \begin{bmatrix} \frac{1}{s+1} & 0 & 1 \\ 0 & \frac{10}{s+1} & 0 \\ \hline 1 & 0 & 0 \end{bmatrix}$$

and $\Delta = \begin{bmatrix} \delta_1 & \\ & \delta_2 \end{bmatrix}$ with $\|\Delta\|_\infty < 1$. Then $\mathcal{F}_u(G, \Delta) = \dfrac{1}{1 - \delta_1 \frac{1}{s+1}} \in \mathcal{RH}_\infty$ for all admissible Δ ($\|\Delta\|_\infty < 1$) but $(I - G_{11}\Delta)^{-1} \in \mathcal{RH}_\infty$ is true only for $\|\Delta\|_\infty < 0.1$. ◇

10.3.2 Robust Performance

Often, stability is not the only property of a closed-loop system that must be robust to perturbations. Typically, there are exogenous disturbances acting on the system (wind gusts, sensor noise) that result in tracking and regulation errors. Under perturbation, the effect that these disturbances have on error signals can greatly increase. In most cases, long before the onset of instability, the closed-loop performance will degrade to the point of unacceptability (hence the need for a "robust performance" test). Such a test will indicate the worst-case level of performance degradation associated with a given level of perturbations.

Assume G_p is a stable, real-rational, proper transfer function with $q_1 + q_2$ inputs and $p_1 + p_2$ outputs. Partition G_p in the obvious manner

$$G_p(s) = \begin{bmatrix} G_{11} & G_{12} \\ G_{21} & G_{22} \end{bmatrix}$$

so that G_{11} has q_1 inputs and p_1 outputs, and so on. Let $\boldsymbol{\Delta} \subset \mathbb{C}^{q_1 \times p_1}$ be a block structure, as in equation (10.1). Define an augmented block structure:

$$\boldsymbol{\Delta}_P := \left\{ \begin{bmatrix} \Delta & 0 \\ 0 & \Delta_f \end{bmatrix} : \Delta \in \boldsymbol{\Delta}, \Delta_f \in \mathbb{C}^{q_2 \times p_2} \right\}.$$

The setup is to address theoretically the robust performance questions about the following loop:

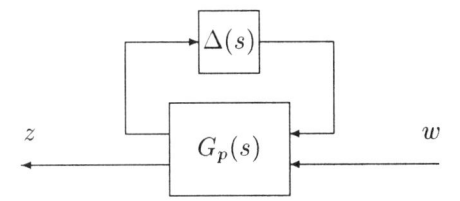

10.3. Structured Robust Stability and Performance

The transfer function from w to z is denoted by $\mathcal{F}_u(G_p, \Delta)$.

Theorem 10.8 *Let $\beta > 0$. For all $\Delta(s) \in \mathcal{M}(\mathbf{\Delta})$ with $\|\Delta\|_\infty < \frac{1}{\beta}$, the loop shown above is well-posed, internally stable, and $\|\mathcal{F}_u(G_p, \Delta)\|_\infty \leq \beta$ if and only if*

$$\sup_{\omega \in \mathbb{R}} \mu_{\mathbf{\Delta}_P}(G_p(j\omega)) \leq \beta.$$

Note that by internal stability, $\sup_{\omega \in \mathbb{R}} \mu_{\mathbf{\Delta}}(G_{11}(j\omega)) \leq \beta$, then the proof of this theorem is exactly along the lines of the earlier proof for Theorem 10.7, but also appeals to Theorem 10.6. This is a remarkably useful theorem. It says that a robust performance problem is equivalent to a robust stability problem with augmented uncertainty Δ, as shown in Figure 10.5.

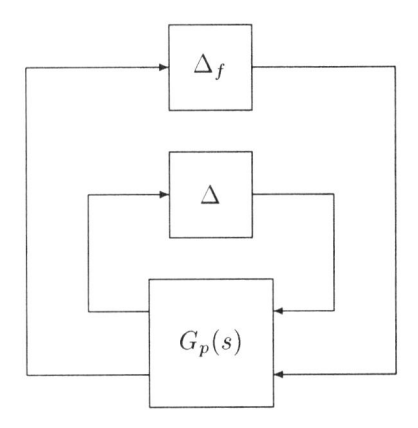

Figure 10.5: Robust performance vs robust stability

Example 10.4 We shall consider again the HIMAT problem from Example 9.1. Use the SIMULINK block diagram in Example 9.1 and run the following commands to get an interconnection model $\hat{G}$, an $\mathcal{H}_\infty$ stabilizing controller K and a closed-loop transfer matrix $G_p(s) = \mathcal{F}_\ell(\hat{G}, K)$. (Do not bother to figure out how **hinfsyn** works; it will be considered in detail in Chapter 14.)

≫ [A, B, C, D] = linmod('aircraft')

≫ $\hat{G}$ = pck(A, B, C, D);

≫ [K, G_p, γ] = hinfsyn($\hat{G}$, 2, 2, 0, 10, 0.001, 2);

which gives $\gamma = 1.8612 = \|G_p\|_\infty$, a stabilizing controller K, and a closed loop transfer matrix G_p:

$$\begin{bmatrix} z_1 \\ z_2 \\ \hdashline e_1 \\ e_2 \end{bmatrix} = G_p(s) \begin{bmatrix} p_1 \\ p_2 \\ \hdashline d_1 \\ d_2 \\ n_1 \\ n_2 \end{bmatrix}, \quad G_p(s) = \begin{bmatrix} G_{p11} & G_{p12} \\ G_{p21} & G_{p22} \end{bmatrix}.$$

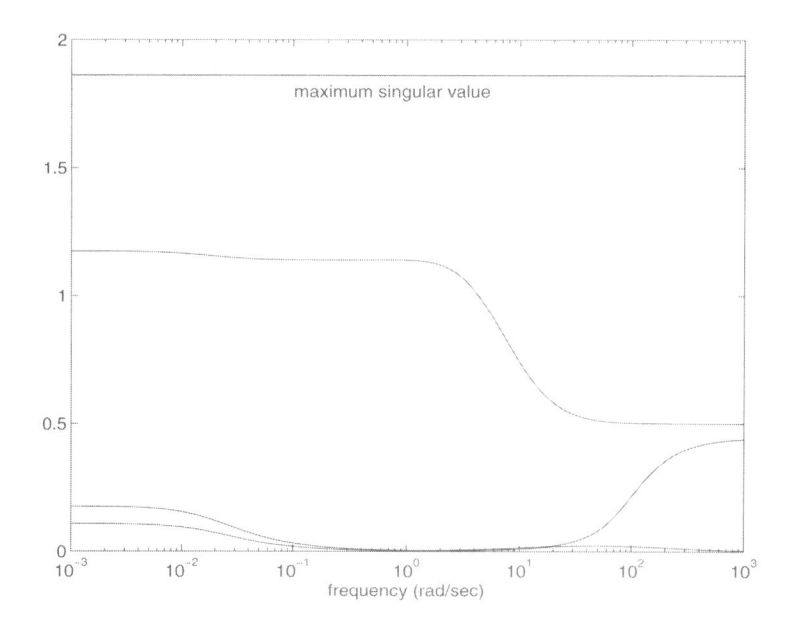

Figure 10.6: Singular values of $G_p(j\omega)$

Now generate the singular value frequency responses of G_p:

≫ w=logspace(-3,3,300);

≫ Gpf = frsp(G_p, w);   % Gpf is the frequency response of G_p;

≫ [u, s, v] = vsvd(Gpf);

10.3. Structured Robust Stability and Performance

≫ vplot('liv, m', s)

The singular value frequency responses of G_p are shown in Figure 10.6. To test the robust stability, we need to compute $\|G_{p11}\|_\infty$:

≫ $\mathbf{G_{p11}}$ = sel($\mathbf{G_p}$, 1 : 2, 1 : 2);

≫ norm_of_$\mathbf{G_{p11}}$ = hinfnorm($\mathbf{G_{p11}}$, 0.001);

which gives $\|G_{P11}\|_\infty = 0.933 < 1$. So the system is robustly stable. To check the robust performance, we shall compute the $\mu_{\Delta_P}(G_p(j\omega))$ for each frequency with

$$\Delta_P = \begin{bmatrix} \Delta & \\ & \Delta_f \end{bmatrix}, \quad \Delta \in \mathbb{C}^{2\times 2}, \quad \Delta_f \in \mathbb{C}^{4\times 2}.$$

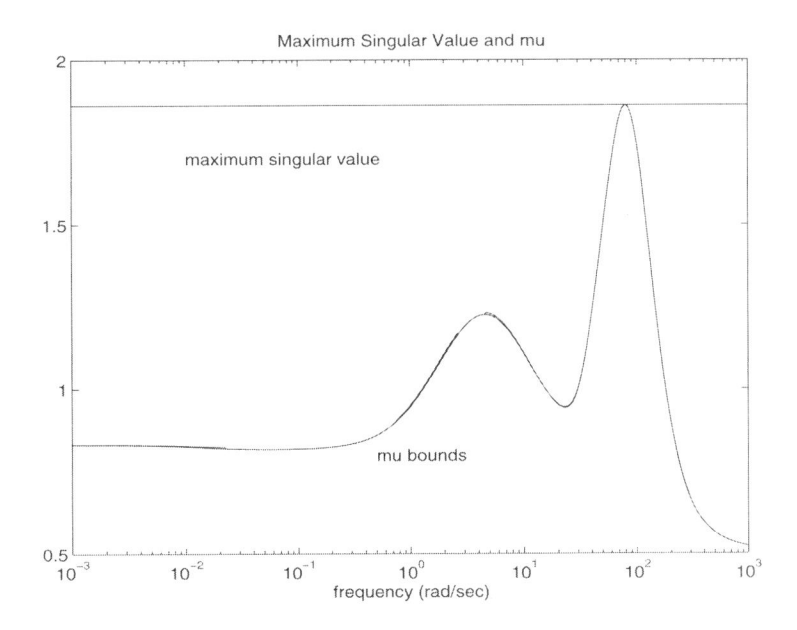

Figure 10.7: $\mu_{\Delta_P}(G_p(j\omega))$ and $\overline{\sigma}(G_p(j\omega))$

≫ blk=[2,2;4,2];

≫ [bnds,dvec,sens,pvec]=mu(Gpf,blk);

≫ vplot('liv, m', vnorm(Gpf), bnds)

≫ title('Maximum Singular Value and mu')

≫ xlabel('frequency(rad/sec)')

≫ **text(0.01, 1.7,' maximum singular value')**

≫ **text(0.5, 0.8,' mu bounds')**

The structured singular value $\mu_{\Delta_P}(G_p(j\omega))$ and $\bar{\sigma}(G_p(j\omega))$ are shown in Figure 10.7. It is clear that the robust performance is not satisfied. Note that

$$\max_{\|\Delta\|_\infty \leq 1} \|\mathcal{F}_u(G_p, \Delta)\|_\infty \leq \gamma \iff \sup_\omega \mu_{\Delta_P}\left(\begin{bmatrix} G_{p11} & G_{p12} \\ G_{p21}/\gamma & G_{p22}/\gamma \end{bmatrix}\right) \leq 1.$$

Using a bisection algorithm, we can also find the worst performance:

$$\max_{\|\Delta\|_\infty \leq 1} \|\mathcal{F}_u(G_p, \Delta)\|_\infty = 12.7824.$$

10.3.3 Two-Block μ: Robust Performance Revisited

Suppose that the uncertainty block is given by

$$\Delta = \begin{bmatrix} \Delta_1 & \\ & \Delta_2 \end{bmatrix} \in \mathcal{RH}_\infty$$

with $\|\Delta\|_\infty < 1$ and that the interconnection model G is given by

$$G(s) = \begin{bmatrix} G_{11}(s) & G_{12}(s) \\ G_{21}(s) & G_{22}(s) \end{bmatrix} \in \mathcal{RH}_\infty.$$

Then the closed-loop system is well-posed and internally stable iff $\sup_\omega \mu_\Delta(G(j\omega)) \leq 1$. Let

$$D_\omega = \begin{bmatrix} d_\omega I & \\ & I \end{bmatrix}, \quad d_\omega \in \mathbb{R}_+.$$

Then

$$D_\omega G(j\omega) D_\omega^{-1} = \begin{bmatrix} G_{11}(j\omega) & d_\omega G_{12}(j\omega) \\ \frac{1}{d_\omega} G_{21}(j\omega) & G_{22}(j\omega) \end{bmatrix}.$$

Hence, by Theorem 10.4, at each frequency ω

$$\mu_\Delta(G(j\omega)) = \inf_{d_\omega \in \mathbb{R}_+} \bar{\sigma}\left(\begin{bmatrix} G_{11}(j\omega) & d_\omega G_{12}(j\omega) \\ \frac{1}{d_\omega} G_{21}(j\omega) & G_{22}(j\omega) \end{bmatrix}\right). \quad (10.20)$$

Since the minimization is convex in $\log d_\omega$ (see, Doyle [1982]), the optimal d_ω can be found by a search; however, two approximations to d_ω can be obtained easily by approximating the right-hand side of equation (10.20):

10.3. Structured Robust Stability and Performance

(1) Note that

$$\mu_\Delta(G(j\omega)) \leq \inf_{d_\omega \in \mathbb{R}_+} \overline{\sigma}\left(\begin{bmatrix} \|G_{11}(j\omega)\| & d_\omega \|G_{12}(j\omega)\| \\ \frac{1}{d_\omega}\|G_{21}(j\omega)\| & \|G_{22}(j\omega)\| \end{bmatrix}\right)$$

$$\leq \sqrt{\inf_{d_\omega \in \mathbb{R}_+}\left(\|G_{11}(j\omega)\|^2 + d_\omega^2\|G_{12}(j\omega)\|^2 + \frac{1}{d_\omega^2}\|G_{21}(j\omega)\|^2 + \|G_{22}(j\omega)\|^2\right)}$$

$$= \sqrt{\|G_{11}(j\omega)\|^2 + \|G_{22}(j\omega)\|^2 + 2\|G_{12}(j\omega)\|\,\|G_{21}(j\omega)\|}$$

with the minimizing d_ω given by

$$\hat{d}_\omega = \begin{cases} \sqrt{\frac{\|G_{21}(j\omega)\|}{\|G_{12}(j\omega)\|}} & \text{if } G_{12} \neq 0 \ \& \ G_{21} \neq 0, \\ 0 & \text{if } G_{21} = 0, \\ \infty & \text{if } G_{12} = 0. \end{cases} \quad (10.21)$$

(2) Alternative approximation can be obtained by using the Frobenius norm:

$$\mu_\Delta(G(j\omega)) \leq \inf_{d_\omega \in \mathbb{R}_+} \left\| \begin{bmatrix} G_{11}(j\omega) & d_\omega G_{12}(j\omega) \\ \frac{1}{d_\omega} G_{21}(j\omega) & G_{22}(j\omega) \end{bmatrix} \right\|_F$$

$$= \sqrt{\inf_{d_\omega \in \mathbb{R}_+}\left(\|G_{11}(j\omega)\|_F^2 + d_\omega^2\|G_{12}(j\omega)\|_F^2 + \frac{1}{d_\omega^2}\|G_{21}(j\omega)\|_F^2 + \|G_{22}(j\omega)\|_F^2\right)}$$

$$= \sqrt{\|G_{11}(j\omega)\|_F^2 + \|G_{22}(j\omega)\|_F^2 + 2\|G_{12}(j\omega)\|_F\,\|G_{21}(j\omega)\|_F}$$

with the minimizing d_ω given by

$$\tilde{d}_\omega = \begin{cases} \sqrt{\frac{\|G_{21}(j\omega)\|_F}{\|G_{12}(j\omega)\|_F}} & \text{if } G_{12} \neq 0 \ \& \ G_{21} \neq 0, \\ 0 & \text{if } G_{21} = 0, \\ \infty & \text{if } G_{12} = 0. \end{cases} \quad (10.22)$$

It can be shown that the approximations for the scalar d_ω obtained previously are exact for a 2×2 matrix G. For higher dimensional G, the approximations for d_ω are still reasonably good. Hence an approximation of μ can be obtained as

$$\mu_\Delta(G(j\omega)) \leq \overline{\sigma}\left(\begin{bmatrix} G_{11}(j\omega) & \hat{d}_\omega G_{12}(j\omega) \\ \frac{1}{\hat{d}_\omega} G_{21}(j\omega) & G_{22}(j\omega) \end{bmatrix}\right) \quad (10.23)$$

or, alternatively, as

$$\mu_\Delta(G(j\omega)) \leq \bar{\sigma}\left(\begin{bmatrix} G_{11}(j\omega) & \tilde{d}_\omega G_{12}(j\omega) \\ \frac{1}{\tilde{d}_\omega} G_{21}(j\omega) & G_{22}(j\omega) \end{bmatrix}\right). \tag{10.24}$$

We can now see how these approximated μ tests are compared with the sufficient conditions obtained in Chapter 8.

Example 10.5 Consider again the robust performance problem of a system with output multiplicative uncertainty in Chapter 8 (see Figure 8.10):

$$P_\Delta = (I + W_1 \Delta W_2)P, \quad \|\Delta\|_\infty < 1.$$

Then it is easy to show that the problem can be put in the general framework by selecting

$$G(s) = \begin{bmatrix} -W_2 T_o W_1 & -W_2 T_o W_d \\ W_e S_o W_1 & W_e S_o W_d \end{bmatrix}$$

and that the robust performance condition is satisfied if and only if

$$\|W_2 T_o W_1\|_\infty \leq 1 \tag{10.25}$$

and

$$\|\mathcal{F}_u(G, \Delta)\|_\infty \leq 1 \tag{10.26}$$

for all $\Delta \in \mathcal{RH}_\infty$ with $\|\Delta\|_\infty < 1$. But equations (10.25) and (10.26) are satisfied iff for each frequency ω

$$\mu_\Delta(G(j\omega)) = \inf_{d_\omega \in \mathbb{R}_+} \bar{\sigma}\left(\begin{bmatrix} -W_2 T_o W_1 & -d_\omega W_2 T_o W_d \\ \frac{1}{d_\omega} W_e S_o W_1 & W_e S_o W_d \end{bmatrix}\right) \leq 1.$$

Note that, in contrast to the sufficient condition obtained in Chapter 8, this condition is an exact test for robust performance. To compare the μ test with the criteria obtained in Chapter 8, some upper-bounds for μ can be derived. Let

$$d_\omega = \sqrt{\frac{\|W_e S_o W_1\|}{\|W_2 T_o W_d\|}}.$$

Then, using the first approximation for μ, we get

$$\begin{aligned} \mu_\Delta(G(j\omega)) &\leq \sqrt{\|W_2 T_o W_1\|^2 + \|W_e S_o W_d\|^2 + 2\|W_2 T_o W_d\|\|W_e S_o W_1\|} \\ &\leq \sqrt{\|W_2 T_o W_1\|^2 + \|W_e S_o W_d\|^2 + 2\kappa(W_1^{-1} W_d)\|W_2 T_o W_1\|\|W_e S_o W_d\|} \\ &\leq \|W_2 T_o W_1\| + \kappa(W_1^{-1} W_d)\|W_e S_o W_d\| \end{aligned}$$

10.3. Structured Robust Stability and Performance

where W_1 is assumed to be invertible in the last two inequalities. The last term is exactly the sufficient robust performance criteria obtained in Chapter 8. It is clear that any term preceding the last forms a tighter test since $\kappa(W_1^{-1}W_d) \geq 1$. Yet another alternative sufficient test can be obtained from the preceding sequence of inequalities:

$$\mu_\Delta(G(j\omega)) \leq \sqrt{\kappa(W_1^{-1}W_d)}(\|W_2T_oW_1\| + \|W_eS_oW_d\|).$$

Note that this sufficient condition is not easy to get from the approach taken in Chapter 8 and is potentially less conservative than the bounds derived there.

Next we consider the skewed specification problem, but first the following lemma is needed in the sequel.

Lemma 10.9 *Suppose* $\bar{\sigma} = \sigma_1 \geq \sigma_2 \geq \ldots \geq \sigma_m = \underline{\sigma} > 0$, *then*

$$\inf_{d \in \mathbb{R}_+} \max_i \left\{ (d\sigma_i)^2 + \frac{1}{(d\sigma_i)^2} \right\} = \frac{\bar{\sigma}}{\underline{\sigma}} + \frac{\underline{\sigma}}{\bar{\sigma}}.$$

Proof. Consider a function $y = x + 1/x$; then y is a convex function and the maximization over a closed interval is achieved at the boundary of the interval. Hence for any fixed d

$$\max_i \left\{ (d\sigma_i)^2 + \frac{1}{(d\sigma_i)^2} \right\} = \max \left\{ (d\bar{\sigma})^2 + \frac{1}{(d\bar{\sigma})^2}, \ (d\underline{\sigma})^2 + \frac{1}{(d\underline{\sigma})^2} \right\}.$$

Then the minimization over d is obtained iff

$$(d\bar{\sigma})^2 + \frac{1}{(d\bar{\sigma})^2} = (d\underline{\sigma})^2 + \frac{1}{(d\underline{\sigma})^2},$$

which gives $d^2 = \frac{1}{\bar{\sigma}\,\underline{\sigma}}$. The result then follows from substituting d. $\square$

Example 10.6 As another example, consider again the skewed specification problem from Chapter 8. Then the corresponding G matrix is given by

$$G = \begin{bmatrix} -W_2T_iW_1 & -W_2KS_oW_d \\ W_eS_oPW_1 & W_eS_oW_d \end{bmatrix}.$$

So the robust performance specification is satisfied iff

$$\mu_\Delta(G(j\omega)) = \inf_{d_\omega \in \mathbb{R}_+} \bar{\sigma}\left(\begin{bmatrix} -W_2T_iW_1 & -d_\omega W_2KS_oW_d \\ \frac{1}{d_\omega}W_eS_oPW_1 & W_eS_oW_d \end{bmatrix} \right) \leq 1$$

for all $\omega \geq 0$. As in the last example, an upper-bound can be obtained by taking

$$d_\omega = \sqrt{\frac{\|W_e S_o P W_1\|}{\|W_2 K S_o W_d\|}}.$$

Then

$$\mu_\Delta(G(j\omega)) \leq \sqrt{\kappa(W_d^{-1} P W_1)}(\|W_2 T_i W_1\| + \|W_e S_o W_d\|).$$

In particular, this suggests that the robust performance margin is inversely proportional to the square root of the plant condition number if $W_d = I$ and $W_1 = I$. This can be further illustrated by considering a plant-inverting control system.

To simplify the exposition, we shall make the following assumptions:

$$W_e = w_s I, \; W_d = I, \; W_1 = I, \; W_2 = w_t I,$$

and P is stable and has a stable inverse (i.e., minimum phase) (P can be strictly proper). Furthermore, we shall assume that the controller has the form

$$K(s) = P^{-1}(s) l(s)$$

where $l(s)$ is a scalar loop transfer function that makes $K(s)$ proper and stabilizes the closed loop. This compensator produces diagonal sensitivity and complementary sensitivity functions with identical diagonal elements; namely,

$$S_o = S_i = \frac{1}{1+l(s)} I, \quad T_o = T_i = \frac{l(s)}{1+l(s)} I.$$

Denote

$$\varepsilon(s) = \frac{1}{1+l(s)}, \quad \tau(s) = \frac{l(s)}{1+l(s)}$$

and substitute these expressions into G; we get

$$G = \begin{bmatrix} -w_t \tau I & -w_t \tau P^{-1} \\ w_s \varepsilon P & w_s \varepsilon I \end{bmatrix}.$$

The structured singular value for G at frequency ω can be computed by

$$\mu_\Delta(G(j\omega)) = \inf_{d \in \mathbb{R}_+} \overline{\sigma}\left(\begin{bmatrix} -w_t \tau I & -w_t \tau (dP)^{-1} \\ w_s \varepsilon dP & w_s \varepsilon I \end{bmatrix}\right).$$

Let the singular value decomposition of $P(j\omega)$ at frequency ω be

$$P(j\omega) = U \Sigma V^*, \quad \Sigma = \mathrm{diag}(\sigma_1, \sigma_2, \ldots, \sigma_m)$$

with $\sigma_1 = \overline{\sigma}$ and $\sigma_m = \underline{\sigma}$, where m is the dimension of P. Then

$$\mu_\Delta(G(j\omega)) = \inf_{d \in \mathbb{R}_+} \overline{\sigma}\left(\begin{bmatrix} -w_t \tau I & -w_t \tau (d\Sigma)^{-1} \\ w_s \varepsilon d\Sigma & w_s \varepsilon I \end{bmatrix}\right)$$

10.3. Structured Robust Stability and Performance

since unitary operations do not change the singular values of a matrix. Note that

$$\begin{bmatrix} -w_t\tau I & -w_t\tau(d\Sigma)^{-1} \\ w_s\varepsilon d\Sigma & w_s\varepsilon I \end{bmatrix} = P_1 \text{diag}(M_1, M_2, \ldots, M_m) P_2$$

where P_1 and P_2 are permutation matrices and where

$$M_i = \begin{bmatrix} -w_t\tau & -w_t\tau(d\sigma_i)^{-1} \\ w_s\varepsilon d\sigma_i & w_s\varepsilon \end{bmatrix}.$$

Hence

$$\begin{aligned}
\mu_\Delta(G(j\omega)) &= \inf_{d\in\mathbb{R}_+} \max_i \bar{\sigma}\left(\begin{bmatrix} -w_t\tau & -w_t\tau(d\sigma_i)^{-1} \\ w_s\varepsilon d\sigma_i & w_s\varepsilon \end{bmatrix}\right) \\
&= \inf_{d\in\mathbb{R}_+} \max_i \bar{\sigma}\left(\begin{bmatrix} -w_t\tau \\ w_s\varepsilon d\sigma_i \end{bmatrix}\begin{bmatrix} 1 & (d\sigma_i)^{-1} \end{bmatrix}\right) \\
&= \inf_{d\in\mathbb{R}_+} \max_i \sqrt{(1+|d\sigma_i|^{-2})(|w_s\varepsilon d\sigma_i|^2 + |w_t\tau|^2)} \\
&= \inf_{d\in\mathbb{R}_+} \max_i \sqrt{|w_s\varepsilon|^2 + |w_t\tau|^2 + |w_s\varepsilon d\sigma_i|^2 + \left|\frac{w_t\tau}{d\sigma_i}\right|^2}.
\end{aligned}$$

Using Lemma 10.9, it is easy to show that the maximum is achieved at either $\bar{\sigma}$ or $\underline{\sigma}$ and that optimal d is given by

$$d^2 = \frac{|w_t\tau|}{|w_s\varepsilon|\underline{\sigma}\bar{\sigma}},$$

so the structured singular value is

$$\mu_\Delta(G(j\omega)) = \sqrt{|w_s\varepsilon|^2 + |w_t\tau|^2 + |w_s\varepsilon||w_t\tau|[\kappa(P) + \frac{1}{\kappa(P)}]}. \tag{10.27}$$

Note that if $|w_s\varepsilon|$ and $|w_t\tau|$ are not too large, which is guaranteed if the nominal performance and robust stability conditions are satisfied, then the structured singular value is proportional to the square root of the plant condition number:

$$\mu_\Delta(G(j\omega)) \approx \sqrt{|w_s\varepsilon||w_t\tau|\kappa(P)}. \tag{10.28}$$

This confirms our intuition that an ill-conditioned plant with skewed specifications is hard to control.

10.3.4 Approximation of Multiple Full Block μ

The approximations given in the last subsection can be generalized to the multiple-block μ problem by assuming that M is partitioned consistently with the structure of

$$\Delta = \text{diag}(\Delta_1, \Delta_2, \ldots, \Delta_F)$$

so that

$$M = \begin{bmatrix} M_{11} & M_{12} & \cdots & M_{1F} \\ M_{21} & M_{22} & \cdots & M_{2F} \\ \vdots & \vdots & & \vdots \\ M_{F1} & M_{F2} & \cdots & M_{FF} \end{bmatrix}$$

and

$$D = \text{diag}(d_1 I, \ldots, d_{F-1} I, I).$$

Now

$$DMD^{-1} = \left[M_{ij} \frac{d_i}{d_j} \right], \; d_F := 1.$$

Hence

$$\mu_\Delta(M) \leq \inf_{D \in \mathcal{D}} \bar{\sigma}(DMD^{-1}) = \inf_{D \in \mathcal{D}} \bar{\sigma}\left[M_{ij} \frac{d_i}{d_j} \right]$$

$$\leq \inf_{D \in \mathcal{D}} \bar{\sigma}\left[\|M_{ij}\| \frac{d_i}{d_j} \right] \leq \inf_{D \in \mathcal{D}} \sqrt{\sum_{i=1}^{F} \sum_{j=1}^{F} \|M_{ij}\|^2 \frac{d_i^2}{d_j^2}}$$

$$\leq \inf_{D \in \mathcal{D}} \sqrt{\sum_{i=1}^{F} \sum_{j=1}^{F} \|M_{ij}\|_F^2 \frac{d_i^2}{d_j^2}}.$$

An approximate D can be found by solving the following minimization problem:

$$\inf_{D \in \mathcal{D}} \sum_{i=1}^{F} \sum_{j=1}^{F} \|M_{ij}\|^2 \frac{d_i^2}{d_j^2}$$

or, more conveniently, by minimizing

$$\inf_{D \in \mathcal{D}} \sum_{i=1}^{F} \sum_{j=1}^{F} \|M_{ij}\|_F^2 \frac{d_i^2}{d_j^2}$$

with $d_F = 1$. The optimal d_i minimizing the preceding two problems satisfies, respectively,

$$d_k^4 = \frac{\sum_{i \neq k} \|M_{ik}\|^2 d_i^2}{\sum_{j \neq k} \|M_{kj}\|^2 / d_j^2}, \quad k = 1, 2, \ldots, F-1 \quad (10.29)$$

10.4. Overview of μ Synthesis

and

$$d_k^4 = \frac{\sum_{i \neq k} \|M_{ik}\|_F^2 d_i^2}{\sum_{j \neq k} \|M_{kj}\|_F^2 / d_j^2}, \quad k = 1, 2, \ldots, F-1. \qquad (10.30)$$

Using these relations, d_k can be obtained by iterations.

Example 10.7 Consider a 3×3 complex matrix

$$M = \begin{bmatrix} 1+j & 10-2j & -20j \\ 5j & 3+j & -1+3j \\ -2 & j & 4-j \end{bmatrix}$$

with structured $\Delta = \text{diag}(\delta_1, \delta_2, \delta_3)$. The largest singular value of M is $\bar{\sigma}(M) = 22.9094$ and the structured singular value of M computed using the μ Analysis and Synthesis Toolbox is equal to its upper-bound:

$$\mu_\Delta(M) = \inf_{D \in \mathcal{D}} \bar{\sigma}(DMD^{-1}) = 11.9636$$

with the optimal scaling $D_{\text{opt}} = \text{diag}(0.3955, 0.6847, 1)$. The optimal D minimizing

$$\inf_{D \in \mathcal{D}} \sum_{i=1}^{F} \sum_{j=1}^{F} \|M_{ij}\|^2 \frac{d_i^2}{d_j^2}$$

is $D_{\text{subopt}} = \text{diag}(0.3212, 0.4643, 1)$, which is solved from equation (10.29). Using this D_{subopt}, we obtain another upper-bound for the structured singular value:

$$\mu_\Delta(M) \leq \bar{\sigma}(D_{\text{subopt}} M D_{\text{subopt}}^{-1}) = 12.2538.$$

One may also use this D_{subopt} as an initial guess for the exact optimization.

10.4 Overview of μ Synthesis

This section briefly outlines various synthesis methods. The details are somewhat complicated and are treated in the other parts of this book. At this point, we simply want to point out how the analysis theory discussed in the previous sections leads naturally to synthesis questions.

From the analysis results, we see that each case eventually leads to the evaluation of

$$\|M\|_\alpha \quad \alpha = 2, \infty, \text{ or } \mu \qquad (10.31)$$

for some transfer matrix M. Thus when the controller is put back into the problem, it involves only a simple linear fractional transformation, as shown in Figure 10.8, with

$$M = \mathcal{F}_\ell(G, K) = G_{11} + G_{12}K(I - G_{22}K)^{-1}G_{21}$$

where $G = \begin{bmatrix} G_{11} & G_{12} \\ G_{21} & G_{22} \end{bmatrix}$ is chosen, respectively, as

- nominal performance only ($\Delta = 0$): $G = \begin{bmatrix} P_{22} & P_{23} \\ P_{32} & P_{33} \end{bmatrix}$

- robust stability only: $G = \begin{bmatrix} P_{11} & P_{13} \\ P_{31} & P_{33} \end{bmatrix}$

- robust performance: $G = P = \begin{bmatrix} P_{11} & P_{12} & P_{13} \\ P_{21} & P_{22} & P_{23} \\ \hline P_{31} & P_{32} & P_{33} \end{bmatrix}$.

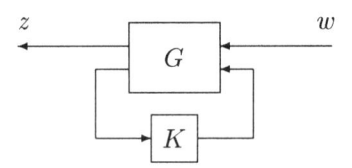

Figure 10.8: Synthesis framework

Each case then leads to the synthesis problem

$$\min_K \|\mathcal{F}_\ell(G, K)\|_\alpha \quad \text{for } \alpha = 2, \infty, \text{ or } \mu, \tag{10.32}$$

which is subject to the internal stability of the nominal.

The solutions of these problems for $\alpha = 2$ and ∞ are the focus of the rest of this book. The $\alpha = 2$ case was already known in the 1960s, and the result presented in this book is simply a new interpretation. The two Riccati solutions for the $\alpha = \infty$ case were new products of the late 1980s.

The synthesis for the $\alpha = \mu$ case is not yet fully solved. Recall that μ may be obtained by scaling and applying $\|\cdot\|_\infty$ (for $F \le 3$ and $S = 0$); a reasonable approach is to "solve"

$$\min_K \inf_{D, D^{-1} \in \mathcal{H}_\infty} \left\| D\mathcal{F}_\ell(G, K)D^{-1} \right\|_\infty \tag{10.33}$$

by iteratively solving for K and D. This is the so-called *D-K iteration*. The stable and minimum phase scaling matrix $D(s)$ is chosen such that $D(s)\Delta(s) = \Delta(s)D(s)$. [Note

10.4. Overview of μ Synthesis

that $D(s)$ is not necessarily belonging to $\mathcal{D}$ since $D(s)$ is not necessarily Hermitian, see Remark 10.2.] For a fixed scaling transfer matrix D, $\min_K \|D\mathcal{F}_\ell(G,K)D^{-1}\|_\infty$ is a standard $\mathcal{H}_\infty$ optimization problem that will be solved later in this book. For a given stabilizing controller K, $\inf_{D,D^{-1}\in\mathcal{H}_\infty} \|D\mathcal{F}_\ell(G,K)D^{-1}\|_\infty$ is a standard convex optimization problem and it can be solved pointwise in the frequency domain:

$$\sup_\omega \inf_{D_\omega \in \mathcal{D}} \bar{\sigma}\left[D_\omega \mathcal{F}_\ell(G,K)(j\omega)D_\omega^{-1}\right].$$

Indeed,

$$\inf_{D,D^{-1}\in\mathcal{H}_\infty} \|D\mathcal{F}_\ell(G,K)D^{-1}\|_\infty = \sup_\omega \inf_{D_\omega \in \mathcal{D}} \bar{\sigma}\left[D_\omega \mathcal{F}_\ell(G,K)(j\omega)D_\omega^{-1}\right].$$

This follows intuitively from the following arguments: The left-hand side is always no smaller than the right-hand side, and, on the other hand, given the minimizing D_ω from the right-hand side across the frequency, there is always a rational function $D(s)$ uniformly approximating the magnitude frequency response D_ω.

Note that when $S = 0$ (no scalar blocks),

$$D_\omega = \text{diag}(d_1^\omega I, \ldots, d_{F-1}^\omega I, I) \in \mathcal{D},$$

which is a block diagonal scaling matrix applied pointwise across frequency to the frequency response $\mathcal{F}_\ell(G,K)(j\omega)$.

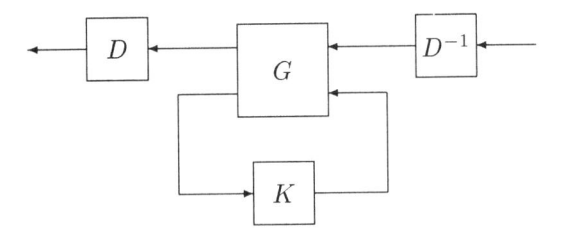

Figure 10.9: μ synthesis via scaling

D-K iterations proceed by performing this two-parameter minimization in sequential fashion: first minimizing over K with D fixed, then minimizing pointwise over D with K fixed, then again over K, and again over D, etc. Details of this process are summarized in the following steps:

(i) Fix an initial estimate of the scaling matrix $D_\omega \in \mathcal{D}$ pointwise across frequency.

(ii) Find scalar transfer functions $d_i(s), d_i^{-1}(s) \in \mathcal{RH}_\infty$ for $i = 1, 2, \ldots, (F-1)$ such that $|d_i(j\omega)| \approx d_i^\omega$. This step can be done using the interpolation theory (Youla and Saito [1967]); however, this will usually result in very high-order transfer functions, which explains why this process is currently done mostly by graphical matching using lower-order transfer functions.

(iii) Let
$$D(s) = \text{diag}(d_1(s)I, \ldots, d_{F-1}(s)I, I).$$
Construct a state-space model for system
$$\hat{G}(s) = \begin{bmatrix} D(s) & \\ & I \end{bmatrix} G(s) \begin{bmatrix} D^{-1}(s) & \\ & I \end{bmatrix},$$
as shown in Figure 10.9.

(iv) Solve an $\mathcal{H}_\infty$-optimization problem to minimize
$$\left\| \mathcal{F}_\ell(\hat{G}, K) \right\|_\infty$$
over all stabilizing K's. Note that this optimization problem uses the scaled version of G. Let its minimizing controller be denoted by $\hat{K}$.

(v) Minimize $\bar{\sigma}[D_\omega \mathcal{F}_\ell(G, \hat{K}) D_\omega^{-1}]$ over D_ω, pointwise across frequency.[1] Note that this evaluation uses the minimizing $\hat{K}$ from the last step, but that G is not scaled. The minimization itself produces a new scaling function. Let this new function be denoted by $\hat{D}_\omega$.

(vi) Compare $\hat{D}_\omega$ with the previous estimate D_ω. Stop if they are close, but otherwise replace D_ω with $\hat{D}_\omega$ and return to step (ii).

With either K or D fixed, the global optimum in the other variable may be found using the μ and $\mathcal{H}_\infty$ solutions. Although the joint optimization of D and K is not convex and the global convergence is not guaranteed, many designs have shown that this approach works very well (see, e.g., Balas [1990]). In fact, this is probably the most effective design methodology available today for dealing with such complicated problems. Detailed treatment of μ analysis is given in Packard and Doyle [1993]. The rest of this book will focus on the $\mathcal{H}_\infty$ optimization, which is a fundamental tool for μ synthesis.

Users are encouraged to try the following demo programs from the μ toolbox:

≫ **himat_x1, himat_x2, himat_x3, himat_x4, himat_x5, himat_x6**

Related MATLAB Commands: musynfit, musynflp, muftbtch, dkit

10.5 Notes and References

This chapter is partially based on the lecture notes given by Doyle [1984] and partially based on the lecture notes by Packard [1991] and the paper by Doyle, Packard, and

---
[1] The approximate solutions given in the preceding section may be used.

Zhou [1991]. Parts of Section 10.3.3 are based on the paper by Stein and Doyle [1991]. The small μ theorem for systems with nonrational plants and uncertainties is proven in Tits [1995]. Connections were established in Poolla and Tikku [1995] between the frequency-dependent D-scaled upper bounds of the structured singular value and the robust performance of a system with arbitrarily slowly varying structured linear perturbations. Robust performance of systems with structured time-varying perturbations was also considered in Shamma [1994] using the constant D-scaled upper bounds of the structured singular value. Other results on μ can be found in Fan and Tits [1986], Fan, Tits, and Doyle [1991], Packard and Doyle [1993], Packard and Pandey [1993], Young [1993], and references therein.

10.6 Problems

Problem 10.1 Let M and N be suitably dimensioned matrices and let Δ be a structured uncertainty. Prove or disprove

(a) $\mu_\Delta(M) = 0 \Longrightarrow M = 0$;

(b) $\mu_\Delta(M_1 + M_2) \leq \mu_\Delta(M_1) + \mu_\Delta(M_2)$.

(c) $\mu_\Delta(\alpha M) = |\alpha|\mu_\Delta(M)$.

(d) $\mu_\Delta(I) = 1$.

(e) $\mu_\Delta(MN) \leq \bar{\sigma}(M)\mu_\Delta(N)$.

(f) $\mu_\Delta(MN) \leq \bar{\sigma}(N)\mu_\Delta(M)$.

Problem 10.2 Let $\Delta = \begin{bmatrix} \Delta_1 & 0 \\ 0 & \Delta_2 \end{bmatrix}$, where Δ_i are structured uncertainties. Show that $\mu_\Delta \left(\begin{bmatrix} M_{11} & M_{12} \\ 0 & M_{22} \end{bmatrix} \right) = \max\{\mu_{\Delta_1}(M_{11}),\ \mu_{\Delta_2}(M_{22})\}$.

Problem 10.3 MATLAB exercise. Let M be a 7×7 random real matrix. Take the perturbation structure to be

$$\Delta = \left\{ \begin{bmatrix} \delta_1 I_3 & 0 & 0 \\ 0 & \Delta_2 & 0 \\ 0 & 0 & \delta_3 I_2 \end{bmatrix} : \delta_1, \delta_3 \in \mathbb{C},\ \Delta_2 \in \mathbb{C}^{2\times 2} \right\}.$$

Compute $\mu(M)$ and a singularizing perturbation.

Problem 10.4 Let $M = \begin{bmatrix} 0 & M_{12} \\ M_{21} & 0 \end{bmatrix}$ be a complex matrix and let $\Delta = \begin{bmatrix} \Delta_1 & \\ & \Delta_2 \end{bmatrix}$. Show that

$$\mu_\Delta(M) = \sqrt{\bar{\sigma}(M_{12})\bar{\sigma}(M_{21})}.$$

Problem 10.5 Let $M = \begin{bmatrix} M_{11} & M_{12} \\ M_{21} & M_{22} \end{bmatrix}$ be a complex matrix and let $\Delta = \begin{bmatrix} \Delta_1 & \\ & \Delta_2 \end{bmatrix}$. Show that

$$\sqrt{\overline{\sigma}(M_{12})\overline{\sigma}(M_{21})} - \max\{\overline{\sigma}(M_{11}),\ \overline{\sigma}(M_{22})\} \leq \mu_\Delta(M)$$
$$\leq \sqrt{\overline{\sigma}(M_{12})\overline{\sigma}(M_{21})} + \max\{\overline{\sigma}(M_{11}),\ \overline{\sigma}(M_{22})\}.$$

Problem 10.6 Let Δ be all diagonal full blocks and M be partitioned as $M = [M_{ij}]$, where M_{ij} are matrices with suitable dimensions. Show that

$$\mu_\Delta(M) \leq \pi\left([\|M_{ij}\|]\right) (= \rho\left([\|M_{ij}\|]\right))$$

where $\pi(\cdot)$ denotes the Perron eigenvalue.

Problem 10.7 Show

$$\inf_{D \in \mathbb{C}^{n \times n}} \|DMD^{-1}\|_p = \rho(M)$$

where $\|\cdot\|_p$ is the induced p-norm, $1 \leq p \leq \infty$.

Problem 10.8 Let D be a nonsingular diagonal matrix $D = \mathrm{diag}(d_1, d_2, \ldots, d_n)$. Show

$$\inf_D \|DMD^{-1}\|_p = \pi(M)$$

if either $M = [|m_{ij}|]$ and $1 \leq p \leq \infty$ or $p = 1$ or ∞. Moreover, the optimal D is given by

$$D = \mathrm{diag}(y_1^{1/p}/x_1^{1/q}, y_2^{1/p}/x_2^{1/q}, \ldots, y_n^{1/p}/x_n^{1/q})$$

where $1/p + 1/q = 1$ and

$$[|m_{ij}|]x = \pi(M)x, \quad y^T[|m_{ij}|] = \pi(M)y^T.$$

Problem 10.9 Let Δ be a structured uncertainty defined in the book. Suppose $M = xy^*$ with $x, y \in \mathbb{C}^n$. Derive an exact expression for $\mu_\Delta(M)$ in terms of the components of x and y. [Note that $\Delta = \mathrm{diag}(\delta_1 I, \delta_2 I, \ldots, \delta_m I, \Delta_1, \ldots, \Delta_F)$ and $\mu_\Delta(M) = \max_{U \in \mathcal{U}} \rho(MU) = \max_{\Delta \in \mathbf{B}\boldsymbol{\Delta}} \rho(M\Delta)$.]

Problem 10.10 Let $\{x_k\}_{k=0}^\infty$, $\{z_k\}_{k=0}^\infty$, and $\{d_k\}_{k=0}^\infty$ be sequences satisfying

$$\|x_{k+1}\|^2 + \|z_k\|^2 \leq \beta^2(\|x_k\|^2 + \|d_k\|^2)$$

for some $\beta < 1$ and all $k \geq 0$. If $d \in \ell_2$, show that both $x \in \ell_2$ and $z \in \ell_2$ and the norms are bounded by

$$\|z\|_2^2 + (1-\beta^2)\|x\|_2^2 \leq \beta^2 \|d\|_2^2 + \|x_0\|^2$$

Give a system interpretation of this result.

10.6. Problems

Problem 10.11 Consider a feedback system shown here with

$$P = P_0(1 + W_1\Delta_1) + W_2\Delta_2, \quad \|\Delta_i\|_\infty < 1, \ i = 1, 2.$$

Suppose W_1 and W_2 are stable, and P and P_0 have the same number of poles in $\text{Re}\{s\} > 0$.

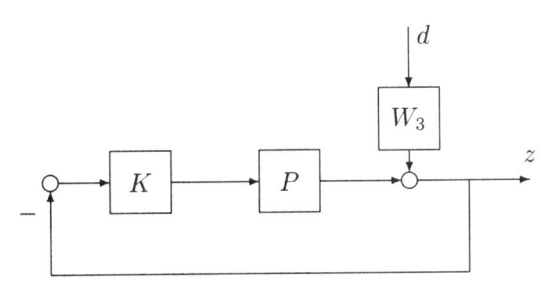

(a) Show that the feedback system is robustly stable if and only if K stabilizes P_0 and

$$\| \, |W_1 T| + |W_2 KS| \, \|_\infty \leq 1$$

where

$$S = \frac{1}{1 + P_0 K}, \quad T = \frac{P_0 K}{1 + P_0 K}.$$

(b) Show that the feedback system has robust performance; that is, $\|T_{zd}\|_\infty \leq 1$, if and only if K stabilizes P_0 and

$$\| \, |W_3 S| + |W_1 T| + |W_2 KS| \, \|_\infty \leq 1.$$

Problem 10.12 In Problem 10.11, find a matrix

$$M = \begin{bmatrix} M_{11} & M_{12} \\ M_{21} & M_{22} \end{bmatrix}$$

such that

$$z = \mathcal{F}_u(M, \Delta)d, \quad \Delta_s = \begin{bmatrix} \Delta_1 & \\ & \Delta_2 \end{bmatrix}.$$

Assume that K stabilizes P_0. Show that at each frequency

$$\mu_{\Delta_s}(M_{11}) = \inf_{D_s \in \mathcal{D}_s} \bar{\sigma}(D_s M_{11} D_s^{-1}) = |W_1 T| + |W_2 KS|$$

with $D_s = \begin{bmatrix} d & \\ & 1 \end{bmatrix}$.

Next let $\Delta = \begin{bmatrix} \Delta_s & \\ & \Delta_p \end{bmatrix}$ and $D = \begin{bmatrix} d_1 & & \\ & d_2 & \\ & & 1 \end{bmatrix}$. Show that

$$\mu_\Delta(M) = \inf_{D \in \mathcal{D}} \overline{\sigma}(DMD^{-1}) = |W_3 S| + |W_1 T| + |W_2 KS|$$

Problem 10.13 Let $\Delta = \text{diag}(\Delta_1, \ldots, \Delta_F)$ be a structured uncertainty and suppose $M = xy^*$ with $x, y \in \mathbb{C}^n$. Let x and y be partitioned compatibly with the Δ:

$$x = \begin{pmatrix} x_1 \\ x_2 \\ \vdots \\ x_{m+F} \end{pmatrix}, \quad y = \begin{pmatrix} y_1 \\ y_2 \\ \vdots \\ y_{m+F} \end{pmatrix}.$$

Show that

$$\mu_\Delta(M) = \inf_{D \in \mathcal{D}} \overline{\sigma}(DMD^{-1})$$

and the minimizing d_i are given by

$$d_i^2 = \frac{\|y_i\| \|x_F\|}{\|x_i\| \|y_F\|}.$$

Problem 10.14 Consider $M \in \mathbb{C}^{2n \times 2n}$ to be given. Let Δ_2 be a $n \times n$ block structure, and suppose that $\mu_2(M_{22}) < 1$. Suppose also that $\mathcal{F}_l(M, \Delta_2)$ is invertible for all $\Delta_2 \in \mathbf{B}\Delta_2$. For each $\alpha > 1$, find a matrix W_α and a block structure $\hat{\Delta}$ such that

$$\max_{\Delta_2 \in \mathbf{B}\Delta_2} \kappa(\mathcal{F}_l(M, \Delta_2)) < \alpha$$

if and only if

$$\mu_{\hat{\Delta}}(W_\alpha) < 1$$

where κ is the condition number.

Problem 10.15 Let $G(s) \in \mathcal{RH}_\infty$ be an $m \times m$ symmetric transfer matrix, i.e., $G^T(s) = G(s)$, and let $\Delta(s) \in \mathcal{RH}_\infty$ be a diagonal perturbation, i.e.,

$$\Delta(s) = \begin{pmatrix} \delta_1(s) & & & \\ & \delta_2(s) & & \\ & & \ddots & \\ & & & \delta_m(s) \end{pmatrix}.$$

Show $(I - G\Delta)^{-1} \in \mathcal{RH}_\infty$ for all $\|\Delta\|_\infty \leq \gamma$ if and only if $\|G\|_\infty < 1/\gamma$. [Hint: Note that for a complex symmetric matrix $M = M^T \in \mathbb{C}^{m \times m}$, there is a unitary matrix U and a diagonal matrix $\Sigma = \text{diag}(\sigma_1, \sigma_2, \ldots, \sigma_m) \geq 0$ such that $M = U\Sigma U^T$, see Horn and Johnson [1990, page 204]. For detailed discussion of this problem, see Qiu [1995].]

Chapter 11

Controller Parameterization

The basic configuration of the feedback systems considered in this chapter is an LFT, as shown in Figure 11.1, where G is the generalized plant with two sets of inputs: the exogenous inputs w, which include disturbances and commands, and control inputs u. The plant G also has two sets of outputs: the measured (or sensor) outputs y and the regulated outputs z. K is the controller to be designed. A control problem in this setup is either to analyze some specific properties (e.g., stability or performance) of the closed loop or to design the feedback control K such that the closed-loop system is stable in some appropriate sense and the error signal z is specified (i.e., some performance specifications are satisfied). In this chapter we are only concerned with the basic internal stabilization problems. We will see again that this setup is very convenient for other general control synthesis problems in the coming chapters.

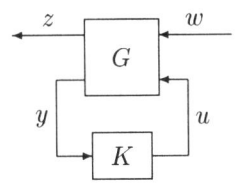

Figure 11.1: General system interconnection

Suppose that a given feedback system is feedback stabilizable. In this chapter, the problem we are mostly interested in is parameterizing all controllers that stabilize the system. The *parameterization* of all internally stabilizing controllers was first introduced by Youla et al. [1976a, 1976b] using the *coprime factorization* technique. We shall, however, focus on the state-space approach in this chapter.

11.1 Existence of Stabilizing Controllers

Consider a system described by the standard block diagram in Figure 11.1. Assume that $G(s)$ has a *stabilizable and detectable* realization of the form

$$G(s) = \begin{bmatrix} G_{11}(s) & G_{12}(s) \\ G_{21}(s) & G_{22}(s) \end{bmatrix} = \left[\begin{array}{c|cc} A & B_1 & B_2 \\ \hline C_1 & D_{11} & D_{12} \\ C_2 & D_{21} & D_{22} \end{array} \right]. \qquad (11.1)$$

The stabilization problem is to find feedback mapping K such that the closed-loop system is internally stable; the well-posedness is required for this interconnection.

Definition 11.1 *A proper system G is said to be* stabilizable *through output feedback if there exists a proper controller K internally stabilizing G in Figure 11.1. Moreover, a proper controller $K(s)$ is said to be* admissible *if it internally stabilizes G.*

The following result is standard and follows from Chapter 3.

Lemma 11.1 *There exists a proper K achieving internal stability iff (A, B_2) is stabilizable and (C_2, A) is detectable. Further, let F and L be such that $A + B_2 F$ and $A + LC_2$ are stable; then an observer-based stabilizing controller is given by*

$$K(s) = \left[\begin{array}{c|c} A + B_2 F + LC_2 + LD_{22}F & -L \\ \hline F & 0 \end{array} \right].$$

Proof. ($\Leftarrow$) By the stabilizability and detectability assumptions, there exist F and L such that $A + B_2 F$ and $A + LC_2$ are stable. Now let $K(s)$ be the observer-based controller given in the lemma, then the closed-loop A matrix is given by

$$\tilde{A} = \begin{bmatrix} A & B_2 F \\ -LC_2 & A + B_2 F + LC_2 \end{bmatrix}.$$

It is easy to check that this matrix is similar to the matrix

$$\begin{bmatrix} A + LC_2 & 0 \\ -LC_2 & A + B_2 F \end{bmatrix}.$$

Thus the spectrum of $\tilde{A}$ equals the union of the spectra of $A + LC_2$ and $A + B_2 F$. In particular, $\tilde{A}$ is stable.

($\Rightarrow$) If (A, B_2) is not stabilizable or if (C_2, A) is not detectable, then there are some eigenvalues of $\tilde{A}$ that are fixed in the right-half plane, no matter what the compensator is. The details are left as an exercise. $\square$

11.1. Existence of Stabilizing Controllers

The stabilizability and detectability conditions of (A, B_2, C_2) are assumed throughout the remainder of this chapter.[1] It follows that the realization for G_{22} is stabilizable and detectable, and these assumptions are enough to yield the following result.

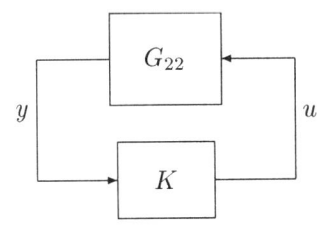

Figure 11.2: Equivalent stabilization diagram

Lemma 11.2 *Suppose the inherited realization* $\left[\begin{array}{c|c} A & B_2 \\ \hline C_2 & D_{22} \end{array}\right]$ *for G_{22} is stabilizable and detectable. Then the system in Figure 11.1 is internally stable iff the one in Figure 11.2 is internally stable.*

In other words, $K(s)$ internally stabilizes $G(s)$ if and only if it internally stabilizes G_{22} [provided that (A, B_2, C_2) is stabilizable and detectable].

Proof. The necessity follows from the definition. To show the sufficiency, it is sufficient to show that the system in Figure 11.1 and that in Figure 11.2 share the same A matrix, which is obvious. □

From Lemma 11.2, we see that the stabilizing controller for G depends only on G_{22}. Hence all stabilizing controllers for G can be obtained by using only G_{22}.

Remark 11.1 There should be no confusion between a given realization for a transfer matrix G_{22} and the inherited realization from G, where G_{22} is a submatrix. A given realization for G_{22} may be stabilizable and detectable while the inherited realization may not be. For instance,

$$G_{22} = \frac{1}{s+1} = \left[\begin{array}{c|c} -1 & 1 \\ \hline 1 & 0 \end{array}\right]$$

is a minimal realization but the inherited realization of G_{22} from

$$\left[\begin{array}{cc} G_{11} & G_{12} \\ G_{21} & G_{22} \end{array}\right] = \left[\begin{array}{cc|cc} -1 & 0 & 0 & 1 \\ 0 & 1 & 1 & 0 \\ \hline 0 & 1 & 0 & 0 \\ 1 & 0 & 0 & 0 \end{array}\right]$$

---

[1] It should be clear that the stabilizability and detectability of a realization for G do not guarantee the stabilizability and/or detectability of the corresponding realization for G_{22}.

is

$$G_{22} = \left[\begin{array}{cc|c} -1 & 0 & 1 \\ 0 & 1 & 0 \\ \hline 1 & 0 & 0 \end{array}\right] \quad \left(= \frac{1}{s+1}\right),$$

which is neither stabilizable nor detectable. ◇

11.2 Parameterization of All Stabilizing Controllers

Consider again the standard system block diagram in Figure 11.1 with

$$G(s) = \left[\begin{array}{c|cc} A & B_1 & B_2 \\ \hline C_1 & D_{11} & D_{12} \\ C_2 & D_{21} & D_{22} \end{array}\right] = \left[\begin{array}{cc} G_{11}(s) & G_{12}(s) \\ G_{21}(s) & G_{22}(s) \end{array}\right].$$

Suppose (A, B_2) is stabilizable and (C_2, A) is detectable. In this section we discuss the following problem:

Given a plant G, parameterize all controllers K that internally stabilize G.

This parameterization for all stabilizing controllers is usually called Youla parameterization. The parameterization of all stabilizing controllers is easy when the plant itself is stable.

Theorem 11.3 *Suppose $G \in \mathcal{RH}_\infty$; then the set of all stabilizing controllers can be described as*

$$K = Q(I + G_{22}Q)^{-1} \tag{11.2}$$

for any $Q \in \mathcal{RH}_\infty$ and $I + D_{22}Q(\infty)$ nonsingular.

Remark 11.2 This result is very natural considering Corollary 5.3, which says that a controller K stabilizes a stable plant G_{22} iff $K(I - G_{22}K)^{-1}$ is stable. Now suppose $Q = K(I - G_{22}K)^{-1}$ is a stable transfer matrix, then K can be solved from this equation which gives exactly the controller parameterization in the preceding theorem. ◇

Proof. Note that $G_{22}(s)$ is stable by the assumptions on G. Then it is straightforward to verify that the controllers given previously stabilize G_{22}. On the other hand, suppose K_0 is a stabilizing controller; then $Q_0 := K_0(I - G_{22}K_0)^{-1} \in \mathcal{RH}_\infty$, so K_0 can be expressed as $K_0 = Q_0(I + G_{22}Q_0)^{-1}$. Note that the invertibility in the last equation is guaranteed by the well-posedness condition of the interconnected system with controller K_0 since $I + D_{22}Q_0(\infty) = (I - D_{22}K_0(\infty))^{-1}$. □

However, if G is not stable, the parameterization is much more complicated. The results can be more conveniently stated using state-space representations.

11.2. Parameterization of All Stabilizing Controllers

Theorem 11.4 *Let F and L be such that $A + LC_2$ and $A + B_2F$ are stable; then all controllers that internally stabilize G can be parameterized as the transfer matrix from y to u:*

$$J = \left[\begin{array}{c|cc} A + B_2F + LC_2 + LD_{22}F & -L & B_2 + LD_{22} \\ \hline F & 0 & I \\ -(C_2 + D_{22}F) & I & -D_{22} \end{array}\right]$$

with any $Q \in \mathcal{RH}_\infty$ and $I + D_{22}Q(\infty)$ nonsingular. Furthermore, the set of all closed-loop transfer matrices from w to z achievable by an internally stabilizing proper controller is equal to

$$\mathcal{F}_\ell(T, Q) = \{T_{11} + T_{12}QT_{21} : Q \in \mathcal{RH}_\infty,\ I + D_{22}Q(\infty)\text{ invertible}\}$$

where T is given by

$$T = \begin{bmatrix} T_{11} & T_{12} \\ T_{21} & T_{22} \end{bmatrix} = \left[\begin{array}{cc|cc} A + B_2F & -B_2F & B_1 & B_2 \\ 0 & A + LC_2 & B_1 + LD_{21} & 0 \\ \hline C_1 + D_{12}F & -D_{12}F & D_{11} & D_{12} \\ 0 & C_2 & D_{21} & 0 \end{array}\right].$$

Proof. Let $K = \mathcal{F}_\ell(J, Q)$. Then it is straightforward to verify, by using the state-space star product formula and some tedious algebra, that $\mathcal{F}_\ell(G, K) = T_{11} + T_{12}QT_{21}$ with the T given in the theorem. Hence the controller $K = \mathcal{F}_\ell(J, Q)$ for any given $Q \in \mathcal{RH}_\infty$ does internally stabilize G. Now let K be any stabilizing controller for G; then $\mathcal{F}_\ell(\hat{J}, K) \in \mathcal{RH}_\infty$, where

$$\hat{J} = \left[\begin{array}{c|cc} A & -L & B_2 \\ \hline -F & 0 & I \\ C_2 & I & 0 \end{array}\right].$$

($\hat{J}$ is stabilized by K since it has the same G_{22} matrix as G.)

Let $Q_0 := \mathcal{F}_\ell(\hat{J}, K) \in \mathcal{RH}_\infty$; then $\mathcal{F}_\ell(J, Q_0) = \mathcal{F}_\ell(J, \mathcal{F}_\ell(\hat{J}, K)) =: \mathcal{F}_\ell(J_{tmp}, K)$, where J_{tmp} can be obtained by using the state-space star product formula given in Chapter 9:

$$J_{tmp} = \left[\begin{array}{cc|cc} A + LC_2 + B_2F & -B_2F & -L & B_2 \\ LC_2 & A & -L & B_2 \\ \hline F & -F & 0 & I \\ -C_2 & C_2 & I & 0 \end{array}\right]$$

$$= \left[\begin{array}{cc|cc} A+LC_2 & -B_2 F & -L & B_2 \\ 0 & A+B_2 F & 0 & 0 \\ \hline 0 & -F & 0 & I \\ 0 & C_2 & I & 0 \end{array}\right]$$

$$= \left[\begin{array}{cc} 0 & I \\ I & 0 \end{array}\right].$$

Hence $\mathcal{F}_\ell(J, Q_0) = \mathcal{F}_\ell(J_{tmp}, K) = K$. This shows that any stabilizing controller can be expressed in the form of $\mathcal{F}_\ell(J, Q_0)$ for some $Q_0 \in \mathcal{RH}_\infty$. □

An important point to note is that the closed-loop transfer matrix is simply an affine function of the controller parameter matrix Q. The proper K's achieving internal stability are precisely those represented in Figure 11.3.

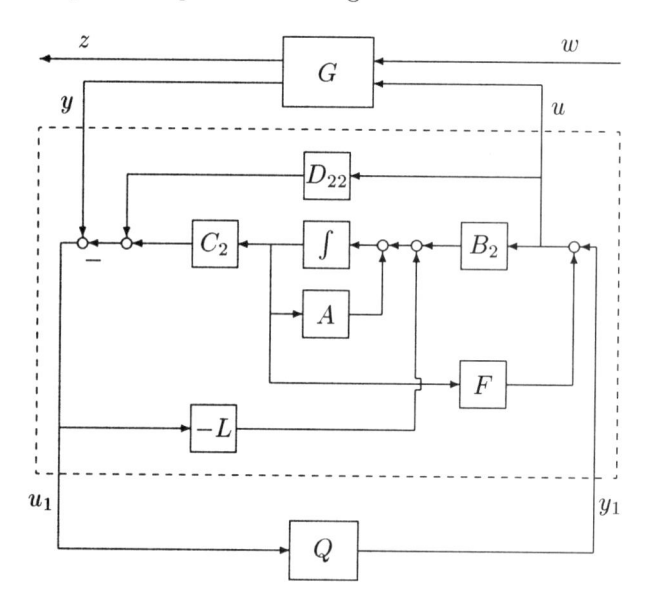

Figure 11.3: Structure of stabilizing controllers

It is interesting to note that the system in the dashed box is an observer-based stabilizing controller for G (or G_{22}). Furthermore, it is easy to show that the transfer function between (y, y_1) and (u, u_1) is J; that is,

$$\left[\begin{array}{c} u \\ u_1 \end{array}\right] = J \left[\begin{array}{c} y \\ y_1 \end{array}\right].$$

It is also easy to show that the transfer matrix from y_1 to u_1 is $T_{22} = 0$.

11.2. Parameterization of All Stabilizing Controllers

This diagram of the parameterization of all stabilizing controllers also suggests an interesting interpretation: Every internal stabilization amounts to adding stable dynamics to the plant and then stabilizing the extended plant by means of an observer. The precise statement is as follows: For simplicity of the formulas, only the cases of strictly proper G_{22} and K are treated.

Theorem 11.5 *Assume that G_{22} and K are strictly proper and the system in Figure 11.1 is internally stable. Then G_{22} can be embedded in a system*

$$\left[\begin{array}{c|c} A_e & B_e \\ \hline C_e & 0 \end{array}\right]$$

where

$$A_e = \left[\begin{array}{cc} A & 0 \\ 0 & A_a \end{array}\right], \quad B_e = \left[\begin{array}{c} B_2 \\ 0 \end{array}\right], \quad C_e = \left[\begin{array}{cc} C_2 & 0 \end{array}\right] \quad (11.3)$$

and where A_a is stable, such that K has the form

$$K = \left[\begin{array}{c|c} A_e + B_e F_e + L_e C_e & -L_e \\ \hline F_e & 0 \end{array}\right] \quad (11.4)$$

where $A_e + B_e F_e$ and $A_e + L_e C_e$ are stable.

Proof. K is representable as in Figure 11.3 for some Q in $\mathcal{RH}_\infty$. For K to be strictly proper, Q must be strictly proper. Take a minimal realization of Q:

$$Q = \left[\begin{array}{c|c} A_a & B_a \\ \hline C_a & 0 \end{array}\right].$$

Since $Q \in \mathcal{RH}_\infty$, A_a is stable. Let x and x_a denote state vectors for J and Q, respectively, and write the equations for the system in Figure 11.3:

$$\begin{aligned} \dot{x} &= (A + B_2 F + L C_2)x - Ly + B_2 y_1 \\ u &= Fx + y_1 \\ u_1 &= -C_2 x + y \\ \dot{x}_a &= A_a x_a + B_a u_1 \\ y_1 &= C_a x_a \end{aligned}$$

These equations yield

$$\begin{aligned} \dot{x}_e &= (A_e + B_e F_e + L_e C_e)x_e - L_e y \\ u &= F_e x_e \end{aligned}$$

where
$$x_e := \begin{bmatrix} x \\ x_a \end{bmatrix}, \quad F_e := \begin{bmatrix} F & C_a \end{bmatrix}, \quad L_e := \begin{bmatrix} L \\ -B_a \end{bmatrix}$$
and where A_e, B_e, C_e are as in equation (11.3). □

Example 11.1 Consider a standard feedback system shown in Figure 5.1 with $P = \dfrac{1}{s-1}$. We shall find all stabilizing controllers for P such that the steady-state errors with respect to the step input and $\sin 2t$ are both zero. It is easy to see that the controller must provide poles at 0 and $\pm 2j$. Now let the set of stabilizing controllers for a modified plant $\dfrac{(s+1)^3}{(s-1)s(s^2+2^2)}$ be K_m. Then the desired set of controllers is given by $K = \dfrac{(s+1)^3}{s(s^2+2^2)} K_m$.

11.3 Coprime Factorization Approach

In this section, all stabilizing controller parameterization will be derived using the conventional coprime factorization approach. Readers should be familiar with the results presented in Section 5.4 of Chapter 5 before preceding further.

Theorem 11.6 *Let $G_{22} = NM^{-1} = \tilde{M}^{-1}\tilde{N}$ be the rcf and lcf of G_{22} over $\mathcal{RH}_\infty$, respectively. Then the set of all proper controllers achieving internal stability is parameterized either by*

$$K = (U_0 + MQ_r)(V_0 + NQ_r)^{-1}, \quad \det(I + V_0^{-1}NQ_r)(\infty) \neq 0 \quad (11.5)$$

for $Q_r \in \mathcal{RH}_\infty$ or by

$$K = (\tilde{V}_0 + Q_l\tilde{N})^{-1}(\tilde{U}_0 + Q_l\tilde{M}), \quad \det(I + Q_l\tilde{N}\tilde{V}_0^{-1})(\infty) \neq 0 \quad (11.6)$$

for $Q_l \in \mathcal{RH}_\infty$, where $U_0, V_0, \tilde{U}_0, \tilde{V}_0 \in \mathcal{RH}_\infty$ satisfy the Bezout identities:

$$\tilde{V}_0 M - \tilde{U}_0 N = I, \quad \tilde{M}V_0 - \tilde{N}U_0 = I.$$

Moreover, if $U_0, V_0, \tilde{U}_0,$ and $\tilde{V}_0$ are chosen such that $U_0 V_0^{-1} = \tilde{V}_0^{-1}\tilde{U}_0$; that is,

$$\begin{bmatrix} \tilde{V}_0 & -\tilde{U}_0 \\ -\tilde{N} & \tilde{M} \end{bmatrix} \begin{bmatrix} M & U_0 \\ N & V_0 \end{bmatrix} = \begin{bmatrix} I & 0 \\ 0 & I \end{bmatrix}$$

11.3. Coprime Factorization Approach

then

$$\begin{aligned} K &= (U_0 + MQ_y)(V_0 + NQ_y)^{-1} \\ &= (\tilde{V}_0 + Q_y\tilde{N})^{-1}(\tilde{U}_0 + Q_y\tilde{M}) \\ &= \mathcal{F}_\ell(J_y, Q_y) \end{aligned} \tag{11.7}$$

where

$$J_y := \begin{bmatrix} U_0 V_0^{-1} & \tilde{V}_0^{-1} \\ V_0^{-1} & -V_0^{-1}N \end{bmatrix} \tag{11.8}$$

and where Q_y ranges over $\mathcal{RH}_\infty$ such that $(I + V_0^{-1}NQ_y)(\infty)$ is invertible.

Proof. We shall prove the parameterization given in equation (11.5) first. Assume that K has the form indicated, and define

$$U := U_0 + MQ_r, \ V := V_0 + NQ_r.$$

Then

$$\tilde{M}V - \tilde{N}U = \tilde{M}(V_0 + NQ_r) - \tilde{N}(U_0 + MQ_r) = \tilde{M}V_0 - \tilde{N}U_0 + (\tilde{M}N - \tilde{N}M)Q_r = I.$$

Thus K achieves internal stability by Lemma 5.7.

Conversely, suppose K is proper and achieves internal stability. Introduce an rcf of K over $\mathcal{RH}_\infty$ as $K = UV^{-1}$. Then by Lemma 5.7, $Z := \tilde{M}V - \tilde{N}U$ is invertible in $\mathcal{RH}_\infty$. Define Q_r by the equation

$$U_0 + MQ_r = UZ^{-1}, \tag{11.9}$$

so

$$Q_r = M^{-1}(UZ^{-1} - U_0).$$

Then, using the Bezout identity, we have

$$\begin{aligned} V_0 + NQ_r &= V_0 + NM^{-1}(UZ^{-1} - U_0) \\ &= V_0 + \tilde{M}^{-1}\tilde{N}(UZ^{-1} - U_0) \\ &= \tilde{M}^{-1}(\tilde{M}V_0 - \tilde{N}U_0 + \tilde{N}UZ^{-1}) \\ &= \tilde{M}^{-1}(I + \tilde{N}UZ^{-1}) \\ &= \tilde{M}^{-1}(Z + \tilde{N}U)Z^{-1} \\ &= \tilde{M}^{-1}\tilde{M}VZ^{-1} \\ &= VZ^{-1}. \end{aligned} \tag{11.10}$$

Thus,

$$\begin{aligned} K &= UV^{-1} \\ &= (U_0 + MQ_r)(V_0 + NQ_r)^{-1}. \end{aligned}$$

230 CONTROLLER PARAMETERIZATION

To see that Q_r belongs to $\mathcal{RH}_\infty$, observe first from equation (11.9) and then from equation (11.10) that both MQ_r and NQ_r belong to $\mathcal{RH}_\infty$. Then

$$Q_r = (\tilde{V}_0 M - \tilde{U}_0 N) Q_r = \tilde{V}_0 (MQ_r) - \tilde{U}_0 (NQ_r) \in \mathcal{RH}_\infty.$$

Finally, since V and Z evaluated at $s = \infty$ are both invertible, so is $V_0 + NQ_r$ from equation (11.10), and hence so is $I + V_0^{-1} NQ_r$.

Similarly, the parameterization given in equation (11.6) can be obtained.

To show that the controller can be written in the form of equation (11.7), note that

$$(U_0 + MQ_y)(V_0 + NQ_y)^{-1} = U_0 V_0^{-1} + (M - U_0 V_0^{-1} N) Q_y (I + V_0^{-1} N Q_y)^{-1} V_0^{-1}$$

and that $U_0 V_0^{-1} = \tilde{V}_0^{-1} \tilde{U}_0$. We have

$$(M - U_0 V_0^{-1} N) = (M - \tilde{V}_0^{-1} \tilde{U}_0 N) = \tilde{V}_0^{-1} (\tilde{V}_0 M - \tilde{U}_0 N) = \tilde{V}_0^{-1}$$

and

$$K = U_0 V_0^{-1} + \tilde{V}_0^{-1} Q_y (I + V_0^{-1} N Q_y)^{-1} V_0^{-1}.$$

$\Box$

Corollary 11.7 *Given an admissible controller K with coprime factorizations $K = UV^{-1} = \tilde{V}^{-1}\tilde{U}$, the free parameter $Q_y \in \mathcal{RH}_\infty$ in Youla parameterization is given by*

$$Q_y = M^{-1}(UZ^{-1} - U_0)$$

where $Z := \tilde{M}V - \tilde{N}U$.

Next, we shall establish the precise relationship between the preceding all stabilizing controller parameterization and the state-space parameterization in the last section. The following theorem follows from some algebraic manipulation.

Theorem 11.8 *Let the doubly coprime factorizations of G_{22} be chosen as*

$$\begin{bmatrix} M & U_0 \\ N & V_0 \end{bmatrix} = \left[\begin{array}{c|cc} A + B_2 F & B_2 & -L \\ \hline F & I & 0 \\ C_2 + D_{22} F & D_{22} & I \end{array}\right]$$

$$\begin{bmatrix} \tilde{V}_0 & -\tilde{U}_0 \\ -\tilde{N} & \tilde{M} \end{bmatrix} = \left[\begin{array}{c|cc} A + LC_2 & -(B_2 + LD_{22}) & L \\ \hline F & I & 0 \\ C_2 & -D_{22} & I \end{array}\right]$$

where F and L are chosen such that $A + B_2 F$ and $A + LC_2$ are both stable. Then J_y can be computed as

$$J_y = \left[\begin{array}{c|cc} A + B_2 F + LC_2 + LD_{22} F & -L & B_2 + LD_{22} \\ \hline F & 0 & I \\ -(C_2 + D_{22} F) & I & -D_{22} \end{array}\right].$$

Remark 11.3 Note that J_y is exactly the same as the J in Theorem 11.4 and that $K_0 := U_0 V_0^{-1}$ is an observer-based stabilizing controller with

$$K_0 := \left[\begin{array}{c|c} A + B_2 F + LC_2 + LD_{22}F & -L \\ \hline F & 0 \end{array} \right].$$

◇

11.4 Notes and References

The conventional Youla parameterization can be found in Youla et al. [1976a, 1976b], Desoer et al. [1980], Doyle [1984], Vidyasagar [1985], and Francis [1987]. The state-space derivation of all stabilizing controllers was reported in Lu, Zhou, and Doyle [1996]. The paper by Moore et al. [1990] contains some other related interesting results. The parameterization of all two-degree-of-freedom stabilizing controllers is given in Youla and Bongiorno [1985] and Vidyasagar [1985].

11.5 Problems

Problem 11.1 Let $P = \dfrac{1}{s-1}$. Find the set of all stabilizing controllers $K = \mathcal{F}_\ell(J, Q)$. Now verify that $K_0 = -4$ is a stabilizing controller and find a $Q_0 \in \mathcal{RH}_\infty$ such that $K_0 = \mathcal{F}_\ell(J, Q_0)$.

Problem 11.2 Suppose that $\{P_i : i = 1, \ldots, n\}$ is a set of MIMO plants and that there is a single controller K that internally stabilizes each P_i in the set. Show that there exists a single transfer function P such that the set

$$\mathcal{P} = \{\mathcal{F}_u(P, \Delta) \mid \Delta \in \mathcal{H}_\infty, \ \|\Delta\|_\infty \leq 1\}$$

is also robustly stabilized by K and that $\{P_i\} \subset \mathcal{P}$.

Problem 11.3 Internal Model Control (IMC): Suppose a plant P is stable. Then it is known that all stabilizing controllers can be parameterized as $K(s) = Q(I - PQ)^{-1}$ for all stable Q. In practice, the exact plant model is not known, only a nominal model P_0 is available. Hence the controller can be implemented as in the following diagram:

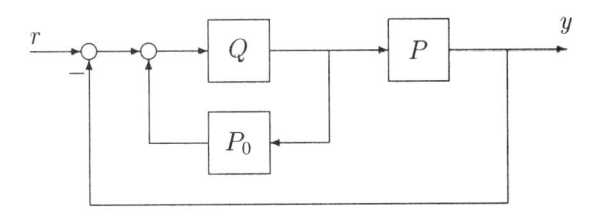

The control diagram can be redrawn as follows:

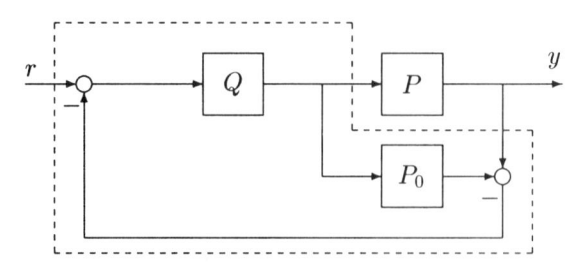

This control implementation is known as internal model control (IMC). Note that no signal is fed back if the model is exact. Discuss the advantage of this implementation and possible generalizations.

Problem 11.4 Use the Youla parameterization (the coprime factor form) to show that a SISO plant cannot be stabilized by a stable controller if the plant does not satisfy the parity interlacing properties. [A SISO plant is said to satisfy the parity interlacing property if the number of unstable real poles between any two unstable real zeros is even; $+\infty$ counts as a unstable zero if the plant is strictly proper. See Youla, Jabr, and Lu [1974] and Vidyasagar [1985].]

Chapter 12

Algebraic Riccati Equations

We studied the Lyapunov equation in Chapter 7 and saw the roles it played in some applications. A more general equation than the Lyapunov equation in control theory is the so-called *algebraic Riccati equation* or ARE for short. Roughly speaking, Lyapunov equations are most useful in system analysis while AREs are most useful in control system synthesis; particularly, they play central roles in $\mathcal{H}_2$ and $\mathcal{H}_\infty$ optimal control.

Let A, Q, and R be real $n \times n$ matrices with Q and R symmetric. Then an algebraic Riccati equation is the following matrix equation:

$$A^*X + XA + XRX + Q = 0. \tag{12.1}$$

Associated with this Riccati equation is a $2n \times 2n$ matrix:

$$H := \begin{bmatrix} A & R \\ -Q & -A^* \end{bmatrix}. \tag{12.2}$$

A matrix of this form is called a *Hamiltonian matrix*. The matrix H in equation (12.2) will be used to obtain the solutions to the equation (12.1). It is useful to note that the spectrum of H is symmetric about the imaginary axis. To see that, introduce the $2n \times 2n$ matrix:

$$J := \begin{bmatrix} 0 & -I \\ I & 0 \end{bmatrix}$$

having the property $J^2 = -I$. Then

$$J^{-1}HJ = -JHJ = -H^*$$

so H and $-H^*$ are similar. Thus λ is an eigenvalue iff $-\bar{\lambda}$ is.

This chapter is devoted to the study of this algebraic Riccati equation.

12.1 Stabilizing Solution and Riccati Operator

Assume that H has no eigenvalues on the imaginary axis. Then it must have n eigenvalues in Re $s < 0$ and n in Re $s > 0$. Consider the n-dimensional invariant spectral subspace, $\mathcal{X}_-(H)$, corresponding to eigenvalues of H in Re $s < 0$. By finding a basis for $\mathcal{X}_-(H)$, stacking the basis vectors up to form a matrix, and partitioning the matrix, we get

$$\mathcal{X}_-(H) = \text{Im} \begin{bmatrix} X_1 \\ X_2 \end{bmatrix}$$

where $X_1, X_2 \in \mathbb{C}^{n \times n}$. ($X_1$ and X_2 can be chosen to be real matrices.) If X_1 is nonsingular or, equivalently, if the two subspaces

$$\mathcal{X}_-(H), \quad \text{Im} \begin{bmatrix} 0 \\ I \end{bmatrix} \tag{12.3}$$

are complementary, we can set $X := X_2 X_1^{-1}$. Then X is uniquely determined by H (i.e., $H \longmapsto X$ is a function, which will be denoted Ric). We will take the domain of Ric, denoted dom(Ric), to consist of Hamiltonian matrices H with two properties: H has no eigenvalues on the imaginary axis and the two subspaces in equation (12.3) are complementary. For ease of reference, these will be called the *stability property* and the *complementarity property*, respectively. This solution will be called the *stabilizing solution*. Thus, $X = \text{Ric}(H)$ and

$$\text{Ric}: \text{dom}(\text{Ric}) \subset \mathbb{R}^{2n \times 2n} \longmapsto \mathbb{R}^{n \times n}.$$

Remark 12.1 It is now clear that to obtain the stabilizing solution to the Riccati equation, it is necessary to construct bases for the stable invariant subspace of H. One way of constructing this invariant subspace is to use eigenvectors and generalized eigenvectors of H. Suppose λ_i is an eigenvalue of H with multiplicity k (then $\lambda_{i+j} = \lambda_i$ for all $j = 1, \ldots, k-1$), and let v_i be a corresponding eigenvector and $v_{i+1}, \ldots, v_{i+k-1}$ be the corresponding generalized eigenvectors associated with v_i and λ_i. Then v_j are related by

$$\begin{aligned} (H - \lambda_i I) v_i &= 0 \\ (H - \lambda_i I) v_{i+1} &= v_i \\ &\vdots \\ (H - \lambda_i I) v_{i+k-1} &= v_{i+k-2}, \end{aligned}$$

and the span$\{v_j, j = i, \ldots, i+k-1\}$ is an invariant subspace of H. The union of all invariant subspaces corresponding to stable eigenvalues is the stable invariant subspace $\mathcal{X}_-(H)$. ◇

12.1. Stabilizing Solution and Riccati Operator

Example 12.1 Let

$$A = \begin{bmatrix} -3 & 2 \\ -2 & 1 \end{bmatrix}, \quad R = \begin{bmatrix} 0 & 0 \\ 0 & -1 \end{bmatrix}, \quad Q = \begin{bmatrix} 0 & 0 \\ 0 & 0 \end{bmatrix}.$$

The eigenvalues of H are $1, 1, -1, -1$, and the corresponding eigenvectors and generalized eigenvectors are

$$v_1 = \begin{bmatrix} 1 \\ 2 \\ 2 \\ -2 \end{bmatrix}, \quad v_2 = \begin{bmatrix} -1 \\ -3/2 \\ 1 \\ 0 \end{bmatrix}, \quad v_3 = \begin{bmatrix} 1 \\ 1 \\ 0 \\ 0 \end{bmatrix}, \quad v_4 = \begin{bmatrix} 1 \\ 3/2 \\ 0 \\ 0 \end{bmatrix}.$$

It is easy to check that $\{v_3, v_4\}$ form a basis for the stable invariant subspace $\mathcal{X}_-(H)$, $\{v_1, v_2\}$ form a basis for the antistable invariant subspace, and $\{v_1, v_3\}$ form a basis for another invariant subspace corresponding to eigenvalues $\{1, -1\}$ so

$$\overline{X} = 0, \quad \tilde{X} = \begin{bmatrix} -10 & 6 \\ 6 & -4 \end{bmatrix}, \quad \hat{X} = \begin{bmatrix} -2 & 2 \\ 2 & -2 \end{bmatrix}$$

are all solutions of the ARE with the property

$$\lambda(A + R\overline{X}) = \{-1, -1\}, \quad \lambda(A + R\tilde{X}) = \{1, 1\}, \quad \lambda(A + R\hat{X}) = \{1, -1\}.$$

Thus only $\overline{X}$ is the stabilizing solution. The stabilizing solution can be found using the following MATLAB command:

≫ [**X₁, X₂**] = ric_schr(**H**), **X** = **X₂/X₁**

The following well-known results give some properties of X as well as verifiable conditions under which H belongs to dom(Ric).

Theorem 12.1 *Suppose $H \in \text{dom}(\text{Ric})$ and $X = \text{Ric}(H)$. Then*

(i) X is real symmetric;

(ii) X satisfies the algebraic Riccati equation

$$A^*X + XA + XRX + Q = 0;$$

(iii) $A + RX$ is stable.

Proof. (i) Let X_1, X_2 be as before. It is claimed that

$$X_1^* X_2 \text{ is Hermitian.} \qquad (12.4)$$

To prove this, note that there exists a stable matrix H_- in $\mathbb{R}^{n \times n}$ such that

$$H \begin{bmatrix} X_1 \\ X_2 \end{bmatrix} = \begin{bmatrix} X_1 \\ X_2 \end{bmatrix} H_-.$$

(H_- is a matrix representation of $H|_{\mathcal{X}_-(H)}$.) Premultiply this equation by

$$\begin{bmatrix} X_1 \\ X_2 \end{bmatrix}^* J$$

to get

$$\begin{bmatrix} X_1 \\ X_2 \end{bmatrix}^* JH \begin{bmatrix} X_1 \\ X_2 \end{bmatrix} = \begin{bmatrix} X_1 \\ X_2 \end{bmatrix}^* J \begin{bmatrix} X_1 \\ X_2 \end{bmatrix} H_-. \qquad (12.5)$$

Since JH is symmetric, so is the left-hand side of equation (12.5) and so is the right-hand side:

$$(-X_1^* X_2 + X_2^* X_1) H_- = H_-^* (-X_1^* X_2 + X_2^* X_1)^*$$
$$= -H_-^* (-X_1^* X_2 + X_2^* X_1).$$

This is a Lyapunov equation. Since H_- is stable, the unique solution is

$$-X_1^* X_2 + X_2^* X_1 = 0.$$

This proves equation (12.4). Hence $X := X_2 X_1^{-1} = (X_1^{-1})^* (X_1^* X_2) X_1^{-1}$ is Hermitian. Since X_1 and X_2 can always be chosen to be real and X is unique, X is real symmetric.

(ii) Start with the equation

$$H \begin{bmatrix} X_1 \\ X_2 \end{bmatrix} = \begin{bmatrix} X_1 \\ X_2 \end{bmatrix} H_-$$

and postmultiply by X_1^{-1} to get

$$H \begin{bmatrix} I \\ X \end{bmatrix} = \begin{bmatrix} I \\ X \end{bmatrix} X_1 H_- X_1^{-1}. \qquad (12.6)$$

Now pre-multiply by $[X \quad -I]$:

$$[X \quad -I] H \begin{bmatrix} I \\ X \end{bmatrix} = 0.$$

12.1. Stabilizing Solution and Riccati Operator

This is precisely the Riccati equation.

(iii) Premultiply equation (12.6) by $[I \quad 0]$ to get

$$A + RX = X_1 H_- X_1^{-1}.$$

Thus $A + RX$ is stable because H_- is. □

Now we are going to state one of the main theorems of this section; it gives the necessary and sufficient conditions for the existence of a unique stabilizing solution of equation (12.1) under certain restrictions on the matrix R.

Theorem 12.2 *Suppose H has no imaginary eigenvalues and R is either positive semidefinite or negative semidefinite. Then $H \in \text{dom}(\text{Ric})$ if and only if (A, R) is stabilizable.*

Proof. ($\Leftarrow$) To prove that $H \in \text{dom}(\text{Ric})$, we must show that

$$\mathcal{X}_-(H), \quad \text{Im} \begin{bmatrix} 0 \\ I \end{bmatrix}$$

are complementary. This requires a preliminary step. As in the proof of Theorem 12.1 define X_1, X_2, H_- so that

$$\mathcal{X}_-(H) = \text{Im} \begin{bmatrix} X_1 \\ X_2 \end{bmatrix}$$

$$H \begin{bmatrix} X_1 \\ X_2 \end{bmatrix} = \begin{bmatrix} X_1 \\ X_2 \end{bmatrix} H_- . \tag{12.7}$$

We want to show that X_1 is nonsingular (i.e., $\text{Ker } X_1 = 0$). First, it is claimed that $\text{Ker } X_1$ is H_- invariant. To prove this, let $x \in \text{Ker } X_1$. Premultiply equation (12.7) by $[I \quad 0]$ to get

$$AX_1 + RX_2 = X_1 H_- . \tag{12.8}$$

Premultiply by $x^* X_2^*$, postmultiply by x, and use the fact that $X_2^* X_1$ is symmetric [see equation (12.4)] to get

$$x^* X_2^* R X_2 x = 0.$$

Since R is semidefinite, this implies that $RX_2 x = 0$. Now postmultiply equation (12.8) by x to get $X_1 H_- x = 0$ (i.e., $H_- x \in \text{Ker } X_1$). This proves the claim.

Now to prove that X_1 is nonsingular, suppose, on the contrary, that $\text{Ker } X_1 \neq 0$. Then $H_-|_{\text{Ker } X_1}$ has an eigenvalue, λ, and a corresponding eigenvector, x:

$$H_- x = \lambda x \tag{12.9}$$

$$\text{Re } \lambda < 0, \quad 0 \neq x \in \text{Ker } X_1.$$

Premultiply equation (12.7) by $[0\ \ I]$:

$$-QX_1 - A^*X_2 = X_2 H_- \ . \tag{12.10}$$

Postmultiply the above equation by x and use equation (12.9):

$$(A^* + \lambda I)X_2 x = 0.$$

Recall that $RX_2 x = 0$; we have

$$x^* X_2^* [A + \overline{\lambda} I \ \ R] = 0.$$

Then the stabilizability of (A, R) implies $X_2 x = 0$. But if both $X_1 x = 0$ and $X_2 x = 0$, then $x = 0$ since $\begin{bmatrix} X_1 \\ X_2 \end{bmatrix}$ has full column rank, which is a contradiction.

($\Rightarrow$) This is obvious since $H \in \text{dom}(\text{Ric})$ implies that X is a stabilizing solution and that $A + RX$ is asymptotically stable. It also implies that (A, R) must be stabilizable. $\square$

The following result is the so-called *bounded real lemma*, which follows immediately from the preceding theorem.

Corollary 12.3 *Let* $\gamma > 0$, $G(s) = \left[\begin{array}{c|c} A & B \\ \hline C & D \end{array}\right] \in \mathcal{RH}_\infty$, *and*

$$H := \begin{bmatrix} A + BR^{-1}D^*C & BR^{-1}B^* \\ -C^*(I + DR^{-1}D^*)C & -(A + BR^{-1}D^*C)^* \end{bmatrix}$$

where $R = \gamma^2 I - D^*D$. *Then the following conditions are equivalent:*

(i) $\|G\|_\infty < \gamma$.

(ii) $\bar{\sigma}(D) < \gamma$ *and* H *has no eigenvalues on the imaginary axis.*

(iii) $\bar{\sigma}(D) < \gamma$ *and* $H \in \text{dom}(\text{Ric})$.

(iv) $\bar{\sigma}(D) < \gamma$ *and* $H \in \text{dom}(\text{Ric})$ *and* $\text{Ric}(H) \geq 0$ *(*$\text{Ric}(H) > 0$ *if* (C, A) *is observable).*

(v) $\bar{\sigma}(D) < \gamma$ *and there exists an* $X \geq 0$ *such that*

$$X(A+BR^{-1}D^*C)+(A+BR^{-1}D^*C)^*X+XBR^{-1}B^*X+C^*(I+DR^{-1}D^*)C = 0$$

and $A + BR^{-1}D^*C + BR^{-1}B^*X$ *has no eigenvalues on the imaginary axis.*

(vi) $\bar{\sigma}(D) < \gamma$ *and there exists an* $X > 0$ *such that*

$$X(A+BR^{-1}D^*C)+(A+BR^{-1}D^*C)^*X+XBR^{-1}B^*X+C^*(I+DR^{-1}D^*)C < 0.$$

12.1. Stabilizing Solution and Riccati Operator

(vii) There exists an $X > 0$ such that

$$\begin{bmatrix} XA + A^*X & XB & C^* \\ B^*X & -\gamma I & D^* \\ C & D & -\gamma I \end{bmatrix} < 0.$$

Proof. The equivalence between (i) and (ii) has been shown in Chapter 4. The equivalence between (ii) and (iv) is obvious. The equivalence between (ii) and (iii) follows from the preceding theorem. It is also obvious that (iv) implies (v). We shall now show that (v) implies (i). Thus suppose that there is an $X \geq 0$ such that

$$X(A + BR^{-1}D^*C) + (A + BR^{-1}D^*C)^*X + XBR^{-1}B^*X + C^*(I + DR^{-1}D^*)C = 0$$

and $A + BR^{-1}(B^*X + D^*C)$ has no eigenvalues on the imaginary axis. Then

$$W(s) := \left[\begin{array}{c|c} A & -B \\ \hline B^*X + D^*C & R \end{array} \right]$$

has no zeros on the imaginary axis since

$$W^{-1}(s) = \left[\begin{array}{c|c} A + BR^{-1}(B^*X + D^*C) & BR^{-1} \\ \hline R^{-1}(B^*X + D^*C) & R^{-1} \end{array} \right]$$

has no poles on the imaginary axis. Next, note that

$$-X(j\omega I - A) - (j\omega I - A)^*X + XBR^{-1}D^*C + C^*DR^{-1}B^*X$$
$$+ XBR^{-1}B^*X + C^*(I + DR^{-1}D^*)C = 0.$$

Multiplying $B^*\{(j\omega I - A)^*\}^{-1}$ on the left and $(j\omega I - A)^{-1}B$ on the right of the preceding equation and completing square, we have

$$G^*(j\omega)G(j\omega) - \gamma^2 I + W^*(j\omega)R^{-1}W(j\omega) = 0$$

or

$$G^*(j\omega)G(j\omega) = \gamma^2 I - W^*(j\omega)R^{-1}W(j\omega).$$

Since $W(s)$ has no zeros on the imaginary axis, we conclude that $\|G\|_\infty < \gamma$.

The equivalence between (vi) and (vii) follows from Schur complement. It is also easy to show that (vi) implies (i) by following the similar procedure as above. To show that (i) implies (vi), let

$$\hat{G} = \left[\begin{array}{c|c} A & B \\ \hline C & D \\ \epsilon I & 0 \end{array} \right].$$

Then there exists an $\epsilon > 0$ such that $\|\hat{G}\|_\infty < \gamma$. Now (vi) follows by applying part (v) to $\hat{G}$. $\square$

Theorem 12.4 *Suppose H has the form*

$$H = \begin{bmatrix} A & -BB^* \\ -C^*C & -A^* \end{bmatrix}.$$

Then $H \in \text{dom(Ric)}$ iff (A, B) is stabilizable and (C, A) has no unobservable modes on the imaginary axis. Furthermore, $X = \text{Ric}(H) \geq 0$ if $H \in \text{dom(Ric)}$, and $\text{Ker}(X) = 0$ if and only if (C, A) has no stable unobservable modes.

Note that $\text{Ker}(X) \subset \text{Ker}(C)$, so that the equation $XM = C^*$ always has a solution for M, and a minimum F-norm solution is given by $X^+ C^*$.

Proof. It is clear from Theorem 12.2 that the stabilizability of (A, B) is necessary, and it is also sufficient if H has no eigenvalues on the imaginary axis. So we only need to show that, assuming (A, B) is stabilizable, H has no imaginary eigenvalues iff (C, A) has no unobservable modes on the imaginary axis. Suppose that $j\omega$ is an eigenvalue and $0 \neq \begin{bmatrix} x \\ z \end{bmatrix}$ is a corresponding eigenvector. Then

$$Ax - BB^* z = j\omega x$$

$$-C^* C x - A^* z = j\omega z.$$

Rearrange:
$$(A - j\omega I)x = BB^* z \tag{12.11}$$

$$-(A - j\omega I)^* z = C^* C x. \tag{12.12}$$

Thus
$$\langle z, (A - j\omega I)x \rangle = \langle z, BB^* z \rangle = \|B^* z\|^2$$

$$-\langle x, (A - j\omega I)^* z \rangle = \langle x, C^* C x \rangle = \|Cx\|^2$$

so $\langle x, (A - j\omega I)^* z \rangle$ is real and

$$-\|Cx\|^2 = \langle (A - j\omega I)x, z \rangle = \overline{\langle z, (A - j\omega I)x \rangle} = \|B^* z\|^2.$$

Therefore, $B^* z = 0$ and $Cx = 0$. So from equations (12.11) and (12.12)

$$(A - j\omega I)x = 0$$

$$(A - j\omega I)^* z = 0.$$

Combine the last four equations to get

$$z^* [A - j\omega I \quad B] = 0$$

12.1. Stabilizing Solution and Riccati Operator

$$\begin{bmatrix} A - j\omega I \\ C \end{bmatrix} x = 0.$$

The stabilizability of (A, B) gives $z = 0$. Now it is clear that $j\omega$ is an eigenvalue of H iff $j\omega$ is an unobservable mode of (C, A).

Next, set $X := \text{Ric}(H)$. We will show that $X \geq 0$. The Riccati equation is

$$A^*X + XA - XBB^*X + C^*C = 0$$

or, equivalently,

$$(A - BB^*X)^*X + X(A - BB^*X) + XBB^*X + C^*C = 0. \tag{12.13}$$

Noting that $A - BB^*X$ is stable (Theorem 12.1), we have

$$X = \int_0^\infty e^{(A-BB^*X)^*t}(XBB^*X + C^*C)e^{(A-BB^*X)t}dt. \tag{12.14}$$

Since $XBB^*X + C^*C$ is positive semidefinite, so is X.

Finally, we will show that $\text{Ker}X$ is nontrivial if and only if (C, A) has stable unobservable modes. Let $x \in \text{Ker}X$, then $Xx = 0$. Premultiply equation (12.13) by x^* and postmultiply by x to get

$$Cx = 0.$$

Now postmultiply equation (12.13) again by x to get

$$XAx = 0.$$

We conclude that $\text{Ker}(X)$ is an A-invariant subspace. Now if $\text{Ker}(X) \neq 0$, then there is a $0 \neq x \in \text{Ker}(X)$ and a λ such that $\lambda x = Ax = (A - BB^*X)x$ and $Cx = 0$. Since $(A - BB^*X)$ is stable, $\text{Re}\lambda < 0$; thus λ is a stable unobservable mode. Conversely, suppose (C, A) has an unobservable stable mode λ (i.e., there is an x such that $Ax = \lambda x, Cx = 0$). By premultiplying the Riccati equation by x^* and postmultiplying by x, we get

$$2\text{Re}\lambda x^*Xx - x^*XBB^*Xx = 0.$$

Hence $x^*Xx = 0$ (i.e., X is singular). □

Example 12.2 This example shows that the observability of (C, A) is not necessary for the existence of a positive definite stabilizing solution. Let

$$A = \begin{bmatrix} 1 & 0 \\ 0 & 2 \end{bmatrix}, \quad B = \begin{bmatrix} 1 \\ 1 \end{bmatrix}, \quad C = \begin{bmatrix} 0 & 0 \end{bmatrix}.$$

Then (A, B) is stabilizable, but (C, A) is not detectable. However,

$$X = \begin{bmatrix} 18 & -24 \\ -24 & 36 \end{bmatrix} > 0$$

is the stabilizing solution.

Corollary 12.5 *Suppose that (A, B) is stabilizable and (C, A) is detectable. Then the Riccati equation*

$$A^*X + XA - XBB^*X + C^*C = 0$$

has a unique positive semidefinite solution. Moreover, the solution is stabilizing.

Proof. It is obvious from the preceding theorem that the Riccati equation has a unique stabilizing solution and that the solution is positive semidefinite. Hence we only need to show that any positive semidefinite solution $X \geq 0$ must also be stabilizing. Then by the uniqueness of the stabilizing solution, we can conclude that there is only one positive semidefinite solution. To achieve that goal, let us assume that $X \geq 0$ satisfies the Riccati equation but that it is not stabilizing. First rewrite the Riccati equation as

$$(A - BB^*X)^*X + X(A - BB^*X) + XBB^*X + C^*C = 0 \tag{12.15}$$

and let λ and x be an unstable eigenvalue and the corresponding eigenvector of $A - BB^*X$, respectively; that is,

$$(A - BB^*X)x = \lambda x.$$

Now premultiply and postmultiply equation (12.15) by x^* and x, respectively, and we have

$$(\bar{\lambda} + \lambda)x^*Xx + x^*(XBB^*X + C^*C)x = 0.$$

This implies

$$B^*Xx = 0, \quad Cx = 0$$

since $\text{Re}(\lambda) \geq 0$ and $X \geq 0$. Finally, we arrive at

$$Ax = \lambda x, \quad Cx = 0.$$

That is, (C, A) is not detectable, which is a contradiction. Hence $\text{Re}(\lambda) < 0$ (i.e., $X \geq 0$ is the stabilizing solution). $\square$

Lemma 12.6 *Suppose D has full column rank and let $R = D^*D > 0$; then the following statements are equivalent:*

(i) $\begin{bmatrix} A - j\omega I & B \\ C & D \end{bmatrix}$ *has full column rank for all ω.*

(ii) $\left((I - DR^{-1}D^*)C, A - BR^{-1}D^*C\right)$ *has no unobservable modes on the $j\omega$ axis.*

12.1. Stabilizing Solution and Riccati Operator

Proof. Suppose $j\omega$ is an unobservable mode of $((I - DR^{-1}D^*)C, A - BR^{-1}D^*C)$; then there is an $x \neq 0$ such that
$$(A - BR^{-1}D^*C)x = j\omega x, \quad (I - DR^{-1}D^*)Cx = 0;$$
that is,
$$\begin{bmatrix} A - j\omega I & B \\ C & D \end{bmatrix} \begin{bmatrix} I & 0 \\ -R^{-1}D^*C & I \end{bmatrix} \begin{bmatrix} x \\ 0 \end{bmatrix} = 0.$$
But this implies that
$$\begin{bmatrix} A - j\omega I & B \\ C & D \end{bmatrix} \tag{12.16}$$
does not have full-column rank. Conversely, suppose equation (12.16) does not have full-column rank for some ω; then there exists $\begin{bmatrix} u \\ v \end{bmatrix} \neq 0$ such that
$$\begin{bmatrix} A - j\omega I & B \\ C & D \end{bmatrix} \begin{bmatrix} u \\ v \end{bmatrix} = 0.$$
Now let
$$\begin{bmatrix} u \\ v \end{bmatrix} = \begin{bmatrix} I & 0 \\ -R^{-1}D^*C & I \end{bmatrix} \begin{bmatrix} x \\ y \end{bmatrix}.$$
Then
$$\begin{bmatrix} x \\ y \end{bmatrix} = \begin{bmatrix} I & 0 \\ R^{-1}D^*C & I \end{bmatrix} \begin{bmatrix} u \\ v \end{bmatrix} \neq 0$$
and
$$(A - BR^{-1}D^*C - j\omega I)x + By = 0 \tag{12.17}$$
$$(I - DR^{-1}D^*)Cx + Dy = 0. \tag{12.18}$$
Premultiply equation (12.18) by D^* to get $y = 0$. Then we have
$$(A - BR^{-1}D^*C)x = j\omega x, \quad (I - DR^{-1}D^*)Cx = 0;$$
that is, $j\omega$ is an unobservable mode of $((I - DR^{-1}D^*)C, A - BR^{-1}D^*C)$. □

Remark 12.2 If D is not square, then there is a $D_\perp$ such that $\begin{bmatrix} D_\perp & DR^{-1/2} \end{bmatrix}$ is unitary and that $D_\perp D_\perp^* = I - DR^{-1}D^*$. Hence, in some cases we will write the condition (ii) in the preceding lemma as $(D_\perp^* C, A - BR^{-1}D^*C)$ having no imaginary unobservable modes. Of course, if D is square, the condition is simplified to $A - BR^{-1}D^*C$ having no imaginary eigenvalues. Note also that if $D^*C = 0$, condition (ii) becomes (C, A) having no imaginary unobservable modes. ◇

Corollary 12.7 *Suppose D has full column rank and denote $R = D^*D > 0$. Let H have the form*

$$H = \begin{bmatrix} A & 0 \\ -C^*C & -A^* \end{bmatrix} - \begin{bmatrix} B \\ -C^*D \end{bmatrix} R^{-1} \begin{bmatrix} D^*C & B^* \end{bmatrix}$$

$$= \begin{bmatrix} A - BR^{-1}D^*C & -BR^{-1}B^* \\ -C^*(I - DR^{-1}D^*)C & -(A - BR^{-1}D^*C)^* \end{bmatrix}.$$

Then $H \in \text{dom}(\text{Ric})$ iff (A, B) is stabilizable and $\begin{bmatrix} A - j\omega I & B \\ C & D \end{bmatrix}$ has full-column rank for all ω. Furthermore, $X = \text{Ric}(H) \geq 0$ if $H \in \text{dom}(\text{Ric})$, and $\text{Ker}(X) = 0$ if and only if $(D_\perp^* C, A - BR^{-1}D^*C)$ has no stable unobservable modes.

Proof. This is the consequence of Lemma 12.6 and Theorem 12.4. □

Remark 12.3 It is easy to see that the detectability (observability) of $(D_\perp^* C, A - BR^{-1}D^*C)$ implies the detectability (observability) of (C, A); however, the converse is, in general, not true. Hence the existence of a stabilizing solution to the Riccati equation in the preceding corollary is not guaranteed by the stabilizability of (A, B) and detectability of (C, A). Furthermore, even if a stabilizing solution exists, the positive definiteness of the solution is not guaranteed by the observability of (C, A) unless $D^*C = 0$. As an example, consider

$$A = \begin{bmatrix} 0 & 1 \\ 0 & 0 \end{bmatrix}, \quad B = \begin{bmatrix} 0 \\ -1 \end{bmatrix}, \quad C = \begin{bmatrix} 1 & 0 \\ 0 & 0 \end{bmatrix}, \quad D = \begin{bmatrix} 1 \\ 0 \end{bmatrix}.$$

Then (C, A) is observable, (A, B) is controllable, and

$$A - BD^*C = \begin{bmatrix} 0 & 1 \\ 1 & 0 \end{bmatrix}, \quad D_\perp^* C = \begin{bmatrix} 0 & 0 \end{bmatrix}.$$

A Riccati equation with the preceding data has a nonnegative definite stabilizing solution since $(D_\perp^* C, A - BR^{-1}D^*C)$ has no unobservable modes on the imaginary axis. However, the solution is not positive definite since $(D_\perp^* C, A - BR^{-1}D^*C)$ has a stable unobservable mode. On the other hand, if the B matrix is changed to

$$B = \begin{bmatrix} 0 \\ 1 \end{bmatrix},$$

then the corresponding Riccati equation has no stabilizing solution since, in this case, $(A - BD^*C)$ has eigenvalues on the imaginary axis although (A, B) is controllable and (C, A) is observable. ◇

Related MATLAB Commands: ric_eig, are

12.2 Inner Functions

A transfer function N is called *inner* if $N \in \mathcal{RH}_\infty$ and $N^\sim N = I$ and *co-inner* if $N \in \mathcal{RH}_\infty$ and $NN^\sim = I$. Note that N need not be square. Inner and co-inner are dual notions (i.e., N is an inner iff N^T is a co-inner). A matrix function $N \in \mathcal{RL}_\infty$ is called *all-pass* if N is square and $N^\sim N = I$; clearly a square inner function is all-pass. We will focus on the characterizations of inner functions here, and the properties of co-inner functions follow by duality.

Note that N inner implies that N has at least as many rows as columns. For N inner and any $q \in \mathbb{C}^m$, $v \in \mathcal{L}_2$, $\|N(j\omega)q\| = \|q\|$, $\forall \omega$ and $\|Nv\|_2 = \|v\|_2$ since $N(j\omega)^*N(j\omega) = I$ for all ω. Because of these norm preserving properties, inner matrices will play an important role in the control synthesis theory in this book. In this section, we present a state-space characterization of inner transfer functions.

Lemma 12.8 *Suppose* $N = \left[\begin{array}{c|c} A & B \\ \hline C & D \end{array}\right] \in \mathcal{RH}_\infty$ *and* $X = X^* \geq 0$ *satisfies*

$$A^*X + XA + C^*C = 0. \qquad (12.19)$$

Then

(a) $D^*C + B^*X = 0$ *implies* $N^\sim N = D^*D$.

(b) (A, B) *is controllable, and* $N^\sim N = D^*D$ *implies that* $D^*C + B^*X = 0$.

Proof. Conjugating the states of

$$N^\sim N = \left[\begin{array}{cc|c} A & 0 & B \\ -C^*C & -A^* & -C^*D \\ \hline D^*C & B^* & D^*D \end{array}\right]$$

by $\begin{bmatrix} I & 0 \\ -X & I \end{bmatrix}$ on the left and $\begin{bmatrix} I & 0 \\ -X & I \end{bmatrix}^{-1} = \begin{bmatrix} I & 0 \\ X & I \end{bmatrix}$ on the right yields

$$N^\sim N = \left[\begin{array}{cc|c} A & 0 & B \\ -(A^*X + XA + C^*C) & -A^* & -(XB + C^*D) \\ \hline B^*X + D^*C & B^* & D^*D \end{array}\right]$$

$$= \left[\begin{array}{cc|c} A & 0 & B \\ 0 & -A^* & -(XB + C^*D) \\ \hline B^*X + D^*C & B^* & D^*D \end{array}\right].$$

Then (a) and (b) follow easily. □

This lemma immediately leads to one characterization of inner matrices in terms of their state-space representations. Simply add the condition that $D^*D = I$ to Lemma 12.8 to get $N^\sim N = I$.

Corollary 12.9 *Suppose $N = \left[\begin{array}{c|c} A & B \\ \hline C & D \end{array}\right]$ is stable and minimal, and X is the observability Gramian. Then N is an inner if and only if*

(a) $D^*C + B^*X = 0$

(b) $D^*D = I$.

A transfer matrix $N_\perp$ is called a *complementary inner factor (CIF)* of N if $[N\ N_\perp]$ is square and is an inner. The dual notion of the complementary co-inner factor is defined in the obvious way. Given an inner N, the following lemma gives a construction of its CIF. The proof of this lemma follows from straightforward calculation and from the fact that $CX^+X = C$ since $\text{Im}(I - X^+X) \subset \text{Ker}(X) \subset \text{Ker}(C)$.

Lemma 12.10 *Let $N = \left[\begin{array}{c|c} A & B \\ \hline C & D \end{array}\right]$ be an inner and X be the observability Gramian. Then a CIF $N_\perp$ is given by*

$$N_\perp = \left[\begin{array}{c|c} A & -X^+C^*D_\perp \\ \hline C & D_\perp \end{array}\right]$$

where $D_\perp$ is an orthogonal complement of D such that $[D\ D_\perp]$ is square and orthogonal.

12.3 Notes and References

The general solutions of a Riccati equation are given by Martensson [1971]. The paper by Willems [1971] contains a comprehensive treatment of ARE and the related optimization problems. Some matrix factorization results are given in Doyle [1984]. Numerical methods for solving ARE can be found in Arnold and Laub [1984], Van Dooren [1981], and references therein. See Zhou, Doyle, and Glover [1996] and Lancaster and Rodman [1995] for a more extensive treatment of this subject.

12.4 Problems

Problem 12.1 Assume that $G(s) := \left[\begin{array}{c|c} A & B \\ \hline C & D \end{array}\right] \in \mathcal{RL}_\infty$ is a stabilizable and detectable realization and $\gamma > \|G(s)\|_\infty$. Show that there exists a transfer matrix $M \in$

12.4. Problems

$\mathcal{RL}_\infty$ such that $M^\sim M = \gamma^2 I - G^\sim G$ and $M^{-1} \in \mathcal{RH}_\infty$. A particular realization of M is

$$M(s) = \left[\begin{array}{c|c} A & B \\ \hline -R^{1/2}F & R^{1/2} \end{array}\right]$$

where

$$\begin{aligned} R &= \gamma^2 I - D^* D \\ F &= R^{-1}(B^* X + D^* C) \\ X &= \mathrm{Ric}\left[\begin{array}{cc} A + BR^{-1}D^*C & BR^{-1}B^* \\ -C^*(I + DR^{-1}D^*)C & -(A + BR^{-1}D^*C)^* \end{array}\right] \end{aligned}$$

and $X \geq 0$ if A is stable.

Problem 12.2 Let $G(s) = \left[\begin{array}{c|c} A & B \\ \hline C & D \end{array}\right]$ be a stabilizable and detectable realization. Suppose $G^\sim(j\omega)G(j\omega) > 0$ for all ω or $\left[\begin{array}{cc} A - j\omega & B \\ C & D \end{array}\right]$ has full-column rank for all ω. Let

$$X = \mathrm{Ric}\left[\begin{array}{cc} A - BR^{-1}D^*C & -BR^{-1}B^* \\ -C^*(I - DR^{-1}D^*)C & -(A - BR^{-1}D^*C)^* \end{array}\right]$$

with $R := D^*D > 0$. Show

$$W^\sim W = G^\sim G$$

where $W^{-1} \in \mathcal{RH}_\infty$ and

$$W = \left[\begin{array}{c|c} A & B \\ \hline R^{-1/2}(D^*C + B^*X) & R^{1/2} \end{array}\right].$$

Problem 12.3 A square $(m \times m)$ matrix function $G(s) \in \mathcal{RH}_\infty$ is said to be *positive real* (PR) if $G(j\omega) + G^*(j\omega) \geq 0$ for all finite ω; and $G(s)$ is said to be *strictly positive real* (SPR) if $G(j\omega) + G^*(j\omega) > 0$ for all $\omega \in \mathbb{R}$. Let $\left[\begin{array}{c|c} A & B \\ \hline C & D \end{array}\right]$ be a state-space realization of $G(s)$ with A stable (not necessarily a minimal realization). Suppose there exist $X \geq 0$, Q, and W such that

$$\begin{aligned} XA + A^*X &= -Q^*Q & (12.20) \\ B^*X + W^*Q &= C & (12.21) \\ D + D^* &= W^*W, & (12.22) \end{aligned}$$

Show that $G(s)$ is positive real and

$$G(s) + G^\sim(s) = M^\sim(s)M(s)$$

with $M(s) = \left[\begin{array}{c|c} A & B \\ \hline Q & W \end{array}\right]$. Furthermore, if $M(j\omega)$ has full-column rank for all $\omega \in \mathbb{R}$, then $G(s)$ is strictly positive real.

Problem 12.4 Suppose (A, B, C, D) is a minimal realization of $G(s)$ with A stable and $G(s)$ positive real. Show that there exist $X \geq 0$, Q, and W such that

$$XA + A^*X = -Q^*Q$$
$$B^*X + W^*Q = C$$
$$D + D^* = W^*W$$

and

$$G(s) + G^\sim(s) = M^\sim(s)M(s)$$

with $M(s) = \left[\begin{array}{c|c} A & B \\ \hline Q & W \end{array}\right]$. Furthermore, if $G(s)$ is strictly positive real, then $M(j\omega)$ has full-column rank for all $\omega \in \mathbb{R}$.

Problem 12.5 Let $\left[\begin{array}{c|c} A & B \\ \hline C & D \end{array}\right]$ be a state-space realization of $G(s) \in \mathcal{RH}_\infty$ with A stable and $R := D + D^* > 0$. Show that $G(s)$ is strictly positive real if and only if there exists a stabilizing solution to the following Riccati equation:

$$X(A - BR^{-1}C) + (A - BR^{-1}C)^*X + XBR^{-1}B^*X + C^*R^{-1}C = 0.$$

Moreover, $M(s) = \left[\begin{array}{c|c} A & B \\ \hline R^{-\frac{1}{2}}(C - B^*X) & R^{\frac{1}{2}} \end{array}\right]$ is minimal phase and

$$G(s) + G^\sim(s) = M^\sim(s)M(s).$$

Problem 12.6 Assume $p \geq m$. Show that there exists an $rcf\ G = NM^{-1}$ such that N is an *inner* if and only if $G^\sim G > 0$ on the $j\omega$ axis, including at ∞. This factorization is unique up to a constant unitary multiple. Furthermore, assume that the realization of $G = \left[\begin{array}{c|c} A & B \\ \hline C & D \end{array}\right]$ is stabilizable and that $\left[\begin{array}{cc} A - j\omega I & B \\ C & D \end{array}\right]$ has full column rank for all $\omega \in \mathbb{R}$. Then a particular realization of the desired coprime factorization is

$$\left[\begin{array}{c} M \\ N \end{array}\right] := \left[\begin{array}{c|c} A + BF & BR^{-1/2} \\ \hline F & R^{-1/2} \\ C + DF & DR^{-1/2} \end{array}\right] \in \mathcal{RH}_\infty$$

where

$$R = D^*D > 0$$

12.4. Problems

$$F = -R^{-1}(B^*X + D^*C)$$

and

$$X = \text{Ric}\begin{bmatrix} A - BR^{-1}D^*C & -BR^{-1}B^* \\ -C^*(I - DR^{-1}D^*)C & -(A - BR^{-1}D^*C)^* \end{bmatrix} \geq 0.$$

Moreover, a complementary inner factor can be obtained as

$$N_\perp = \left[\begin{array}{c|c} A + BF & -X^\dagger C^* D_\perp \\ \hline C + DF & D_\perp \end{array}\right]$$

if $p > m$.

Problem 12.7 Assume that $G = \left[\begin{array}{c|c} A & B \\ \hline C & D \end{array}\right] \in \mathcal{R}_p(s)$ and (A, B) is stabilizable. Show that there exists a right coprime factorization $G = NM^{-1}$ such that $M \in \mathcal{RH}_\infty$ is an inner if and only if G has no poles on the $j\omega$ axis. A particular realization is

$$\begin{bmatrix} M \\ N \end{bmatrix} := \left[\begin{array}{c|c} A + BF & B \\ \hline F & I \\ C + DF & D \end{array}\right] \in \mathcal{RH}_\infty$$

where

$$F = -B^*X$$

$$X = \text{Ric}\begin{bmatrix} A & -BB^* \\ 0 & -A^* \end{bmatrix} \geq 0.$$

Problem 12.8 A right coprime factorization of $G = NM^{-1}$ with $N, M \in \mathcal{RH}_\infty$ is called a *normalized right coprime factorization* if $M^\sim M + N^\sim N = I$; that is, if $\begin{bmatrix} M \\ N \end{bmatrix}$ is an inner. Similarly, an $lcf\ G = \tilde{M}^{-1}\tilde{N}$ is called a *normalized left coprime factorization* if $\begin{bmatrix} \tilde{M} & \tilde{N} \end{bmatrix}$ is a co-inner. Let a realization of G be given by $G = \left[\begin{array}{c|c} A & B \\ \hline C & D \end{array}\right]$ and define $R = I + D^*D > 0$ and $\tilde{R} = I + DD^* > 0$.

(a) Suppose (A, B) is stabilizable and (C, A) has no unobservable modes on the imaginary axis. Show that there is a normalized right coprime factorization $G = NM^{-1}$

$$\begin{bmatrix} M \\ N \end{bmatrix} := \left[\begin{array}{c|c} A + BF & BR^{-1/2} \\ \hline F & R^{-1/2} \\ C + DF & DR^{-1/2} \end{array}\right] \in \mathcal{RH}_\infty$$

where
$$F = -R^{-1}(B^*X + D^*C)$$
and
$$X = \operatorname{Ric}\begin{bmatrix} A - BR^{-1}D^*C & -BR^{-1}B^* \\ -C^*\tilde{R}^{-1}C & -(A - BR^{-1}D^*C)^* \end{bmatrix} \geq 0.$$

(b) Suppose (C, A) is detectable and (A, B) has no uncontrollable modes on the imaginary axis. Show that there is a normalized left coprime factorization $G = \tilde{M}^{-1}\tilde{N}$

$$\begin{bmatrix} \tilde{M} & \tilde{N} \end{bmatrix} := \left[\begin{array}{c|cc} A + LC & L & B + LD \\ \hline \tilde{R}^{-1/2}C & \tilde{R}^{-1/2} & \tilde{R}^{-1/2}D \end{array}\right]$$

where
$$L = -(BD^* + YC^*)\tilde{R}^{-1}$$
and
$$Y = \operatorname{Ric}\begin{bmatrix} (A - BD^*\tilde{R}^{-1}C)^* & -C^*\tilde{R}^{-1}C \\ -BR^{-1}B^* & -(A - BD^*\tilde{R}^{-1}C) \end{bmatrix} \geq 0.$$

(c) Show that the controllability Gramian P and the observability Gramian Q of $\begin{bmatrix} M \\ N \end{bmatrix}$ are given by
$$P = (I + YX)^{-1}Y, \quad Q = X$$
while the controllability Gramian $\tilde{P}$ and observability Gramian $\tilde{Q}$ of $\begin{bmatrix} \tilde{M} & \tilde{N} \end{bmatrix}$ are given by
$$\tilde{P} = Y, \quad \tilde{Q} = (I + XY)^{-1}X.$$

Problem 12.9 Let $G(s) = \left[\begin{array}{c|c} A & B \\ \hline C & D \end{array}\right]$. Find M_1 and M_2 such that $M_1^{-1}, M_2^{-1} \in \mathcal{RH}_\infty$ and
$$M_1 M_1^\sim = I + GG^\sim, \quad M_2^\sim M_2 = I + G^\sim G.$$

Problem 12.10 Let $A \in \mathbb{R}^{m \times m}$, $B \in \mathbb{R}^{n \times n}$, $C \in \mathbb{R}^{m \times n}$, and consider the Sylvester equation
$$AX + XB = C$$
for an unknown matrix $X \in \mathbb{R}^{m \times n}$. Let
$$M = \begin{bmatrix} B & 0 \\ C & -A \end{bmatrix}, \quad N = \begin{bmatrix} B & 0 \\ 0 & -A \end{bmatrix}.$$

12.4. Problems

1. Let the columns of $\begin{bmatrix} U \\ V \end{bmatrix} \in \mathbb{C}^{n+m \times n}$ be the eigenvectors of M associated with the eigenvalues of B and suppose U is nonsingular. Show that
$$X = VU^{-1}$$
solves the Sylvester equation. Moreover, every solution of the Sylvester equation can be written in the above form.

2. Show that the Sylvester equation has a solution if and only if M and N are similar. (See Lancaster and Tismenetsky [1985, page 423].)

Problem 12.11 Let $A \in \mathbb{R}^{n \times n}$. Show that
$$P(t) = \int_0^t e^{A^*\tau} Q e^{A\tau} d\tau$$
satisfies
$$\dot{P}(t) = A^* P(t) + P(t)A + Q, \quad P(0) = 0.$$

Problem 12.12 A more general case of the above problem is when the given matrices are time varying and the initial condition is not zero. Let $A(t), Q(t), P_0 \in \mathbb{R}^{n \times n}$. Show that
$$P(t) = \Phi^T(t, t_0) P_0 \Phi(t, t_0) + \int_{t_0}^t \Phi^T(t, \tau) Q(\tau) \Phi(t, \tau) d\tau$$
satisfies
$$\dot{P}(t) = A^* P(t) + P(t)A + Q(t), \quad P(t_0) = P_0$$
where $\Phi(t, \tau)$ is the state transition matrix for the system $\dot{x} = A(t)x$.

Problem 12.13 Let $A \in \mathbb{R}^{n \times n}$, $R = R^*$, $Q = Q^*$. Define
$$H = \begin{bmatrix} A & R \\ -Q & -A^* \end{bmatrix}.$$
Let
$$\Theta(t) = \begin{bmatrix} \Theta_{11}(t) & \Theta_{12}(t) \\ \Theta_{21}(t) & \Theta_{22}(t) \end{bmatrix} = e^{Ht}.$$
Show that
$$P(t) = (\Theta_{21}(t) + \Theta_{22} P_0)(\Theta_{11}(t) + \Theta_{12}(t) P_0)^{-1}$$
is the solution to the following differential Riccati equation:
$$-\dot{P}(t) = A^* P(t) + P(t)A + PRP + Q, \quad P(0) = P_0.$$

Problem 12.14 Let $A \in \mathbb{R}^{n \times n}$, $R = R^*$, $Q = Q^*$. Define

$$H = \begin{bmatrix} A & R \\ -Q & -A^* \end{bmatrix}.$$

Let

$$\Theta(t) = \begin{bmatrix} \Theta_{11}(t) & \Theta_{12}(t) \\ \Theta_{21}(t) & \Theta_{22}(t) \end{bmatrix} = e^{H(t-T)}.$$

Show that

$$P(t) = \Theta_{21}(t)\Theta_{11}^{-1}(t)$$

is the solution to the following differential Riccati equation:

$$-\dot{P}(t) = A^*P(t) + P(t)A + PRP + Q, \quad P(T) = 0.$$

Chapter 13

$\mathcal{H}_2$ Optimal Control

In this chapter we treat the optimal control of linear time-invariant systems with a quadratic performance criterion.

13.1 Introduction to Regulator Problem

Consider the following dynamical system:

$$\dot{x} = Ax + B_2 u, \quad x(t_0) = x_0 \tag{13.1}$$

where x_0 is given but arbitrary. Our objective is to find a control function $u(t)$ defined on $[t_0, T]$, which can be a function of the state $x(t)$, such that the state $x(t)$ is driven to a (small) neighborhood of origin at time T. This is the so-called *regulator problem*. One might suggest that this regulator problem can be trivially solved for any $T > t_0$ if the system is controllable. This is indeed the case if the controller can provide arbitrarily large amount of energy since, by the definition of controllability, one can immediately construct a control function that will drive the state to zero in an arbitrarily short time. However, this is not practical since any physical system has energy limitation (i.e., the actuator will eventually saturate). Furthermore, large control action can easily drive the system out of the region, where the given linear model is valid. Hence certain limitations have to be imposed on the control in practical engineering implementation. The constraints on control u may be measured in many different ways; for example,

$$\int_{t_0}^T \|u\| \, dt, \quad \int_{t_0}^T \|u\|^2 \, dt, \quad \sup_{t \in [t_0, T]} \|u\|;$$

That is, in terms of $\mathcal{L}_1$ norm, $\mathcal{L}_2$ norm, and $\mathcal{L}_\infty$ norm or, more generally, weighted $\mathcal{L}_1$ norm, $\mathcal{L}_2$ norm, and $\mathcal{L}_\infty$ norm

$$\int_{t_0}^T \|W_u u\| \, dt, \quad \int_{t_0}^T \|W_u u\|^2 \, dt, \quad \sup_{t \in [t_0, T]} \|W_u u\|$$

for some constant weighting matrix W_u.

Similarly, one might also want to impose some constraints on the transient response $x(t)$ in a similar fashion:

$$\int_{t_0}^T \|W_x x\| \, dt, \quad \int_{t_0}^T \|W_x x\|^2 \, dt, \quad \sup_{t \in [t_0, T]} \|W_x x\|$$

for some weighting matrix W_x. Hence the regulator problem can be posed as an optimal control problem with certain combined performance index on u and x. In this chapter, we shall be concerned exclusively with the $\mathcal{L}_2$ performance problem or quadratic performance problem. Moreover, we shall focus on the infinite time regulator problem (i.e., $T \to \infty$) and, without loss of generality, we shall assume $t_0 = 0$. In this case, our problem is as follows: Find a control $u(t)$ defined on $[0, \infty)$ such that the state $x(t)$ is driven to the origin as $t \to \infty$ and the following performance index is minimized:

$$\min_u \int_0^\infty \begin{bmatrix} x(t) \\ u(t) \end{bmatrix}^* \begin{bmatrix} Q & S \\ S^* & R \end{bmatrix} \begin{bmatrix} x(t) \\ u(t) \end{bmatrix} dt \quad (13.2)$$

for some $Q = Q^*$, S, and $R = R^* > 0$. This problem is traditionally called a *linear quadratic regulator* problem or, simply, an LQR problem. Here we have assumed $R > 0$ to emphasize that the control energy has to be finite (i.e., $u(t) \in \mathcal{L}_2[0, \infty)$). So $\mathcal{L}_2[0, \infty)$ is the space over which the integral is minimized. Moreover, it is also generally assumed that

$$\begin{bmatrix} Q & S \\ S^* & R \end{bmatrix} \geq 0. \quad (13.3)$$

Since R is positive definite, it has a square root, $R^{1/2}$, which is also positive-definite. By the substitution

$$u \leftarrow R^{1/2} u,$$

we may as well assume at the start that $R = I$. In fact, we can even assume $S = 0$ by using a pre-state feedback $u = -S^* x + v$ provided some care is exercised; however, this will not be assumed in the sequel. Since the matrix in equation (13.3) is positive semi-definite with $R = I$, it can be factored as

$$\begin{bmatrix} Q & S \\ S^* & I \end{bmatrix} = \begin{bmatrix} C_1^* \\ D_{12}^* \end{bmatrix} \begin{bmatrix} C_1 & D_{12} \end{bmatrix}.$$

Then equation (13.2) can be rewritten as

$$\min_{u \in \mathcal{L}_2[0,\infty)} \|C_1 x + D_{12} u\|_2^2.$$

13.2. Standard LQR Problem

In fact, the LQR problem is posed traditionally as the minimization problem:

$$\min_{u \in \mathcal{L}_2[0,\infty)} \|C_1 x + D_{12} u\|_2^2 \tag{13.4}$$

$$\text{subject to: } \dot{x} = Ax + Bu, \quad x(0) = x_0 \tag{13.5}$$

without explicitly mentioning the condition that the control should drive the state to the origin. Instead some assumptions are imposed on Q, S, and R (or, equivalently, on C_1 and D_{12}) to ensure that the optimal control law u has this property. To see what assumption one needs to make to ensure that the minimization problem formulated in equations (13.4) and (13.5) has a sensible solution, let us consider a simple example with $A = 1$, $B = 1, Q = 0, S = 0$, and $R = 1$:

$$\min_{u \in \mathcal{L}_2[0,\infty)} \int_0^\infty u^2 dt, \quad \dot{x} = x + u, \quad x(0) = x_0.$$

It is clear that $u = 0$ is the optimal solution. However, the system with $u = 0$ is unstable and $x(t)$ diverges exponentially to infinity since $x(t) = e^t x_0$. The problem with this example is that this performance index does not "see" the unstable state x. Thus, to ensure that the minimization problem in equations (13.4) and (13.5) is sensible, we must assume that all unstable states can be "seen" from the performance index; that is, (C_1, A) must be detectable. An LQR problem with such an assumption will be called a *standard LQR problem*.

On the other hand, if the closed-loop stability is imposed on the preceding minimization, then it can be shown that $\min_{u \in \mathcal{L}_2[0,\infty)} \int_0^\infty u^2 dt = 2x_0^2$ and $u(t) = -2x(t)$ is the optimal control. This can also be generalized to a more general case where (C_1, A) is not necessarily detectable. Such a LQR problem will be referred to as an *Extended LQR problem*.

13.2 Standard LQR Problem

In this section, we shall consider the LQR problem as traditionally formulated.

Standard LQR Problem

Let a dynamical system be described by

$$\dot{x} = Ax + B_2 u, \quad x(0) = x_0 \text{ given but arbitrary} \tag{13.6}$$
$$z = C_1 x + D_{12} u \tag{13.7}$$

and suppose that the system parameter matrices satisfy the following assumptions:

(A1) (A, B_2) is stabilizable;

(A2) D_{12} has full column rank with $\begin{bmatrix} D_{12} & D_\perp \end{bmatrix}$ unitary;

(A3) (C_1, A) is detectable;

(A4) $\begin{bmatrix} A - j\omega I & B_2 \\ C_1 & D_{12} \end{bmatrix}$ has full column rank for all ω.

Find an optimal control law $u \in \mathcal{L}_2[0, \infty)$ such that the performance criterion $\|z\|_2^2$ is minimized.

Remark 13.1 Assumption (A1) is clearly necessary for the existence of a stabilizing control function u. The assumption (A2) is made for simplicity of notation and is actually a restatement that $R = D_{12}^* D_{12} = I$. Note also that $D_\perp$ drops out when D_{12} is square. It is interesting to point out that (A3) is not needed in the Extended LQR problem. The assumption (A3) enforces that the unconditional optimization problem will result in a stabilizing control law. In fact, the assumption (A3) together with (A1) guarantees that the input/output stability implies the internal stability; that is, $u \in \mathcal{L}_2$ and $z \in \mathcal{L}_2$ imply $x \in \mathcal{L}_2$, which will be shown in Lemma 13.1. Finally note that (A4) is equivalent to the condition that $(D_\perp^* C_1, A - B_2 D_{12}^* C_1)$ has no unobservable modes on the imaginary axis and is weaker than the popular assumption of detectability of $(D_\perp^* C_1, A - B_2 D_{12}^* C_1)$. (A4), together with the stabilizability of (A, B_2), guarantees by Corollary 12.7 that the following Hamiltonian matrix belongs to dom(Ric) and that $X = \text{Ric}(H) \geq 0$:

$$H = \begin{bmatrix} A & 0 \\ -C_1^* C_1 & -A^* \end{bmatrix} - \begin{bmatrix} B_2 \\ -C_1^* D_{12} \end{bmatrix} \begin{bmatrix} D_{12}^* C_1 & B_2^* \end{bmatrix}$$

$$= \begin{bmatrix} A - B_2 D_{12}^* C_1 & -B_2 B_2^* \\ -C_1^* D_\perp D_\perp^* C_1 & -(A - B_2 D_{12}^* C_1)^* \end{bmatrix}. \tag{13.8}$$

Note also that if $D_{12}^* C_1 = 0$, then (A4) is implied by the detectability of (C_1, A). ◇

Note that the Riccati equation corresponding to equation (13.8) is

$$(A - B_2 D_{12}^* C_1)^* X + X(A - B_2 D_{12}^* C_1) - X B_2 B_2^* X + C_1^* D_\perp D_\perp^* C_1 = 0. \tag{13.9}$$

Now let X be the corresponding stabilizing solution and define

$$F := -(B_2^* X + D_{12}^* C_1). \tag{13.10}$$

Then $A + B_2 F$ is stable. Denote

$$A_F := A + B_2 F, \quad C_F := C_1 + D_{12} F$$

13.2. Standard LQR Problem

and rearrange equation (13.9) to get

$$A_F^* X + X A_F + C_F^* C_F = 0. \tag{13.11}$$

Thus X is the observability Gramian of (C_F, A_F).

Consider applying the control law $u = Fx$ to the system equations (13.6) and (13.7). The controlled system becomes

$$\dot{x} = A_F x, \quad x(0) = x_0$$
$$z = C_F x$$

or, equivalently,

$$\dot{x} = A_F x + x_0 \delta(t), \quad x(0_-) = 0$$
$$z = C_F x.$$

The associated transfer matrix is

$$G_c(s) = \left[\begin{array}{c|c} A_F & I \\ \hline C_F & 0 \end{array} \right]$$

and

$$\|G_c x_0\|_2^2 = x_0^* X x_0.$$

The proof of the following theorem requires a preliminary result about internal stability given input-output stability.

Lemma 13.1 *If $u, z \in \mathcal{L}_2[0, \infty)$ and (C_1, A) is detectable in the system described by equations (13.6) and (13.7), then $x \in \mathcal{L}_2[0, \infty)$. Furthermore, $x(t) \to 0$ as $t \to \infty$.*

Proof. Since (C_1, A) is detectable, there exists L such that $A + LC_1$ is stable. Let $\hat{x}$ be the state estimate of x given by

$$\dot{\hat{x}} = (A + LC_1)\hat{x} + (LD_{12} + B_2)u - Lz.$$

Then $\hat{x} \in \mathcal{L}_2[0, \infty)$ since z and u are in $\mathcal{L}_2[0, \infty)$. Now let $e = x - \hat{x}$; then

$$\dot{e} = (A + LC_1)e$$

and $e \in \mathcal{L}_2[0, \infty)$. Therefore, $x = e + \hat{x} \in \mathcal{L}_2[0, \infty)$. It is easy to see that $e(t) \to 0$ as $t \to \infty$ for any initial condition $e(0)$. Finally, $x(t) \to 0$ since $\hat{x} \to 0$. □

Theorem 13.2 *There exists a unique optimal control for the LQR problem, namely $u = Fx$. Moreover,*

$$\min_{u \in \mathcal{L}_2[0, \infty)} \|z\|_2 = \|G_c x_0\|_2.$$

Note that the optimal control strategy is a constant gain state feedback, and this gain is independent of the initial condition x_0.

Proof. With the change of variable $v = u - Fx$, the system can be written as

$$\begin{bmatrix} \dot{x} \\ z \end{bmatrix} = \begin{bmatrix} A_F & B_2 \\ C_F & D_{12} \end{bmatrix} \begin{bmatrix} x \\ v \end{bmatrix}, \qquad x(0) = x_0. \tag{13.12}$$

Now if $v \in \mathcal{L}_2[0, \infty)$, then $x, z \in \mathcal{L}_2[0, \infty)$ and $x(\infty) = 0$ since A_F is stable. Hence $u = Fx + v \in \mathcal{L}_2[0, \infty)$. Conversely, if $u, z \in \mathcal{L}_2[0, \infty)$, then from Lemma 13.1 $x \in \mathcal{L}_2[0, \infty)$. So $v \in \mathcal{L}_2[0, \infty)$. Thus the mapping $v = u - Fx$ between $v \in \mathcal{L}_2[0, \infty)$ and those $u \in \mathcal{L}_2[0, \infty)$ that make $z \in \mathcal{L}_2[0, \infty)$ is one-to-one and onto. Therefore,

$$\min_{u \in \mathcal{L}_2[0,\infty)} \|z\|_2 = \min_{v \in \mathcal{L}_2[0,\infty)} \|z\|_2.$$

By differentiating $x(t)^* X x(t)$ with respect to t along a solution of the differential equation (13.12) and by using equation (13.9) and the fact that $C_F^* D_{12} = -XB_2$, we see that

$$\begin{aligned} \frac{d}{dt} x^* X x &= \dot{x}^* X x + x^* X \dot{x} = x^* (A_F^* X + X A_F) x + 2x^* X B_2 v \\ &= -x^* C_F^* C_F x + 2x^* X B_2 v \\ &= -(C_F x + D_{12} v)^* (C_F x + D_{12} v) + 2x^* C_F^* D_{12} v + v^* v + 2x^* X B_2 v \\ &= -\|z\|^2 + \|v\|^2. \end{aligned} \tag{13.13}$$

Now integrate equation (13.13) from 0 to ∞ to get

$$\|z\|_2^2 = x_0^* X x_0 + \|v\|_2^2.$$

Clearly, the unique optimal control is $v = 0$, i.e., $u = Fx$. $\square$

13.3 Extended LQR Problem

This section considers the extended LQR problem where no detectability assumption is made for (C_1, A).

Extended LQR Problem

Let a dynamical system be given by

$$\begin{aligned} \dot{x} &= Ax + B_2 u, \quad x(0) = x_0 \text{ given but arbitrary} \\ z &= C_1 x + D_{12} u \end{aligned}$$

with the following assumptions:

(A1) (A, B_2) is stabilizable;

(A2) D_{12} has full column rank with $\begin{bmatrix} D_{12} & D_\perp \end{bmatrix}$ unitary;

(A3) $\begin{bmatrix} A - j\omega I & B_2 \\ C_1 & D_{12} \end{bmatrix}$ has full column rank for all ω.

Find an optimal control law $u \in \mathcal{L}_2[0, \infty)$ such that the system is internally stable (i.e., $x \in \mathcal{L}_2[0, \infty)$) and the performance criterion $\|z\|_2^2$ is minimized.

Assume the same notation as in the last section, we have:

Theorem 13.3 *There exists a unique optimal control for the extended LQR problem, namely $u = Fx$. Moreover,*

$$\min_{u \in \mathcal{L}_2[0,\infty)} \|z\|_2 = \|G_c x_0\|_2.$$

Proof. The proof of this theorem is very similar to the proof of the standard LQR problem except that, in this case, the input/output stability may not necessarily imply the internal stability. Instead, the internal stability is guaranteed by the way of choosing control law.

Suppose that $u \in \mathcal{L}_2[0, \infty)$ is such a control law that the system is stable, i.e., $x \in \mathcal{L}_2[0, \infty)$. Then $v = u - Fx \in \mathcal{L}_2[0, \infty)$. On the other hand, let $v \in \mathcal{L}_2[0, \infty)$ and consider

$$\begin{bmatrix} \dot{x} \\ z \end{bmatrix} = \begin{bmatrix} A_F & B_2 \\ C_F & D_{12} \end{bmatrix} \begin{bmatrix} x \\ v \end{bmatrix}, \quad x(0) = x_0.$$

Then $x, z \in \mathcal{L}_2[0, \infty)$ and $x(\infty) = 0$ since A_F is stable. Hence $u = Fx + v \in \mathcal{L}_2[0, \infty)$. Again the mapping $v = u - Fx$ between $v \in \mathcal{L}_2[0, \infty)$ and those $u \in \mathcal{L}_2[0, \infty)$ that make $z \in \mathcal{L}_2[0, \infty)$ and $x \in \mathcal{L}_2[0, \infty)$ is one to one and onto. Therefore,

$$\min_{u \in \mathcal{L}_2[0,\infty)} \|z\|_2 = \min_{v \in \mathcal{L}_2[0,\infty)} \|z\|_2.$$

Using the same technique as in the proof of the standard LQR problem, we have

$$\|z\|_2^2 = x_0^* X x_0 + \|v\|_2^2.$$

Thus, the unique optimal control is $v = 0$, i.e., $u = Fx$. □

13.4 Guaranteed Stability Margins of LQR

Now we will consider the system described by equation (13.6) with the LQR control law $u = Fx$. The closed-loop block diagram is as shown in Figure 13.1.

The following result is the key to stability margins of an LQR control law.

Lemma 13.4 *Let $F = -(B_2^* X + D_{12}^* C_1)$ and define $G_{12} = D_{12} + C_1(sI - A)^{-1}B_2$. Then*

$$\left(I - B_2^*(-sI - A^*)^{-1}F^*\right)\left(I - F(sI - A)^{-1}B_2\right) = G_{12}^\sim(s) G_{12}(s).$$

```
         u                          x
    ┌─ F ──→ ẋ = Ax + B₂u ──→
    │                      │
    └──────────────────────┘
```

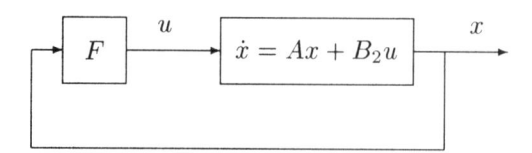

Figure 13.1: LQR closed-loop system

Proof. Note that the Riccati equation (13.9) can be written as

$$XA + A^*X - F^*F + C_1^*C_1 = 0.$$

Add and subtract sX to the above equation to get

$$-X(sI - A) - (-sI - A^*)X - F^*F + C_1^*C_1 = 0.$$

Now multiply the above equation from the left by $B_2^*(-sI - A^*)^{-1}$ and from the right by $(sI - A)^{-1}B_2$ to get

$$-B_2^*(-sI - A^*)^{-1}XB_2 - B_2^*X(sI - A)^{-1}B_2 - B_2^*(-sI - A^*)^{-1}F^*F(sI - A)^{-1}B_2$$
$$+ B_2^*(-sI - A^*)^{-1}C_1^*C_1(sI - A)^{-1}B_2 = 0.$$

Using $-B_2^*X = F + D_{12}^*C_1$ in the above equation, we have

$$B_2^*(-sI - A^*)^{-1}F^* + F(sI - A)^{-1}B_2 - B_2^*(-sI - A^*)^{-1}F^*F(sI - A)^{-1}B_2$$
$$+ B_2^*(-sI - A^*)^{-1}C_1^*D_{12} + D_{12}^*C_1(sI - A)^{-1}B_2$$
$$+ B_2^*(-sI - A^*)^{-1}C_1^*C_1(sI - A)^{-1}B_2 = 0.$$

Then the result follows from completing the square and from the fact that $D_{12}^*D_{12} = I$. □

Corollary 13.5 *Suppose* $D_{12}^*C_1 = 0$. *Then*

$$\left(I - B_2^*(-sI - A^*)^{-1}F^*\right)\left(I - F(sI - A)^{-1}B_2\right) = I + B_2^*(-sI - A^*)^{-1}C_1^*C_1(sI - A)^{-1}B_2.$$

In particular,

$$\left(I - B_2^*(-j\omega I - A^*)^{-1}F^*\right)\left(I - F(j\omega I - A)^{-1}B_2\right) \geq I \qquad (13.14)$$

and

$$\left(I + B_2^*(-j\omega I - A^* - F^*B_2^*)^{-1}F^*\right)\left(I + F(j\omega I - A - B_2F)^{-1}B_2\right) \leq I. \qquad (13.15)$$

13.5. Standard $\mathcal{H}_2$ Problem

Note that the inequality (13.15) follows from taking the inverse of inequality (13.14).

Define $G(s) = -F(sI - A)^{-1}B_2$ and assume for the moment that the system is single-input. Then the inequality (13.14) shows that the open-loop Nyquist diagram of the system $G(s)$ in Figure 13.1 never enters the unit disk centered at $(-1,0)$ of the complex plane. Hence the system has at least a 6 dB ($= 20\log 2$) gain margin and a $60°$ phase margin in both directions. A similar interpretation may be generalized to multiple-input systems.

Next, it is noted that the inequality (13.15) can also be given some robustness interpretation. In fact, it implies that the closed-loop system in Figure 13.1 is stable even if the open-loop system $G(s)$ is perturbed additively by a $\Delta \in \mathcal{RH}_\infty$ as long as $\|\Delta\|_\infty < 1$. This can be seen from the following block diagram and the small gain theorem, where the transfer matrix from w to z is exactly $I + F(j\omega I - A - B_2 F)^{-1}B_2$.

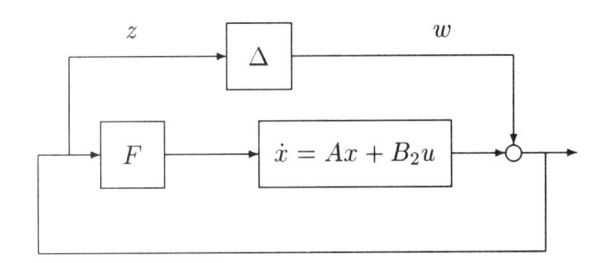

13.5 Standard $\mathcal{H}_2$ Problem

The system considered in this section is described by the following standard block diagram:

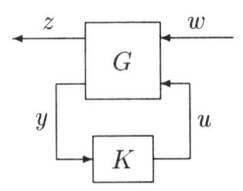

The realization of the transfer matrix G is taken to be of the form

$$G(s) = \left[\begin{array}{c|cc} A & B_1 & B_2 \\ \hline C_1 & 0 & D_{12} \\ C_2 & D_{21} & 0 \end{array} \right].$$

Notice the special off-diagonal structure of D: D_{22} is assumed to be zero so that G_{22} is strictly proper;[1] also, D_{11} is assumed to be zero in order to guarantee that the $\mathcal{H}_2$

---
[1] This assumption is made without loss of generality since a substitution of $K_D = K(I + D_{22}K)^{-1}$ would give the controller for $D_{22} \neq 0$.

problem is properly posed.[2]

The following additional assumptions are made for the output feedback $\mathcal{H}_2$ problem in this chapter:

(i) (A, B_2) is stabilizable and (C_2, A) is detectable;

(ii) $R_1 = D_{12}^* D_{12} > 0$ and $R_2 = D_{21} D_{21}^* > 0$;

(iii) $\begin{bmatrix} A - j\omega I & B_2 \\ C_1 & D_{12} \end{bmatrix}$ has full column rank for all ω;

(iv) $\begin{bmatrix} A - j\omega I & B_1 \\ C_2 & D_{21} \end{bmatrix}$ has full row rank for all ω.

The first assumption is for the stabilizability of G by output feedback, and the third and the fourth assumptions together with the first guarantee that the two Hamiltonian matrices associated with the following $\mathcal{H}_2$ problem belong to dom(Ric). The assumptions in (ii) guarantee that the $\mathcal{H}_2$ optimal control problem is nonsingular.

$\mathcal{H}_2$ **Problem** *The $\mathcal{H}_2$ control problem is to find a proper, real rational controller K that stabilizes G internally and minimizes the $\mathcal{H}_2$ norm of the transfer matrix T_{zw} from w to z.*

In the following discussions we shall assume that we have state models of G and K. Recall that a controller is said to be admissible if it is internally stabilizing and proper. By Corollary 12.7 the two Hamiltonian matrices

$$H_2 := \begin{bmatrix} A - B_2 R_1^{-1} D_{12}^* C_1 & -B_2 R_1^{-1} B_2^* \\ -C_1^*(I - D_{12} R_1^{-1} D_{12}^*) C_1 & -(A - B_2 R_1^{-1} D_{12}^* C_1)^* \end{bmatrix}$$

$$J_2 := \begin{bmatrix} (A - B_1 D_{21}^* R_2^{-1} C_2)^* & -C_2^* R_2^{-1} C_2 \\ -B_1(I - D_{21}^* R_2^{-1} D_{21}) B_1^* & -(A - B_1 D_{21}^* R_2^{-1} C_2) \end{bmatrix}$$

belong to dom(Ric), and, moreover, $X_2 := \text{Ric}(H_2) \geq 0$ and $Y_2 := \text{Ric}(J_2) \geq 0$. Define

$$F_2 := -R_1^{-1}(B_2^* X_2 + D_{12}^* C_1), \quad L_2 := -(Y_2 C_2^* + B_1 D_{21}^*) R_2^{-1}$$

and

$$A_{F_2} := A + B_2 F_2, \quad C_{1F_2} := C_1 + D_{12} F_2$$
$$A_{L_2} := A + L_2 C_2, \quad B_{1L_2} := B_1 + L_2 D_{21}$$
$$\hat{A}_2 := A + B_2 F_2 + L_2 C_2$$

$$G_c(s) := \left[\begin{array}{c|c} A_{F_2} & I \\ \hline C_{1F_2} & 0 \end{array}\right], \quad G_f(s) := \left[\begin{array}{c|c} A_{L_2} & B_{1L_2} \\ \hline I & 0 \end{array}\right].$$

Before stating the main theorem, we note the following fact:

---

[2] Recall that a rational proper stable transfer function is an $\mathcal{RH}_2$ function iff it is strictly proper.

13.5. Standard $\mathcal{H}_2$ Problem

Lemma 13.6 *Let $U, V \in \mathcal{RH}_\infty$ be defined as*

$$U := \left[\begin{array}{c|c} A_{F_2} & B_2 R_1^{-1/2} \\ \hline C_{1F_2} & D_{12} R_1^{-1/2} \end{array}\right], \quad V := \left[\begin{array}{c|c} A_{L_2} & B_{1L_2} \\ \hline R_2^{-1/2} C_2 & R_2^{-1/2} D_{21} \end{array}\right].$$

Then U is an inner and V is a co-inner, $U^\sim G_c \in \mathcal{RH}_2^\perp$, and $G_f V^\sim \in \mathcal{RH}_2^\perp$.

Proof. The proof uses standard manipulations of state-space realizations. From U we get

$$U^\sim(s) = \left[\begin{array}{c|c} -A_{F_2}^* & -C_{1F_2}^* \\ \hline R_1^{-1/2} B_2^* & R_1^{-1/2} D_{12}^* \end{array}\right].$$

Then it is easy to compute

$$U^\sim U = \left[\begin{array}{cc|c} -A_{F_2}^* & -C_{1F_2}^* C_{1F_2} & -C_{1F_2}^* D_{12} R_1^{-1/2} \\ 0 & A_{F_2} & B_2 R_1^{-1/2} \\ \hline R_1^{-1/2} B_2^* & R_1^{-1/2} D_{12}^* C_{1F_2} & I \end{array}\right]$$

$$U^\sim G_c = \left[\begin{array}{cc|c} -A_{F_2}^* & -C_{1F_2}^* C_{1F_2} & 0 \\ 0 & A_{F_2} & I \\ \hline R_1^{-1/2} B_2^* & R_1^{-1/2} D_{12}^* C_{1F_2} & 0 \end{array}\right].$$

Now do the similarity transformation

$$\begin{bmatrix} I & -X_2 \\ 0 & I \end{bmatrix}$$

on the states of the preceding transfer matrices and note that

$$A_{F_2}^* X_2 + X_2 A_{F_2} + C_{1F_2}^* C_{1F_2} = 0.$$

We get

$$U^\sim U = \left[\begin{array}{cc|c} -A_{F_2}^* & 0 & 0 \\ 0 & A_{F_2} & B_2 R_1^{-1/2} \\ \hline R_1^{-1/2} B_2^* & 0 & I \end{array}\right] = I$$

$$U^\sim G_c = \left[\begin{array}{cc|c} -A_{F_2}^* & 0 & -X_2 \\ 0 & A_{F_2} & I \\ \hline R_1^{-1/2} B_2^* & 0 & 0 \end{array}\right] = \left[\begin{array}{c|c} -A_{F_2}^* & -X_2 \\ \hline R_1^{-1/2} B_2^* & 0 \end{array}\right] \in \mathcal{RH}_2^\perp.$$

It follows by duality that $G_f V^\sim \in \mathcal{RH}_2^\perp$ and V is a co-inner. □

Theorem 13.7 *There exists a unique optimal controller*

$$K_{\text{opt}}(s) := \left[\begin{array}{c|c} \hat{A}_2 & -L_2 \\ \hline F_2 & 0 \end{array}\right].$$

Moreover,

$$\min \|T_{zw}\|_2^2 = \|G_c B_1\|_2^2 + \|R_1^{1/2} F_2 G_f\|_2^2 = \text{trace}\,(B_1^* X_2 B_1) + \text{trace}\,(R_1 F_2 Y_2 F_2^*).$$

Proof. Consider the all-stabilizing controller parameterization $K(s) = \mathcal{F}_\ell(M_2, Q)$, $Q \in \mathcal{RH}_\infty$ with

$$M_2(s) = \left[\begin{array}{c|cc} \hat{A}_2 & -L_2 & B_2 \\ \hline F_2 & 0 & I \\ -C_2 & I & 0 \end{array}\right]$$

and consider the following system diagram:

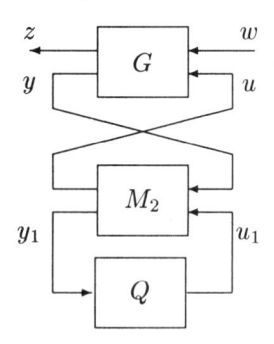

Then $T_{zw} = \mathcal{F}_\ell(N, Q)$ with

$$N = \left[\begin{array}{cc|cc} A_{F_2} & -B_2 F_2 & B_1 & B_2 \\ 0 & A_{L_2} & B_{1 L_2} & 0 \\ \hline C_{1 F_2} & -D_{12} F_2 & 0 & D_{12} \\ 0 & C_2 & D_{21} & 0 \end{array}\right]$$

and

$$T_{zw} = G_c B_1 - U R_1^{1/2} F_2 G_f + U R_1^{1/2} Q R_2^{1/2} V.$$

It follows from Lemma 13.6 that $G_c B_1$ and U are orthogonal. Thus

$$\begin{aligned} \|T_{zw}\|_2^2 &= \|G_c B_1\|_2^2 + \left\| U R_1^{1/2} F_2 G_f - U R_1^{1/2} Q R_2^{1/2} V \right\|_2^2 \\ &= \|G_c B_1\|_2^2 + \left\| R_1^{1/2} F_2 G_f - R_1^{1/2} Q R_2^{1/2} V \right\|_2^2. \end{aligned}$$

13.6. Stability Margins of $\mathcal{H}_2$ Controllers

Since G_f and V are also orthogonal by Lemma 13.6, we have

$$\begin{aligned}
\|T_{zw}\|_2^2 &= \|G_c B_1\|_2^2 + \left\|R_1^{1/2} F_2 G_f - R_1^{1/2} Q R_2^{1/2} V\right\|_2^2 \\
&= \|G_c B_1\|_2^2 + \left\|R_1^{1/2} F_2 G_f\right\|_2^2 + \left\|R_1^{1/2} Q R_2^{1/2}\right\|_2^2.
\end{aligned}$$

This shows clearly that $Q = 0$ gives the unique optimal control, so $K = \mathcal{F}_\ell(M_2, 0)$ is the unique optimal controller. □

The optimal $\mathcal{H}_2$ controller, K_{opt}, and the closed-loop transfer matrix, T_{zw}, can be obtained by the following MATLAB program:

≫ [**K**, **T**$_{\mathbf{zw}}$] = h2syn(**G**, n$_y$, n$_u$)

where n_y and n_u are the dimensions of y and u, respectively.

Related MATLAB Commands: lqg, lqr, lqr2, lqry, reg, lqe

13.6 Stability Margins of $\mathcal{H}_2$ Controllers

It is well-known that a system with LQR controller has at least 60° phase margin and 6 dB gain margin. However, it is not clear whether these stability margins will be preserved if the states are not available and the output feedback $\mathcal{H}_2$ (or LQG) controller has to be used. The answer is provided here through a counterexample from Doyle [1978]: There are no guaranteed stability margins for a $\mathcal{H}_2$ controller.

Consider a single-input and single-output two-state generalized dynamical system:

$$G(s) = \left[\begin{array}{c|c|c} \begin{bmatrix} 1 & 1 \\ 0 & 1 \end{bmatrix} & \begin{bmatrix} \sqrt{\sigma} & 0 \\ \sqrt{\sigma} & 0 \end{bmatrix} & \begin{bmatrix} 0 \\ 1 \end{bmatrix} \\ \hline \begin{bmatrix} \sqrt{q} & \sqrt{q} \\ 0 & 0 \end{bmatrix} & 0 & \begin{bmatrix} 0 \\ 1 \end{bmatrix} \\ \hline \begin{bmatrix} 1 & 0 \end{bmatrix} & \begin{bmatrix} 0 & 1 \end{bmatrix} & 0 \end{array}\right].$$

It can be shown analytically that

$$X_2 = \begin{bmatrix} 2\alpha & \alpha \\ \alpha & \alpha \end{bmatrix}, \quad Y_2 = \begin{bmatrix} 2\beta & \beta \\ \beta & \beta \end{bmatrix}$$

and

$$F_2 = -\alpha \begin{bmatrix} 1 & 1 \end{bmatrix}, \quad L_2 = -\beta \begin{bmatrix} 1 \\ 1 \end{bmatrix}$$

where
$$\alpha = 2 + \sqrt{4+q}, \quad \beta = 2 + \sqrt{4+\sigma}.$$
Then the optimal output $\mathcal{H}_2$ controller is given by
$$K_{\text{opt}} = \left[\begin{array}{cc|c} 1-\beta & 1 & \beta \\ -(\alpha+\beta) & 1-\alpha & \beta \\ \hline -\alpha & -\alpha & 0 \end{array}\right].$$

Suppose that the resulting closed-loop controller (or plant G_{22}) has a scalar gain k with a nominal value $k = 1$. Then the controller implemented in the system is actually
$$K = kK_{\text{opt}},$$
and the closed-loop system A matrix becomes
$$\tilde{A} = \begin{bmatrix} 1 & 1 & 0 & 0 \\ 0 & 1 & -k\alpha & -k\alpha \\ \beta & 0 & 1-\beta & 1 \\ \beta & 0 & -\alpha-\beta & 1-\alpha \end{bmatrix}.$$

It can be shown that the characteristic polynomial has the form
$$\det(sI - \tilde{A}) = s^4 + a_3 s^3 + a_2 s^2 + a_1 s + a_0$$
with
$$a_1 = \alpha + \beta - 4 + 2(k-1)\alpha\beta, \quad a_0 = 1 + (1-k)\alpha\beta.$$

Note that for closed-loop stability it is necessary to have $a_0 > 0$ and $a_1 > 0$. Note also that $a_0 \approx (1-k)\alpha\beta$ and $a_1 \approx 2(k-1)\alpha\beta$ for sufficiently large α and β if $k \neq 1$. It is easy to see that for sufficiently large α and β (or q and σ), the system is unstable for arbitrarily small perturbations in k in either direction. Thus, by choice of q and σ, the gain margins may be made arbitrarily small.

It is interesting to note that the margins deteriorate as control weight $(1/q)$ gets small (large q) and/or system driving noise gets large (large σ). In modern control folklore, these have often been considered ad hoc means of improving sensitivity.

It is also important to recognize that vanishing margins are not only associated with open-loop unstable systems. It is easy to construct minimum phase, open-loop stable counterexamples for which the margins are arbitrarily small.

The point of this example is that $\mathcal{H}_2$ (LQG) solutions, unlike LQR solutions, provide no global system-independent guaranteed robustness properties. Like their more classical colleagues, modern LQG designers are obliged to test their margins for each specific design.

It may, however, be possible to improve the robustness of a given design by relaxing the optimality of the filter with respect to error properties. A successful approach in this direction is the so called LQG loop transfer recovery (LQG/LTR) design technique. The idea is to design a filtering gain, L_2, in such way so that the LQG (or $\mathcal{H}_2$) control law will approximate the loop properties of the regular LQR control. This will not be explored further here; interested readers may consult related references.

13.7 Notes and References

Detailed treatment of $\mathcal{H}_2$ related theory, LQ optimal control, Kalman filtering, etc., can be found in Anderson and Moore [1989] or Kwakernaak and Sivan [1972]. The LQG/LTR control design was first introduced by Doyle and Stein [1981], and much work has been reported in this area since then. Additional results on the LQR stability margins can be found in Zhang and Fu [1996].

13.8 Problems

Problem 13.1 Parameterize all stabilizing controllers satisfying $\|T_{zw}\|_2 \leq \gamma$ for a given $\gamma > 0$.

Problem 13.2 Consider the feedback system in Figure 6.3 and suppose

$$P = \frac{s-10}{(s+1)(s+10)}, \quad W_e = \frac{1}{s+0.001}, \quad W_u = \frac{s+2}{s+10}.$$

Design a controller that minimizes

$$\left\| \begin{bmatrix} W_e S_o \\ W_u K S_o \end{bmatrix} \right\|_2.$$

Simulate the time response of the system when r is a step.

Problem 13.3 Repeat Problem 13.2 when $W_e = 1/s$. (Note that the solution given in this chapter cannot be applied directly.)

Problem 13.4 Consider the model matching (or reference) control problem shown here:

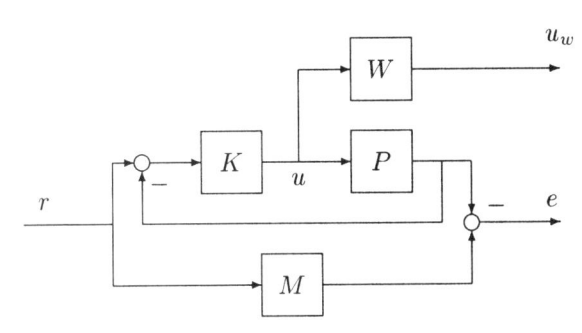

Let $M(s) \in \mathcal{H}_\infty$ be a strictly proper transfer matrix and $W(s), W^{-1}(s) \in \mathcal{RH}_\infty$. Formulate an $\mathcal{H}_2$ control problem that minimizes u_w and the error e through minimizing the $\mathcal{H}_2$ norm of the transfer matrix from r to (e, u_w). Apply your formula to

$$M(s) = \frac{4}{s^2 + 2s + 4}, \quad P(s) = \frac{10(s+2)}{(s+1)^3}, \quad W(s) = \frac{0.1(s+1)}{s+10}.$$

Problem 13.5 Repeat Problem 13.4 with $W = \epsilon$ for $\epsilon = 0.01$ and 0.0001. Study the behavior of the controller when $\epsilon \to 0$.

Problem 13.6 Repeat Problem 13.4 and Problem 13.5 with

$$P = \frac{10(2-s)}{(s+1)^3}.$$

Chapter 14

$\mathcal{H}_\infty$ Control

In this chapter we consider $\mathcal{H}_\infty$ control theory. Specifically, we formulate the optimal and suboptimal $\mathcal{H}_\infty$ control problems in Section 14.1. However, we will focus on the suboptimal case in this book and discuss why we do so. In Section 14.2 a suboptimal controller is characterized together with an algebraic proof for a class of simplified problems while leaving the more general problems to a later section. The behavior of the $\mathcal{H}_\infty$ controller as a function of performance level γ is considered in Section 14.3. The optimal controllers are also briefly considered in this section. Some other interpretations of the $\mathcal{H}_\infty$ controllers are given in Section 14.4. Section 14.5 presents the formulas for an optimal $\mathcal{H}_\infty$ controller. Section 14.6 considers again the standard $\mathcal{H}_\infty$ control problem but with some assumptions in the previous sections relaxed. Since the proof techniques in Section 14.2 can, in principle, be applied to this general case except with some more involved algebra, the detailed proof for the general case will not be given; only the formulas are presented. We shall indicate how the assumptions in the general case can be relaxed further to accommodate other more complicated problems in Section 14.7. Section 14.8 considers the integral control in the $\mathcal{H}_2$ and $\mathcal{H}_\infty$ theory and Section 14.9 considers how the general $\mathcal{H}_\infty$ solution can be used to solve the $\mathcal{H}_\infty$ filtering problem.

14.1 Problem Formulation

Consider the system described by the block diagram

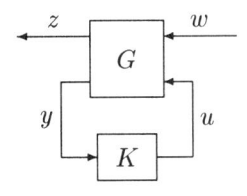

where the plant G and controller K are assumed to be real rational and proper. It will be assumed that state-space models of G and K are available and that their realizations are assumed to be stabilizable and detectable. Recall again that a controller is said to be *admissible* if it internally stabilizes the system. Clearly, stability is the most basic requirement for a practical system to work. Hence any sensible controller has to be admissible.

Optimal $\mathcal{H}_\infty$ Control: *Find all admissible controllers $K(s)$ such that $\|T_{zw}\|_\infty$ is minimized.*

It should be noted that the optimal $\mathcal{H}_\infty$ controllers as just defined are generally not unique for MIMO systems. Furthermore, finding an optimal $\mathcal{H}_\infty$ controller is often both numerically and theoretically complicated, as shown in Glover and Doyle [1989]. This is certainly in contrast with the standard $\mathcal{H}_2$ theory, in which the optimal controller is unique and can be obtained by solving two Riccati equations without iterations. Knowing the achievable optimal (minimum) $\mathcal{H}_\infty$ norm may be useful theoretically since it sets a limit on what we can achieve. However, in practice it is often not necessary and sometimes even undesirable to design an optimal controller, and it is usually much cheaper to obtain controllers that are very close in the norm sense to the optimal ones, which will be called *suboptimal controllers*. A suboptimal controller may also have other nice properties (e.g., lower bandwidth) over the optimal ones.

Suboptimal $\mathcal{H}_\infty$ Control: *Given $\gamma > 0$, find all admissible controllers $K(s)$, if there are any, such that $\|T_{zw}\|_\infty < \gamma$.*

For the reasons mentioned above, we focus our attention in this book on suboptimal control. When appropriate, we briefly discuss what will happen when γ approaches the optimal value.

14.2 A Simplified $\mathcal{H}_\infty$ Control Problem

The realization of the transfer matrix G is taken to be of the form

$$G(s) = \left[\begin{array}{c|cc} A & B_1 & B_2 \\ \hline C_1 & 0 & D_{12} \\ C_2 & D_{21} & 0 \end{array}\right].$$

The following assumptions are made:

(i) (A, B_1) is controllable and (C_1, A) is observable;

(ii) (A, B_2) is stabilizable and (C_2, A) is detectable;

(iii) $D_{12}^* \begin{bmatrix} C_1 & D_{12} \end{bmatrix} = \begin{bmatrix} 0 & I \end{bmatrix}$;

(iv) $\begin{bmatrix} B_1 \\ D_{21} \end{bmatrix} D_{21}^* = \begin{bmatrix} 0 \\ I \end{bmatrix}$.

14.2. A Simplified $\mathcal{H}_\infty$ Control Problem

Two additional assumptions that are implicit in the assumed realization for $G(s)$ are that $D_{11} = 0$ and $D_{22} = 0$. As we mentioned in the last chapter, $D_{22} \neq 0$ does not pose any problem since it is easy to form an equivalent problem with $D_{22} = 0$ by a linear fractional transformation on the controller $K(s)$. However, relaxing the assumption $D_{11} = 0$ complicates the formulas substantially.

The $\mathcal{H}_\infty$ solution involves the following two Hamiltonian matrices:

$$H_\infty := \begin{bmatrix} A & \gamma^{-2}B_1 B_1^* - B_2 B_2^* \\ -C_1^* C_1 & -A^* \end{bmatrix}, \quad J_\infty := \begin{bmatrix} A^* & \gamma^{-2}C_1^* C_1 - C_2^* C_2 \\ -B_1 B_1^* & -A \end{bmatrix}.$$

The important difference here from the $\mathcal{H}_2$ problem is that the (1,2)-blocks are not sign definite, so we cannot use Theorem 12.4 in Chapter 12 to guarantee that $H_\infty \in \text{dom}(\text{Ric})$ or $\text{Ric}(H_\infty) \geq 0$. Indeed, these conditions are intimately related to the existence of $\mathcal{H}_\infty$ suboptimal controllers. Note that the (1,2)-blocks are a suggestive combination of expressions from the $\mathcal{H}_\infty$ norm characterization in Chapter 4 (or bounded real lemma in Chapter 12) and from the $\mathcal{H}_2$ synthesis of Chapter 13. It is also clear that if γ approaches infinity, then these two Hamiltonian matrices become the corresponding $\mathcal{H}_2$ control Hamiltonian matrices. The reasons for the form of these expressions should become clear through the discussions and proofs for the following theorem.

Theorem 14.1 *There exists an admissible controller such that $\|T_{zw}\|_\infty < \gamma$ iff the following three conditions hold:*

(i) $H_\infty \in \text{dom}(\text{Ric})$ *and* $X_\infty := \text{Ric}(H_\infty) > 0$;

(ii) $J_\infty \in \text{dom}(\text{Ric})$ *and* $Y_\infty := \text{Ric}(J_\infty) > 0$;

(iii) $\rho(X_\infty Y_\infty) < \gamma^2$.

Moreover, when these conditions hold, one such controller is

$$K_{\text{sub}}(s) := \left[\begin{array}{c|c} \hat{A}_\infty & -Z_\infty L_\infty \\ \hline F_\infty & 0 \end{array} \right]$$

where

$$\hat{A}_\infty := A + \gamma^{-2} B_1 B_1^* X_\infty + B_2 F_\infty + Z_\infty L_\infty C_2$$
$$F_\infty := -B_2^* X_\infty, \quad L_\infty := -Y_\infty C_2^*, \quad Z_\infty := (I - \gamma^{-2} Y_\infty X_\infty)^{-1}.$$

Furthermore, the set of all admissible controllers such that $\|T_{zw}\|_\infty < \gamma$ equals the set of all transfer matrices from y to u in

$$M_\infty(s) = \left[\begin{array}{c|cc} \hat{A}_\infty & -Z_\infty L_\infty & Z_\infty B_2 \\ \hline F_\infty & 0 & I \\ -C_2 & I & 0 \end{array} \right]$$

where $Q \in \mathcal{RH}_\infty$, $\|Q\|_\infty < \gamma$.

We shall only give a proof of the first part of the theorem; the proof for the all-controller parameterization needs much more work and is omitted (see Zhou, Doyle, and Glover [1996] for a comprehensive treatment of the related topics). We shall first show some preliminary results.

Lemma 14.2 *Suppose that $X \in \mathbb{R}^{n \times n}$, $Y \in \mathbb{R}^{n \times n}$, with $X = X^* > 0$, and $Y = Y^* > 0$. Let r be a positive integer. Then there exist matrices $X_{12} \in \mathbb{R}^{n \times r}$, $X_2 \in \mathbb{R}^{r \times r}$ such that $X_2 = X_2^*$*

$$\begin{bmatrix} X & X_{12} \\ X_{12}^* & X_2 \end{bmatrix} > 0 \quad \text{and} \quad \begin{bmatrix} X & X_{12} \\ X_{12}^* & X_2 \end{bmatrix}^{-1} = \begin{bmatrix} Y & \star \\ \star & \star \end{bmatrix}$$

if and only if

$$\begin{bmatrix} X & I_n \\ I_n & Y \end{bmatrix} \geq 0 \quad \text{and} \quad \text{rank} \begin{bmatrix} X & I_n \\ I_n & Y \end{bmatrix} \leq n + r.$$

Proof. ($\Leftarrow$) By the assumption, there is a matrix $X_{12} \in \mathbb{R}^{n \times r}$ such that $X - Y^{-1} = X_{12} X_{12}^*$. Defining $X_2 := I_r$ completes the construction.

($\Rightarrow$) Using Schur complements,

$$Y = X^{-1} + X^{-1} X_{12} (X_2 - X_{12}^* X^{-1} X_{12})^{-1} X_{12}^* X^{-1}.$$

Inverting, using the matrix inversion lemma, gives

$$Y^{-1} = X - X_{12} X_2^{-1} X_{12}^*.$$

Hence, $X - Y^{-1} = X_{12} X_2^{-1} X_{12}^* \geq 0$, and, indeed, $\text{rank}(X - Y^{-1}) = \text{rank}(X_{12} X_2^{-1} X_{12}^*) \leq r$. $\square$

Lemma 14.3 *There exists an rth-order admissible controller such that $\|T_{zw}\|_\infty < \gamma$ only if the following three conditions hold:*

(i) There exists a $Y_1 > 0$ such that

$$AY_1 + Y_1 A^* + Y_1 C_1^* C_1 Y_1/\gamma^2 + B_1 B_1^* - \gamma^2 B_2 B_2^* < 0. \tag{14.1}$$

(ii) There exists an $X_1 > 0$ such that

$$X_1 A + A^* X_1 + X_1 B_1 B_1^* X_1/\gamma^2 + C_1^* C_1 - \gamma^2 C_2^* C_2 < 0. \tag{14.2}$$

(iii) $\begin{bmatrix} X_1/\gamma & I_n \\ I_n & Y_1/\gamma \end{bmatrix} \geq 0 \quad \text{rank} \begin{bmatrix} X_1/\gamma & I_n \\ I_n & Y_1/\gamma \end{bmatrix} \leq n + r.$

14.2. A Simplified $\mathcal{H}_\infty$ Control Problem

Proof. Suppose that there exists an rth-order controller $K(s)$ such that $\|T_{zw}\|_\infty < \gamma$. Let $K(s)$ have a state-space realization

$$K(s) = \left[\begin{array}{c|c} \hat{A} & \hat{B} \\ \hline \hat{C} & \hat{D} \end{array}\right].$$

Then

$$T_{zw} = \mathcal{F}_\ell(G,K) = \left[\begin{array}{cc|c} A + B_2\hat{D}C_2 & B_2\hat{C} & B_1 + B_2\hat{D}D_{21} \\ \hat{B}C_2 & \hat{A} & \hat{B}D_{21} \\ \hline C_1 + D_{12}\hat{D}C_2 & D_{12}\hat{C} & D_{12}\hat{D}D_{21} \end{array}\right] =: \left[\begin{array}{c|c} A_c & B_c \\ \hline C_c & D_c \end{array}\right].$$

Denote

$$R = \gamma^2 I - D_c^* D_c, \qquad \tilde{R} = \gamma^2 I - D_c D_c^*.$$

By Corollary 12.3, there exists an $\tilde{X} = \begin{bmatrix} X_1 & X_{12} \\ X_{12}^* & X_2 \end{bmatrix} > 0$ such that

$$\tilde{X}(A_c + B_c R^{-1} D_c^* C_c) + (A_c + B_c R^{-1} D_c^* C_c)^* \tilde{X} + \tilde{X} B_c R^{-1} B_c^* \tilde{X} + C_c^* \tilde{R}^{-1} C_c < 0. \quad (14.3)$$

This gives, after much algebraic manipulation,

$$X_1 A + A^* X_1 + X_1 B_1 B_1^* X_1/\gamma^2 + C_1^* C_1 - \gamma^2 C_2^* C_2$$
$$+ (X_1 B_1 \hat{D} + X_{12} \hat{B} + \gamma^2 C_2^*)(\gamma^2 I - \hat{D}^* \hat{D})^{-1}(X_1 B_1 \hat{D} + X_{12} \hat{B} + \gamma^2 C_2^*)^* < 0,$$

which implies that

$$X_1 A + A^* X_1 + X_1 B_1 B_1^* X_1/\gamma^2 + C_1^* C_1 - \gamma^2 C_2^* C_2 < 0.$$

On the other hand, let

$$\tilde{Y} = \gamma^2 \tilde{X}^{-1}$$

and partition $\tilde{Y}$ as $\tilde{Y} = \begin{bmatrix} Y_1 & Y_{12} \\ Y_{12}^* & Y_2 \end{bmatrix} > 0$. Then

$$(A_c + B_c R^{-1} D_c^* C_c)\tilde{Y} + \tilde{Y}(A_c + B_c R^{-1} D_c^* C_c)^* + \tilde{Y} C_c^* \tilde{R}^{-1} C_c \tilde{Y} + B_c R^{-1} B_c^* < 0. \quad (14.4)$$

This gives

$$AY_1 + Y_1 A^* + B_1 B_1^* - \gamma^2 B_2 B_2^* + Y_1 C_1^* C_1 Y_1/\gamma^2$$
$$+ (Y_1 C_1^* \hat{D}^* + Y_{12} \hat{C}^* + \gamma^2 B_2)(\gamma^2 I - \hat{D}\hat{D}^*)^{-1}(Y_1 C_1^* \hat{D}^* + Y_{12} \hat{C}^* + \gamma^2 B_2)^* < 0,$$

which implies that

$$AY_1 + Y_1 A^* + B_1 B_1^* - \gamma^2 B_2 B_2^* + Y_1 C_1^* C_1 Y_1/\gamma^2 < 0.$$

By Lemma 14.2, given $X_1 > 0$ and $Y_1 > 0$, there exists X_{12} and X_2 such that $\tilde{Y} = \gamma^2 \tilde{X}^{-1}$ or $\tilde{Y}/\gamma = (\tilde{X}/\gamma)^{-1}$:

$$\begin{bmatrix} X_1/\gamma & X_{12}/\gamma \\ X_{12}^*/\gamma & X_2/\gamma \end{bmatrix}^{-1} = \begin{bmatrix} Y_1/\gamma & \star \\ \star & \star \end{bmatrix}$$

if and only if

$$\begin{bmatrix} X_1/\gamma & I_n \\ I_n & Y_1/\gamma \end{bmatrix} \geq 0 \qquad \text{rank} \begin{bmatrix} X_1/\gamma & I_n \\ I_n & Y_1/\gamma \end{bmatrix} \leq n + r.$$

$\square$

To show that the inequalities in the preceding lemma imply the existence of the stabilizing solutions to the Riccati equations of X_∞ and Y_∞, we need the following theorem.

Theorem 14.4 *Let $R \geq 0$ and suppose (A, R) is controllable and there is an $X = X^*$ such that*

$$\mathcal{Q}(X) := XA + A^*X + XRX + Q < 0. \tag{14.5}$$

Then there exists a solution $X_+ > X$ to the Riccati equation

$$X_+ A + A^* X_+ + X_+ R X_+ + Q = 0 \tag{14.6}$$

such that $A + RX_+$ is antistable.

Proof. Let $R = BB^*$ for some B. Note the fact that (A, R) is controllable iff (A, B) is. Let X be such that $\mathcal{Q}(X) < 0$. Since (A, B) is controllable, there is an F_0 such that

$$A_0 := A - BF_0$$

is antistable. Now let $X_0 = X_0^*$ be the unique solution to the Lyapunov equation

$$X_0 A_0 + A_0^* X_0 - F_0^* F_0 + Q = 0.$$

Define

$$\hat{F}_0 := F_0 + B^*X,$$

and we have the following equation:

$$(X_0 - X)A_0 + A_0^*(X_0 - X) = \hat{F}_0^* \hat{F}_0 - \mathcal{Q}(X) > 0.$$

The antistability of A_0 implies that

$$X_0 > X.$$

Starting with X_0, we shall define a nonincreasing sequence of Hermitian matrices $\{X_i\}$. Associated with $\{X_i\}$, we shall also define a sequence of antistable matrices $\{A_i\}$ and a

14.2. A Simplified $\mathcal{H}_\infty$ Control Problem

sequence of matrices $\{F_i\}$. Assume inductively that we have already defined matrices $\{X_i\}$, $\{A_i\}$, and $\{F_i\}$ for i up to $n-1$ such that X_i is Hermitian and

$$X_0 \geq X_1 \geq \cdots \geq X_{n-1} > X,$$

$$A_i = A - BF_i \text{ is antistable}, \ i = 0, \ldots, n-1;$$

$$F_i = -B^* X_{i-1}, \ i = 1, \ldots, n-1;$$

$$X_i A_i + A_i^* X_i = F_i^* F_i - Q, \ i = 0, 1, \ldots, n-1. \tag{14.7}$$

Next, introduce

$$F_n = -B^* X_{n-1},$$

$$A_n = A - BF_n.$$

First we show that A_n is antistable. Using equation (14.7), with $i = n-1$, we get

$$X_{n-1} A_n + A_n^* X_{n-1} + Q - F_n^* F_n - (F_n - F_{n-1})^* (F_n - F_{n-1}) = 0. \tag{14.8}$$

Let

$$\hat{F}_n := F_n + B^* X;$$

then

$$(X_{n-1} - X) A_n + A_n^* (X_{n-1} - X) = -\mathcal{Q}(X) + \hat{F}_n^* \hat{F}_n + (F_n - F_{n-1})^* (F_n - F_{n-1}) > 0, \tag{14.9}$$

which implies that A_n is antistable by Lyapunov theorem since $X_{n-1} - X > 0$.

Now we introduce X_n as the unique solution of the Lyapunov equation:

$$X_n A_n + A_n^* X_n = F_n^* F_n - Q. \tag{14.10}$$

Then X_n is Hermitian. Next, we have

$$(X_n - X) A_n + A_n^* (X_n - X) = -\mathcal{Q}(X) + \hat{F}_n^* \hat{F}_n > 0,$$

and, by using equation (14.8),

$$(X_{n-1} - X_n) A_n + A_n^* (X_{n-1} - X_n) = (F_n - F_{n-1})^* (F_n - F_{n-1}) \geq 0.$$

Since A_n is antistable, we have

$$X_{n-1} \geq X_n > X.$$

We have a nonincreasing sequence $\{X_i\}$, and the sequence is bounded below by $X_i > X$. Hence the limit

$$X_+ := \lim_{n \to \infty} X_n$$

exists and is Hermitian, and we have $X_+ \geq X$. Passing the limit $n \to \infty$ in equation (14.10), we get $\mathcal{Q}(X_+) = 0$. So X_+ is a solution of equation (14.6).

Note that $X_+ - X \geq 0$ and

$$(X_+ - X) A_+ + A_+^* (X_+ - X) = -\mathcal{Q}(X) + (X_+ - X) R (X_+ - X) > 0. \tag{14.11}$$

Hence, $X_+ - X > 0$ and $A_+ = A + RX_+$ is antistable. $\square$

Lemma 14.5 *There exists an admissible controller such that $\|T_{zw}\|_\infty < \gamma$ only if the following three conditions hold:*

(i) There exists a stabilizing solution $X_\infty > 0$ to

$$X_\infty A + A^* X_\infty + X_\infty(B_1 B_1^*/\gamma^2 - B_2 B_2^*) X_\infty + C_1^* C_1 = 0. \tag{14.12}$$

(ii) There exists a stabilizing solution $Y_\infty > 0$ to

$$AY_\infty + Y_\infty A^* + Y_\infty(C_1^* C_1/\gamma^2 - C_2^* C_2) Y_\infty + B_1 B_1^* = 0. \tag{14.13}$$

(iii) $\begin{bmatrix} \gamma Y_\infty^{-1} & I_n \\ I_n & \gamma X_\infty^{-1} \end{bmatrix} > 0$ or $\rho(X_\infty Y_\infty) < \gamma^2.$

Proof. Applying Theorem 14.4 to part (i) of Lemma 14.3, we conclude that there exists a $Y > Y_1 > 0$ such that

$$AY + YA^* + YC_1^* C_1 Y/\gamma^2 + B_1 B_1^* - \gamma^2 B_2 B_2^* = 0$$

and $A + C_1^* C_1 Y/\gamma^2$ is antistable. Let $X_\infty := \gamma^2 Y^{-1}$; we have

$$X_\infty A + A^* X_\infty + X_\infty(B_1 B_1^*/\gamma^2 - B_2 B_2^*) X_\infty + C_1^* C_1 = 0 \tag{14.14}$$

and

$$A + (B_1 B_1^*/\gamma^2 - B_2 B_2^*) X_\infty = -X_\infty^{-1}(A + C_1^* C_1 X_\infty^{-1}) X_\infty = -X_\infty^{-1}(A + C_1^* C_1 Y/\gamma^2) X_\infty$$

is stable.

Similarly, applying Theorem 14.4 to part (ii) of Lemma 14.3, we conclude that there exists an $X > X_1 > 0$ such that

$$XA + A^* X + XB_1 B_1^* X/\gamma^2 + C_1^* C_1 - \gamma^2 C_2^* C_2 = 0$$

and $A + B_1 B_1^* X/\gamma^2$ is antistable. Let $Y_\infty := \gamma^2 X^{-1}$, we have

$$AY_\infty + Y_\infty A^* + Y_\infty(C_1^* C_1/\gamma^2 - C_2^* C_2) Y_\infty + B_1 B_1^* = 0 \tag{14.15}$$

and $A + (C_1^* C_1/\gamma^2 - C_2^* C_2) Y_\infty$ is stable.

Finally, note that the rank condition in part (iii) of Lemma 14.3 is automatically satisfied by $r \geq n$, and

$$\begin{bmatrix} \gamma Y_\infty^{-1} & I_n \\ I_n & \gamma X_\infty^{-1} \end{bmatrix} = \begin{bmatrix} X/\gamma & I_n \\ I_n & Y/\gamma \end{bmatrix} > \begin{bmatrix} X_1/\gamma & I_n \\ I_n & Y_1/\gamma \end{bmatrix} \geq 0$$

or $\rho(X_\infty Y_\infty) < \gamma^2$. □

14.2. A Simplified $\mathcal{H}_\infty$ Control Problem

Proof of Theorem 14.1: To complete the proof, we only need to show that the controller K_{sub} given in Theorem 14.1 renders $\|T_{zw}\|_\infty < \gamma$. Note that the closed-loop transfer function with K_{sub} is given by

$$T_{zw} = \left[\begin{array}{cc|c} A & B_2 F_\infty & B_1 \\ -Z_\infty L_\infty C_2 & \hat{A}_\infty & -Z_\infty L_\infty D_{21} \\ \hline C_1 & D_{12} F_\infty & 0 \end{array}\right] =: \left[\begin{array}{c|c} A_c & B_c \\ \hline C_c & D_c \end{array}\right].$$

Define

$$P = \begin{bmatrix} \gamma^2 Y_\infty^{-1} & -\gamma^2 Y_\infty^{-1} Z_\infty^{-1} \\ -\gamma^2 (Z_\infty^*)^{-1} Y_\infty^{-1} & \gamma^2 Y_\infty^{-1} Z_\infty^{-1} \end{bmatrix}.$$

Then it is easy to show that $P > 0$ and

$$PA_c + A_c^* P + PB_c B_c^* P/\gamma^2 + C_c^* C_c = 0.$$

Moreover,

$$A_c + B_c B_c^* P/\gamma^2 = \begin{bmatrix} A + B_1 B_1^* Y_\infty^{-1} & B_2 F_\infty - B_1 B_1^* Y_\infty^{-1} Z_\infty^{-1} \\ 0 & A + B_1 B_1^* X_\infty/\gamma^2 + B_2 F_\infty \end{bmatrix}$$

has no eigenvalues on the imaginary axis since $A + B_1 B_1^* X_\infty/\gamma^2 + B_2 F_\infty$ is stable and $A + B_1 B_1^* Y_\infty^{-1}$ is antistable. Thus, by Corollary 12.3, $\|T_{zw}\|_\infty < \gamma$. □

Remark 14.1 It is appropriate to point out that the conditions stated in Lemma 14.3 are, in fact, necessary and sufficient; see Gahinet and Apkarian [1994] and Gahinet [1996] for a linear matrix inequality (LMI) approach to the $\mathcal{H}_\infty$ problem. But the necessity should be suitably interpreted. For example, if one finds an $X_1 > 0$ and a $Y_1 > 0$ satisfying conditions (i) and (ii) but not condition (iii), this does not imply that there is no admissible $\mathcal{H}_\infty$ controller since there might be other $X_1 > 0$ and $Y_1 > 0$ that satisfy all three conditions. For example, consider $\gamma = 1$ and

$$G(s) = \left[\begin{array}{c|cc} -1 & \begin{bmatrix} 1 & 0 \end{bmatrix} & 1 \\ \hline \begin{bmatrix} 1 \\ 0 \end{bmatrix} & 0 & \begin{bmatrix} 0 \\ 1 \end{bmatrix} \\ 1 & \begin{bmatrix} 0 & 1 \end{bmatrix} & 0 \end{array}\right].$$

It is easy to check that $X_1 = Y_1 = 0.5$ satisfy (i) and (ii) but not (iii). Nevertheless, we shall show in the next section that $\gamma_{\text{opt}} = 0.7321$ and thus a suboptimal controller exists for $\gamma = 1$. In fact, we can check that $1 < X_1 < 2$, $1 < Y_1 < 2$ also satisfy (i), (ii) and (iii). ◇

Example 14.1 Consider the feedback system shown in Figure 6.3 with

$$P = \frac{50(s+1.4)}{(s+1)(s+2)}, \quad W_e = \frac{2}{s+0.2}, \quad W_u = \frac{s+1}{s+10}.$$

We shall design a controller so that the $\mathcal{H}_\infty$ norm from $w = \begin{bmatrix} d \\ d_i \end{bmatrix}$ to $z = \begin{bmatrix} e \\ \tilde{u} \end{bmatrix}$ is minimized. Note that

$$\begin{bmatrix} e \\ \tilde{u} \end{bmatrix} = \begin{bmatrix} W_e(I+PK)^{-1} & W_e(I+PK)^{-1}P \\ -W_uK(I+PK)^{-1} & -W_uK(I+PK)^{-1}P \end{bmatrix} \begin{bmatrix} d \\ d_i \end{bmatrix} =: T_{zw} \begin{bmatrix} d \\ d_i \end{bmatrix}.$$

Then the problem can be set up in an LFT framework with

$$G(s) = \begin{bmatrix} W_e & W_eP & -W_eP \\ 0 & 0 & -W_u \\ \hline I & P & -P \end{bmatrix} = \left[\begin{array}{cccc|cc|c} -0.2 & 2 & 2 & 0 & 2 & 0 & 0 \\ 0 & -1 & 0 & 0 & 0 & 20 & -20 \\ 0 & 0 & -2 & 0 & 0 & 30 & -30 \\ 0 & 0 & 0 & -10 & 0 & 0 & -3 \\ \hline 1 & 0 & 0 & 0 & 0 & 0 & 0 \\ 0 & 0 & 0 & -3 & 0 & 0 & -1 \\ \hline 0 & 1 & 1 & 0 & 1 & 0 & 0 \end{array}\right].$$

A suboptimal $\mathcal{H}_\infty$ controller can be computed by using the following command:

≫ $[\mathbf{K}, \mathbf{T_{zw}}, \gamma_{\mathrm{subopt}}] = \mathbf{hinfsyn}(\mathbf{G}, \mathbf{n_y}, \mathbf{n_u}, \gamma_{\min}, \gamma_{\max}, \mathbf{tol})$

where n_y and n_u are the dimensions of y and u; $\gamma_{\min}$ and $\gamma_{\max}$ are, respectively a lower bound and an upper bound for γ_{opt}; and tol is a tolerance to the optimal value. Set $n_y = 1, n_u = 1, \gamma_{\min} = 0, \gamma_{\max} = 10,$ tol $= 0.0001$; we get $\gamma_{\mathrm{subopt}} = 0.7849$ and a suboptimal controller

$$K = \frac{12.82(s/10+1)(s/7.27+1)(s/1.4+1)}{(s/32449447.67+1)(s/22.19+1)(s/1.4+1)(s/0.2+1)}.$$

If we set tol $= 0.01$, we would get $\gamma_{\mathrm{subopt}} = 0.7875$ and a suboptimal controller

$$\tilde{K} = \frac{12.78(s/10+1)(s/7.27+1)(s/1.4+1)}{(s/2335.59+1)(s/21.97+1)(s/1.4+1)(s/0.2+1)}.$$

The only significant difference between K and $\tilde{K}$ is the exact location of the far-away stable controller pole. Figure 14.1 shows the closed-loop frequency response of $\bar{\sigma}(T_{zw})$ and Figure 14.2 shows the frequency responses of S, T, KS, and SP.

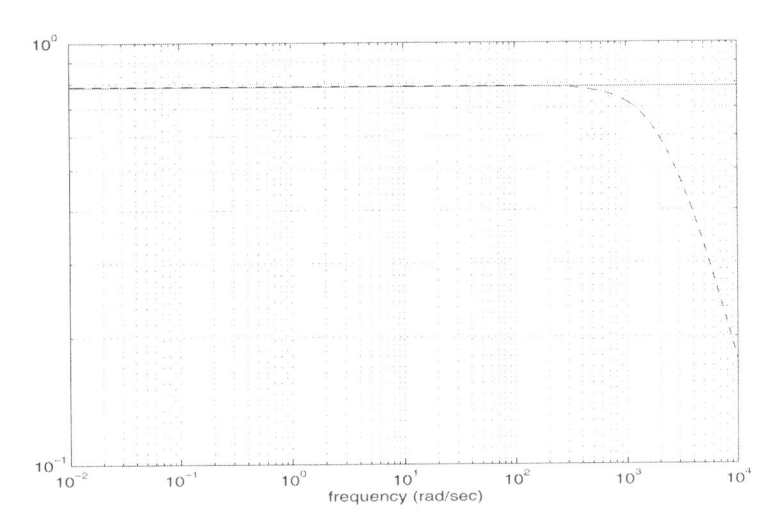

14.2. A Simplified $\mathcal{H}_\infty$ Control Problem

Figure 14.1: The closed-loop frequency responses of $\overline{\sigma}(T_{zw})$ with K (solid line) and $\tilde{K}$ (dashed line)

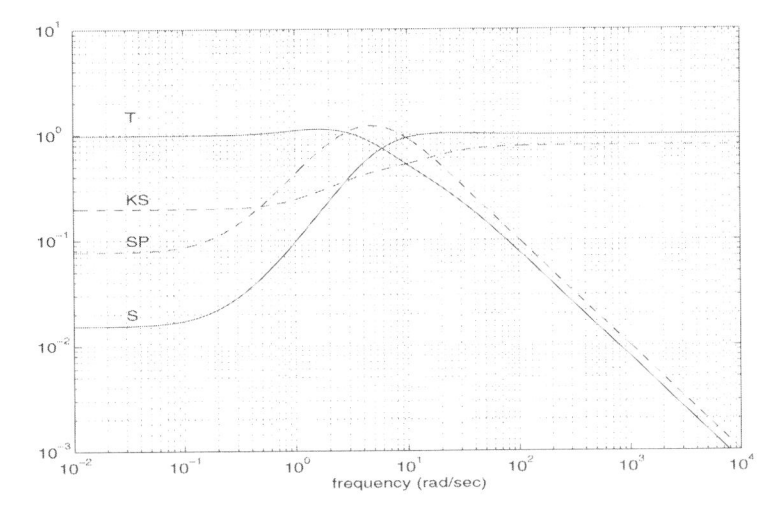

Figure 14.2: The frequency responses of S, T, KS, and SP with K

Example 14.2 Consider again the two-mass/spring/damper system shown in Figure 4.2. Assume that F_1 is the control force, F_2 is the disturbance force, and the measurements of x_1 and x_2 are corrupted by measurement noise:

$$y = \begin{bmatrix} y_1 \\ y_2 \end{bmatrix} = \begin{bmatrix} x_1 \\ x_2 \end{bmatrix} + W_n \begin{bmatrix} n_1 \\ n_2 \end{bmatrix}, \quad W_n = \begin{bmatrix} \dfrac{0.01(s+10)}{s+100} & 0 \\ 0 & \dfrac{0.01(s+10)}{s+100} \end{bmatrix}.$$

Our objective is to design a control law so that the effect of the disturbance force F_2 on the positions of the two masses, x_1 and x_2, are reduced in a frequency range $0 \leq \omega \leq 2$. The problem can be set up as shown in Figure 14.3, where $W_e = \begin{bmatrix} W_1 & 0 \\ 0 & W_2 \end{bmatrix}$ is the performance weight and W_u is the control weight. In order to limit the control force, we shall choose

$$W_u = \frac{s+5}{s+50}.$$

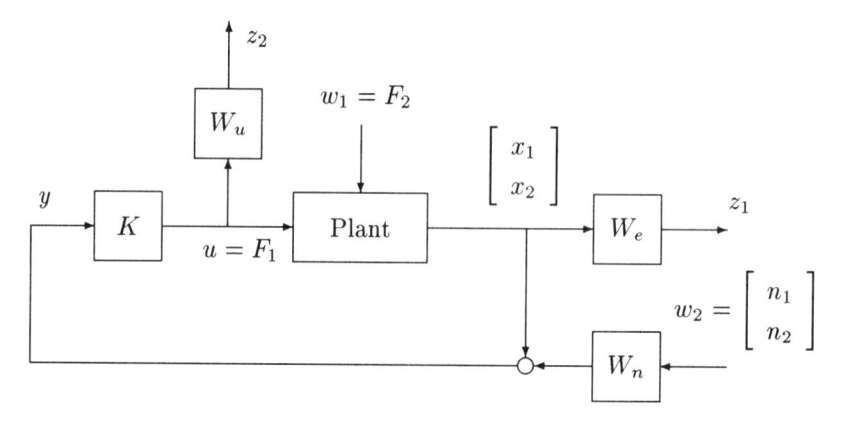

Figure 14.3: Rejecting the disturbance force F_2 by a feedback control

Now let $u = F_1$, $w = \begin{bmatrix} F_2 \\ n_1 \\ n_2 \end{bmatrix}$; then the problem can be formulated in an LFT form with

$$G(s) = \begin{bmatrix} \begin{bmatrix} W_e P_1 & 0 \\ 0 & 0 \\ P_1 & W_n \end{bmatrix} & \begin{bmatrix} W_e P_2 \\ W_u \\ P_2 \end{bmatrix} \end{bmatrix}$$

14.2. A Simplified $\mathcal{H}_\infty$ Control Problem

where P_1 and P_2 denote the transfer matrices from F_1 and F_2 to $\begin{bmatrix} x_1 \\ x_2 \end{bmatrix}$, respectively.

Let
$$W_1 = \frac{5}{s/2+1}, \quad W_2 = 0.$$

That is, we only want to reject the effect of the disturbance force F_2 on the position x_1. Then the optimal $\mathcal{H}_2$ performance is $\|\mathcal{F}_\ell(G, K_2)\|_2 = 2.6584$ and the $\mathcal{H}_\infty$ performance with the optimal $\mathcal{H}_2$ controller is $\|\mathcal{F}_\ell(G, K_2)\|_\infty = 2.6079$ while the optimal $\mathcal{H}_\infty$ performance with an $\mathcal{H}_\infty$ controller is $\|\mathcal{F}_\ell(G, K_\infty)\|_\infty = 1.6101$. This means that the effect of the disturbance force F_2 in the desired frequency rang $0 \leq \omega \leq 2$ will be effectively reduced with the $\mathcal{H}_\infty$ controller K_∞ by $5/1.6101 = 3.1054$ times at x_1. On the other hand, let
$$W_1 = 0, \quad W_2 = \frac{5}{s/2+1}.$$

That is, we only want to reject the effect of the disturbance force F_2 on the position x_2. Then the optimal $\mathcal{H}_2$ performance is $\|\mathcal{F}_\ell(G, K_2)\|_2 = 0.1659$ and the $\mathcal{H}_\infty$ performance with the optimal $\mathcal{H}_2$ controller is $\|\mathcal{F}_\ell(G, K_2)\|_\infty = 0.5202$ while the optimal $\mathcal{H}_\infty$ performance with an $\mathcal{H}_\infty$ controller is $\|\mathcal{F}_\ell(G, K_\infty)\|_\infty = 0.5189$. This means that the effect of the disturbance force F_2 in the desired frequency rang $0 \leq \omega \leq 2$ will be effectively reduced with the $\mathcal{H}_\infty$ controller K_∞ by $5/0.5189 = 9.6358$ times at x_2.

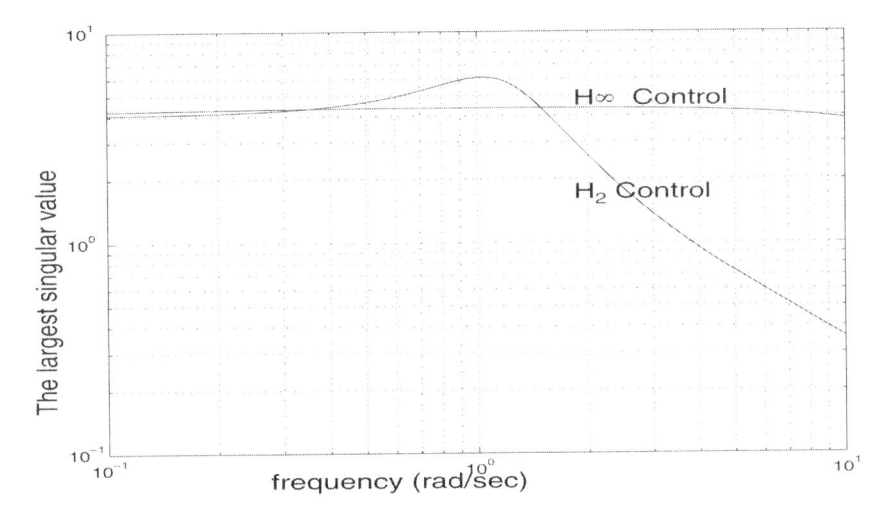

Figure 14.4: The largest singular value plot of the closed-loop system T_{zw} with an $\mathcal{H}_2$ controller and an $\mathcal{H}_\infty$ controller

Finally, set
$$W_1 = W_2 = \frac{5}{s/2+1}.$$

That is, we want to reject the effect of the disturbance force F_2 on both x_1 and x_2. Then the optimal $\mathcal{H}_2$ performance is $\|\mathcal{F}_\ell(G, K_2)\|_2 = 4.087$ and the $\mathcal{H}_\infty$ performance with the optimal $\mathcal{H}_2$ controller is $\|\mathcal{F}_\ell(G, K_2)\|_\infty = 6.0921$ while the optimal $\mathcal{H}_\infty$ performance with an $\mathcal{H}_\infty$ controller is $\|\mathcal{F}_\ell(G, K_\infty)\|_\infty = 4.3611$. This means that the effect of the disturbance force F_2 in the desired frequency rang $0 \leq \omega \leq 2$ will only be effectively reduced with the $\mathcal{H}_\infty$ controller K_∞ by $5/4.3611 = 1.1465$ times at both x_1 and x_2. This result shows clearly that it is very hard to reject the disturbance effect on both positions at the same time. The largest singular value Bode plots of the closed-loop system are shown in Figure 14.4. We note that the $\mathcal{H}_\infty$ controller typically gives a relatively flat frequency response since it tries to minimize the peak of the frequency response. On the other hand, the $\mathcal{H}_2$ controller would typically produce a frequency response that rolls off fast in the high-frequency range but with a large peak in the low-frequency range.

14.3 Optimality and Limiting Behavior

In this section, we will discuss, without proof, the behavior of the $\mathcal{H}_\infty$ suboptimal solution as γ varies, especially as γ approaches the infima achievable norm, denoted by γ_{opt}. Since Theorem 14.1 gives necessary and sufficient conditions for the existence of an admissible controller such that $\|T_{zw}\|_\infty < \gamma$, γ_{opt} is the infimum over all γ such that conditions (i)–(iii) are satisfied. Theorem 14.1 does not give an explicit formula for γ_{opt}, but, just as for the $\mathcal{H}_\infty$ norm calculation, it can be computed as closely as desired by a search technique.

Although we have not focused on the problem of $\mathcal{H}_\infty$ *optimal* controllers, the assumptions in this book make them relatively easy to obtain in most cases. In addition to describing the qualitative behavior of suboptimal solutions as γ varies, we will indicate why the descriptor version of the controller formulas below can usually provide formulas for the optimal controller when $\gamma = \gamma_{\text{opt}}$.

As $\gamma \to \infty$, $H_\infty \to H_2$, $X_\infty \to X_2$, etc., and $K_{\text{sub}} \to K_2$. This fact is the result of the particular choice of the suboptimal controller. While it could be argued that K_{sub} is a natural choice, this connection with $\mathcal{H}_2$ actually hints at deeper interpretations. In fact, K_{sub} is the minimum entropy solution (see Section 14.4) as well as the minimax controller for $\|z\|_2^2 - \gamma^2\|w\|_2^2$.

If $\gamma_2 \geq \gamma_1 > \gamma_{\text{opt}}$, then $X_\infty(\gamma_1) \geq X_\infty(\gamma_2)$ and $Y_\infty(\gamma_1) \geq Y_\infty(\gamma_2)$. Thus X_∞ and Y_∞ are decreasing functions of γ, as is $\rho(X_\infty Y_\infty)$. At $\gamma = \gamma_{\text{opt}}$, any one of the three conditions in Theorem 14.1 can fail. If only condition (iii) fails, then it is relatively straightforward to show that the descriptor formulas below for $\gamma = \gamma_{\text{opt}}$ are optimal; that is, the optimal controller is given by

$$(I - \gamma_{\text{opt}}^{-2} Y_\infty X_\infty)\dot{\hat{x}} = A_s \hat{x} - L_\infty y \qquad (14.16)$$

$$u = F_\infty \hat{x} \qquad (14.17)$$

14.3. Optimality and Limiting Behavior

where $A_s := A + B_2 F_\infty + L_\infty C_2 + \gamma_{opt}^{-2} Y_\infty A^* X_\infty + \gamma_{opt}^{-2} B_1 B_1^* X_\infty + \gamma_{opt}^{-2} Y_\infty C_1^* C_1$. (See Example 14.3.)

The formulas in Theorem 14.1 are not well-defined in the optimal case because the term $(I - \gamma_{opt}^{-2} X_\infty Y_\infty)$ is not invertible. It is possible but far less likely that conditions (i) or (ii) would fail before (iii). To see this, consider (i) and let γ_1 be the largest γ for which H_∞ fails to be in dom(Ric) because the H_∞ matrix fails to have either the stability property or the complementarity property. The same remarks will apply to (ii) by duality.

If complementarity fails at $\gamma = \gamma_1$, then $\rho(X_\infty) \to \infty$ as $\gamma \to \gamma_1$. For $\gamma < \gamma_1$, H_∞ may again be in dom(Ric), but X_∞ will be indefinite. For such γ, the controller $u = -B_2^* X_\infty x$ would make $\|T_{zw}\|_\infty < \gamma$ but would not be stabilizing. (See part 1 of Example 14.3.) If the stability property fails at $\gamma = \gamma_1$, then $H_\infty \notin$ dom(Ric) but Ric can be extended to obtain X_∞ so that a controller can be obtained to make $\|T_{zw}\|_\infty = \gamma_1$. The stability property will also not hold for any $\gamma \leq \gamma_1$, and no controller whatsoever exists that makes $\|T_{zw}\|_\infty < \gamma_1$. In other words, if stability breaks down first, then the infimum over stabilizing controllers equals the infimum over all controllers, stabilizing or otherwise. (See part 2 of Example 14.3.) In view of this, we would typically expect that complementarity would fail first.

Complementarity failing at $\gamma = \gamma_1$ means $\rho(X_\infty) \to \infty$ as $\gamma \to \gamma_1$, so condition (iii) would fail at even larger values of γ, unless the eigenvectors associated with $\rho(X_\infty)$ as $\gamma \to \gamma_1$ are in the null space of Y_∞. Thus condition (iii) is the most likely of all to fail first. If condition (i) or (ii) fails first because the stability property fails, the formulas in Theorem 14.1 as well as their descriptor versions are optimal at $\gamma = \gamma_{opt}$. This is illustrated in Example 14.3. If the complementarity condition fails first, [but (iii) does not fail], then obtaining formulas for the optimal controllers is a more subtle problem.

Example 14.3 Let an interconnected dynamical system realization be given by

$$G(s) = \left[\begin{array}{c|cc|c} a & \begin{array}{cc} 1 & 0 \end{array} & 1 \\ \hline \begin{array}{c} 1 \\ 0 \end{array} & 0 & \begin{array}{c} 0 \\ 1 \end{array} \\ \hline 1 & \begin{array}{cc} 0 & 1 \end{array} & 0 \end{array}\right].$$

Then all assumptions for output feedback problem are satisfied and

$$H_\infty = \begin{bmatrix} a & \frac{1-\gamma^2}{\gamma^2} \\ -1 & -a \end{bmatrix}, \quad J_\infty = \begin{bmatrix} a & \frac{1-\gamma^2}{\gamma^2} \\ -1 & -a \end{bmatrix}.$$

The eigenvalues of H_∞ and J_∞ are given by

$$\left\{ \pm \frac{\sqrt{(a^2+1)\gamma^2 - 1}}{\gamma} \right\}.$$

If $\gamma > \dfrac{1}{\sqrt{a^2+1}}$, then $\mathcal{X}_-(H_\infty)$ and $\mathcal{X}_-(J_\infty)$ exist and

$$\mathcal{X}_-(H_\infty) = \text{Im}\begin{bmatrix} \dfrac{\sqrt{(a^2+1)\gamma^2-1}-a\gamma}{\gamma} \\ 1 \end{bmatrix}$$

$$\mathcal{X}_-(J_\infty) = \text{Im}\begin{bmatrix} \dfrac{\sqrt{(a^2+1)\gamma^2-1}-a\gamma}{\gamma} \\ 1 \end{bmatrix}.$$

We shall consider two cases:

1) $a > 0$: In this case, the complementary property of dom(Ric) will fail before the stability property fails since

$$\sqrt{(a^2+1)\gamma^2-1} - a\gamma = 0$$

when $\gamma = 1$.

Nevertheless, if $\gamma > \dfrac{1}{\sqrt{a^2+1}}$ and $\gamma \neq 1$, then $H_\infty \in \text{dom(Ric)}$ and

$$X_\infty = \dfrac{\gamma}{\sqrt{(a^2+1)\gamma^2-1}-a\gamma} = \begin{cases} > 0; & \text{if } \gamma > 1 \\ < 0; & \text{if } \dfrac{1}{\sqrt{a^2+1}} < \gamma < 1. \end{cases}$$

Note that if $\gamma > 1$, then $H_\infty \in \text{dom(Ric)}$, $J_\infty \in \text{dom(Ric)}$, and

$$X_\infty = \dfrac{\gamma}{\sqrt{(a^2+1)\gamma^2-1}-a\gamma} > 0$$

$$Y_\infty = \dfrac{\gamma}{\sqrt{(a^2+1)\gamma^2-1}-a\gamma} > 0.$$

Hence conditions (i) and (ii) in Theorem 14.1 are satisfied, and we need to check condition (iii). Since

$$\rho(X_\infty Y_\infty) = \dfrac{\gamma^2}{(\sqrt{(a^2+1)\gamma^2-1}-a\gamma)^2},$$

it is clear that $\rho(X_\infty Y_\infty) \to \infty$ when $\gamma \to 1$. So condition (iii) will fail before condition (i) or (ii) fails.

2) $a < 0$: In this case, the complementary property is always satisfied, and, furthermore, $H_\infty \in \text{dom(Ric)}$, $J_\infty \in \text{dom(Ric)}$, and

$$X_\infty = \dfrac{\gamma}{\sqrt{(a^2+1)\gamma^2-1}-a\gamma} > 0$$

14.3. Optimality and Limiting Behavior

$$Y_\infty = \frac{\gamma}{\sqrt{(a^2+1)\gamma^2 - 1} - a\gamma} > 0$$

for $\gamma > \frac{1}{\sqrt{a^2+1}}$.

However, for $\gamma \leq \frac{1}{\sqrt{a^2+1}}$, $H_\infty \notin \mathrm{dom}(\mathrm{Ric})$ since stability property fails. Nevertheless, in this case, if $\gamma_0 = \frac{1}{\sqrt{a^2+1}}$, we can extend the dom(Ric) to include those matrices H_∞ with imaginary axis eigenvalues as

$$\overline{\mathcal{X}_-}(H_\infty) = \mathrm{Im}\begin{bmatrix} -a \\ 1 \end{bmatrix}$$

such that $X_\infty = -\frac{1}{a}$ is a solution to the Riccati equation

$$A^* X_\infty + X_\infty A + C_1^* C_1 + \gamma_0^{-2} X_\infty B_1 B_1^* X_\infty - X_\infty B_2 B_2^* X_\infty = 0$$

and $A + \gamma_0^{-2} B_1 B_1^* X_\infty - B_2 B_2^* X_\infty = 0$. It can be shown that

$$\rho(X_\infty Y_\infty) = \frac{\gamma^2}{(\sqrt{(a^2+1)\gamma^2 - 1} - a\gamma)^2} < \gamma^2$$

is satisfied if and only if

$$\gamma > \sqrt{a^2+2} + a \quad \left(> \frac{1}{\sqrt{a^2+1}} \right).$$

So condition (iii) of Theorem 14.1 will fail before either (i) or (ii) fails.

In both $a > 0$ and $a < 0$ cases, the optimal γ for the output feedback is given by

$$\gamma_{\mathrm{opt}} = \sqrt{a^2+2} + a$$

and the optimal controller given by the descriptor formula in equations (14.16) and (14.17) is a constant. In fact,

$$u_{\mathrm{opt}} = -\frac{\gamma_{\mathrm{opt}}}{\sqrt{(a^2+1)\gamma_{\mathrm{opt}}^2 - 1} - a\gamma_{\mathrm{opt}}} y.$$

For instance, let $a = -1$ then $\gamma_{\mathrm{opt}} = \sqrt{3} - 1 = 0.7321$ and $u_{\mathrm{opt}} = -0.7321\, y$. Further,

$$T_{zw} = \left[\begin{array}{c|cc} -1.7321 & 1 & -0.7321 \\ \hline 1 & 0 & 0 \\ -0.7321 & 0 & -0.7321 \end{array}\right].$$

It is easy to check that $\|T_{zw}\|_\infty = 0.7321$.

14.4 Minimum Entropy Controller

Let T be a transfer matrix with $\|T\|_\infty < \gamma$. Then the entropy of $T(s)$ is defined by

$$I(T,\gamma) = -\frac{\gamma^2}{2\pi}\int_{-\infty}^{\infty} \ln\left|\det\left(I - \gamma^{-2}T^*(j\omega)T(j\omega)\right)\right| d\omega.$$

It is easy to see that

$$I(T,\gamma) = -\frac{\gamma^2}{2\pi}\int_{-\infty}^{\infty} \sum_i \ln\left|1 - \gamma^{-2}\sigma_i^2\left(T(j\omega)\right)\right| d\omega$$

and $I(T,\gamma) \geq 0$, where $\sigma_i\left(T(j\omega)\right)$ is the ith singular value of $T(j\omega)$. It is also easy to show that

$$\lim_{\gamma \to \infty} I(T,\gamma) = \frac{1}{2\pi}\int_{-\infty}^{\infty} \sum_i \sigma_i^2\left(T(j\omega)\right) d\omega = \|T\|_2^2.$$

Thus the entropy $I(T,\gamma)$ is, in fact, a performance index measuring the tradeoff between the $\mathcal{H}_\infty$ optimality ($\gamma \to \|T\|_\infty$) and the $\mathcal{H}_2$ optimality ($\gamma \to \infty$).

It has been shown in Glover and Mustafa [1989] that the suboptimal controller given in Theorem 14.1 is actually the controller that satisfies the norm condition $\|T_{zw}\|_\infty < \gamma$ and minimizes the following entropy:

$$-\frac{\gamma^2}{2\pi}\int_{-\infty}^{\infty} \ln\left|\det\left(I - \gamma^{-2}T_{zw}^*(j\omega)T_{zw}(j\omega)\right)\right| d\omega.$$

Therefore, the given suboptimal controller is also called the minimum entropy controller [maximum entropy controller if the entropy is defined as $\tilde{I}(T,\gamma) = -I(T,\gamma)$].

Related MATLAB Commands: hinfsyne, hinffi

14.5 An Optimal Controller

To offer a general idea about the appearance of an optimal controller, we shall give in the following (without proof) the conditions under which an optimal controller exists and an explicit formula for an optimal controller.

Theorem 14.6 *There exists an admissible controller such that $\|T_{zw}\|_\infty \leq \gamma$ iff the following three conditions hold:*

(i) There exists a full column rank matrix

$$\begin{bmatrix} X_{\infty 1} \\ X_{\infty 2} \end{bmatrix} \in \mathbb{R}^{2n \times n}$$

14.5. An Optimal Controller

such that
$$H_\infty \begin{bmatrix} X_{\infty 1} \\ X_{\infty 2} \end{bmatrix} = \begin{bmatrix} X_{\infty 1} \\ X_{\infty 2} \end{bmatrix} T_X, \quad \text{Re } \lambda_i(T_X) \leq 0 \;\forall i$$

and
$$X_{\infty 1}^* X_{\infty 2} = X_{\infty 2}^* X_{\infty 1};$$

(ii) *There exists a full column rank matrix*
$$\begin{bmatrix} Y_{\infty 1} \\ Y_{\infty 2} \end{bmatrix} \in \mathbb{R}^{2n \times n}$$

such that
$$J_\infty \begin{bmatrix} Y_{\infty 1} \\ Y_{\infty 2} \end{bmatrix} = \begin{bmatrix} Y_{\infty 1} \\ Y_{\infty 2} \end{bmatrix} T_Y, \quad \text{Re } \lambda_i(T_Y) \leq 0 \;\forall i$$

and
$$Y_{\infty 1}^* Y_{\infty 2} = Y_{\infty 2}^* Y_{\infty 1};$$

(iii)
$$\begin{bmatrix} X_{\infty 2}^* X_{\infty 1} & \gamma^{-1} X_{\infty 2}^* Y_{\infty 2} \\ \gamma^{-1} Y_{\infty 2}^* X_{\infty 2} & Y_{\infty 2}^* Y_{\infty 1} \end{bmatrix} \geq 0.$$

Moreover, when these conditions hold, one such controller is
$$K_{\text{opt}}(s) := C_K (sE_K - A_K)^+ B_K$$

where
$$\begin{aligned} E_K &:= Y_{\infty 1}^* X_{\infty 1} - \gamma^{-2} Y_{\infty 2}^* X_{\infty 2} \\ B_K &:= Y_{\infty 2}^* C_2^* \\ C_K &:= -B_2^* X_{\infty 2} \\ A_K &:= E_K T_X - B_K C_2 X_{\infty 1} = T_Y^* E_K + Y_{\infty 1}^* B_2 C_K. \end{aligned}$$

Remark 14.2 It is simple to show that if $X_{\infty 1}$ and $Y_{\infty 1}$ are nonsingular and if $X_\infty = X_{\infty 2} X_{\infty 1}^{-1}$ and $Y_\infty = Y_{\infty 2} Y_{\infty 1}^{-1}$, then condition (iii) in the preceding theorem is equivalent to $X_\infty \geq 0$, $Y_\infty \geq 0$, and $\rho(Y_\infty X_\infty) \leq \gamma^2$. So, in this case, the conditions for the existence of an optimal controller can be obtained from "taking the limit" of the corresponding conditions in Theorem 14.1. Moreover, the controller given above is reduced to the descriptor form given in equations (14.16) and (14.17). ◇

14.6 General $\mathcal{H}_\infty$ Solutions

Consider the system described by the block diagram

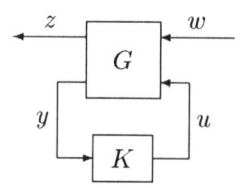

```
      z ┌─────┐ w
      ←─┤  G  ├←─
         └─────┘
      y │     │ u
      ←─┤     ├←─
         └─────┘
            │
         ┌─────┐
         │  K  │
         └─────┘
```

where, as usual, G and K are assumed to be real rational and proper with K constrained to provide internal stability. The controller is said to be admissible if it is real rational, proper, and stabilizing. Although we are taking everything to be real, the results presented here are still true for the complex case with some obvious modifications. We will again only be interested in characterizing all suboptimal $\mathcal{H}_\infty$ controllers.

The realization of the transfer matrix G is taken to be of the form

$$G(s) = \left[\begin{array}{c|cc} A & B_1 & B_2 \\ \hline C_1 & D_{11} & D_{12} \\ C_2 & D_{21} & 0 \end{array}\right] = \left[\begin{array}{c|c} A & B \\ \hline C & D \end{array}\right],$$

which is compatible with the dimensions of $z(t) \in \mathbb{R}^{p_1}$, $y(t) \in \mathbb{R}^{p_2}$, $w(t) \in \mathbb{R}^{m_1}$, $u(t) \in \mathbb{R}^{m_2}$, and the state $x(t) \in \mathbb{R}^n$. The following assumptions are made:

(A1) (A, B_2) is stabilizable and (C_2, A) is detectable;

(A2) $D_{12} = \begin{bmatrix} 0 \\ I \end{bmatrix}$ and $D_{21} = \begin{bmatrix} 0 & I \end{bmatrix}$;

(A3) $\begin{bmatrix} A - j\omega I & B_2 \\ C_1 & D_{12} \end{bmatrix}$ has full column rank for all ω;

(A4) $\begin{bmatrix} A - j\omega I & B_1 \\ C_2 & D_{21} \end{bmatrix}$ has full row rank for all ω.

Assumption (A1) is necessary for the existence of stabilizing controllers. The assumptions in (A2) mean that the penalty on $z = C_1 x + D_{12} u$ includes a nonsingular, normalized penalty on the control u, that the exogenous signal w includes both plant disturbance and sensor noise, and that the sensor noise weighting is normalized and nonsingular. Relaxation of (A2) leads to singular control problems; see Stroorvogel [1992]. For those problems that have D_{12} full column rank and D_{21} full row rank but do not satisfy assumption (A2), a normalizing procedure is given in the next section so that an equivalent new system will satisfy this assumption.

14.6. General $\mathcal{H}_\infty$ Solutions

Assumptions (A3) and (A4) are made for a technical reason: Together with (A1) they guarantee that the two Hamiltonian matrices in the corresponding $\mathcal{H}_2$ problem belong to dom(Ric), as we have seen in Chapter 13. Dropping (A3) and (A4) would make the solution very complicated. A further discussion of the assumptions and their possible relaxation will be provided in Section 14.7.

The main result is now stated in terms of the solutions of the X_∞ and Y_∞ Riccati equations together with the "state feedback" and "output injection" matrices F and L.

$$R := D_{1\bullet}^* D_{1\bullet} - \begin{bmatrix} \gamma^2 I_{m_1} & 0 \\ 0 & 0 \end{bmatrix}, \quad \text{where} \quad D_{1\bullet} := [D_{11} \ D_{12}]$$

$$\tilde{R} := D_{\bullet 1} D_{\bullet 1}^* - \begin{bmatrix} \gamma^2 I_{p_1} & 0 \\ 0 & 0 \end{bmatrix}, \quad \text{where} \quad D_{\bullet 1} := \begin{bmatrix} D_{11} \\ D_{21} \end{bmatrix}$$

$$H_\infty := \begin{bmatrix} A & 0 \\ -C_1^* C_1 & -A^* \end{bmatrix} - \begin{bmatrix} B \\ -C_1^* D_{1\bullet} \end{bmatrix} R^{-1} \begin{bmatrix} D_{1\bullet}^* C_1 & B^* \end{bmatrix}$$

$$J_\infty := \begin{bmatrix} A^* & 0 \\ -B_1 B_1^* & -A \end{bmatrix} - \begin{bmatrix} C^* \\ -B_1 D_{\bullet 1}^* \end{bmatrix} \tilde{R}^{-1} \begin{bmatrix} D_{\bullet 1} B_1^* & C \end{bmatrix}$$

$$X_\infty := \text{Ric}(H_\infty) \qquad Y_\infty := \text{Ric}(J_\infty)$$

$$F := \begin{bmatrix} F_{1\infty} \\ F_{2\infty} \end{bmatrix} := -R^{-1}[D_{1\bullet}^* C_1 + B^* X_\infty]$$

$$L := \begin{bmatrix} L_{1\infty} & L_{2\infty} \end{bmatrix} := -[B_1 D_{\bullet 1}^* + Y_\infty C^*]\tilde{R}^{-1}$$

Partition D, $F_{1\infty}$, and $L_{1\infty}$ are as follows:

$$\left[\begin{array}{c|c} & F' \\ \hline L' & D \end{array}\right] = \left[\begin{array}{c|ccc} & F_{11\infty}^* & F_{12\infty}^* & F_{2\infty}^* \\ \hline L_{11\infty}^* & D_{1111} & D_{1112} & 0 \\ L_{12\infty}^* & D_{1121} & D_{1122} & I \\ L_{2\infty}^* & 0 & I & 0 \end{array}\right].$$

Remark 14.3 In the above matrix partitioning, some matrices may not exist depending on whether D_{12} or D_{21} is square. This issue will be discussed further later. For the time being, we shall assume that all matrices in the partition exist. ◇

Theorem 14.7 *Suppose G satisfies the assumptions (A1)–(A4).*

(a) There exists an admissible controller $K(s)$ such that $\|\mathcal{F}_\ell(G, K)\|_\infty < \gamma$ (i.e., $\|T_{zw}\|_\infty < \gamma$) if and only if

(i) $\gamma > \max(\bar{\sigma}[D_{1111}, D_{1112},], \bar{\sigma}[D_{1111}^*, D_{1121}^*])$;
(ii) $H_\infty \in \text{dom}(\text{Ric})$ with $X_\infty = \text{Ric}(H_\infty) \geq 0$;
(iii) $J_\infty \in \text{dom}(\text{Ric})$ with $Y_\infty = \text{Ric}(J_\infty) \geq 0$;
(iv) $\rho(X_\infty Y_\infty) < \gamma^2$.

(b) *Given that the conditions of part (a) are satisfied, then all rational internally stabilizing controllers $K(s)$ satisfying $\|\mathcal{F}_\ell(G,K)\|_\infty < \gamma$ are given by*

$$K = \mathcal{F}_\ell(M_\infty, Q) \quad \text{for arbitrary } Q \in \mathcal{RH}_\infty \quad \text{such that} \quad \|Q\|_\infty < \gamma$$

where

$$M_\infty = \left[\begin{array}{c|cc} \hat{A} & \hat{B}_1 & \hat{B}_2 \\ \hline \hat{C}_1 & \hat{D}_{11} & \hat{D}_{12} \\ \hat{C}_2 & \hat{D}_{21} & 0 \end{array}\right]$$

$$\hat{D}_{11} = -D_{1121} D_{1111}^* (\gamma^2 I - D_{1111} D_{1111}^*)^{-1} D_{1112} - D_{1122}$$

$\hat{D}_{12} \in \mathbb{R}^{m_2 \times m_2}$ and $\hat{D}_{21} \in \mathbb{R}^{p_2 \times p_2}$ are any matrices (e.g., Cholesky factors) satisfying

$$\hat{D}_{12} \hat{D}_{12}^* = I - D_{1121}(\gamma^2 I - D_{1111}^* D_{1111})^{-1} D_{1121}^*,$$
$$\hat{D}_{21}^* \hat{D}_{21} = I - D_{1112}^*(\gamma^2 I - D_{1111} D_{1111}^*)^{-1} D_{1112},$$

and

$$\begin{aligned} \hat{B}_2 &= Z_\infty (B_2 + L_{12\infty}) \hat{D}_{12}, \\ \hat{C}_2 &= -\hat{D}_{21}(C_2 + F_{12\infty}), \\ \hat{B}_1 &= -Z_\infty L_{2\infty} + \hat{B}_2 \hat{D}_{12}^{-1} \hat{D}_{11}, \\ \hat{C}_1 &= F_{2\infty} + \hat{D}_{11} \hat{D}_{21}^{-1} \hat{C}_2, \\ \hat{A} &= A + BF + \hat{B}_1 \hat{D}_{21}^{-1} \hat{C}_2 \end{aligned}$$

where

$$Z_\infty = (I - \gamma^{-2} Y_\infty X_\infty)^{-1}.$$

(Note that if $D_{11} = 0$ then the formulas are considerably simplified.)

Some Special Cases:

Case 1: $D_{12} = I$

In this case

1. In part (a), (i) becomes $\gamma > \bar{\sigma}(D_{1121})$.

2. In part (b)
$$\hat{D}_{11} = -D_{1122}$$
$$\hat{D}_{12}\hat{D}_{12}^* = I - \gamma^{-2}D_{1121}D_{1121}^*$$
$$\hat{D}_{21}^*\hat{D}_{21} = I.$$

Case 2: $D_{21} = I$

In this case

1. In part (a), (i) becomes $\gamma > \overline{\sigma}(D_{1112})$.

2. In part (b)
$$\hat{D}_{11} = -D_{1122}$$
$$\hat{D}_{12}\hat{D}_{12}^* = I$$
$$\hat{D}_{21}^*\hat{D}_{21} = I - \gamma^{-2}D_{1112}^*D_{1112}.$$

Case 3: $D_{12} = I$ & $D_{21} = I$

In this case

1. In part (a), (i) drops out.

2. In part (b)
$$\hat{D}_{11} = -D_{1122}$$
$$\hat{D}_{12}\hat{D}_{12}^* = I$$
$$\hat{D}_{21}^*\hat{D}_{21} = I.$$

14.7 Relaxing Assumptions

In this section, we indicate how the results of Section 14.6 can be extended to more general cases. Let a given problem have the following diagram, where $z_p(t) \in \mathbb{R}^{p_1}$, $y_p(t) \in \mathbb{R}^{p_2}$, $w_p(t) \in \mathbb{R}^{m_1}$, and $u_p(t) \in \mathbb{R}^{m_2}$:

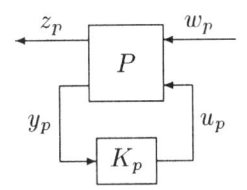

The plant P has the following state-space realization with D_{p12} *full column rank and* D_{p21} *full row rank*:

$$P(s) = \left[\begin{array}{c|cc} A_p & B_{p1} & B_{p2} \\ \hline C_{p1} & D_{p11} & D_{p12} \\ C_{p2} & D_{p21} & D_{p22} \end{array} \right].$$

The objective is to find all rational proper controllers $K_p(s)$ that stabilize P and $\|\mathcal{F}_\ell(P, K_p)\|_\infty < \gamma$. To solve this problem, we first transform it to the standard one treated in the last section. *Note that the following procedure can also be applied to the* $\mathcal{H}_2$ *problem (except the procedure for the case* $D_{11} \neq 0$).

Normalize D_{12} and D_{21}

Perform singular value decompositions to obtain the matrix factorizations

$$D_{p12} = U_p \begin{bmatrix} 0 \\ I \end{bmatrix} R_p, \quad D_{p21} = \tilde{R}_p \begin{bmatrix} 0 & I \end{bmatrix} \tilde{U}_p$$

such that U_p and $\tilde{U}_p$ are square and unitary. Now let

$$z_p = U_p z, \quad w_p = \tilde{U}_p^* w, \quad y_p = \tilde{R}_p y, \quad u_p = R_p u$$

and

$$K(s) = R_p K_p(s) \tilde{R}_p$$

$$\begin{aligned} G(s) &= \begin{bmatrix} U_p^* & 0 \\ 0 & \tilde{R}_p^{-1} \end{bmatrix} P(s) \begin{bmatrix} \tilde{U}_p^* & 0 \\ 0 & R_p^{-1} \end{bmatrix} \\ &= \left[\begin{array}{c|cc} A_p & B_{p1} \tilde{U}_p^* & B_{p2} R_p^{-1} \\ \hline U_p^* C_{p1} & U_p^* D_{p11} \tilde{U}_p^* & U_p^* D_{p12} R_p^{-1} \\ \tilde{R}_p^{-1} C_{p2} & \tilde{R}_p^{-1} D_{p21} \tilde{U}_p^* & \tilde{R}_p^{-1} D_{p22} R_p^{-1} \end{array} \right] \\ &=: \left[\begin{array}{c|cc} A & B_1 & B_2 \\ \hline C_1 & D_{11} & D_{12} \\ C_2 & D_{21} & D_{22} \end{array} \right] = \left[\begin{array}{c|c} A & B \\ \hline C & D \end{array} \right]. \end{aligned}$$

Then

$$D_{12} = \begin{bmatrix} 0 \\ I \end{bmatrix} \quad D_{21} = \begin{bmatrix} 0 & I \end{bmatrix},$$

and the new system is as follows:

14.7. Relaxing Assumptions

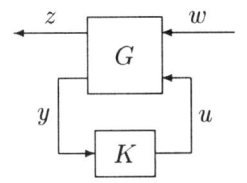

Furthermore, $\|\mathcal{F}_\ell(P, K_p)\|_\alpha = \|U_p\mathcal{F}_\ell(G,K)\tilde{U}_p\|_\alpha = \|\mathcal{F}_\ell(G,K)\|_\alpha$ for $\alpha = 2$ or ∞ since U_p and $\tilde{U}_p$ are unitary.

Remove the Assumption $D_{22} = 0$

Suppose $K(s)$ is a controller for G with D_{22} set to zero. Then the controller for $D_{22} \neq 0$ is $K(I + D_{22}K)^{-1}$. Hence there is no loss of generality in assuming $D_{22} = 0$.

Relax (A3) and (A4)

Suppose that

$$G = \left[\begin{array}{c|cc} 0 & 0 & 1 \\ \hline 0 & 0 & 1 \\ 1 & 1 & 0 \end{array}\right]$$

which violates both (A3) and (A4) and corresponds to the robust stabilization of an integrator. If the controller $u = -\epsilon x$, where $\epsilon > 0$ is used, then

$$T_{zw} = \frac{-\epsilon s}{s + \epsilon}, \quad \text{with } \|T_{zw}\|_\infty = \epsilon.$$

Hence the norm can be made arbitrarily small as $\epsilon \to 0$, but $\epsilon = 0$ is not admissible since it is not stabilizing. This may be thought of as a case where the $\mathcal{H}_\infty$ optimum is not achieved on the set of admissible controllers. Of course, for this system, $\mathcal{H}_\infty$ *optimal control is a silly problem*, although the suboptimal case is not obviously so.

Relax (A1)

If assumption (A1) is violated, then it is obvious that no admissible controllers exist. Suppose (A1) is relaxed to allow unstabilizable and/or undetectable modes on the $j\omega$ axis and internal stability is also relaxed to also allow closed-loop $j\omega$ axis poles, but (A2)–(A4) is still satisfied. It can be easily shown that under these conditions the closed-loop $\mathcal{H}_\infty$ norm cannot be made finite and, in particular, that the unstabilizable and/or undetectable modes on the $j\omega$ axis must show up as poles in the closed-loop system.

Violate (A1) and either or both (A3) and (A4)

Sensible control problems can be posed that violate (A1) *and* either or both (A3) and (A4). For example, cases when A has modes at $s = 0$ that are unstabilizable through B_2 and/or undetectable through C_2 arise when an integrator is included in a weight on a disturbance input or an error term. In these cases, either (A3) or (A4) are also violated, or the closed-loop $\mathcal{H}_\infty$ norm cannot be made finite. In many applications, such problems can be reformulated so that the integrator occurs inside the loop (essentially using the internal model principle) and is hence detectable and stabilizable. We will show this process in the next section.

An alternative is to introduce ϵ perturbations so that (A1), (A3), and (A4) are satisfied. Roughly speaking, this would produce sensible answers for sensible problems, but the behavior as $\epsilon \to 0$ could be problematic.

Relax (A2)

In the cases that either D_{12} is not full column rank or that D_{21} is not full row rank, improper controllers can give a bounded $\mathcal{H}_\infty$ norm for T_{zw}, although the controllers will not be admissible by our definition. Such singular filtering and control problems have been well-studied in $\mathcal{H}_2$ theory and many of the same techniques go over to the $\mathcal{H}_\infty$ case (e.g., Willems [1981] and Willems, Kitapci, and Silverman [1986]). A complete solution to the singular problem can be found in Stroorvogel [1992].

14.8 $\mathcal{H}_2$ and $\mathcal{H}_\infty$ Integral Control

It is interesting to note that the $\mathcal{H}_2$ and $\mathcal{H}_\infty$ design frameworks do not, in general, produce integral control. In this section we show how to introduce integral control into the $\mathcal{H}_2$ and $\mathcal{H}_\infty$ design framework through a simple disturbance rejection problem. We consider a feedback system shown in Figure 14.5. We shall assume that the frequency contents of the disturbance w are effectively modeled by the weighting $W_d \in \mathcal{RH}_\infty$ and the constraints on control signal are limited by an appropriate choice of $W_u \in \mathcal{RH}_\infty$. In order to let the output y track the reference signal r, we require K to contain an integrator [i.e., $K(s)$ has a pole at $s = 0$]. (In general, K is required to have poles on the imaginary axis.)

There are several ways to achieve the integral design. One approach is to introduce an integral in the performance weight W_e. Then the transfer function between w and z_1 is given by

$$z_1 = W_e(I + PK)^{-1}W_d w.$$

Now if the resulting controller K stabilizes the plant P and makes the norm (2-norm or ∞-norm) between w and z_1 finite, then K must have a pole at $s = 0$ that is the zero of the sensitivity function (assuming W_d has no zeros at $s = 0$). (This follows from the well-known internal model principle.) The problem with this approach is that the $\mathcal{H}_\infty$ (or $\mathcal{H}_2$) control theory presented in this chapter and in the previous chapters cannot be

14.8. $\mathcal{H}_2$ and $\mathcal{H}_\infty$ Integral Control

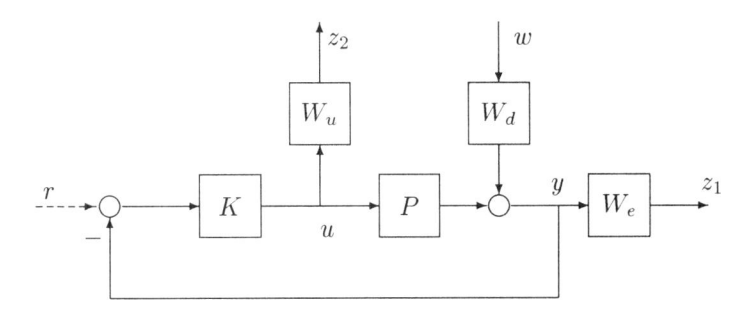

Figure 14.5: A simple disturbance rejection problem

applied to this problem formulation directly because the pole $s = 0$ of W_e becomes an uncontrollable pole of the feedback system and the very first assumption for the $\mathcal{H}_\infty$ (or $\mathcal{H}_2$) theory is violated.

However, the obstacles can be overcome by appropriately reformulating the problem. Suppose W_e can be factorized as follows:

$$W_e = \tilde{W}_e(s)M(s)$$

where $M(s)$ is proper, containing all the imaginary axis poles of W_e, and $M^{-1}(s) \in \mathcal{RH}_\infty$, $\tilde{W}_e(s)$ is stable and minimum phase. Now suppose there exists a controller $K(s)$ that contains the same imaginary axis poles that achieves the performance specifications. Then, without loss of generality, K can be factorized as

$$K(s) = -\hat{K}(s)M(s)$$

such that there is no unstable pole/zero cancellation in forming the product $\hat{K}(s)M(s)$. Now the problem can be reformulated as in Figure 14.6. Figure 14.6 can be put in the general LFT framework as in Figure 14.7 with

$$G(s) = \begin{bmatrix} \begin{bmatrix} \tilde{W}_e M W_d \\ 0 \\ M W_d \end{bmatrix} & \begin{bmatrix} \tilde{W}_e M P \\ W_u \\ M P \end{bmatrix} \end{bmatrix}.$$

We shall illustrate this design through a simple numerical example. Let

$$P = \frac{s-2}{(s+1)(s-3)} = \left[\begin{array}{cc|c} 0 & 1 & 0 \\ 3 & 2 & 1 \\ \hline -2 & 1 & 0 \end{array}\right], \quad W_d = 1,$$

$$W_u = \frac{s+10}{s+100} = \left[\begin{array}{c|c} -100 & -90 \\ \hline 1 & 1 \end{array}\right], \quad W_e = \frac{1}{s}.$$

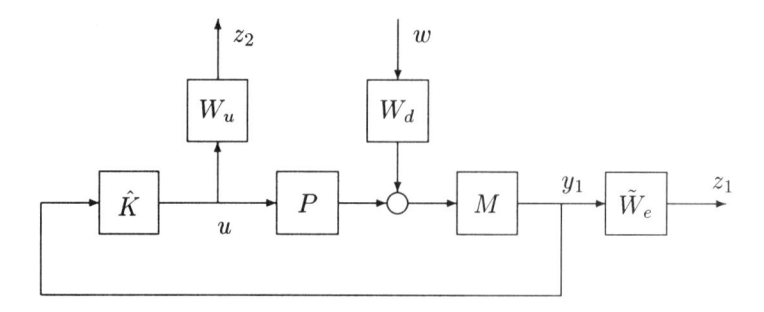

Figure 14.6: Disturbance rejection with imaginary axis poles

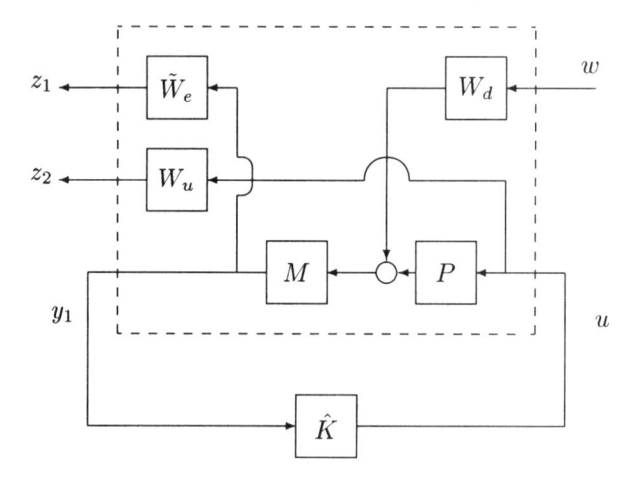

Figure 14.7: LFT framework for the disturbance rejection problem

Then we can choose without loss of generality that

$$M = \frac{s+\alpha}{s}, \quad \tilde{W}_e = \frac{1}{s+\alpha}, \quad \alpha > 0.$$

This gives the following generalized system:

$$G(s) = \left[\begin{array}{ccccc|cc} -\alpha & 0 & 1 & -2 & 1 & 1 & 0 \\ 0 & -100 & 0 & 0 & 0 & 0 & -90 \\ 0 & 0 & 0 & -2\alpha & \alpha & \alpha & 0 \\ 0 & 0 & 0 & 0 & 1 & 0 & 0 \\ 0 & 0 & 0 & 3 & 2 & 0 & 1 \\ \hline 1 & 0 & 0 & 0 & 0 & 0 & 0 \\ 0 & 1 & 0 & 0 & 0 & 0 & 1 \\ \hline 0 & 0 & 1 & -2 & 1 & 1 & 0 \end{array}\right].$$

14.9. $\mathcal{H}_\infty$ Filtering

The suboptimal $\mathcal{H}_\infty$ controller $\hat{K}_\infty$ for each α can be computed easily as

$$\hat{K}_\infty = \frac{-2060381.4(s+1)(s+\alpha)(s+100)(s-0.1557)}{(s+\alpha)^2(s+32.17)(s+262343)(s-19.89)},$$

which gives the closed-loop $\mathcal{H}_\infty$ norm 7.854. Hence the controller $K_\infty = -\hat{K}_\infty(s)M(s)$ is given by

$$K_\infty(s) = \frac{2060381.4(s+1)(s+100)(s-0.1557)}{s(s+32.17)(s+262343)(s-19.89)} \approx \frac{7.85(s+1)(s+100)(s-0.1557)}{s(s+32.17)(s-19.89)},$$

which is independent of α as expected. Similarly, we can solve an optimal $\mathcal{H}_2$ controller

$$\hat{K}_2 = \frac{-43.487(s+1)(s+\alpha)(s+100)(s-0.069)}{(s+\alpha)^2(s^2+30.94s+411.81)(s-7.964)}$$

and

$$K_2(s) = -\hat{K}_2(s)M(s) = \frac{43.487(s+1)(s+100)(s-0.069)}{s(s^2+30.94s+411.81)(s-7.964)}.$$

An approximate integral control can also be achieved without going through the preceding process by letting

$$W_e = \tilde{W}_e = \frac{1}{s+\epsilon}, \quad M(s) = 1$$

for a sufficiently small $\epsilon > 0$. For example, a controller for $\epsilon = 0.001$ is given by

$$K_\infty = \frac{316880(s+1)(s+100)(s-0.1545)}{(s+0.001)(s+32)(s+40370)(s-20)} \approx \frac{7.85(s+1)(s+100)(s-0.1545)}{s(s+32)(s-20)},$$

which gives the closed-loop $\mathcal{H}_\infty$ norm of 7.85. Similarly, an approximate $\mathcal{H}_2$ integral controller is obtained as

$$K_2 = \frac{43.47(s+1)(s+100)(s-0.0679)}{(s+0.001)(s^2+30.93s+411.7)(s-7.972)} \approx \frac{43.47(s+1)(s+100)(s-0.0679)}{s(s^2+30.93s+411.7)(s-7.972)}.$$

14.9 $\mathcal{H}_\infty$ Filtering

In this section we show how the filtering problem can be solved using the $\mathcal{H}_\infty$ theory developed earlier. Suppose a dynamic system is described by the following equations:

$$\dot{x} = Ax + B_1 w(t), \quad x(0) = 0 \qquad (14.18)$$
$$y = C_2 x + D_{21} w(t) \qquad (14.19)$$
$$z = C_1 x + D_{11} w(t) \qquad (14.20)$$

The filtering problem is to find an estimate $\hat{z}$ of z in some sense using the measurement of y. The restriction on the filtering problem is that the filter has to be causal so that it can be realized (i.e., $\hat{z}$ has to be generated by a causal system acting on the measurements). We will further restrict our filter to be *unbiased*; that is, given $T > 0$ the estimate $\hat{z}(t) = 0$ $\forall t \in [0, T]$ if $y(t) = 0$, $\forall t \in [0, T]$. Now we can state our $\mathcal{H}_\infty$ filtering problem.

$\mathcal{H}_\infty$ **Filtering**: Given a $\gamma > 0$, find a causal filter $F(s) \in \mathcal{RH}_\infty$ if it exists such that

$$J := \sup_{w \in \mathcal{L}_2[0,\infty)} \frac{\|z - \hat{z}\|_2^2}{\|w\|_2^2} < \gamma^2$$

with $\hat{z} = F(s)y$.

A diagram for the filtering problem is shown in Figure 14.8.

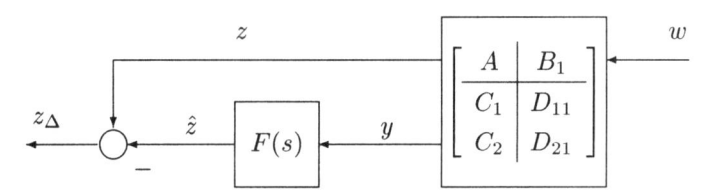

Figure 14.8: Filtering problem formulation

The preceding filtering problem can also be formulated in an LFT framework: Given a system shown below

$$G(s) = \left[\begin{array}{c|cc} A & B_1 & 0 \\ \hline C_1 & D_{11} & -I \\ C_2 & D_{21} & 0 \end{array}\right]$$

find a filter $F(s) \in \mathcal{RH}_\infty$ such that

$$\sup_{w \in \mathcal{L}_2} \frac{\|z_\Delta\|_2^2}{\|w\|_2^2} < \gamma^2. \tag{14.21}$$

Hence the filtering problem can be regarded as a special $\mathcal{H}_\infty$ problem. However, compared with control problems, there is no internal stability requirement in the filtering problem. Hence the solution to the above filtering problem can be obtained from the $\mathcal{H}_\infty$ solution in the previous sections by setting $B_2 = 0$ and dropping the internal stability requirement.

Theorem 14.8 *Suppose* (C_2, A) *is detectable and*

$$\left[\begin{array}{cc} A - j\omega I & B_1 \\ C_2 & D_{21} \end{array}\right]$$

has full row rank for all ω. Let D_{21} be normalized and D_{11} partitioned conformably as

$$\begin{bmatrix} D_{11} \\ D_{21} \end{bmatrix} = \begin{bmatrix} D_{111} & D_{112} \\ \hline 0 & I \end{bmatrix}.$$

Then there exists a causal $F(s) \in \mathcal{RH}_\infty$ such that $J < \gamma^2$ if and only if $\bar{\sigma}(D_{111}) < \gamma$ and $J_\infty \in \text{dom}(\text{Ric})$ with $Y_\infty = \text{Ric}(J_\infty) \geq 0$, where

$$\tilde{R} := \begin{bmatrix} D_{11} \\ D_{21} \end{bmatrix} \begin{bmatrix} D_{11} \\ D_{21} \end{bmatrix}^* - \begin{bmatrix} \gamma^2 I & 0 \\ 0 & 0 \end{bmatrix}$$

$$J_\infty := \begin{bmatrix} A^* & 0 \\ -B_1 B_1^* & -A \end{bmatrix} - \begin{bmatrix} C_1^* & C_2^* \\ -B_1 D_{11}^* & -B_1 D_{21}^* \end{bmatrix} \tilde{R}^{-1} \begin{bmatrix} D_{11} B_1^* & C_1 \\ D_{21} B_1^* & C_2 \end{bmatrix}.$$

Moreover, if the above conditions are satisfied, then a rational causal filter $F(s)$ satisfying $J < \gamma^2$ is given by

$$\hat{z} = F(s)y = \left[\begin{array}{c|c} A + L_{2\infty} C_2 + L_{1\infty} D_{112} C_2 & -L_{2\infty} - L_{1\infty} D_{112} \\ \hline C_1 - D_{112} C_2 & D_{112} \end{array} \right] y$$

where

$$\begin{bmatrix} L_{1\infty} & L_{2\infty} \end{bmatrix} := - \begin{bmatrix} B_1 D_{11}^* + Y_\infty C_1^* & B_1 D_{21}^* + Y_\infty C_2^* \end{bmatrix} \tilde{R}^{-1}.$$

In the case where $D_{11} = 0$ and $B_1 D_{21}^* = 0$ the filter becomes much simpler:

$$\hat{z} = \left[\begin{array}{c|c} A - Y_\infty C_2^* C_2 & Y_\infty C_2^* \\ \hline C_1 & 0 \end{array} \right] y$$

where Y_∞ is the stabilizing solution to

$$Y_\infty A^* + A Y_\infty + Y_\infty (\gamma^{-2} C_1^* C_1 - C_2^* C_2) Y_\infty + B_1 B_1^* = 0.$$

14.10 Notes and References

The first part of this chapter is based on Sampei, Mita, and Nakamichi [1990], Packard [1994], Doyle, Glover, Khargonekar, and Francis [1989], and Chen and Zhou [1996]. The proof of Theorem 14.4 comes from Ran and Vreugdenhil [1988]. The detailed derivation of the $\mathcal{H}_\infty$ solution for the general case is treated in Glover and Doyle [1988, 1989]. A fairly complete solution to the singular $\mathcal{H}_\infty$ problem is obtained in Stoorvogel [1992]. The $\mathcal{H}_\infty$ filtering and smoothing problems are considered in detail in Nagpal and Khargonekar [1991]. There is a rich body of literature on the LMI approach to $\mathcal{H}_\infty$ control and related problems. In particular, readers are referred to the monograph by Boyd, Ghaoui, Feron, and Balakrishnan [1994] for a comprehensive treatment of LMIs and their applications in control. See also the paper by Chilali and Gahinet [1996] for an application of LMIs in $\mathcal{H}_\infty$ control with closed-loop pole constraints.

14.11 Problems

Problem 14.1 Figure 14.9 shows a single-loop analog feedback system. The plant is

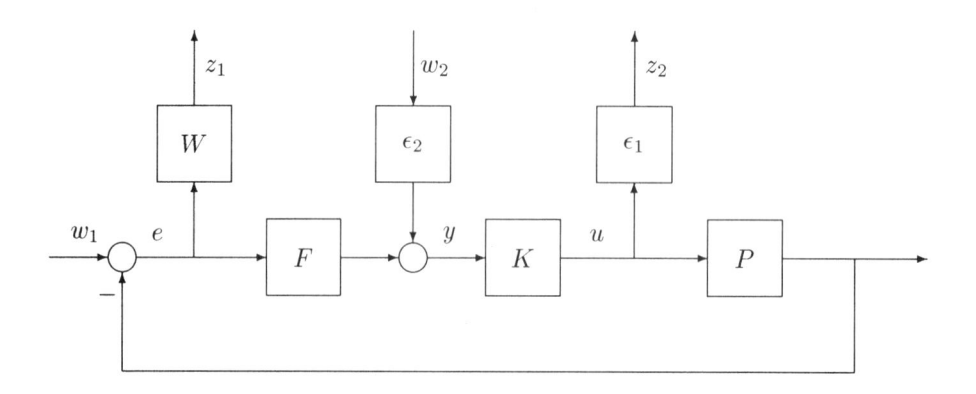

Figure 14.9: Analog feedback system

P and the controller K; F is an antialiasing filter for future digital implementation of the controller (it is a good idea to include F at the start of the analog design so that there are no surprises later due to additional phase lag). The basic control specification is to get good tracking over a certain frequency range, say $[0, \omega_1]$; that is, to make the magnitude of the transfer function from w_1 to e small over this frequency range. The weighted tracking error is z_1 in the figure, where the weight W is selected to be a low-pass filter with bandwidth ω_1. We could attempt to minimize the $\mathcal{H}_\infty$ norm from w_1 to z_1, but this problem is not regular. To regularize it, another input, w_2, is added and another signal, z_2, is penalized. The two weights ϵ_1 and ϵ_2 are small positive scalars. The design problem is to minimize the $\mathcal{H}_\infty$ norm

$$\text{from } w = \begin{bmatrix} w_1 \\ w_2 \end{bmatrix} \text{ to } z = \begin{bmatrix} z_1 \\ z_2 \end{bmatrix}.$$

Figure 14.9 can then be converted to the standard setup by stacking the states of P, F, and W to form the state of G.

The plant transfer function is taken to be

$$P(s) = \frac{20 - s}{(s + 0.01)(s + 20)}.$$

With a view toward subsequent digital control with $h = 0.5$, the filter F is taken to have bandwidth $\pi/0.5$, and Nyquist frequency ω_N:

$$F(s) = \frac{1}{(0.5/\pi)s + 1}.$$

14.11. Problems

The weight W is then taken to have bandwidth one-fifth the Nyquist frequency:

$$W(s) = \left[\frac{1}{(2.5/\pi)s + 1}\right]^2.$$

Finally, ϵ_1 and ϵ_2 are both set to 0.01.

Run *hinfsyn* and show your plots of the closed-loop frequency responses.

Problem 14.2 Make the same assumptions as in Chapter 13 for $\mathcal{H}_2$ control and derive the $\mathcal{H}_\infty$ controller parameterization by using the normalization procedure and Theorem 14.7.

Problem 14.3 Consider the feedback system in Figure 6.3 and suppose

$$P = \frac{s-10}{(s+1)(s+10)}, \quad W_e = \frac{1}{s+0.001}, \quad W_u = \frac{s+2}{s+10}.$$

Design a controller that minimizes

$$\left\| \begin{bmatrix} W_e S_o \\ W_u K S_o \end{bmatrix} \right\|_\infty.$$

Simulate the time response of the system when r is a step.

Problem 14.4 Repeat Problem 14.3 when $W_e = 1/s$.

Problem 14.5 Consider again Problem 13.4 and design a controller that minimizes the $\mathcal{H}_\infty$ norm of the transfer matrix from r to (e, u_w).

Problem 14.6 Repeat Problem 14.5 with $W = \epsilon$ for $\epsilon = 0.01$ and 0.0001. Study the behavior of the controller when $\epsilon \to 0$.

Problem 14.7 Repeat Problem 14.5 and Problem 14.6 with

$$P = \frac{10(2-s)}{(s+1)^3}.$$

Problem 14.8 Let $N \in \mathcal{RH}_\infty^-$. The Nehari problem is the following approximation problem:

$$\inf_{Q \in \mathcal{RH}_\infty} \|N - Q\|_\infty.$$

Formulate the Nehari problem as a standard $\mathcal{H}_\infty$ control problem.

Problem 14.9 Consider a generalized plant

$$P = \left[\begin{array}{cc|cc} -4 & 25 & 0.8 & -1 \\ -10 & 29 & 0.9 & -1 \\ \hline 10 & -25 & 0 & 1 \\ 13 & 25 & 1 & 0 \end{array}\right]$$

with an SISO controller K. Find the optimal $\mathcal{H}_\infty$ performance γ_{opt}. Calculate the central controller for each $\gamma \in (\gamma_{opt}, 2)$ and the corresponding $\mathcal{H}_\infty$ performance, γ_∞. Plot γ_∞ versus γ. Is γ_∞ monotonic with respect to γ? (See Ushida and Kimura [1996] for a detailed discussion.)

Problem 14.10 Let a satellite model be given by $P_o(s) = \left[\begin{array}{c|c} A & B \\ \hline C & 0 \end{array}\right]$, where

$$A = \begin{bmatrix} 0 & 1 & 0 & 0 \\ 0 & 0 & 0 & 0 \\ 0 & 0 & 0 & 1 \\ 0 & 0 & -1.539^2 & -2 \times 0.003 \times 1.539 \end{bmatrix}, \quad B = \begin{bmatrix} 0 \\ 1.7319 \times 10^{-5} \\ 0 \\ 3.7859 \times 10^{-4} \end{bmatrix},$$

$$C = \begin{bmatrix} 1 & 0 & 1 & 0 \end{bmatrix}, \quad D = 0.$$

Suppose the true system is described by

$$P = (N + \Delta_N)(M + \Delta_M)^{-1}$$

where $P_o = NM^{-1}$ is a normalized coprime factorization. Design a controller so that the controller stabilizes the largest

$$\Delta = \begin{bmatrix} \Delta_N \\ \Delta_M \end{bmatrix}.$$

Problem 14.11 Consider the feedback system shown below and let

$$P = \frac{0.5(1-s)}{(s+2)(s+0.5)}, \quad W_1 = 50\frac{s/1.245 + 1}{s/0.007 + 1}, \quad W_2 = 0.1256\frac{s/0.502 + 1}{s/2 + 1}.$$

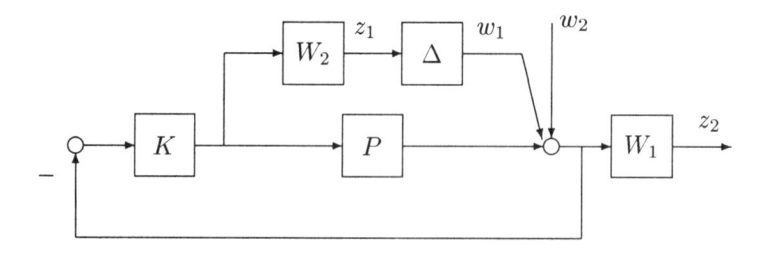

Figure 14.10: System with additive uncertainty

Then

$$\begin{bmatrix} z_1 \\ z_2 \end{bmatrix} = \begin{bmatrix} -W_2 KS & -W_2 KS \\ W_1 S & W_1 S \end{bmatrix} \begin{bmatrix} w_1 \\ w_2 \end{bmatrix} = M \begin{bmatrix} w_1 \\ w_2 \end{bmatrix}$$

where $S = (I + PK)^{-1}$.

14.11. Problems

(a) Design a controller K such that

$$\inf_{K \text{ stabilizing}} \|M\|_\infty.$$

(b) Design a controller K so that

$$\inf_{K \text{ stabilizing}} \sup_\omega \mu_\Delta(M), \quad \Delta = \begin{bmatrix} \Delta_1 & \\ & \Delta_2 \end{bmatrix}.$$

Note that $\mu_\Delta(M) = |W_1 S| + |W_2 K S|$.

Problem 14.12 Design a μ-synthesis controller for the HIMAT control problem in Example 9.1.

Problem 14.13 Let $G(s) \in \mathcal{H}_\infty$ be a square transfer matrix and $\alpha > 0$. Show that G is (extended) strictly positive real (i.e., $G^*(j\omega) + G(j\omega) > 0, \ \forall \ \omega \in \mathbb{R} \cup \{\infty\}$) if and only if $\left\|(\alpha I - G)(\alpha I + G)^{-1}\right\|_\infty < 1$.

Problem 14.14 Consider a generalized system

$$G(s) = \left[\begin{array}{c|cc} A & B_1 & B_2 \\ \hline C_1 & D_{11} & D_{12} \\ C_2 & D_{21} & D_{22} \end{array}\right]$$

and suppose we want to find a controller K such that $\mathcal{F}_\ell(G, K)$ is (extended) strictly postive real. Show that the problem can be converted to a standard $\mathcal{H}_\infty$ control problem by using the transformation in the last problem.

Problem 14.15 (Synthesis Using Popov Criterion) A stability criterion by Popov involves finding a controller and a multiplier matrix N such that

$$Q + (I + sN)\mathcal{F}_\ell(G, K)$$

is strictly positive real where Q is a constant matrix. Assume $D_{11} = 0$ and $D_{12} = 0$ in the realization of G. Find the $\tilde{G}$ so that

$$Q + (I + sN)\mathcal{F}_\ell(G, K) = \mathcal{F}_\ell(\tilde{G}, K)$$

Hence the results in the last problem can be applied.

Chapter 15

Controller Reduction

We have shown in the previous chapters that the $\mathcal{H}_\infty$ control theory and μ synthesis can be used to design robust performance controllers for highly complex uncertain systems. However, since a great many physical plants are modeled as high-order dynamical systems, the controllers designed with these methodologies typically have orders comparable to those of the plants. Simple linear controllers are normally preferred over complex linear controllers in control system designs for some obvious reasons: They are easier to understand and computationally less demanding; they are also easier to implement and have higher reliability since there are fewer things to go wrong in the hardware or bugs to fix in the software. Therefore, a lower-order controller should be sought whenever the resulting performance degradation is kept within an acceptable level. There are usually three ways to arrive at a lower-order controller. A seemingly obvious approach is to design lower-order controllers directly based on the high-order models. However, this is still largely an open research problem. Another approach is first to reduce the order of a high-order plant and, second, based on the reduced plant model, design a lower-order controller. A potential problem associated with this approach is that such a lower-order controller may not even stabilize the full-order plant since the error information between the full-order model and the reduced-order model is not considered in the design of the controller. On the other hand, one may seek to design first a high-order, high-performance controller and subsequently proceed with a reduction of the designed controller. This approach is usually referred to as controller reduction. A crucial consideration in controller order reduction is to take into account the closed loop so that closed-loop stability is guaranteed and the performance degradation is minimized with the reduced-order controllers. The purpose of this chapter is to introduce several controller reduction methods that can guarantee closed-loop stability and possibly closed-loop performance as well.

15.1 $\mathcal{H}_\infty$ Controller Reductions

In this section, we consider an $\mathcal{H}_\infty$ performance-preserving controller order reduction problem. We consider the feedback system shown in Figure 15.1 with a generalized plant realization given by

$$G(s) = \left[\begin{array}{c|cc} A & B_1 & B_2 \\ \hline C_1 & D_{11} & D_{12} \\ C_2 & D_{21} & D_{22} \end{array}\right].$$

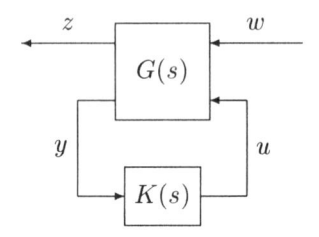

Figure 15.1: Closed-loop system diagram

The following assumptions are made:

(A1) (A, B_2) is stabilizable and (C_2, A) is detectable;

(A2) D_{12} has full column rank and D_{21} has full row rank;

(A3) $\begin{bmatrix} A - j\omega I & B_2 \\ C_1 & D_{12} \end{bmatrix}$ has full column rank for all ω;

(A4) $\begin{bmatrix} A - j\omega I & B_1 \\ C_2 & D_{21} \end{bmatrix}$ has full row rank for all ω.

As stated in Chapter 14, all stabilizing controllers satisfying $\|T_{zw}\|_\infty < \gamma$ can be parameterized as

$$K = \mathcal{F}_\ell(M_\infty, Q), \quad Q \in \mathcal{RH}_\infty, \quad \|Q\|_\infty < \gamma \qquad (15.1)$$

where M_∞ is of the form

$$M_\infty = \begin{bmatrix} M_{11}(s) & M_{12}(s) \\ M_{21}(s) & M_{22}(s) \end{bmatrix} = \left[\begin{array}{c|cc} \hat{A} & \hat{B}_1 & \hat{B}_2 \\ \hline \hat{C}_1 & \hat{D}_{11} & \hat{D}_{12} \\ \hat{C}_2 & \hat{D}_{21} & \hat{D}_{22} \end{array}\right]$$

such that $\hat{D}_{12}$ and $\hat{D}_{21}$ are invertible and $\hat{A} - \hat{B}_2\hat{D}_{12}^{-1}\hat{C}_1$ and $\hat{A} - \hat{B}_1\hat{D}_{21}^{-1}\hat{C}_2$ are both stable (i.e., M_{12}^{-1} and M_{21}^{-1} are both stable).

15.1. $\mathcal{H}_\infty$ Controller Reductions

The problem to be considered here is to find a controller $\hat{K}$ with a minimal possible order such that the $\mathcal{H}_\infty$ performance requirement $\left\|\mathcal{F}_\ell(G, \hat{K})\right\|_\infty < \gamma$ is satisfied. This is clearly equivalent to finding a Q so that it satisfies the preceding constraint and the order of $\hat{K}$ is minimized. Instead of choosing Q directly, we shall approach this problem from a different perspective. The following lemma is useful in the subsequent development and can be regarded as a special case of Theorem 10.6 (main loop theorem).

Lemma 15.1 *Consider a feedback system shown below*

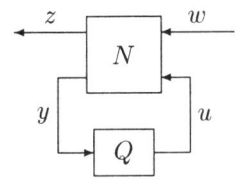

where N is a suitably partitioned transfer matrix

$$N(s) = \begin{bmatrix} N_{11} & N_{12} \\ N_{21} & N_{22} \end{bmatrix}.$$

Then the closed-loop transfer matrix from w to z is given by

$$T_{zw} = \mathcal{F}_\ell(N, Q) = N_{11} + N_{12} Q (I - N_{22} Q)^{-1} N_{21}.$$

Assume that the feedback loop is well-posed [i.e., $\det(I - N_{22}(\infty)Q(\infty)) \neq 0$] and either $N_{21}(j\omega)$ has full row rank for all $\omega \in \mathbb{R} \cup \infty$ or $N_{12}(j\omega)$ has full column rank for all $\omega \in \mathbb{R} \cup \infty$ and $\|N\|_\infty \leq 1$; then $\|\mathcal{F}_\ell(N, Q)\|_\infty < 1$ if $\|Q\|_\infty < 1$.

Proof. We shall assume that N_{21} has full row rank. The case when N_{12} has full column rank can be shown in the same way.

To show that $\|T_{zw}\|_\infty < 1$, consider the closed-loop system at any frequency $s = j\omega$ with the signals fixed as complex constant vectors. Let $\|Q\|_\infty =: \epsilon < 1$ and note that $T_{wy} = N_{21}^+(I - N_{22}Q)$, where N_{21}^+ is a right inverse of N_{21}. Also let $\kappa := \|T_{wy}\|_\infty$. Then $\|w\|_2 \leq \kappa \|y\|_2$, and $\|N\|_\infty \leq 1$ implies that $\|z\|_2^2 + \|y\|_2^2 \leq \|w\|_2^2 + \|u\|_2^2$. Therefore,

$$\|z\|_2^2 \leq \|w\|_2^2 + (\epsilon^2 - 1)\|y\|_2^2 \leq [1 - (1 - \epsilon^2)\kappa^{-2}]\|w\|_2^2,$$

which implies $\|T_{zw}\|_\infty < 1$. □

15.1.1 Additive Reduction

Consider the class of (reduced-order) controllers that can be represented in the form

$$\hat{K} = K_0 + W_2 \Delta W_1,$$

where K_0 may be interpreted as a nominal, higher-order controller, and Δ is a stable perturbation with stable, minimum phase and invertible weighting functions W_1 and W_2. Suppose that $\|\mathcal{F}_\ell(G, K_0)\|_\infty < \gamma$. A natural question is whether it is possible to obtain a reduced-order controller $\hat{K}$ in this class such that the $\mathcal{H}_\infty$ performance bound remains valid when $\hat{K}$ is in place of K_0. Note that this is somewhat a special case of the preceding general problem: The specific form of $\hat{K}$ means that $\hat{K}$ and K_0 must possess the same right-half plane poles, thus to a certain degree limiting the set of attainable reduced-order controllers.

Suppose $\hat{K}$ is a suboptimal $\mathcal{H}_\infty$ controller; that is, there is a $Q \in \mathcal{RH}_\infty$ with $\|Q\|_\infty < \gamma$ such that $\hat{K} = \mathcal{F}_\ell(M_\infty, Q)$. It follows from simple algebra that

$$Q = \mathcal{F}_\ell(\bar{K}_a^{-1}, \hat{K})$$

where

$$\bar{K}_a^{-1} := \begin{bmatrix} 0 & I \\ I & 0 \end{bmatrix} M_\infty^{-1} \begin{bmatrix} 0 & I \\ I & 0 \end{bmatrix}.$$

Furthermore, it follows from straightforward manipulations that

$$\|Q\|_\infty < \gamma \iff \left\|\mathcal{F}_\ell(\bar{K}_a^{-1}, \hat{K})\right\|_\infty < \gamma$$

$$\iff \left\|\mathcal{F}_\ell(\bar{K}_a^{-1}, K_0 + W_2 \Delta W_1)\right\|_\infty < \gamma$$

$$\iff \left\|\mathcal{F}_\ell(\tilde{R}, \Delta)\right\|_\infty < 1$$

where

$$\tilde{R} = \begin{bmatrix} \gamma^{-1/2} I & 0 \\ 0 & W_1 \end{bmatrix} \begin{bmatrix} R_{11} & R_{12} \\ R_{21} & R_{22} \end{bmatrix} \begin{bmatrix} \gamma^{-1/2} I & 0 \\ 0 & W_2 \end{bmatrix}$$

and R is given by the star product

$$\begin{bmatrix} R_{11} & R_{12} \\ R_{21} & R_{22} \end{bmatrix} = \mathcal{S}(\bar{K}_a^{-1}, \begin{bmatrix} K_o & I \\ I & 0 \end{bmatrix}).$$

It is easy to see that $\tilde{R}_{12}$ and $\tilde{R}_{21}$ are both minimum phase and invertible and hence have full column and full row rank, respectively, for all $\omega \in \mathbb{R} \cup \infty$. Consequently, by invoking Lemma 15.1, we conclude that if $\tilde{R}$ is a contraction and $\|\Delta\|_\infty < 1$, then $\left\|\mathcal{F}_\ell(\tilde{R}, \Delta)\right\|_\infty < 1$. This guarantees the existence of a Q such that $\|Q\|_\infty < \gamma$ or, equivalently, the existence of a $\hat{K}$ such that $\left\|\mathcal{F}_\ell(G, \hat{K})\right\|_\infty < \gamma$. This observation leads to the following theorem.

15.1. $\mathcal{H}_\infty$ Controller Reductions

Theorem 15.2 *Suppose W_1 and W_2 are stable, minimum phase and invertible transfer matrices such that $\tilde{R}$ is a contraction. Let K_0 be a stabilizing controller such that $\|\mathcal{F}_\ell(G, K_0)\|_\infty < \gamma$. Then $\hat{K}$ is also a stabilizing controller such that $\left\|\mathcal{F}_\ell(G, \hat{K})\right\|_\infty < \gamma$ if*

$$\|\Delta\|_\infty = \left\|W_2^{-1}(\hat{K} - K_0)W_1^{-1}\right\|_\infty < 1.$$

Since $\tilde{R}$ can always be made contractive for sufficiently small W_1 and W_2, there are infinite many W_1 and W_2 that satisfy the conditions in the theorem. It is obvious that to make $\left\|W_2^{-1}(\hat{K} - K_0)W_1^{-1}\right\|_\infty < 1$ for some $\hat{K}$, one would like to select the "largest" W_1 and W_2 such that $\tilde{R}$ is a contraction.

Lemma 15.3 *Assume that $\|R_{22}\|_\infty < \gamma$ and define*

$$L = \begin{bmatrix} L_1 & L_2 \\ L_2^\sim & L_3 \end{bmatrix} = \mathcal{F}_\ell\left(\left[\begin{array}{cc|cc} 0 & -R_{11} & 0 & R_{12} \\ -R_{11}^\sim & 0 & R_{21}^\sim & 0 \\ \hline 0 & R_{21} & 0 & -R_{22} \\ R_{12}^\sim & 0 & -R_{22}^\sim & 0 \end{array}\right], \gamma^{-1}I\right).$$

Then $\tilde{R}$ is a contraction if W_1 and W_2 satisfy

$$\begin{bmatrix} (W_1^\sim W_1)^{-1} & 0 \\ 0 & (W_2 W_2^\sim)^{-1} \end{bmatrix} \geq \begin{bmatrix} L_1 & L_2 \\ L_2^\sim & L_3 \end{bmatrix}.$$

Proof. See Goddard and Glover [1993]. □

An algorithm that maximizes $\det(W_1^\sim W_1)\det(W_2 W_2^\sim)$ has been developed by Goddard and Glover [1993]. The procedure below, devised directly from the preceding theorem, can be used to generate a required reduced-order controller that will preserve the closed-loop $\mathcal{H}_\infty$ performance bound $\left\|\mathcal{F}_\ell(G, \hat{K})\right\|_\infty < \gamma$.

1. Let K_0 be a full-order controller such that $\|\mathcal{F}_\ell(G, K_0)\|_\infty < \gamma$;

2. Compute W_1 and W_2 so that $\tilde{R}$ is a contraction;

3. Use the weighted model reduction method in Chapter 7 or any other methods to find a $\hat{K}$ so that $\left\|W_2^{-1}(\hat{K} - K_0)W_1^{-1}\right\|_\infty < 1$.

Note that all controller reduction methods introduced in this book are only sufficient; that is, there may be desired reduced-order controllers that cannot be found from the proposed procedures.

15.1.2 Coprime Factor Reduction

The $\mathcal{H}_\infty$ controller reduction problem can also be considered in the coprime factor framework. For that purpose, we need the following alternative representation of all admissible $\mathcal{H}_\infty$ controllers:

Lemma 15.4 *The family of all admissible controllers such that $\|T_{zw}\|_\infty < \gamma$ can also be written as*

$$\begin{aligned}K(s) = \mathcal{F}_\ell(M_\infty, Q) &= (\Theta_{11}Q + \Theta_{12})(\Theta_{21}Q + \Theta_{22})^{-1} := UV^{-1} \\ &= (Q\tilde{\Theta}_{12} + \tilde{\Theta}_{22})^{-1}(Q\tilde{\Theta}_{11} + \tilde{\Theta}_{21}) := \tilde{V}^{-1}\tilde{U}\end{aligned}$$

where $Q \in \mathcal{RH}_\infty$, $\|Q\|_\infty < \gamma$, and UV^{-1} and $\tilde{V}^{-1}\tilde{U}$ are, respectively, right and left coprime factorizations over $\mathcal{RH}_\infty$, and

$$\Theta = \begin{bmatrix} \Theta_{11} & \Theta_{12} \\ \Theta_{21} & \Theta_{22} \end{bmatrix} = \left[\begin{array}{c|cc} \hat{A} - \hat{B}_1\hat{D}_{21}^{-1}\hat{C}_2 & \hat{B}_2 - \hat{B}_1\hat{D}_{21}^{-1}\hat{D}_{22} & \hat{B}_1\hat{D}_{21}^{-1} \\ \hline \hat{C}_1 - \hat{D}_{11}\hat{D}_{21}^{-1}\hat{C}_2 & \hat{D}_{12} - \hat{D}_{11}\hat{D}_{21}^{-1}\hat{D}_{22} & \hat{D}_{11}\hat{D}_{21}^{-1} \\ -\hat{D}_{21}^{-1}\hat{C}_2 & -\hat{D}_{21}^{-1}\hat{D}_{22} & \hat{D}_{21}^{-1} \end{array}\right]$$

$$\tilde{\Theta} = \begin{bmatrix} \tilde{\Theta}_{11} & \tilde{\Theta}_{12} \\ \tilde{\Theta}_{21} & \tilde{\Theta}_{22} \end{bmatrix} = \left[\begin{array}{c|cc} \hat{A} - \hat{B}_2\hat{D}_{12}^{-1}\hat{C}_1 & \hat{B}_1 - \hat{B}_2\hat{D}_{12}^{-1}\hat{D}_{11} & -\hat{B}_2\hat{D}_{12}^{-1} \\ \hline \hat{C}_2 - \hat{D}_{22}\hat{D}_{12}^{-1}\hat{C}_1 & \hat{D}_{21} - \hat{D}_{22}\hat{D}_{12}^{-1}\hat{D}_{11} & -\hat{D}_{22}\hat{D}_{12}^{-1} \\ \hat{D}_{12}^{-1}\hat{C}_1 & \hat{D}_{12}^{-1}\hat{D}_{11} & \hat{D}_{12}^{-1} \end{array}\right]$$

$$\Theta^{-1} = \left[\begin{array}{c|ccc} \hat{A} - \hat{B}_2\hat{D}_{12}^{-1}\hat{C}_1 & \hat{B}_2\hat{D}_{12}^{-1} & \hat{B}_1 - \hat{B}_2\hat{D}_{12}^{-1}\hat{D}_{11} \\ \hline -\hat{D}_{12}^{-1}\hat{C}_1 & \hat{D}_{12}^{-1} & -\hat{D}_{12}^{-1}\hat{D}_{11} \\ \hat{C}_2 - \hat{D}_{22}\hat{D}_{12}^{-1}\hat{C}_1 & \hat{D}_{22}\hat{D}_{12}^{-1} & \hat{D}_{21} - \hat{D}_{22}\hat{D}_{12}^{-1}\hat{D}_{11} \end{array}\right]$$

$$\tilde{\Theta}^{-1} = \left[\begin{array}{c|ccc} \hat{A} - \hat{B}_1\hat{D}_{21}^{-1}\hat{C}_2 & -\hat{B}_1\hat{D}_{21}^{-1} & \hat{B}_2 - \hat{B}_1\hat{D}_{21}^{-1}\hat{D}_{22} \\ \hline \hat{D}_{21}^{-1}\hat{C}_2 & \hat{D}_{21}^{-1} & \hat{D}_{21}^{-1}\hat{D}_{22} \\ \hat{C}_1 - \hat{D}_{11}\hat{D}_{21}^{-1}\hat{C}_2 & -\hat{D}_{11}\hat{D}_{21}^{-1} & \hat{D}_{12} - \hat{D}_{11}\hat{D}_{21}^{-1}\hat{D}_{22} \end{array}\right].$$

Proof. The results follow immediately from Lemma 9.2. $\square$

Theorem 15.5 *Let $K_0 = \Theta_{12}\Theta_{22}^{-1}$ be the central $\mathcal{H}_\infty$ controller such that $\|\mathcal{F}_\ell(G, K_0)\|_\infty < \gamma$ and let $\hat{U}, \hat{V} \in \mathcal{RH}_\infty$ with $\det \hat{V}(\infty) \neq 0$ be such that*

$$\left\| \begin{bmatrix} \gamma^{-1}I & 0 \\ 0 & I \end{bmatrix} \Theta^{-1} \left(\begin{bmatrix} \Theta_{12} \\ \Theta_{22} \end{bmatrix} - \begin{bmatrix} \hat{U} \\ \hat{V} \end{bmatrix} \right) \right\|_\infty < 1/\sqrt{2}. \qquad (15.2)$$

Then $\hat{K} = \hat{U}\hat{V}^{-1}$ is also a stabilizing controller such that $\|\mathcal{F}_\ell(G, \hat{K})\|_\infty < \gamma$.

15.1. $\mathcal{H}_\infty$ Controller Reductions

Proof. Note that by Lemma 15.4, K is an admissible controller such that $\|T_{zw}\|_\infty < \gamma$ if and only if there exists a $Q \in \mathcal{RH}_\infty$ with $\|Q\|_\infty < \gamma$ such that

$$\begin{bmatrix} U \\ V \end{bmatrix} := \begin{bmatrix} \Theta_{11}Q + \Theta_{12} \\ \Theta_{21}Q + \Theta_{22} \end{bmatrix} = \Theta \begin{bmatrix} Q \\ I \end{bmatrix} \quad (15.3)$$

and
$$K = UV^{-1}.$$

Hence, to show that $\hat{K} = \hat{U}\hat{V}^{-1}$ with $\hat{U}$ and $\hat{V}$ satisfying equation (15.2) is also a stabilizing controller such that $\|\mathcal{F}_\ell(G, \hat{K})\|_\infty < \gamma$, we need to show that there is another coprime factorization for $\hat{K} = UV^{-1}$ and a $Q \in \mathcal{RH}_\infty$ with $\|Q\|_\infty < \gamma$ such that equation (15.3) is satisfied.

Define
$$\Delta := \begin{bmatrix} \gamma^{-1}I & 0 \\ 0 & I \end{bmatrix} \Theta^{-1} \left(\begin{bmatrix} \Theta_{12} \\ \Theta_{22} \end{bmatrix} - \begin{bmatrix} \hat{U} \\ \hat{V} \end{bmatrix} \right)$$

and partition Δ as
$$\Delta := \begin{bmatrix} \Delta_U \\ \Delta_V \end{bmatrix}.$$

Then
$$\begin{bmatrix} \hat{U} \\ \hat{V} \end{bmatrix} = \begin{bmatrix} \Theta_{12} \\ \Theta_{22} \end{bmatrix} - \Theta \begin{bmatrix} \gamma I & 0 \\ 0 & I \end{bmatrix} \Delta = \Theta \begin{bmatrix} -\gamma \Delta_U \\ I - \Delta_V \end{bmatrix}$$

and
$$\begin{bmatrix} \hat{U}(I - \Delta_V)^{-1} \\ \hat{V}(I - \Delta_V)^{-1} \end{bmatrix} = \Theta \begin{bmatrix} -\gamma \Delta_U (I - \Delta_V)^{-1} \\ I \end{bmatrix}.$$

Define $U := \hat{U}(I-\Delta_V)^{-1}$, $V := \hat{V}(I-\Delta_V)^{-1}$, and $Q := -\gamma \Delta_U(I-\Delta_V)^{-1}$. Then UV^{-1} is another coprime factorization for $\hat{K}$. To show that $\hat{K} = UV^{-1} = \hat{U}\hat{V}^{-1}$ is a stabilizing controller such that $\|\mathcal{F}_\ell(G,\hat{K})\|_\infty < \gamma$, we need to show that $\left\|\gamma \Delta_U(I-\Delta_V)^{-1}\right\|_\infty < \gamma$ or, equivalently, $\left\|\Delta_U(I-\Delta_V)^{-1}\right\|_\infty < 1$. Now

$$\begin{aligned}
\Delta_U(I - \Delta_V)^{-1} &= \begin{bmatrix} I & 0 \end{bmatrix} \Delta \left(I - \begin{bmatrix} 0 & I \end{bmatrix} \Delta \right)^{-1} \\
&= \mathcal{F}_\ell \left(\begin{bmatrix} 0 & \begin{bmatrix} I & 0 \end{bmatrix} \\ I/\sqrt{2} & \begin{bmatrix} 0 & I/\sqrt{2} \end{bmatrix} \end{bmatrix}, \sqrt{2}\Delta \right)
\end{aligned}$$

and, by Lemma 15.1, $\left\|\Delta_U(I - \Delta_V)^{-1}\right\|_\infty < 1$ since

$$\begin{bmatrix} 0 & \begin{bmatrix} I & 0 \end{bmatrix} \\ I/\sqrt{2} & \begin{bmatrix} 0 & I/\sqrt{2} \end{bmatrix} \end{bmatrix}$$

is a contraction and $\left\|\sqrt{2}\Delta\right\|_\infty < 1$. □

Similarly, we have the following theorem:

Theorem 15.6 *Let $K_0 = \tilde{\Theta}_{22}^{-1}\tilde{\Theta}_{21}$ be the central $\mathcal{H}_\infty$ controller such that $\|\mathcal{F}_\ell(G, K_0)\|_\infty < \gamma$ and let $\hat{\tilde{U}}, \hat{\tilde{V}} \in \mathcal{RH}_\infty$ with $\det \hat{\tilde{V}}(\infty) \neq 0$ be such that*

$$\left\|\left(\begin{bmatrix} \tilde{\Theta}_{21} & \tilde{\Theta}_{22} \end{bmatrix} - \begin{bmatrix} \hat{\tilde{U}} & \hat{\tilde{V}} \end{bmatrix}\right)\tilde{\Theta}^{-1}\begin{bmatrix} \gamma^{-1}I & 0 \\ 0 & I \end{bmatrix}\right\|_\infty < 1/\sqrt{2}.$$

Then $\hat{K} = \hat{\tilde{V}}^{-1}\hat{\tilde{U}}$ is also a stabilizing controller such that $\|\mathcal{F}_\ell(G, \hat{K})\|_\infty < \gamma$.

The preceding two theorems show that the sufficient conditions for $\mathcal{H}_\infty$ controller reduction problem are equivalent to frequency-weighted $\mathcal{H}_\infty$ model reduction problems.

$\mathcal{H}_\infty$ Controller Reduction Procedures

(i) Let $K_0 = \Theta_{12}\Theta_{22}^{-1}(= \tilde{\Theta}_{22}^{-1}\tilde{\Theta}_{21})$ be a suboptimal $\mathcal{H}_\infty$ central controller ($Q = 0$) such that $\|T_{zw}\|_\infty < \gamma$.

(ii) Find a reduced-order controller $\hat{K} = \hat{U}\hat{V}^{-1}$ (or $\hat{\tilde{V}}^{-1}\hat{\tilde{U}}$) such that

$$\left\|\begin{bmatrix} \gamma^{-1}I & 0 \\ 0 & I \end{bmatrix}\Theta^{-1}\left(\begin{bmatrix} \Theta_{12} \\ \Theta_{22} \end{bmatrix} - \begin{bmatrix} \hat{U} \\ \hat{V} \end{bmatrix}\right)\right\|_\infty < 1/\sqrt{2}$$

or

$$\left\|\left(\begin{bmatrix} \tilde{\Theta}_{21} & \tilde{\Theta}_{22} \end{bmatrix} - \begin{bmatrix} \hat{\tilde{U}} & \hat{\tilde{V}} \end{bmatrix}\right)\tilde{\Theta}^{-1}\begin{bmatrix} \gamma^{-1}I & 0 \\ 0 & I \end{bmatrix}\right\|_\infty < 1/\sqrt{2}.$$

Then the closed-loop system with the reduced-order controller $\hat{K}$ is stable and the performance is maintained with the reduced-order controller; that is,

$$\|T_{zw}\|_\infty = \left\|\mathcal{F}_\ell(G, \hat{K})\right\|_\infty < \gamma.$$

15.2 Notes and References

The main results presented in this chapter are based on the work of Goddard and Glover [1993, 1994]. Other controller reduction methods include the stability-oriented controller reduction criterion proposed by Enns [1984a, 1984b]; the weighted and unweighted coprime factor controller reduction methods studied by Liu and Anderson [1986, 1990]; Liu, Anderson, and Ly [1990]; Anderson and Liu [1989]; and Anderson [1993]; the normalized $\mathcal{H}_\infty$ controller reduction studied by Mustafa and Glover [1991]; the normalized coprime factor method studied by McFarlane and Glover [1990] in the $\mathcal{H}_\infty$ loop-shaping setup; and the controller reduction in the ν-gap metric setup studied by Vinnicombe [1993b]. Lenz, Khargonekar, and Doyle [1987] have also proposed another $\mathcal{H}_\infty$ controller reduction method with guaranteed performance for a class of $\mathcal{H}_\infty$ problems.

15.3 Problems

Problem 15.1 Find a lower-order controller for the system in Example 10.4 when $\gamma = 2$.

Problem 15.2 Find a lower-order controller for Problem 14.3 when $\gamma = 1.1\gamma_{opt}$ where γ_{opt} is the optimal norm.

Problem 15.3 Find a lower-order controller for the HIMAT control problem in Problem 14.12 when $\gamma = 1.1\gamma_{opt}$ where γ_{opt} is the optimal norm. Compare the controller reduction methods presented in this chapter with other available methods.

Problem 15.4 Let G be a generalized plant and K be a stabilizing controller. Let $\Delta = \text{diag}(\Delta_p, \Delta_k)$ be a suitably dimensioned perturbation and let $T_{\hat{z}\hat{w}}$ be the transfer matrix from $\hat{w} = \begin{bmatrix} w \\ w_1 \end{bmatrix}$ to $\hat{z} = \begin{bmatrix} z \\ z_1 \end{bmatrix}$ in the following diagram:

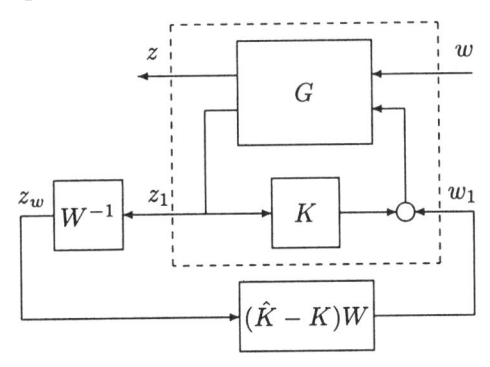

Let $W, W^{-1} \in \mathcal{H}_\infty$ be a given transfer matrix. Show that the following statements are equivalent:

1. $\mu_\Delta \left(\begin{bmatrix} I & 0 \\ 0 & W^{-1} \end{bmatrix} T_{\hat{z}\hat{w}} \right) < 1$;

2. $\|\mathcal{F}_\ell(G, K)\|_\infty < 1$ and $\|W^{-1}\mathcal{F}_u(T_{\hat{z}\hat{w}}, \Delta_p)\|_\infty < 1$ for all $\overline{\sigma}(\Delta_p) \leq 1$;

3. $\|W^{-1}T_{z_1 w_1}\|_\infty < 1$ and $\left\| \mathcal{F}_\ell \left(\begin{bmatrix} I & 0 \\ 0 & W^{-1} \end{bmatrix} T_{\hat{z}\hat{w}}, \Delta_k \right) \right\|_\infty < 1$ for all $\overline{\sigma}(\Delta_k) \leq 1$.

Problem 15.5 In the part 3 of Problem 15.4, if we let $\Delta_k = (\hat{K} - K)W$, then $T_{z_1 w_1} = G_{22}(I - KG_{22})^{-1}$ and $\mathcal{F}_\ell \left(\begin{bmatrix} I & 0 \\ 0 & W^{-1} \end{bmatrix} T_{\hat{z}\hat{w}}, \Delta_k \right) = \mathcal{F}_\ell(G, \hat{K})$. Thus $\hat{K}$ stabilizes the system and satisfies $\left\| \mathcal{F}_\ell(G, \hat{K}) \right\|_\infty < 1$ if $\|\Delta_k\|_\infty = \left\| (\hat{K} - K)W \right\|_\infty \leq 1$ and part 2 of

Problem 15.4 is satisfied by a controller K. Hence to reduce the order of the controller K, it is sufficient to solve a frequency-weighted model reduction problem if W can be calculated. In the single-input and single-output case, a "smallest" weighting function $W(s)$ can be calculated using part 2 of Problem 15.4 as follows:

$$|W(j\omega)| \geq \sup_{\overline{\sigma}(\Delta_p) \leq 1} |\mathcal{F}_u(T_{\hat{z}\hat{w}}(j\omega), \Delta_p)|.$$

Repeat Problems 15.1 and 15.2 using the foregoing method. (Hint: W can be computed frequency by frequency using μ software and then fitted by a stable and minimum phase transfer function.)

Problem 15.6 One way to generalize the method in Problem 15.5 to the MIMO case is to take a diagonal W

$$W = \text{diag}(W_1, W_2, \ldots, W_m)$$

and let $\hat{W}_i$ be computed from

$$|\hat{W}_i(j\omega)| \geq \sup_{\overline{\sigma}(\Delta_p) \leq 1} \left\| e_i^T \mathcal{F}_u(T_{\hat{z}\hat{w}}(j\omega), \Delta_p) \right\|$$

where e_i is the ith unit vector. Next let $\alpha(s)$ be computed from

$$|\alpha(j\omega)| \geq \sup_{\overline{\sigma}(\Delta_p) \leq 1} \left\| \hat{W}^{-1} \mathcal{F}_u(T_{\hat{z}\hat{w}}(j\omega), \Delta_p) \right\|$$

where $\hat{W} = \text{diag}(\hat{W}_1, \hat{W}_2, \ldots, \hat{W}_m)$. Then a suitable W can be taken as

$$W = \alpha \hat{W}.$$

Apply this method to Problem 15.3.

Problem 15.7 Generalize the procedures in Problems 15.5 and 15.6 to problems with additional structured uncertainty cases. (A more general case can be found in Yang and Packard [1995].)

Chapter 16

$\mathcal{H}_\infty$ Loop Shaping

This chapter introduces a design technique that incorporates loop-shaping methods to obtain performance/robust stability tradeoffs, and a particular $\mathcal{H}_\infty$ optimization problem to guarantee closed-loop stability and a level of robust stability at all frequencies. The proposed technique uses only the basic concept of loop-shaping methods, and then a robust stabilization controller for the normalized coprime factor perturbed system is used to construct the final controller. This chapter is arranged as follows: The $\mathcal{H}_\infty$ theory is applied to solve the stabilization problem of a normalized coprime factor perturbed system in Section 16.1. The loop-shaping design procedure is described in Section 16.2. The theoretical justification for the loop-shaping design procedure is given in Section 16.3. Some further loop-shaping guidelines are given in Section 16.4.

16.1 Robust Stabilization of Coprime Factors

In this section, we use the $\mathcal{H}_\infty$ control theory developed in previous chapters to solve the robust stabilization of a left coprime factor perturbed plant given by

$$P_\Delta = (\tilde{M} + \tilde{\Delta}_M)^{-1}(\tilde{N} + \tilde{\Delta}_N)$$

with $\tilde{M}, \tilde{N}, \tilde{\Delta}_M, \tilde{\Delta}_N \in \mathcal{RH}_\infty$ and $\left\| \begin{bmatrix} \tilde{\Delta}_N & \tilde{\Delta}_M \end{bmatrix} \right\|_\infty < \epsilon$ (see Figure 16.1). The transfer matrices $(\tilde{M}, \tilde{N})$ are assumed to be a left coprime factorization of P (i.e., $P = \tilde{M}^{-1}\tilde{N}$), and K internally stabilizes the nominal system.

It has been shown in Chapter 8 that the system is robustly stable iff

$$\left\| \begin{bmatrix} K \\ I \end{bmatrix} (I + PK)^{-1} \tilde{M}^{-1} \right\|_\infty \leq 1/\epsilon.$$

Finding a controller such that the above norm condition holds is an $\mathcal{H}_\infty$ norm minimization problem that can be solved using $\mathcal{H}_\infty$ theory developed in previous chapters.

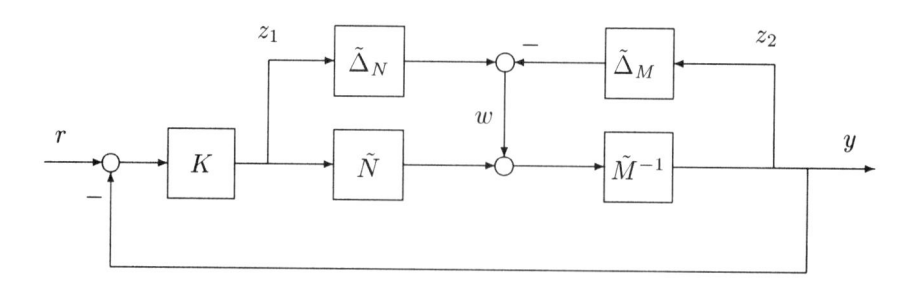

Figure 16.1: Left coprime factor perturbed systems

Suppose P has a stabilizable and detectable state-space realization given by

$$P = \left[\begin{array}{c|c} A & B \\ \hline C & D \end{array}\right]$$

and let L be a matrix such that $A + LC$ is stable. Then a left coprime factorization of $P = \tilde{M}^{-1}\tilde{N}$ is given by

$$\left[\begin{array}{cc} \tilde{N} & \tilde{M} \end{array}\right] = \left[\begin{array}{c|cc} A + LC & B + LD & L \\ \hline ZC & ZD & Z \end{array}\right]$$

where Z can be any nonsingular matrix. In particular, we shall choose $Z = (I + DD^*)^{-1/2}$ if $P = \tilde{M}^{-1}\tilde{N}$ is chosen to be a normalized left coprime factorization. Denote

$$\hat{K} = -K.$$

Then the system diagram can be put in an LFT form, as in Figure 16.2, with the generalized plant

$$G(s) = \left[\begin{array}{cc} \begin{bmatrix} 0 \\ \tilde{M}^{-1} \\ \tilde{M}^{-1} \end{bmatrix} & \begin{bmatrix} I \\ P \\ P \end{bmatrix} \end{array}\right] = \left[\begin{array}{c|cc} A & -LZ^{-1} & B \\ \hline \begin{bmatrix} 0 \\ C \\ C \end{bmatrix} & \begin{bmatrix} 0 \\ Z^{-1} \\ Z^{-1} \end{bmatrix} & \begin{bmatrix} I \\ D \\ D \end{bmatrix} \end{array}\right]$$

$$=: \left[\begin{array}{c|cc} A & B_1 & B_2 \\ \hline C_1 & D_{11} & D_{12} \\ C_2 & D_{21} & D_{22} \end{array}\right].$$

To apply the $\mathcal{H}_\infty$ control formulas in Chapter 14, we need to normalize the D_{12} and D_{21} first. Note that

$$\begin{bmatrix} I \\ D \end{bmatrix} = U \begin{bmatrix} 0 \\ I \end{bmatrix} (I + D^*D)^{\frac{1}{2}}, \text{ where } U = \begin{bmatrix} D^*(I + DD^*)^{-\frac{1}{2}} & (I + D^*D)^{-\frac{1}{2}} \\ -(I + DD^*)^{-\frac{1}{2}} & D(I + D^*D)^{-\frac{1}{2}} \end{bmatrix}$$

16.1. Robust Stabilization of Coprime Factors

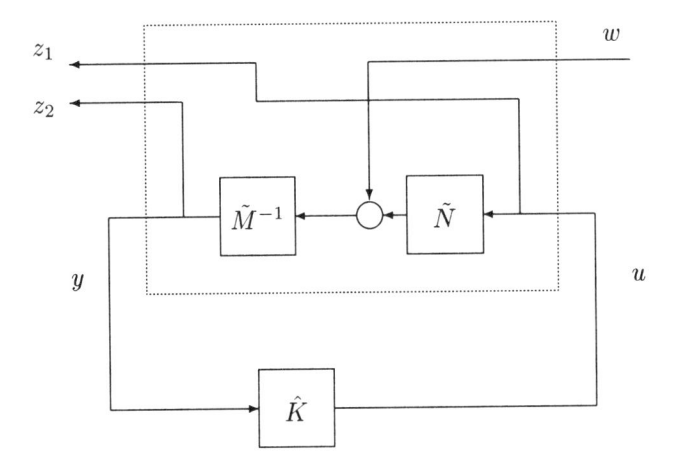

Figure 16.2: LFT diagram for coprime factor stabilization

and U is a unitary matrix. Let
$$\hat{K} = (I+D^*D)^{-\frac{1}{2}}\tilde{K}Z$$
$$\begin{bmatrix} z_1 \\ z_2 \end{bmatrix} = U \begin{bmatrix} \hat{z}_1 \\ \hat{z}_2 \end{bmatrix}.$$

Then $\|T_{zw}\|_\infty = \|U^*T_{zw}\|_\infty = \|T_{\hat{z}w}\|_\infty$ and the problem becomes one of finding a controller $\tilde{K}$ so that $\|T_{\hat{z}w}\|_\infty < \gamma$ with the following generalized plant:

$$\hat{G} = \begin{bmatrix} U^* & 0 \\ 0 & Z \end{bmatrix} G \begin{bmatrix} I & 0 \\ 0 & (I+D^*D)^{-\frac{1}{2}} \end{bmatrix}$$

$$= \left[\begin{array}{c|cc} A & -LZ^{-1} & B \\ \hline \begin{bmatrix} -(I+DD^*)^{-\frac{1}{2}}C \\ (I+D^*D)^{-\frac{1}{2}}D^*C \end{bmatrix} & \begin{bmatrix} -(I+DD^*)^{-\frac{1}{2}}Z^{-1} \\ (I+D^*D)^{-\frac{1}{2}}D^*Z^{-1} \end{bmatrix} & \begin{bmatrix} 0 \\ I \end{bmatrix} \\ ZC & I & ZD(I+D^*D)^{-\frac{1}{2}} \end{array} \right].$$

Now the formulas in Chapter 14 can be applied to $\hat{G}$ to obtain a controller $\tilde{K}$ and then the K can be obtained from $K = -(I+D^*D)^{-\frac{1}{2}}\tilde{K}Z$. We shall leave the detail to the reader. In the sequel, we shall consider the case $D = 0$ and $Z = I$. In this case, we have $\gamma > 1$ and

$$X_\infty(A - \frac{LC}{\gamma^2-1}) + (A - \frac{LC}{\gamma^2-1})^* X_\infty - X_\infty(BB^* - \frac{LL^*}{\gamma^2-1})X_\infty + \frac{\gamma^2 C^*C}{\gamma^2-1} = 0$$

$$Y_\infty(A+LC)^* + (A+LC)Y_\infty - Y_\infty C^*CY_\infty = 0.$$

It is clear that $Y_\infty = 0$ is the stabilizing solution. Hence by the formulas in Chapter 14 we have
$$\begin{bmatrix} L_{1\infty} & L_{2\infty} \end{bmatrix} = \begin{bmatrix} 0 & L \end{bmatrix}$$
and
$$Z_\infty = I, \quad \hat{D}_{11} = 0, \quad \hat{D}_{12} = I, \quad \hat{D}_{21} = \frac{\sqrt{\gamma^2 - 1}}{\gamma} I.$$
The results are summarized in the following theorem.

Theorem 16.1 *Let $D = 0$ and let L be such that $A + LC$ is stable. Then there exists a controller K such that*
$$\left\| \begin{bmatrix} K \\ I \end{bmatrix} (I + PK)^{-1} \tilde{M}^{-1} \right\|_\infty < \gamma$$
iff $\gamma > 1$ and there exists a stabilizing solution $X_\infty \geq 0$ solving
$$X_\infty \left(A - \frac{LC}{\gamma^2 - 1} \right) + \left(A - \frac{LC}{\gamma^2 - 1} \right)^* X_\infty - X_\infty \left(BB^* - \frac{LL^*}{\gamma^2 - 1} \right) X_\infty + \frac{\gamma^2 C^* C}{\gamma^2 - 1} = 0.$$
Moreover, if the above conditions hold a central controller is given by
$$K = \left[\begin{array}{c|c} A - BB^* X_\infty + LC & L \\ \hline -B^* X_\infty & 0 \end{array} \right].$$

It is clear that the existence of a robust stabilizing controller depends on the choice of the stabilizing matrix L (i.e., the choice of the coprime factorization). Now let $Y \geq 0$ be the stabilizing solution to
$$AY + YA^* - YC^*CY + BB^* = 0$$
and let $L = -YC^*$. Then the left coprime factorization $(\tilde{M}, \tilde{N})$ given by
$$\begin{bmatrix} \tilde{N} & \tilde{M} \end{bmatrix} = \left[\begin{array}{c|cc} A - YC^*C & B & -YC^* \\ \hline C & 0 & I \end{array} \right]$$
is a normalized left coprime factorization (see Chapter 12). Let $\|\cdot\|_H$ denote the Hankel norm (i.e., the largest Hankel singular value). Then we have the following result.

Corollary 16.2 *Let $D = 0$ and $L = -YC^*$, where $Y \geq 0$ is the stabilizing solution to*
$$AY + YA^* - YC^*CY + BB^* = 0.$$
Then $P = \tilde{M}^{-1}\tilde{N}$ is a normalized left coprime factorization and
$$\inf_{K \text{ stabilizing}} \left\| \begin{bmatrix} K \\ I \end{bmatrix} (I + PK)^{-1} \tilde{M}^{-1} \right\|_\infty = \frac{1}{\sqrt{1 - \lambda_{\max}(YQ)}}$$
$$= \left(1 - \left\| \begin{bmatrix} \tilde{N} & \tilde{M} \end{bmatrix} \right\|_H^2 \right)^{-1/2} =: \gamma_{\min}$$

16.1. Robust Stabilization of Coprime Factors

where Q is the solution to the following Lyapunov equation:

$$Q(A - YC^*C) + (A - YC^*C)^*Q + C^*C = 0.$$

Moreover, if the preceding conditions hold, then for any $\gamma > \gamma_{\min}$ a controller achieving

$$\left\| \begin{bmatrix} K \\ I \end{bmatrix} (I + PK)^{-1} \tilde{M}^{-1} \right\|_\infty < \gamma$$

is given by

$$K(s) = \left[\begin{array}{c|c} A - BB^*X_\infty - YC^*C & -YC^* \\ \hline -B^*X_\infty & 0 \end{array} \right]$$

where

$$X_\infty = \frac{\gamma^2}{\gamma^2 - 1} Q \left(I - \frac{\gamma^2}{\gamma^2 - 1} YQ \right)^{-1}.$$

Proof. Note that the Hamiltonian matrix associated with X_∞ is given by

$$H_\infty = \begin{bmatrix} A + \frac{1}{\gamma^2-1}YC^*C & -BB^* + \frac{1}{\gamma^2-1}YC^*CY \\ -\frac{\gamma^2}{\gamma^2-1}C^*C & -(A + \frac{1}{\gamma^2-1}YC^*C)^* \end{bmatrix}.$$

Straightforward calculation shows that

$$H_\infty = \begin{bmatrix} I & -\frac{\gamma^2}{\gamma^2-1}Y \\ 0 & \frac{\gamma^2}{\gamma^2-1}I \end{bmatrix} H_q \begin{bmatrix} I & -\frac{\gamma^2}{\gamma^2-1}Y \\ 0 & \frac{\gamma^2}{\gamma^2-1}I \end{bmatrix}^{-1}$$

where

$$H_q = \begin{bmatrix} A - YC^*C & 0 \\ -C^*C & -(A - YC^*C)^* \end{bmatrix}.$$

It is clear that the stable invariant subspace of H_q is given by

$$\mathcal{X}_-(H_q) = \mathrm{Im} \begin{bmatrix} I \\ Q \end{bmatrix}$$

and the stable invariant subspace of H_∞ is given by

$$\mathcal{X}_-(H_\infty) = \begin{bmatrix} I & -\frac{\gamma^2}{\gamma^2-1}Y \\ 0 & \frac{\gamma^2}{\gamma^2-1}I \end{bmatrix} \mathcal{X}_-(H_q) = \mathrm{Im} \begin{bmatrix} I - \frac{\gamma^2}{\gamma^2-1}YQ \\ \frac{\gamma^2}{\gamma^2-1}Q \end{bmatrix}.$$

Hence there is a nonnegative definite stabilizing solution to the algebraic Riccati equation of X_∞ if and only if

$$I - \frac{\gamma^2}{\gamma^2 - 1} YQ > 0$$

or
$$\gamma > \frac{1}{\sqrt{1 - \lambda_{\max}(YQ)}}$$
and the solution, if it exists, is given by
$$X_\infty = \frac{\gamma^2}{\gamma^2 - 1} Q \left(I - \frac{\gamma^2}{\gamma^2 - 1} YQ \right)^{-1}.$$

Note that Y and Q are the controllability Gramian and the observability Gramian of $\begin{bmatrix} \tilde{N} & \tilde{M} \end{bmatrix}$ respectively. Therefore, we also have that the Hankel norm of $\begin{bmatrix} \tilde{N} & \tilde{M} \end{bmatrix}$ is $\sqrt{\lambda_{\max}(YQ)}$. □

Corollary 16.3 *Let $P = \tilde{M}^{-1}\tilde{N}$ be a normalized left coprime factorization and*
$$P_\Delta = (\tilde{M} + \tilde{\Delta}_M)^{-1}(\tilde{N} + \tilde{\Delta}_N)$$
with
$$\left\| \begin{bmatrix} \tilde{\Delta}_N & \tilde{\Delta}_M \end{bmatrix} \right\|_\infty < \epsilon.$$
Then there is a robustly stabilizing controller for P_Δ if and only if
$$\epsilon \leq \sqrt{1 - \lambda_{\max}(YQ)} = \sqrt{1 - \left\| \begin{bmatrix} \tilde{N} & \tilde{M} \end{bmatrix} \right\|_H^2}.$$

The solutions to the normalized left coprime factorization stabilization problem are also solutions to a related $\mathcal{H}_\infty$ problem, which is shown in the following lemma.

Lemma 16.4 *Let $P = \tilde{M}^{-1}\tilde{N}$ be a normalized left coprime factorization. Then*
$$\left\| \begin{bmatrix} K \\ I \end{bmatrix} (I + PK)^{-1} \tilde{M}^{-1} \right\|_\infty = \left\| \begin{bmatrix} K \\ I \end{bmatrix} (I + PK)^{-1} \begin{bmatrix} I & P \end{bmatrix} \right\|_\infty.$$

Proof. Since $(\tilde{M}, \tilde{N})$ is a normalized left coprime factorization of P, we have
$$\begin{bmatrix} \tilde{M} & \tilde{N} \end{bmatrix} \begin{bmatrix} \tilde{M} & \tilde{N} \end{bmatrix}^\sim = I$$
and
$$\left\| \begin{bmatrix} \tilde{M} & \tilde{N} \end{bmatrix} \right\|_\infty = \left\| \begin{bmatrix} \tilde{M} & \tilde{N} \end{bmatrix}^\sim \right\|_\infty = 1.$$
Using these equations, we have
$$\left\| \begin{bmatrix} K \\ I \end{bmatrix} (I + PK)^{-1} \tilde{M}^{-1} \right\|_\infty$$

16.1. Robust Stabilization of Coprime Factors

$$= \left\| \begin{bmatrix} K \\ I \end{bmatrix} (I+PK)^{-1} \tilde{M}^{-1} \begin{bmatrix} \tilde{M} & \tilde{N} \end{bmatrix} \begin{bmatrix} \tilde{M} & \tilde{N} \end{bmatrix}^{\sim} \right\|_{\infty}$$

$$\leq \left\| \begin{bmatrix} K \\ I \end{bmatrix} (I+PK)^{-1} \tilde{M}^{-1} \begin{bmatrix} \tilde{M} & \tilde{N} \end{bmatrix} \right\|_{\infty} \left\| \begin{bmatrix} \tilde{M} & \tilde{N} \end{bmatrix}^{\sim} \right\|_{\infty}$$

$$= \left\| \begin{bmatrix} K \\ I \end{bmatrix} (I+PK)^{-1} \begin{bmatrix} I & P \end{bmatrix} \right\|_{\infty}$$

$$\leq \left\| \begin{bmatrix} K \\ I \end{bmatrix} (I+PK)^{-1} \tilde{M}^{-1} \right\|_{\infty} \left\| \begin{bmatrix} \tilde{M} & \tilde{N} \end{bmatrix} \right\|_{\infty}$$

$$= \left\| \begin{bmatrix} K \\ I \end{bmatrix} (I+PK)^{-1} \tilde{M}^{-1} \right\|_{\infty}.$$

This implies

$$\left\| \begin{bmatrix} K \\ I \end{bmatrix} (I+PK)^{-1} \tilde{M}^{-1} \right\|_{\infty} = \left\| \begin{bmatrix} K \\ I \end{bmatrix} (I+PK)^{-1} \begin{bmatrix} I & P \end{bmatrix} \right\|_{\infty}.$$

$\square$

Combining Corollary 16.3 and Lemma 16.4, we have the following result.

Corollary 16.5 *A controller solves the normalized left coprime factor robust stabilization problem if and only if it solves the following $\mathcal{H}_\infty$ control problem:*

$$\left\| \begin{bmatrix} I \\ K \end{bmatrix} (I+PK)^{-1} \begin{bmatrix} I & P \end{bmatrix} \right\|_{\infty} < \gamma$$

and

$$\inf_{K \text{ stabilizing}} \left\| \begin{bmatrix} I \\ K \end{bmatrix} (I+PK)^{-1} \begin{bmatrix} I & P \end{bmatrix} \right\|_{\infty} = \frac{1}{\sqrt{1 - \lambda_{\max}(YQ)}}$$

$$= \left(1 - \left\| \begin{bmatrix} \tilde{N} & \tilde{M} \end{bmatrix} \right\|_{H}^{2} \right)^{-1/2}.$$

The solution Q can also be obtained in other ways. Let $X \geq 0$ be the stabilizing solution to

$$XA + A^*X - XBB^*X + C^*C = 0.$$

Then it is easy to verify that

$$Q = (I+XY)^{-1}X.$$

Hence

$$\gamma_{\min} = \frac{1}{\sqrt{1 - \lambda_{\max}(YQ)}} = \left(1 - \left\|\begin{bmatrix} \tilde{N} & \tilde{M} \end{bmatrix}\right\|_H^2\right)^{-1/2} = \sqrt{1 + \lambda_{\max}(XY)}.$$

Similar results can be obtained if one starts with a normalized right coprime factorization. In fact, a rather strong relation between the normalized left and right coprime factorization problems can be established using the following matrix fact.

Lemma 16.6 *Let M be a square matrix such that $M^2 = M$. Then $\sigma_i(M) = \sigma_i(I - M)$ for all i such that $0 < \sigma_i(M) \neq 1$.*

Proof. We first show that the eigenvalues of M are either 0 or 1 and M is diagonalizable. In fact, assume that λ is an eigenvalue of M and x is a corresponding eigenvector. Then $\lambda x = Mx = MMx = M(Mx) = \lambda Mx = \lambda^2 x$; that is, $\lambda(1 - \lambda)x = 0$. This implies that either $\lambda = 0$ or $\lambda = 1$. To show that M is diagonalizable, assume that $M = TJT^{-1}$, where J is a Jordan canonical form. It follows immediately that J must be diagonal by the condition $M = M^2$.

Next, assume that M is diagonalized by a nonsingular matrix T such that

$$M = T \begin{bmatrix} I & 0 \\ 0 & 0 \end{bmatrix} T^{-1}.$$

Then

$$N := I - M = T \begin{bmatrix} 0 & 0 \\ 0 & I \end{bmatrix} T^{-1}.$$

Define

$$\begin{bmatrix} A & B \\ B^* & D \end{bmatrix} := T^*T$$

and assume $0 < \lambda \neq 1$. Then $A > 0$ and

$$\det(M^*M - \lambda I) = 0$$

$$\Leftrightarrow \quad \det\left(\begin{bmatrix} I & 0 \\ 0 & 0 \end{bmatrix} T^*T \begin{bmatrix} I & 0 \\ 0 & 0 \end{bmatrix} - \lambda T^*T\right) = 0$$

$$\Leftrightarrow \quad \det\begin{bmatrix} (1-\lambda)A & -\lambda B \\ -\lambda B^* & -\lambda D \end{bmatrix} = 0$$

$$\Leftrightarrow \quad \det(-\lambda D - \frac{\lambda^2}{1-\lambda} B^* A^{-1} B) = 0$$

$$\Leftrightarrow \quad \det(\frac{1-\lambda}{\lambda} D + B^* A^{-1} B) = 0$$

16.1. Robust Stabilization of Coprime Factors

$$\Leftrightarrow \quad \det \begin{bmatrix} -\lambda A & -\lambda B \\ -\lambda B^* & (1-\lambda)D \end{bmatrix} = 0$$

$$\Leftrightarrow \quad \det(N^*N - \lambda I) = 0.$$

This implies that all nonzero eigenvalues of M^*M and N^*N that are not equal to 1 are equal; that is, $\sigma_i(M) = \sigma_i(I - M)$ for all i such that $0 < \sigma_i(M) \neq 1$. □

Using this matrix fact, we have the following corollary.

Corollary 16.7 *Let K and P be any compatibly dimensioned complex matrices. Then*

$$\left\| \begin{bmatrix} I \\ K \end{bmatrix} (I + PK)^{-1} \begin{bmatrix} I & P \end{bmatrix} \right\| = \left\| \begin{bmatrix} I \\ P \end{bmatrix} (I + KP)^{-1} \begin{bmatrix} I & K \end{bmatrix} \right\|.$$

Proof. Define

$$M = \begin{bmatrix} I \\ K \end{bmatrix} (I + PK)^{-1} \begin{bmatrix} I & P \end{bmatrix}, \quad N = \begin{bmatrix} -P \\ I \end{bmatrix} (I + KP)^{-1} \begin{bmatrix} -K & I \end{bmatrix}.$$

Then it is easy to verify that $M^2 = M$ and $N = I - M$. By Lemma 16.6, we have $\|M\| = \|N\|$. The corollary follows by noting that

$$\begin{bmatrix} I \\ P \end{bmatrix} (I + KP)^{-1} \begin{bmatrix} I & K \end{bmatrix} = \begin{bmatrix} 0 & I \\ -I & 0 \end{bmatrix} N \begin{bmatrix} 0 & -I \\ I & 0 \end{bmatrix}.$$

□

Corollary 16.8 *Let $P = \tilde{M}^{-1}\tilde{N} = NM^{-1}$ be, respectively, the normalized left and right coprime factorizations. Then*

$$\left\| \begin{bmatrix} K \\ I \end{bmatrix} (I + PK)^{-1} \tilde{M}^{-1} \right\|_\infty = \left\| M^{-1}(I + KP)^{-1} \begin{bmatrix} I & K \end{bmatrix} \right\|_\infty.$$

Proof. This follows from Corollary 16.7 and the fact that

$$\left\| M^{-1}(I + KP)^{-1} \begin{bmatrix} I & K \end{bmatrix} \right\|_\infty = \left\| \begin{bmatrix} I \\ P \end{bmatrix} (I + KP)^{-1} \begin{bmatrix} I & K \end{bmatrix} \right\|_\infty.$$

□

This corollary says that any $\mathcal{H}_\infty$ controller for the normalized left coprime factorization is also an $\mathcal{H}_\infty$ controller for the normalized right coprime factorization. Hence one can work with either factorization.

For future reference, we shall define

$$b_{P,K} := \begin{cases} \left(\left\|\begin{bmatrix} I \\ K \end{bmatrix}(I+PK)^{-1}\begin{bmatrix} I & P \end{bmatrix}\right\|_\infty\right)^{-1} & \text{if } K \text{ stabilizes } P \\ 0 & \text{otherwise} \end{cases}$$

and

$$b_{\text{opt}} := \sup_K b_{P,K}.$$

Then $b_{P,K} = b_{K,P}$ and

$$b_{\text{opt}} = \sqrt{1 - \lambda_{\max}(YQ)} = \sqrt{1 - \left\|\begin{bmatrix} \tilde{N} & \tilde{M} \end{bmatrix}\right\|_H^2}.$$

The number $b_{P,K}$ can be related to the classical gain and phase margins as shown in Vinnicombe [1993b].

Theorem 16.9 *Let P be a SISO plant and K be a stabilizing controller. Then*

$$\text{gain margin} \geq \frac{1+b_{P,K}}{1-b_{P,K}}$$

and

$$\text{phase margin} \geq 2\arcsin(b_{P,K}).$$

Proof. Note that for SISO system

$$b_{P,K} \leq \frac{|1+P(j\omega)K(j\omega)|}{\sqrt{1+|P(j\omega)|^2}\sqrt{1+|K(j\omega)|^2}}, \quad \forall \omega.$$

So, at frequencies where $k := -P(j\omega)K(j\omega) \in \mathbb{R}^+$,

$$b_{P,K} \leq \frac{|1-k|}{\sqrt{(1+|P|^2)(1+\frac{k^2}{|P|^2})}} \leq \frac{|1-k|}{\sqrt{\min_P\left\{(1+|P|^2)(1+\frac{k^2}{|P|^2})\right\}}} = \left|\frac{1-k}{1+k}\right|,$$

which implies that

$$k \leq \frac{1-b_{P,K}}{1+b_{P,K}}, \quad \text{or} \quad k \geq \frac{1+b_{P,K}}{1-b_{P,K}}$$

from which the gain margin result follows. Similarly, at frequencies where $P(j\omega)K(j\omega) = -e^{j\theta}$,

$$b_{P,K} \leq \frac{|1-e^{j\theta}|}{\sqrt{(1+|P|^2)(1+\frac{1}{|P|^2})}} \leq \frac{|2\sin\frac{\theta}{2}|}{\sqrt{\min_P\left\{(1+|P|^2)(1+\frac{1}{|P|^2})\right\}}} = \frac{|2\sin\frac{\theta}{2}|}{2},$$

16.2. Loop-Shaping Design

which implies $\theta \geq 2\arcsin b_{P,K}$. □

For example, $b_{P,K} = 1/2$ guarantees a gain margin of 3 and a phase margin of 60°.

Illustrative MATLAB Commands:

≫ **b**$_{\mathbf{p,k}}$ = **emargin**(**P**, **K**); % given P and K, compute $b_{P,K}$.

≫ [**K**$_{\mathrm{opt}}$, **b**$_{\mathbf{p,k}}$] = **ncfsyn**(**P**, **1**); % find the optimal controller K_{opt}.

≫ [**K**$_{\mathrm{sub}}$, **b**$_{\mathbf{p,k}}$] = **ncfsyn**(**P**, **2**); % find a suboptimal controller K_{sub}.

16.2 Loop-Shaping Design

This section considers the $\mathcal{H}_\infty$ loop-shaping design. The objective of this approach is to incorporate the simple performance/robustness tradeoff obtained in the loop-shaping with the guaranteed stability properties of $\mathcal{H}_\infty$ design methods. Recall from Section 6.1 of Chapter 6 that good performance controller design requires that

$$\bar{\sigma}\left((I+PK)^{-1}\right),\ \bar{\sigma}\left((I+PK)^{-1}P\right),\ \bar{\sigma}\left((I+KP)^{-1}\right),\ \bar{\sigma}\left(K(I+PK)^{-1}\right) \quad (16.1)$$

be made small, particularly in some low-frequency range. Good robustness requires that

$$\bar{\sigma}\left(PK(I+PK)^{-1}\right),\ \bar{\sigma}\left(KP(I+KP)^{-1}\right) \quad (16.2)$$

be made small, particularly in some high-frequency range. These requirements, in turn, imply that good controller design boils down to achieving the desired loop (and controller) gains in the appropriate frequency range:

$$\underline{\sigma}(PK) \gg 1,\ \underline{\sigma}(KP) \gg 1,\ \underline{\sigma}(K) \gg 1$$

in some low-frequency range and

$$\bar{\sigma}(PK) \ll 1,\ \bar{\sigma}(KP) \ll 1,\ \bar{\sigma}(K) \leq M$$

in some high-frequency range where M is not too large.

The $\mathcal{H}_\infty$ loop-shaping design procedure is developed by McFarlane and Glover [1990, 1992] and is stated next.

Loop-Shaping Design Procedure

(1) Loop-Shaping: The singular values of the nominal plant, as shown in Figure 16.3, are shaped, using a precompensator W_1 and/or a postcompensator W_2, to give a desired open-loop shape. The nominal plant P and the shaping functions W_1, W_2 are combined to form the shaped plant, P_s, where $P_s = W_2 P W_1$. We assume that W_1 and W_2 are such that P_s contains no hidden modes.

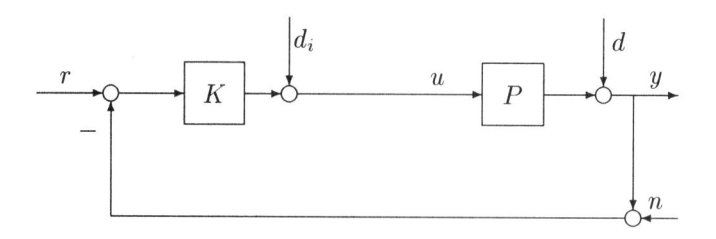

Figure 16.3: Standard feedback configuration

(2) Robust Stabilization: a) Calculate $\epsilon_{\max}$ (i.e., $b_{\text{opt}}(P_s)$), where

$$\epsilon_{\max} = \left(\inf_{K \text{ stabilizing}} \left\| \begin{bmatrix} I \\ K \end{bmatrix} (I + P_s K)^{-1} \tilde{M}_s^{-1} \right\|_\infty \right)^{-1}$$

$$= \sqrt{1 - \left\| \begin{bmatrix} \tilde{N}_s & \tilde{M}_s \end{bmatrix} \right\|_H^2} < 1$$

and $\tilde{M}_s, \tilde{N}_s$ define the normalized coprime factors of P_s such that $P_s = \tilde{M}_s^{-1} \tilde{N}_s$ and

$$\tilde{M}_s \tilde{M}_s^\sim + \tilde{N}_s \tilde{N}_s^\sim = I.$$

If $\epsilon_{\max} \ll 1$ return to (1) and adjust W_1 and W_2.

b) Select $\epsilon \leq \epsilon_{\max}$; then synthesize a stabilizing controller K_∞ that satisfies

$$\left\| \begin{bmatrix} I \\ K_\infty \end{bmatrix} (I + P_s K_\infty)^{-1} \tilde{M}_s^{-1} \right\|_\infty \leq \epsilon^{-1}.$$

(3) The final feedback controller K is then constructed by combining the $\mathcal{H}_\infty$ controller K_∞ with the shaping functions W_1 and W_2, as shown in Figure 16.4, such that

$$K = W_1 K_\infty W_2.$$

A typical design works as follows: the designer inspects the open-loop singular values of the nominal plant and shapes these by pre- and/or postcompensation until nominal performance (and possibly robust stability) specifications are met. (Recall that the open-loop shape is related to closed-loop objectives.) A feedback controller K_∞ with associated stability margin (for the shaped plant) $\epsilon \leq \epsilon_{\max}$, is then synthesized. If $\epsilon_{\max}$ is small, then the specified loop shape is incompatible with robust stability requirements and should be adjusted accordingly; then K_∞ is reevaluated.

In the preceding design procedure we have specified the desired loop shape by $W_2 P W_1$. But after Stage (2) of the design procedure, the actual loop shape achieved

16.2. Loop-Shaping Design

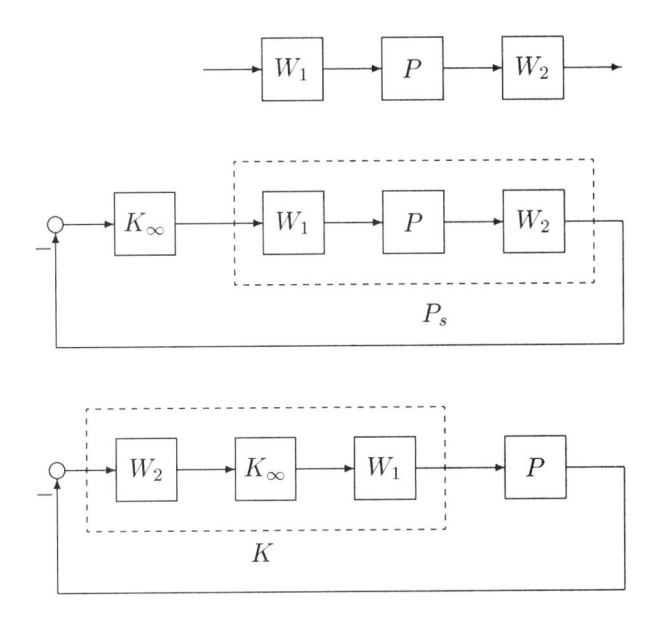

Figure 16.4: The loop-shaping design procedure

is, in fact, given by $W_1 K_\infty W_2 P$ at plant input and $P W_1 K_\infty W_2$ at plant output. It is therefore possible that the inclusion of K_∞ in the open-loop transfer function will cause deterioration in the open-loop shape specified by P_s. In the next section, we will show that the degradation in the loop shape caused by the $\mathcal{H}_\infty$ controller K_∞ is limited at frequencies where the specified loop shape is sufficiently large or sufficiently small. In particular, we show in the next section that ϵ can be interpreted as an indicator of the success of the loop-shaping in addition to providing a robust stability guarantee for the closed-loop systems. A small value of ϵ_{max} ($\epsilon_{max} \ll 1$) in Stage (2) always indicates incompatibility between the specified loop shape, the nominal plant phase, and robust closed-loop stability.

Remark 16.1 Note that, in contrast to the classical loop-shaping approach, the loop-shaping here is done without explicit regard for the nominal plant phase information. That is, closed-loop stability requirements are disregarded at this stage. Also, in contrast with conventional $\mathcal{H}_\infty$ design, the robust stabilization is done without frequency weighting. The design procedure described here is both simple and systematic and only assumes knowledge of elementary loop-shaping principles on the part of the designer.
◇

Remark 16.2 The preceding robust stabilization objective can also be interpreted as the more standard $\mathcal{H}_\infty$ problem formulation of minimizing the $\mathcal{H}_\infty$ norm of the frequency-weighted gain from disturbances on the plant input and output to the controller input and output as follows:

$$\left\| \begin{bmatrix} I \\ K_\infty \end{bmatrix} (I + P_s K_\infty)^{-1} \tilde{M}_s^{-1} \right\|_\infty = \left\| \begin{bmatrix} I \\ K_\infty \end{bmatrix} (I + P_s K_\infty)^{-1} \begin{bmatrix} I & P_s \end{bmatrix} \right\|_\infty$$

$$= \left\| \begin{bmatrix} W_2 \\ W_1^{-1} K \end{bmatrix} (I + PK)^{-1} \begin{bmatrix} W_2^{-1} & PW_1 \end{bmatrix} \right\|_\infty$$

$$= \left\| \begin{bmatrix} I \\ P_s \end{bmatrix} (I + K_\infty P_s)^{-1} \begin{bmatrix} I & K_\infty \end{bmatrix} \right\|_\infty$$

$$= \left\| \begin{bmatrix} W_1^{-1} \\ W_2 P \end{bmatrix} (I + KP)^{-1} \begin{bmatrix} W_1 & KW_2^{-1} \end{bmatrix} \right\|_\infty$$

This shows how all the closed-loop objectives in equations (16.1) and (16.2) are incorporated. As an example, it is easy to see that the signal relationship in Figure 16.5 is given by

$$\begin{bmatrix} z_1 \\ z_2 \end{bmatrix} = \begin{bmatrix} W_2 \\ W_1^{-1} K \end{bmatrix} (I + PK)^{-1} \begin{bmatrix} W_2^{-1} & PW_1 \end{bmatrix} \begin{bmatrix} w_1 \\ w_2 \end{bmatrix}.$$

$\diamond$

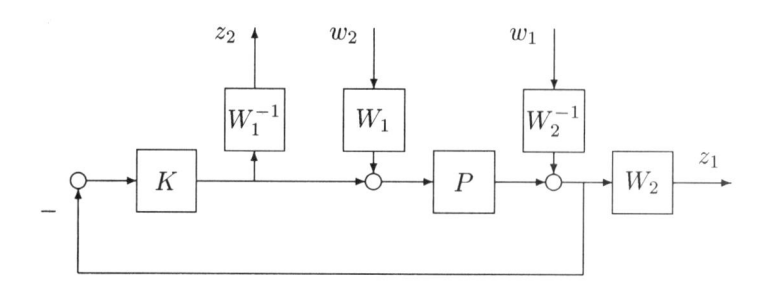

Figure 16.5: An equivalent $\mathcal{H}_\infty$ formulation

16.3 Justification for $\mathcal{H}_\infty$ Loop Shaping

The objective of this section is to provide justification for the use of parameter ϵ as a design indicator. We will show that ϵ is a measure of both closed-loop robust stability and the success of the design in meeting the loop-shaping specifications. Readers are encouraged to consult the original reference by McFarlane and Glover [1990] for further details.

We first examine the possibility of loop shape deterioration at frequencies of high loop gain (typically low-frequency). At low-frequency [in particular, $\omega \in (0, \omega_l)$], the deterioration in loop shape at plant output can be obtained by comparing $\underline{\sigma}(PW_1 K_\infty W_2)$

16.3. Justification for $\mathcal{H}_\infty$ Loop Shaping

to $\underline{\sigma}(P_s) = \underline{\sigma}(W_2 P W_1)$. Note that

$$\underline{\sigma}(PK) = \underline{\sigma}(PW_1 K_\infty W_2) = \underline{\sigma}(W_2^{-1} W_2 P W_1 K_\infty W_2) \geq \underline{\sigma}(W_2 P W_1)\underline{\sigma}(K_\infty)/\kappa(W_2) \quad (16.3)$$

where $\kappa(\cdot)$ denotes condition number. Similarly, for loop shape deterioration at plant input, we have

$$\underline{\sigma}(KP) = \underline{\sigma}(W_1 K_\infty W_2 P) = \underline{\sigma}(W_1 K_\infty W_2 P W_1 W_1^{-1}) \geq \underline{\sigma}(W_2 P W_1)\underline{\sigma}(K_\infty)/\kappa(W_1). \quad (16.4)$$

In each case, $\underline{\sigma}(K_\infty)$ is required to obtain a bound on the deterioration in the loop shape at low-frequency. Note that the condition numbers $\kappa(W_1)$ and $\kappa(W_2)$ are selected by the designer.

Next, recalling that P_s denotes the shaped plant and that K_∞ robustly stabilizes the normalized coprime factorization of P_s with stability margin ϵ, we have

$$\left\| \begin{bmatrix} I \\ K_\infty \end{bmatrix} (I + P_s K_\infty)^{-1} \tilde{M}_s^{-1} \right\|_\infty \leq \epsilon^{-1} := \gamma \quad (16.5)$$

where $(\tilde{N}_s, \tilde{M}_s)$ is a normalized left coprime factorization of P_s, and the parameter γ is defined to simplify the notation to follow. The following result shows that $\underline{\sigma}(K_\infty)$ is explicitly bounded by functions of ϵ and $\underline{\sigma}(P_s)$, the minimum singular value of the shaped plant, and hence by equation (16.3) and (16.4) K_∞ will only have a limited effect on the specified loop shape at low-frequency.

Theorem 16.10 *Any controller K_∞ satisfying equation (16.5), where P_s is assumed square, also satisfies*

$$\underline{\sigma}(K_\infty(j\omega)) \geq \frac{\underline{\sigma}(P_s(j\omega)) - \sqrt{\gamma^2 - 1}}{\sqrt{\gamma^2 - 1}\underline{\sigma}(P_s(j\omega)) + 1}$$

for all ω such that

$$\underline{\sigma}(P_s(j\omega)) > \sqrt{\gamma^2 - 1}.$$

Furthermore, if $\underline{\sigma}(P_s) \gg \sqrt{\gamma^2 - 1}$, then $\underline{\sigma}(K_\infty(j\omega)) \gtrapprox 1/\sqrt{\gamma^2 - 1}$, where $\gtrapprox$ denotes asymptotically greater than or equal to as $\underline{\sigma}(P_s) \to \infty$.

Proof. First note that $\underline{\sigma}(P_s) > \sqrt{\gamma^2 - 1}$ implies that

$$I + P_s P_s^* > \gamma^2 I.$$

Further, since $(\tilde{N}_s, \tilde{M}_s)$ is a normalized left coprime factorization of P_s, we have

$$\tilde{M}_s \tilde{M}_s^* = I - \tilde{N}_s \tilde{N}_s^* = I - \tilde{M}_s P_s P_s^* \tilde{M}_s^*.$$

Then

$$\tilde{M}_s^* \tilde{M}_s = (I + P_s P_s^*)^{-1} < \gamma^{-2} I.$$

Now
$$\left\| \begin{bmatrix} I \\ K_\infty \end{bmatrix} (I + P_s K_\infty)^{-1} \tilde{M}_s^{-1} \right\|_\infty \leq \gamma$$
can be rewritten as
$$(I + K_\infty^* K_\infty) \leq \gamma^2 (I + K_\infty^* P_s^*)(\tilde{M}_s^* \tilde{M}_s)(I + P_s K_\infty). \tag{16.6}$$

We will next show that K_∞ is invertible. Suppose that there exists an x such that $K_\infty x = 0$, then $x^* \times$ equation (16.6) $\times x$ gives
$$\gamma^{-2} x^* x \leq x^* \tilde{M}_s^* \tilde{M}_s x,$$
which implies that $x = 0$ since $\tilde{M}_s^* \tilde{M}_s < \gamma^{-2} I$, and hence K_∞ is invertible. Equation (16.6) can now be written as
$$(K_\infty^{-*} K_\infty^{-1} + I) \leq \gamma^2 (K_\infty^{-*} + P_s^*) \tilde{M}_s^* \tilde{M}_s (K_\infty^{-1} + P_s). \tag{16.7}$$

Define W such that
$$(WW^*)^{-1} = I - \gamma^2 \tilde{M}_s^* \tilde{M}_s = I - \gamma^2 (I + P_s P_s^*)^{-1}.$$

Completing the square in equation (16.7) with respect to K_∞^{-1} yields
$$(K_\infty^{-*} + N^*)(WW^*)^{-1}(K_\infty^{-1} + N) \leq (\gamma^2 - 1) R^* R$$
where
$$\begin{aligned} N &= \gamma^2 P_s((1-\gamma^2)I + P_s^* P_s)^{-1} \\ R^* R &= (I + P_s^* P_s)((1-\gamma^2)I + P_s^* P_s)^{-1}. \end{aligned}$$

Hence we have
$$R^{-*}(K_\infty^{-*} + N^*)(WW^*)^{-1}(K_\infty^{-1} + N)R^{-1} \leq (\gamma^2 - 1) I$$
and
$$\bar{\sigma}\left(W^{-1}(K_\infty^{-1} + N)R^{-1}\right) \leq \sqrt{\gamma^2 - 1}.$$
Use $\bar{\sigma}\left(W^{-1}(K_\infty^{-1} + N)R^{-1}\right) \geq \underline{\sigma}(W^{-1}) \bar{\sigma}(K_\infty^{-1} + N) \underline{\sigma}(R^{-1})$ to get
$$\bar{\sigma}(K_\infty^{-1} + N) \leq \sqrt{\gamma^2 - 1} \bar{\sigma}(W) \bar{\sigma}(R)$$
and use $\underline{\sigma}(K_\infty^{-1} + N) \geq \underline{\sigma}(K_\infty) - \bar{\sigma}(N)$ to get
$$\underline{\sigma}(K_\infty) \geq \left\{ (\gamma^2 - 1)^{1/2} \bar{\sigma}(W) \bar{\sigma}(R) + \bar{\sigma}(N) \right\}^{-1}. \tag{16.8}$$

16.3. Justification for $\mathcal{H}_\infty$ Loop Shaping

Next, note that the eigenvalues of WW^*, N^*N, and R^*R can be computed as follows:

$$\lambda(WW^*) = \frac{1 + \lambda(P_s P_s^*)}{1 - \gamma^2 + \lambda(P_s P_s^*)}$$

$$\lambda(N^*N) = \frac{\gamma^4 \lambda(P_s P_s^*)}{(1 - \gamma^2 + \lambda(P_s P_s^*))^2}$$

$$\lambda(R^*R) = \frac{1 + \lambda(P_s P_s^*)}{1 - \gamma^2 + \lambda(P_s P_s^*)}.$$

Therefore,

$$\bar{\sigma}(W) = \sqrt{\lambda_{\max}(WW^*)} = \left(\frac{1 + \lambda_{\min}(P_s P_s^*)}{1 - \gamma^2 + \lambda_{\min}(P_s P_s^*)} \right)^{1/2} = \left(\frac{1 + \underline{\sigma}^2(P_s)}{1 - \gamma^2 + \underline{\sigma}^2(P_s)} \right)^{1/2}$$

$$\bar{\sigma}(N) = \sqrt{\lambda_{\max}(N^*N)} = \frac{\gamma^2 \sqrt{\lambda_{\min}(P_s P_s^*)}}{1 - \gamma^2 + \lambda_{\min}(P_s P_s^*)} = \frac{\gamma^2 \underline{\sigma}(P_s)}{1 - \gamma^2 + \underline{\sigma}^2(P_s)}$$

$$\bar{\sigma}(R) = \sqrt{\lambda_{\max}(R^*R)} = \left(\frac{1 + \lambda_{\min}(P_s P_s^*)}{1 - \gamma^2 + \lambda_{\min}(P_s P_s^*)} \right)^{1/2} = \left(\frac{1 + \underline{\sigma}^2(P_s)}{1 - \gamma^2 + \underline{\sigma}^2(P_s)} \right)^{1/2}.$$

Substituting these formulas into equation (16.8), we have

$$\underline{\sigma}(K_\infty) \geq \left\{ \frac{(\gamma^2 - 1)^{1/2}(1 + \underline{\sigma}^2(P_s)) + \gamma^2 \underline{\sigma}(P_s)}{\underline{\sigma}^2(P_s) - (\gamma^2 - 1)} \right\}^{-1} = \frac{\underline{\sigma}(P_s)) - \sqrt{\gamma^2 - 1}}{\sqrt{\gamma^2 - 1}\underline{\sigma}(P_s) + 1}.$$

□

The main implication of Theorem 16.10 is that the bound on $\underline{\sigma}(K_\infty)$ depends only on the selected loop shape and the stability margin of the shaped plant. The value of $\gamma (= \epsilon^{-1})$ directly determines the frequency range over which this result is valid – a small γ (large ϵ) is desirable, as we would expect. Further, P_s has a sufficiently large loop gain; then so also will $P_s K_\infty$ provided that $\gamma (= \epsilon^{-1})$ is sufficiently small.

In an analogous manner, we now examine the possibility of deterioration in the loop shape at high-frequency due to the inclusion of K_∞. Note that at high frequency [in particular, $\omega \in (\omega_h, \infty)$] the deterioration in plant output loop shape can be obtained by comparing $\bar{\sigma}(PW_1 K_\infty W_2)$ to $\bar{\sigma}(P_s) = \bar{\sigma}(W_2 P W_1)$. Note that, analogous to equation (16.3) and (16.4), we have

$$\bar{\sigma}(PK) = \bar{\sigma}(PW_1 K_\infty W_2) \leq \bar{\sigma}(W_2 P W_1) \bar{\sigma}(K_\infty) \kappa(W_2).$$

Similarly, the corresponding deterioration in plant input loop shape is obtained by comparing $\bar{\sigma}(W_1 K_\infty W_2 P)$ to $\bar{\sigma}(W_2 P W_1)$, where

$$\bar{\sigma}(KP) = \bar{\sigma}(W_1 K_\infty W_2 P) \leq \bar{\sigma}(W_2 P W_1) \bar{\sigma}(K_\infty) \kappa(W_1).$$

Hence, in each case, $\overline{\sigma}(K_\infty)$ is required to obtain a bound on the deterioration in the loop shape at high-frequency. In an identical manner to Theorem 16.10, we now show that $\overline{\sigma}(K_\infty)$ is explicitly bounded by functions of γ and $\overline{\sigma}(P_s)$, the maximum singular value of the shaped plant.

Theorem 16.11 *Any controller K_∞ satisfying equation (16.5) also satisfies*

$$\overline{\sigma}(K_\infty(j\omega)) \leq \frac{\sqrt{\gamma^2 - 1} + \overline{\sigma}(P_s(j\omega))}{1 - \sqrt{\gamma^2 - 1}\ \overline{\sigma}(P_s(j\omega))}$$

for all ω such that

$$\overline{\sigma}(P_s(j\omega)) < \frac{1}{\sqrt{\gamma^2 - 1}}.$$

Furthermore, if $\overline{\sigma}(P_s) \ll 1/\sqrt{\gamma^2 - 1}$, then $\overline{\sigma}(K_\infty(j\omega)) \lesssim \sqrt{\gamma^2 - 1}$, where $\lesssim$ denotes asymptotically less than or equal to as $\overline{\sigma}(P_s) \to 0$.

Proof. The proof of Theorem 16.11 is similar to that of Theorem 16.10 and is only sketched here: As in the proof of Theorem 16.10, we have $\tilde{M}_s^* \tilde{M}_s = (I + P_s P_s^*)^{-1}$ and

$$(I + K_\infty^* K_\infty) \leq \gamma^2 (I + K_\infty^* P_s^*)(\tilde{M}_s^* \tilde{M}_s)(I + P_s K_\infty). \tag{16.9}$$

Since $\overline{\sigma}(P_s) < \frac{1}{\sqrt{\gamma^2-1}}$,

$$I - \gamma^2 P_s^* (I + P_s P_s^*)^{-1} P_s > 0$$

and there exists a spectral factorization

$$V^* V = I - \gamma^2 P_s^* (I + P_s P_s^*)^{-1} P_s.$$

Now, completing the square in equation (16.9) with respect to K_∞ yields

$$(K_\infty^* + M^*) V^* V (K_\infty + M) \leq (\gamma^2 - 1) Y^* Y$$

where

$$\begin{aligned} M &= \gamma^2 P_s^* (I + (1 - \gamma^2) P_s P_s^*)^{-1} \\ Y^* Y &= (\gamma^2 - 1)(I + P_s P_s^*)(I + (1 - \gamma^2) P_s P_s^*)^{-1}. \end{aligned}$$

Hence we have

$$\overline{\sigma}\left(V(K_\infty + M)Y^{-1}\right) \leq \sqrt{\gamma^2 - 1},$$

which implies

$$\overline{\sigma}(K_\infty) \leq \frac{\sqrt{\gamma^2 - 1}}{\underline{\sigma}(V)\underline{\sigma}(Y^{-1})} + \overline{\sigma}(M). \tag{16.10}$$

16.3. Justification for $\mathcal{H}_\infty$ Loop Shaping

As in the proof of Theorem 16.10, it is easy to show that

$$\underline{\sigma}(V) = \underline{\sigma}(Y^{-1}) = \left(\frac{1-(\gamma^2-1)\overline{\sigma}^2(P_s)}{1+\overline{\sigma}^2(P_s)}\right)^{1/2}$$

$$\overline{\sigma}(M) = \frac{\gamma^2 \overline{\sigma}(P_s)}{1-(\gamma^2-1)\overline{\sigma}^2(P_s)}.$$

Substituting these formulas into equation (16.10), we have

$$\overline{\sigma}(K_\infty) \le \frac{(\gamma^2-1)^{1/2}(1+\overline{\sigma}^2(P_s)) + \gamma^2\overline{\sigma}(P_s)}{1-(\gamma^2-1)\overline{\sigma}^2(P_s)} = \frac{\sqrt{\gamma^2-1}+\overline{\sigma}(P_s)}{1-\sqrt{\gamma^2-1}\overline{\sigma}(P_s)}.$$

$\square$

The results in Theorems 16.10 and 16.11 confirm that γ (alternatively ϵ) indicates the compatibility between the specified loop shape and closed-loop stability requirements.

Theorem 16.12 *Let P be the nominal plant and let $K = W_1 K_\infty W_2$ be the associated controller obtained from loop-shaping design procedure in the last section. Then if*

$$\left\| \begin{bmatrix} I \\ K_\infty \end{bmatrix} (I + P_s K_\infty)^{-1} \tilde{M}_s^{-1} \right\|_\infty \le \gamma$$

we have

$$\overline{\sigma}\left(K(I+PK)^{-1}\right) \le \gamma\overline{\sigma}(\tilde{M}_s)\overline{\sigma}(W_1)\overline{\sigma}(W_2) \tag{16.11}$$

$$\overline{\sigma}\left((I+PK)^{-1}\right) \le \min\left\{\gamma\overline{\sigma}(\tilde{M}_s)\kappa(W_2),\ 1+\gamma\overline{\sigma}(N_s)\kappa(W_2)\right\} \tag{16.12}$$

$$\overline{\sigma}\left(K(I+PK)^{-1}P\right) \le \min\left\{\gamma\overline{\sigma}(\tilde{N}_s)\kappa(W_1),\ 1+\gamma\overline{\sigma}(M_s)\kappa(W_1)\right\} \tag{16.13}$$

$$\overline{\sigma}\left((I+PK)^{-1}P\right) \le \frac{\gamma\overline{\sigma}(\tilde{N}_s)}{\underline{\sigma}(W_1)\underline{\sigma}(W_2)} \tag{16.14}$$

$$\overline{\sigma}\left((I+KP)^{-1}\right) \le \min\left\{1+\gamma\overline{\sigma}(\tilde{N}_s)\kappa(W_1),\ \gamma\overline{\sigma}(M_s)\kappa(W_1)\right\} \tag{16.15}$$

$$\overline{\sigma}\left(G(I+KP)^{-1}K\right) \le \min\left\{1+\gamma\overline{\sigma}(\tilde{M}_s)\kappa(W_2),\ \gamma\overline{\sigma}(N_s)\kappa(W_2)\right\} \tag{16.16}$$

where

$$\overline{\sigma}(\tilde{N}_s) = \overline{\sigma}(N_s) = \left(\frac{\overline{\sigma}^2(W_2 P W_1)}{1+\overline{\sigma}^2(W_2 P W_1)}\right)^{1/2} \tag{16.17}$$

$$\overline{\sigma}(\tilde{M}_s) = \overline{\sigma}(M_s) = \left(\frac{1}{1+\underline{\sigma}^2(W_2 P W_1)}\right)^{1/2} \tag{16.18}$$

and $(\tilde{N}_s, \tilde{M}_s)$, (N_s, M_s) is a normalized left coprime factorization and right coprime factorization, respectively, of $P_s = W_2 P W_1$.

Proof. Note that
$$\tilde{M}_s^* \tilde{M}_s = (I + P_s P_s^*)^{-1}$$
and
$$\tilde{M}_s \tilde{M}_s^* = I - \tilde{N}_s \tilde{N}_s^*.$$
Then
$$\bar{\sigma}^2(\tilde{M}_s) = \lambda_{\max}(\tilde{M}_s^* \tilde{M}_s) = \frac{1}{1 + \lambda_{\min}(P_s P_s^*)} = \frac{1}{1 + \underline{\sigma}^2(P_s)}$$
$$\bar{\sigma}^2(\tilde{N}_s) = 1 - \underline{\sigma}^2(\tilde{M}_s) = \frac{\bar{\sigma}^2(P_s)}{1 + \bar{\sigma}^2(P_s)}.$$

The proof for the normalized right coprime factorization is similar. All other inequalities follow from noting that

$$\left\| \begin{bmatrix} I \\ K_\infty \end{bmatrix} (I + P_s K_\infty)^{-1} \tilde{M}_s^{-1} \right\|_\infty \leq \gamma$$

and

$$\left\| \begin{bmatrix} I \\ K_\infty \end{bmatrix} (I + P_s K_\infty)^{-1} \tilde{M}_s^{-1} \right\|_\infty = \left\| \begin{bmatrix} W_2 \\ W_1^{-1} K \end{bmatrix} (I + PK)^{-1} \begin{bmatrix} W_2^{-1} & PW_1 \end{bmatrix} \right\|_\infty$$
$$= \left\| \begin{bmatrix} W_1^{-1} \\ W_2 P \end{bmatrix} (I + KP)^{-1} \begin{bmatrix} W_1 & PW_2^{-1} \end{bmatrix} \right\|_\infty$$

$\square$

This theorem shows that all closed-loop objectives are guaranteed to have bounded magnitude and the bounds depend only on γ, W_1, W_2, and P.

16.4 Further Guidelines for Loop Shaping

Let $P = NM^{-1}$ be a normalized right coprime factorization. It was shown in Georgiou and Smith [1990] that

$$b_{\text{opt}}(P) \leq \lambda(P) := \inf_{\text{Re}\,s > 0} \underline{\sigma}\left(\begin{bmatrix} M(s) \\ N(s) \end{bmatrix} \right).$$

Hence a small $\lambda(P)$ would necessarily imply a small $b_{\text{opt}}(P)$. We shall now discuss the performance limitations implied by this relationship for a scalar system. The following argument is based on Vinnicombe [1993b], to which the reader is referred for further discussions. Let $z_1, z_2, \ldots, z_m$ and $p_1, p_2, \ldots, p_k$ be the open right-half plane zeros and poles of the plant P. Define

$$N_z(s) = \frac{z_1 - s}{z_1 + s} \frac{z_2 - s}{z_2 + s} \cdots \frac{z_m - s}{z_m + s}, \quad N_p(s) = \frac{p_1 - s}{p_1 + s} \frac{p_2 - s}{p_2 + s} \cdots \frac{p_k - s}{p_k + s}.$$

16.4. Further Guidelines for Loop Shaping

Then P can be written as
$$P(s) = P_0(s)N_z(s)/N_p(s)$$
where $P_0(s)$ has no open right-half plane poles or zeros. Let $N_0(s)$ and $M_0(s)$ be stable and minimum phase spectral factors:

$$N_0(s)N_0^\sim(s) = \left(1 + \frac{1}{P(s)P^\sim(s)}\right)^{-1}, \quad M_0(s)M_0^\sim(s) = (1 + P(s)P^\sim(s))^{-1}.$$

Then $P_0 = N_0/M_0$ is a normalized coprime factorization and $(N_0 N_z)$ and $(M_0 N_p)$ form a pair of normalized coprime factorizations of P. Thus

$$b_{opt}(P) \leq \sqrt{|N_0(s)N_z(s)|^2 + |M_0(s)N_p(s)|^2}, \quad \forall \text{Re}(s) > 0. \tag{16.19}$$

Since N_0 and M_0 are both stable and have no zeros in the open right-half plane, $\ln(N_0(s))$ and $\ln(M_0(s))$ are both analytic in $\text{Re}(s) > 0$ and so can be determined from their boundary values on $\text{Re}(s) = 0$ via Poisson integrals (see also Problem 16.15):

$$\ln|N_0(re^{j\theta})| = \int_{-\infty}^{\infty} \ln\left(\frac{1}{\sqrt{1 + 1/|P(j\omega)|^2}}\right) K_\theta(\omega/r) \, d(\ln \omega)$$

$$\ln|M_0(re^{j\theta})| = \int_{-\infty}^{\infty} \ln\left(\frac{1}{\sqrt{1 + |P(j\omega)|^2}}\right) K_\theta(\omega/r) \, d(\ln \omega)$$

where $r > 0$, $-\pi/2 < \theta < \pi/2$, and

$$K_\theta(\omega/r) = \frac{1}{\pi} \frac{2(\omega/r)[1 + (\omega/r)^2] \cos \theta}{[1 - (\omega/r)^2]^2 + 4(\omega/r)^2 \cos^2 \theta}$$

The function $K_\theta(\omega/r)$ is plotted in Figure 16.6 against logarithmic frequency for various values of θ. Note that the function is symmetric to $\omega = r$ in $\log \omega$ and it attends the maximum at $\omega = r$. The function converges to an impulse function at $\omega = r$ when θ approaches $90°$; that is, when $|N_0(s)|$ or $|M_0(s)|$ is evaluated close to the imaginary axis.

Since the kernel function $K_\theta(\omega/r)$ has the greatest weighting near $\omega = r$, the Poisson integral is largely determined by the frequency response near that frequency. Thus it is clear that $|N_0(re^{j\theta})|$ will be small if $|P(j\omega)|$ is small near $\omega = r$. Similarly, $|M_0(re^{j\theta})|$ will be small if $|P(j\omega)|$ is large near $\omega = r$.

It is also important to note that a very large percentage of weighting is concentrated in a very narrow frequency range for a large θ (i.e., when $s = re^{j\theta}$ has a much larger imaginary part than the real part). Hence $|N_0(re^{j\theta})|$ and $|M_0(re^{j\theta})|$ will essentially be determined by $|P(j\omega)|$ in a very narrow frequency range near $\omega = r$ when θ is large. On the other hand, when θ is small, a larger range of frequency response $|P(j\omega)|$ around $\omega = r$ will have affect on the value $|N_0(re^{j\theta})|$ and $|M_0(re^{j\theta})|$. (This, in fact, will imply

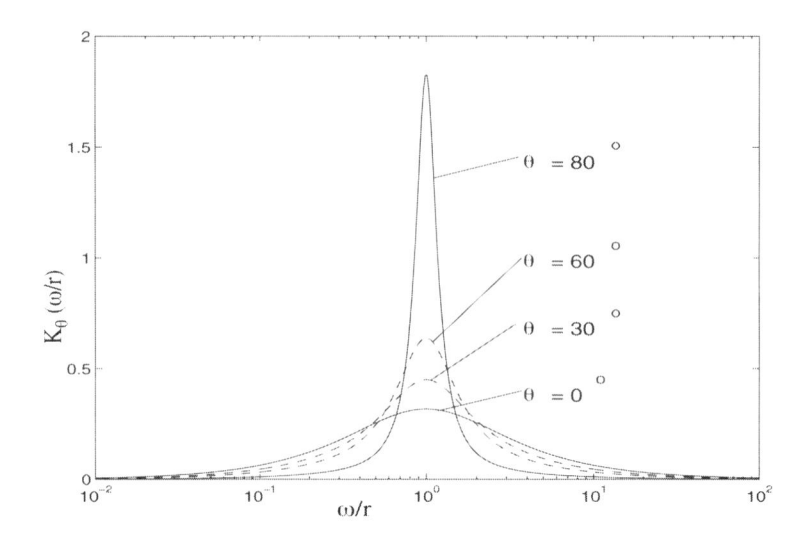

Figure 16.6: $K_\theta(\omega/r)$ vs. normalized frequency ω/r

that a right-plane zero (pole) with a much larger real part than the imaginary part will have much worse effect on the performance than a right-plane zero (pole) with a much larger imaginary part than the real part.)

Let $s = re^{j\theta}$. Consider again the bound of equation (16.19) and note that $N_z(z_i) = 0$ and $N_p(p_j) = 0$, we see that there are several ways in which the bound may be small (i.e., $b_{\text{opt}}(P)$ is small).

▷ $|N_z(s)|$ and $|N_p(s)|$ are both small for some s. That is, $|N_z(s)| \approx 0$ (i.e., s is close to a right-half plane zero of P) and $|N_p(s)| \approx 0$ (i.e., s is close to a right-half plane pole of P). This is only possible if $P(s)$ has a right-half plane zero near a right-half plane pole. (See Example 16.1.)

▷ $|N_z(s)|$ and $|M_0(s)|$ are both small for some s. That is, $|N_z(s)| \approx 0$ (i.e., s is close to a right-half plane zero of P) and $|M_0(s)| \approx 0$ (i.e., $|P(j\omega)|$ is large around $\omega = |s| = r$). This is only possible if $|P(j\omega)|$ is large around $\omega = r$, where r is the modulus of a right-half plane zero of P. (See Example 16.2.)

▷ $|N_p(s)|$ and $|N_0(s)|$ are both small for some s. That is, $|N_p(s)| \approx 0$ (i.e., s is close to a right-half plane pole of P) and $|N_0(s)| \approx 0$ (i.e., $|P(j\omega)|$ is small around $\omega = |s| = r$). This is only possible if $|P(j\omega)|$ is small around $\omega = r$, where r is the modulus of a right-half plane pole of P. (See Example 16.3.)

▷ $|N_0(s)|$ and $|M_0(s)|$ are both small for some s. That is, $|N_0(s)| \approx 0$ (i.e., $|P(j\omega)|$ is small around $\omega = |s| = r$) and $|M_0(s)| \approx 0$ (i.e., $|P(j\omega)|$ is large around $\omega = |s| = r$). The only way in which $|P(j\omega)|$ can be both small and large

16.4. Further Guidelines for Loop Shaping

at frequencies near $\omega = r$ is that $|P(j\omega)|$ is approximately equal to 1 and the absolute value of the slope of $|P(j\omega)|$ is large. (See Example 16.4.)

Example 16.1 Consider an unstable and nonminimum phase system

$$P_1(s) = \frac{K(s-r)}{(s+1)(s-1)}.$$

The frequency responses of $P_1(s)$ with $r = 0.9$ and $K = 0.1, 1$, and 10 are shown in Figure 16.7. The following table shows that $b_{\text{opt}}(P_1)$ will be very small for all K whenever r is close to 1 (i.e., whenever there is an unstable pole close to an unstable zero).

	r	0.5	0.7	0.9	1.1	1.3	1.5
$K = 0.1$	$b_{\text{opt}}(P_1)$	0.0125	0.0075	0.0025	0.0025	0.0074	0.0124
	r	0.5	0.7	0.9	1.1	1.3	1.5
$K = 1$	$b_{\text{opt}}(P_1)$	0.1036	0.0579	0.0179	0.0165	0.0457	0.0706
	r	0.5	0.7	0.9	1.1	1.3	1.5
$K = 10$	$b_{\text{opt}}(P_1)$	0.0658	0.0312	0.0088	0.0077	0.0208	0.0318

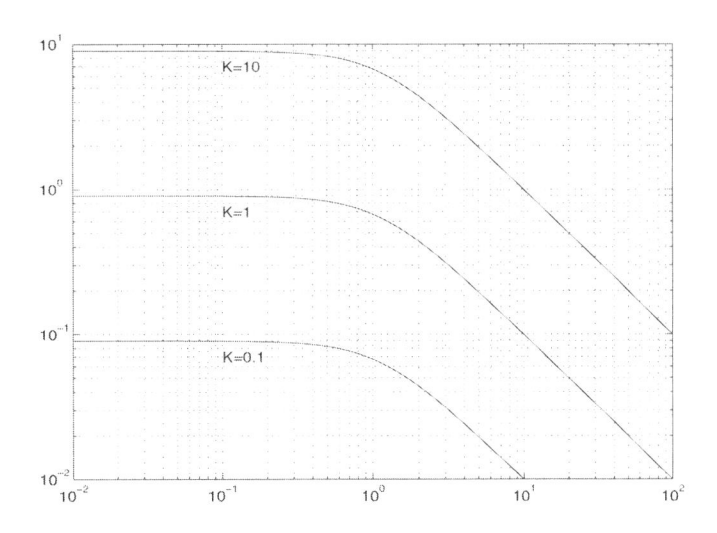

Figure 16.7: Frequency responses of P_1 for $r = 0.9$ and $K = 0.1, 1$, and 10

Example 16.2 Consider a nonminimum phase plant

$$P_2(s) = \frac{K(s-1)}{s(s+1)}.$$

The frequency responses of $P_2(s)$ with $K = 0.1, 1$, and 10 are shown in Figure 16.8. The following table shows clearly that $b_{\text{opt}}(P_2)$ will be small if $|P_2(j\omega)|$ is large around $\omega = 1$, the modulus of the right-half plane zero.

K	0.01	0.1	1	10	100
$b_{\text{opt}}(P_2)$	0.7001	0.6451	0.3827	0.0841	0.0098

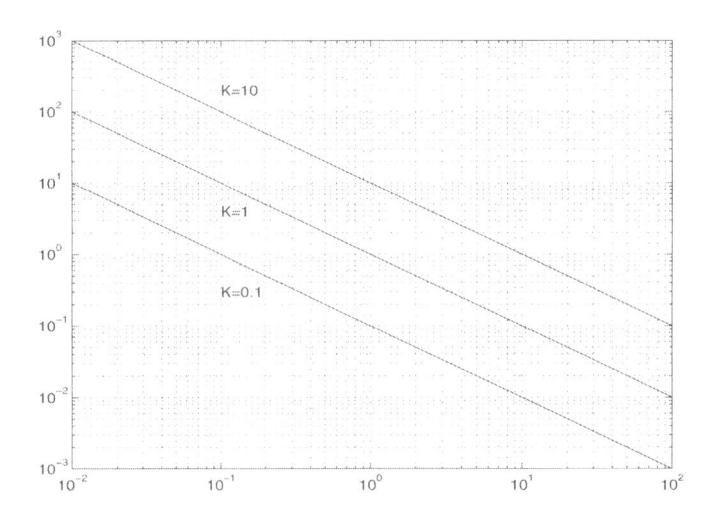

Figure 16.8: Frequency responses of P_2 for $K = 0.1, 1$, and 10

Note that $b_{\text{opt}}(L/s) = 0.707$ for any L and $b_{\text{opt}}(P_2) \longrightarrow 0.707$ as $K \longrightarrow 0$. This is because $|P_2(j\omega)|$ around the frequency of the right-half plane zero is very small as $K \longrightarrow 0$.

Next consider a plant with a pair of complex right-half plane zeros:

$$P_3(s) = \frac{K[(s - \cos\theta)^2 + \sin^2\theta]}{s[(s + \cos\theta)^2 + \sin^2\theta]}.$$

The magnitude frequency response of P_3 is the same as that of P_2 for the same K. The optimal $b_{\text{opt}}(P_3)$ for various θ's are listed in the following table:

16.4. Further Guidelines for Loop Shaping

	θ (degree)	0	45	60	80	85
$K = 0.1$	$b_{\text{opt}}(P_3)$	0.5952	0.6230	0.6447	0.6835	0.6950
	θ (degree)	0	45	60	80	85
$K = 1$	$b_{\text{opt}}(P_3)$	0.2588	0.3078	0.3568	0.4881	0.5512
	θ (degree)	0	45	60	80	85
$K = 10$	$b_{\text{opt}}(P_3)$	0.0391	0.0488	0.0584	0.0813	0.0897

It can also be concluded from the table that $b_{\text{opt}}(P_3)$ will be small if $|P_3(j\omega)|$ is large around the frequency of $\omega = 1$ (the modulus of the right-half plane zero). It can be further concluded that, for zeros with the same modulus, $b_{\text{opt}}(P_3)$ will be smaller for a plant with relatively larger real part zeros than for a plant with relatively larger imaginary part zeros (i.e., a pair of real right-half plane zeros has a much worse effect on the performance than a pair of almost pure imaginary axis right-half plane zeros of the same modulus).

Example 16.3 Consider an unstable plant

$$P_4(s) = \frac{K(s+1)}{s(s-1)}.$$

The magnitude frequency response of P_4 is again the same as that of P_2 for the same K. The following table shows that $b_{\text{opt}}(P_4)$ will be small if $|P_4(j\omega)|$ is small around $\omega = 1$ (the modulus of the right-half plane pole).

K	0.01	0.1	1	10	100
$b_{\text{opt}}(P_4)$	0.0098	0.0841	0.3827	0.6451	0.7001

Note that $b_{\text{opt}}(P_4) \longrightarrow 0.707$ as $K \longrightarrow \infty$. This is because $|P_4(j\omega)|$ is very large around the frequency of the modulus of the right-half plane pole as $K \longrightarrow \infty$.

Next consider a plant with complex right-half plane poles:

$$P_5(s) = \frac{K[(s+\cos\theta)^2 + \sin^2\theta]}{s[(s-\cos\theta)^2 + \sin^2\theta]}.$$

The optimal $b_{\text{opt}}(P_5)$ for various θ's are listed in the following table:

	θ (degree)	0	45	60	80	85
$K = 0.1$	$b_{\rm opt}(P_5)$	0.0391	0.0488	0.0584	0.0813	0.0897
	θ (degree)	0	45	60	80	85
$K = 1$	$b_{\rm opt}(P_5)$	0.2588	0.3078	0.3568	0.4881	0.5512
	θ (degree)	0	45	60	80	85
$K = 10$	$b_{\rm opt}(P_5)$	0.5952	0.6230	0.6447	0.6835	0.6950

It can also be concluded from the table that $b_{\rm opt}(P_5)$ will be small if $|P_5(j\omega)|$ is small around the frequency of the modulus of the right-half plane pole. It can be further concluded that, for poles with the same modulus, $b_{\rm opt}(P_5)$ will be smaller for a plant with relatively larger real part poles than for a plant with relatively larger imaginary part poles (i.e., a pair of real right-half plane poles has a much worse effect on the performance than a pair of almost pure imaginary axis right-half plane poles of the same modulus).

Example 16.4 Let a stable and minimum phase transfer function be

$$P_6(s) = \frac{K(0.2s + 1)^4}{s(s + 1)^4}.$$

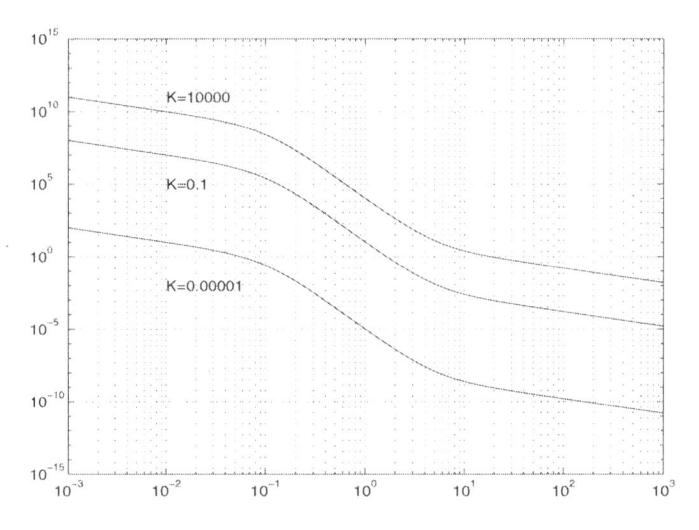

Figure 16.9: Frequency response of P_6 for $K = 10^{-5}, 10^{-1}$ and 10^4

The frequency responses of P_6 with $K = 10^{-5}, 10^{-1}$, and 10^4 are shown in Figure 16.9. It is clear that the slope of the frequency response near the crossover frequency for $K = 10^{-5}$ is not too large, which implies a reasonably good loop shape. Thus we should expect $b_{\text{opt}}(P_6)$ to be not too small. A similar conclusion applies to $K = 10^4$. On the other hand, the slope of the frequency response near the crossover frequency for $K = 0.1$ is quite large which implies a bad loop shape. Thus we should expect $b_{\text{opt}}(P_6)$ to be quite small. This is clear from the following table:

K	10^{-5}	10^{-3}	0.1	1	10	10^2	10^4
$b_{\text{opt}}(P_6)$	0.3566	0.0938	0.0569	0.0597	0.0765	0.1226	0.4933

Based on the preceding discussion, we can give some guidelines for the loop-shaping design.

- ♡ The loop transfer function should be shaped in such a way that it has low gain around the frequency of the modulus of any right-half plane zero z. Typically, it requires that the crossover frequency be much smaller than the modulus of the right-half plane zero; say, $\omega_c < |z|/2$ for any real zero and $\omega_c < |z|$ for any complex zero with a much larger imaginary part than the real part (see Figure 16.6).

- ♡ The loop transfer function should have a large gain around the frequency of the modulus of any right-half plane pole.

- ♡ The loop transfer function should not have a large slope near the crossover frequencies.

These guidelines are consistent with the rules used in classical control theory (see Bode [1945] and Horowitz [1963]).

16.5 Notes and References

The $\mathcal{H}_\infty$ loop-shaping using normalized coprime factorization was developed by McFarlane and Glover [1990, 1992], on which most parts of this chapter is based. In the same references, some design examples were also shown. The method has been applied to the design of scheduled controllers for a VSTOL aircraft in Hyde and Glover [1993]. Some limitations of this loop-shaping design are discussed in detail in Vinnicombe [1993b] (on which Section 16.4 is based) and Christian and Freudenberg [1994]. The robust stabilization of normalized coprime factors is closely related to the robustness in the gap metric and ν-gap metric, which will be discussed in the next chapter, see El-Sakkary [1985], Georgiou and Smith [1990], Glover and McFarlane [1989], McFarlane, Glover, and Vidyasagar [1990], Qiu and Davison [1992a, 1992b], Vinnicombe [1993a, 1993b], Vidyasagar [1984, 1985], Zhu [1989], and references therein.

16.6 Problems

Problem 16.1 Consider a feedback system with $G(s) = \dfrac{1}{s-2}$. Compute by hand ϵ_{max}, the maximum stability radius for a normalized coprime factor perturbation of $G(s)$.

Problem 16.2 In Corollary 16.2, find the parameterization of all $\mathcal{H}_\infty$ controllers.

Problem 16.3 Let $P_\Delta = (N + \Delta_N)(M + \Delta_M)^{-1}$ be a right coprime factor perturbed plant with a nominal plant $P = NM^{-1}$, where (N, M) is a pair of normalized right coprime factorization. Formulate the corresponding robust stabilization problem as an $\mathcal{H}_\infty$ control problem and find a stabilizing controller using the $\mathcal{H}_\infty$ formulas in Chapter 14. Are there any connections between this stabilizing controller and the controller obtained in this chapter for left coprime stabilization?

Problem 16.4 Let P have coprime factorizations $P = NM^{-1} = \tilde{M}^{-1}\tilde{N}$. Then there exist $U, V, \tilde{U}, \tilde{V} \in \mathcal{H}_\infty$ such that

$$\begin{bmatrix} M & U \\ N & V \end{bmatrix} \begin{bmatrix} \tilde{V} & -\tilde{U} \\ \tilde{N} & \tilde{M} \end{bmatrix} = I.$$

Furthermore, all stabilizing controllers for P can be written as

$$K = (U + MQ)(V + NQ)^{-1}, \quad Q \in \mathcal{H}_\infty.$$

Show that

$$\begin{bmatrix} K \\ I \end{bmatrix} (I + PK)^{-1} \tilde{M}^{-1} = \begin{bmatrix} U + MQ \\ V + NQ \end{bmatrix}.$$

Suppose $P = NM^{-1} = \tilde{M}^{-1}\tilde{N}$ are normalized coprime factorizations. Show that

$$b_{P,K}^{-1} = \left\| \begin{bmatrix} U \\ V \end{bmatrix} + \begin{bmatrix} M \\ N \end{bmatrix} Q \right\|_\infty = \left\| \begin{bmatrix} R + Q \\ I \end{bmatrix} \right\|_\infty$$

where $R = M^\sim U + N^\sim V$.

Problem 16.5 Let K be a controller that stabilizes the plant P. Show that

1. K stabilizes $\tilde{P} = P + \Delta_a$ such that $\Delta_a \in \mathcal{H}_\infty$ and $\|\Delta_a\| < b_{P,K}$;

2. K stabilizes $\tilde{P} = P(I + \Delta_m)$ such that $\Delta_m \in \mathcal{H}_\infty$ and $\|\Delta_m\| < b_{P,K}$;

3. K stabilizes $\tilde{P} = (I + \Delta_m)P$ such that $\Delta_m \in \mathcal{H}_\infty$ and $\|\Delta_m\| < b_{P,K}$;

4. K stabilizes $\tilde{P} = P(I + \Delta_f)^{-1}$ such that $\Delta_f \in \mathcal{H}_\infty$ and $\|\Delta_f\| < b_{P,K}$;

5. K stabilizes $\tilde{P} = (I + \Delta_f)^{-1}P$ such that $\Delta_f \in \mathcal{H}_\infty$ and $\|\Delta_f\| < b_{P,K}$.

16.6. Problems

Discuss the possible implications of the preceding results.

Problem 16.6 Let K be a controller that stabilizes the plant P. Show that

1. any controller in the form of $\tilde{K} = (U + \Delta_U)(V + \Delta_V)^{-1}$ such that $\Delta_U, \Delta_V \in \mathcal{H}_\infty$ and $\left\| \begin{bmatrix} \Delta_U \\ \Delta_V \end{bmatrix} \right\|_\infty < b_{P,K}$ also stabilizes P;

2. any controller $\tilde{K} = K + \Delta_a$ such that $\Delta_a \in \mathcal{H}_\infty$ and $\|\Delta_a\| < b_{P,K}$ also stabilizes P;

3. any controller $\tilde{K} = K(I + \Delta_m)$ such that $\Delta_m \in \mathcal{H}_\infty$ and $\|\Delta_m\| < b_{P,K}$ also stabilizes P;

4. any controller $\tilde{K} = (I + \Delta_m)K$ such that $\Delta_m \in \mathcal{H}_\infty$ and $\|\Delta_m\| < b_{P,K}$ also stabilizes P;

5. any controller $\tilde{K} = K(I + \Delta_f)^{-1}$ such that $\Delta_f \in \mathcal{H}_\infty$ and $\|\Delta_f\| < b_{P,K}$ also stabilizes P;

6. any controller $\tilde{K} = (I + \Delta_f)^{-1}K$ such that $\Delta_f \in \mathcal{H}_\infty$ and $\|\Delta_f\| < b_{P,K}$ also stabilizes P.

Discuss the possible implications of the preceding results.

Problem 16.7 Let an uncertain plant be given by

$$P_\delta = \frac{s + \alpha}{s^2 + 2\zeta s + 1}, \quad \alpha \in [1, 3], \quad \zeta \in [0.2, 0.4]$$

and let a nominal model be

$$P = \frac{s + \alpha_0}{s^2 + 2\zeta_0 s + 1}.$$

1. Let $\alpha_0 = 2$ and $\zeta_0 = 0.3$. Find the largest possible $\|\Delta_{\text{add}}\|_\infty$ and $\|\Delta_{\text{mul}}\|_\infty$ where

$$\Delta_{\text{add}} = P_\delta - P, \quad \Delta_{\text{mul}} = (P_\delta - P)/P.$$

2. Let $\alpha_0 = 2$ and $\zeta_0 = 0.3$. Show that $P = N/M$ with

$$N = \frac{s + 2}{s^2 + 1.9576s + 2.2361}, \quad M = \frac{s^2 + 0.6s + 1}{s^2 + 1.9576s + 2.2361}$$

is a normalized coprime factorization. Now let

$$N_\delta = \frac{s + \alpha}{s^2 + 1.9576s + 2.2361}, \quad M_\delta = \frac{s^2 + 2\zeta s + 1}{s^2 + 1.9576s + 2.2361}$$

$$\Delta_n = N_\delta - N, \quad \Delta_m = M_\delta - M.$$

Find the largest possible $\left\| \begin{bmatrix} \Delta_n & \Delta_m \end{bmatrix} \right\|_\infty$.

3. In part 2, let (N_δ, M_δ) be a normalized coprime factorization of P_δ. Find the largest possible $\left\| \begin{bmatrix} \Delta_n & \Delta_m \end{bmatrix} \right\|_\infty$.

4. Find the optimal nominal α_0 and ζ_0 such that the largest possible $\|\Delta_{\text{add}}\|_\infty$, $\|\Delta_{\text{mul}}\|_\infty$, and $\left\| \begin{bmatrix} \Delta_n & \Delta_m \end{bmatrix} \right\|_\infty$ are minimized, respectively.

Discuss the advantages of each uncertainty modeling method in terms of robust stabilizations.

Problem 16.8 Let $P = \dfrac{-10}{s(s-1)}$. Design (a) a precompensator W of order no greater than 2 such that the crossover frequency $\omega_c \leq 2$ and $b_{\text{opt}}(WP)$ is as large as possible; (b) find the optimal loop-shaping controller $K = K_\infty W$ with the W obtained in part (a).

Problem 16.9 Let $P = \dfrac{100(1-s)}{s(s+10)}$. Design (a) a precompensator W of order no greater than 2 such that the crossover frequency $\omega_c \geq 1$ and $b_{\text{opt}}(WP)$ is as large as possible; (b) find the optimal loop-shaping controller $K = K_\infty W$ with the W obtained in part (a).

Problem 16.10 Let $G(s) = \left[\begin{array}{c|c} A & B \\ \hline C & 0 \end{array} \right]$ and $G(s) = NM^{-1}$ with

$$\begin{bmatrix} N \\ M \end{bmatrix} = \left[\begin{array}{c|c} A+BF & B \\ C & 0 \\ F & I \end{array} \right]$$

where F is chosen such that NM^{-1} is a normalized right coprime factorization. Let $\begin{bmatrix} \hat{N} \\ \hat{M} \end{bmatrix}$ be an rth order balanced truncation of $\begin{bmatrix} N \\ M \end{bmatrix}$. Show that $\begin{bmatrix} \hat{N} \\ \hat{M} \end{bmatrix}$ is also a normalized right coprime factorization.

Problem 16.11 (Reduced-Order Controllers by Controller Model Reduction; see McFarlane and Glover [1990], Zhou and Chen [1995].) Let $G(s) = \left[\begin{array}{c|c} A & B \\ \hline C & 0 \end{array} \right] = \tilde{M}^{-1}\tilde{N}$ be a normalized left coprime factorization and let $K(s)$ be a suboptimal controller given in Corollary 18.2 (with performance γ). Let $K = UV^{-1}$ be a right coprime factorization

$$\begin{bmatrix} U \\ V \end{bmatrix} = \left[\begin{array}{c|c} A - BB^*X_\infty & -YC^* \\ \hline -C & I \\ -B^*X_\infty & 0 \end{array} \right]$$

16.6. Problems

and $\hat{U}, \hat{V} \in \mathcal{RH}_\infty$ be approximations of U and V. Define

$$\epsilon := \left\| \begin{bmatrix} U \\ V \end{bmatrix} - \begin{bmatrix} \hat{U} \\ \hat{V} \end{bmatrix} \right\|_\infty$$

and $K_r = \hat{U}\hat{V}^{-1}$. Show that K_r is a stabilizing controller for G if $\epsilon < 1$ and

$$\left\| \begin{bmatrix} K_r \\ I \end{bmatrix} (I + GK_r)^{-1} \tilde{M}^{-1} \right\|_\infty = \left\| \begin{bmatrix} K_r \\ I \end{bmatrix} (I + GK_r)^{-1} \begin{bmatrix} I & G \end{bmatrix} \right\|_\infty < \frac{\gamma}{1 - \epsilon}.$$

Problem 16.12 (Reduced Order Controllers by Plant Model Reduction; see McFarlane and Glover [1990].) Let $G = \tilde{M}^{-1}\tilde{N}$ be a normalized left coprime factorization and let K be a stabilizing controller such that

$$\left\| \begin{bmatrix} K \\ I \end{bmatrix} (I + GK)^{-1} \tilde{M}^{-1} \right\|_\infty \leq \delta^{-1}.$$

Let $G_r := \tilde{M}_r^{-1}\tilde{N}_r$ be an approximation of G and

$$\epsilon := \left\| \begin{bmatrix} \tilde{M} - \tilde{M}_r & \tilde{N} - \tilde{N}_r \end{bmatrix} \right\|_\infty < \delta.$$

(a) Show that K stabilizes G_r and

$$\left\| \begin{bmatrix} K \\ I \end{bmatrix} (I + G_r K)^{-1} \tilde{M}_r^{-1} \right\|_\infty \leq (\delta - \epsilon)^{-1}.$$

(b) Let $W(s), W^{-1}(s) \in \mathcal{RH}_\infty$ be obtained from the following spectral factorization:

$$W^{-1}W^{-*} = \tilde{M}_r \tilde{M}_r^* + \tilde{N}_r \tilde{N}_r^*.$$

Show that $\|W\|_\infty \leq \dfrac{1}{1-\epsilon}$ and $\|W^{-1}\|_\infty \leq 1 + \epsilon$.

(c) Show that

$$\delta_{rn}^{-1} := \inf_{K_1} \left\| \begin{bmatrix} K_1 \\ I \end{bmatrix} (I + G_r K_1)^{-1} (W \tilde{M}_r)^{-1} \right\|_\infty$$

$$= \inf_{K_1} \left\| \begin{bmatrix} K_1 \\ I \end{bmatrix} (I + G_r K_1)^{-1} \begin{bmatrix} I & G_r \end{bmatrix} \right\|_\infty \leq \frac{\|W^{-1}\|_\infty}{\delta - \epsilon} \leq \frac{1 + \epsilon}{\delta - \epsilon}.$$

and

$$\left\| \begin{bmatrix} K_1 \\ I \end{bmatrix} (I + G_r K_1)^{-1} \tilde{M}_r^{-1} \right\|_\infty \leq \delta_{rn}^{-1} \|W\|_\infty.$$

(d) With the controller K_1 given in (c), show that

$$\left\| \begin{bmatrix} K_1 \\ I \end{bmatrix} (I+GK_1)^{-1} \tilde{M}^{-1} \right\|_\infty = \left\| \begin{bmatrix} K_1 \\ I \end{bmatrix} (I+GK_1)^{-1} \begin{bmatrix} I & G \end{bmatrix} \right\|_\infty \leq \delta_{\text{red}}^{-1}$$

where

$$\delta_{\text{red}} := \frac{\delta_{rn}}{\|W\|_\infty} - \epsilon \leq \frac{\delta - \epsilon}{\|W^{-1}\|_\infty \|W\|_\infty} - \epsilon \leq \frac{1-\epsilon}{1+\epsilon}(\delta - \epsilon) - \epsilon.$$

Note that if $\tilde{N}_r$ and $\tilde{M}_r$ are the kth-order balanced truncation of $\tilde{N}$ and $\tilde{M}$, then $\delta = \delta_{rn} = \sqrt{1-\sigma_1^2}$, $\delta_{\text{red}} = \delta - \epsilon$, and $\epsilon \leq 2\sum_{i=k+1}^n \sigma_i$, where σ_i are the Hankel singular values of $\begin{bmatrix} \tilde{M} & \tilde{N} \end{bmatrix}$.

(e) Show that

$$\tilde{\delta}_r^{-1} := \inf_{K_2} \left\| \begin{bmatrix} K_2 \\ I \end{bmatrix} (I+G_r K_2)^{-1} \tilde{M}_r^{-1} \right\|_\infty \leq (\delta - \epsilon)^{-1}.$$

(f) With the controller K_2 given in (e), show that

$$\left\| \begin{bmatrix} K_2 \\ I \end{bmatrix} (I+GK_2)^{-1} \tilde{M}^{-1} \right\|_\infty = \left\| \begin{bmatrix} K_2 \\ I \end{bmatrix} (I+GK_2)^{-1} \begin{bmatrix} I & G \end{bmatrix} \right\|_\infty$$

$$\leq (\tilde{\delta}_r - \epsilon)^{-1} \leq (\delta - 2\epsilon)^{-1}.$$

Again note that if $\tilde{N}_r$ and $\tilde{M}_r$ are the kth-order balanced truncation of $\tilde{N}$ and $\tilde{M}$, then $\tilde{\delta}_r = \delta$.

(Note that K_1 and K_2 are reduced-order controllers.)

Problem 16.13 Let $G(s) = \left[\begin{array}{c|c} A & B \\ \hline C & 0 \end{array}\right] = \tilde{M}^{-1}\tilde{N}$ be a normalized left coprime factorization and let $K(s)$ be a suboptimal controller given in Corollary 18.2 (with performance γ):

$$K(s) = \left[\begin{array}{c|c} A - BB^* X_\infty - YC^*C & -YC^* \\ \hline -B^* X_\infty & 0 \end{array}\right]$$

where

$$X_\infty = \frac{\gamma^2}{\gamma^2 - 1} Q \left(I - \frac{\gamma^2}{\gamma^2 - 1} YQ\right)^{-1}$$

and

$$AY + YA^* - YC^*CY + BB^* = 0$$

16.6. Problems

$$Q(A - YC^*C) + (A - YC^*C)^*Q + C^*C = 0.$$

Suppose Y and Q are balanced; that is,

$$Y = Q = \text{diag}(\sigma_1, \ldots, \sigma_r, \sigma_{r+1}, \ldots, \sigma_n) = \text{diag}(\Sigma_1, \Sigma_2)$$

and let $G(s)$ be partitioned accordingly as

$$G(s) = \left[\begin{array}{cc|c} A_{11} & A_{12} & B_1 \\ A_{21} & A_{22} & B_2 \\ \hline C_1 & C_2 & 0 \end{array}\right].$$

Denote $Y_1 = \Sigma_1$ and $X_1 = \frac{\gamma^2}{\gamma^2-1}\Sigma_1\left(I - \frac{\gamma^2}{\gamma^2-1}\Sigma_1^2\right)^{-1}$. Show that

$$K_r(s) = \left[\begin{array}{c|c} A_{11} - B_1 B_1^* X_1 - Y_1 C_1^* C_1 & -Y_1 C_1^* \\ \hline -B_1^* X_1 & 0 \end{array}\right]$$

is exactly the reduced-order controller obtained from the last problem with balanced model reduction procedure. (It is also interesting to note that

$$Q = X(I + YX)^{-1}$$

where $X = X^* \geq 0$ is the stabilizing solution to

$$XA + A^*X - XBB^*X + C^*C = 0.$$

Hence balancing Y and Q is equivalent to balancing X and Y. This is called Riccati balancing; see Jonckheere and Silverman [1983].)

Problem 16.14 Apply the controller reduction methods in the last three problems, respectively, to a satellite system $G(s) = \left[\begin{array}{c|c} A & B \\ \hline C & 0 \end{array}\right]$ where

$$A = \begin{bmatrix} 0 & 1 & 0 & 0 \\ 0 & 0 & 0 & 0 \\ 0 & 0 & 0 & 1 \\ 0 & 0 & -1.539^2 & -2 \times 0.003 \times 1.539 \end{bmatrix}, \quad B = \begin{bmatrix} 0 \\ 1.7319 \times 10^{-5} \\ 0 \\ 3.7859 \times 10^{-4} \end{bmatrix},$$

$$C = \begin{bmatrix} 1 & 0 & 1 & 0 \end{bmatrix}, \quad D = 0.$$

Compare the results (see McFarlane and Glover [1990] for further details).

Problem 16.15 Let $f(s)$ be analytic in the closed right-half plane and suppose

$$\lim_{r \to \infty} \max_{\theta \in [-\pi/2, \pi/2]} \frac{|f(re^{j\theta})|}{r} = 0.$$

Then the Poisson integral formula (see, for example, Freudenberg and Looze [1988], page 37) says that $f(s)$ at any point $s = x + jy$ in the open right-half plane can be recovered from $f(j\omega)$ via the integral relation:

$$f(s) = \frac{1}{\pi} \int_{-\infty}^{\infty} f(j\omega) \frac{x}{x^2 + (y-\omega)^2} d\omega.$$

Let $s = re^{j\theta}$ (i.e., $x = r\cos\theta$ and $y = r\sin\theta$) with $r > 0$ and $-\pi/2 < \theta < \pi/2$. Suppose $f(j\omega) = f(-j\omega)$. Show that

$$f(re^{j\theta}) = \int_{-\infty}^{\infty} f(j\omega) K_\theta(\omega/r) \, d(\ln \omega)$$

where

$$K_\theta(\omega/r) = \frac{1}{\pi} \frac{2(\omega/r)[1 + (\omega/r)^2]\cos\theta}{[1 - (\omega/r)^2]^2 + 4(\omega/r)^2 \cos^2\theta}$$

Chapter 17

Gap Metric and ν-Gap Metric

In the previous chapters, we have seen that all of robust control design techniques assume that we have some description of the model uncertainties (i.e., we have a measure of the distance from the nominal plant to the set of uncertainty systems). This measure is usually chosen to be a metric or a norm. However, the operator norm can be a poor measure of the distance between systems in respect to feedback control system design. For example, consider

$$P_1(s) = \frac{1}{s}, \quad P_2(s) = \frac{1}{s+0.1}.$$

The closed-loop complementary sensitivity functions corresponding to P_1 and P_2 with unity feedback are relatively close and their difference is

$$\left\| P_1(I+P_1)^{-1} - P_2(I+P_2)^{-1} \right\|_\infty = 0.0909,$$

but the difference between P_1 and P_2 is

$$\|P_1 - P_2\|_\infty = \infty.$$

This shows that the closed-loop behavior of two systems can be very close even though the norm of the difference between the two open-loop systems can be arbitrarily large.

To deal with such problems, the gap metric and the ν-gap metric were introduced into the control literature by Zames and El-Sakkary [1980] (see also El-Sakkary [1985] and Vinnicombe [1993]) as being appropriate for the study of uncertainty in feedback systems. An alternative metric, the graph metric, was also introduced by Vidyasagar [1984] in terms of normalized coprime factorizations. All of these metrics are equivalent, and thus induce the same topology. This topology is the weakest in which feedback stability is a robust property. The metrics define notions of distance in the space of (possible) unstable systems that do not assume that the plants have the same number of poles in the right-half plane.

We shall briefly introduce the gap metric in Section 17.1 and study some of its applications in robust control. Our focus in this chapter is Sections 17.2–17.4, which study in some detail the ν-gap metric. In particular, we introduce the ν-gap metric in Section 17.2 and explore its frequency domain interpretation and applications in Section 17.3 and Section 17.4. Finally, we consider controller order reduction in the gap or ν-gap metric framework in Section 17.5.

17.1 Gap Metric

In this section we briefly introduce the gap metric and discuss some of its applications in controller design.

Let $P(s)$ be a $p \times m$ rational transfer matrix and let P have the following normalized right and left stable coprime factorizations:

$$P = NM^{-1} = \tilde{M}^{-1}\tilde{N}.$$

That is,

$$M^{\sim}M + N^{\sim}N = I, \quad \tilde{M}\tilde{M}^{\sim} + \tilde{N}\tilde{N}^{\sim} = I.$$

The graph of the operator P is the subspace of $\mathcal{H}_2$ consisting of all pairs (u, y) such that $y = Pu$. This is given by

$$\begin{bmatrix} M \\ N \end{bmatrix} \mathcal{H}_2$$

and is a closed subspace of $\mathcal{H}_2$. The gap between two systems P_1 and P_2 is defined by

$$\delta_g(P_1, P_2) = \left\| \Pi \begin{bmatrix} M_1 \\ N_1 \end{bmatrix}_{\mathcal{H}_2} - \Pi \begin{bmatrix} M_2 \\ N_2 \end{bmatrix}_{\mathcal{H}_2} \right\|$$

where Π_K denotes the orthogonal projection onto K and $P_1 = N_1 M_1^{-1}$ and $P_2 = N_2 M_2^{-1}$ are normalized right coprime factorizations.

It is shown by Georgiou [1988] that the gap metric can be computed as follows:

Theorem 17.1 *Let $P_1 = N_1 M_1^{-1}$ and $P_2 = N_2 M_2^{-1}$ be normalized right coprime factorizations. Then*

$$\delta_g(P_1, P_2) = \max\left\{ \vec{\delta}(P_1, P_2), \vec{\delta}(P_2, P_1) \right\}$$

where $\vec{\delta}_g(P_1, P_2)$ is the directed gap and can be computed by

$$\vec{\delta}_g(P_1, P_2) = \inf_{Q \in \mathcal{H}_\infty} \left\| \begin{bmatrix} M_1 \\ N_1 \end{bmatrix} - \begin{bmatrix} M_2 \\ N_2 \end{bmatrix} Q \right\|_\infty.$$

17.1. Gap Metric

The following procedures can be used in computing the directed gap $\vec{\delta}_g(P_1, P_2)$.

Computing $\vec{\delta}_g(P_1, P_2)$: Let

$$P_1 = \left[\begin{array}{c|c} A_1 & B_1 \\ \hline C_1 & D_1 \end{array}\right], \quad P_2 = \left[\begin{array}{c|c} A_2 & B_2 \\ \hline C_2 & D_2 \end{array}\right].$$

1. Let $P_i = N_i M_i^{-1}$, $i = 1, 2$ be normalized right coprime factorizations. Then

$$\left[\begin{array}{c} M_i \\ N_i \end{array}\right] = \left[\begin{array}{c|c} A_i + B_i F_i & B_i R_i^{-1/2} \\ \hline F_i & R_i^{-1/2} \\ C_i + D_i F_i & D_i R_i^{-1/2} \end{array}\right], \quad \begin{array}{l} R_i = I + D_i^* D_i \\ \tilde{R}_i = I + D_i D_i^* \\ F_i = -R_i^{-1}(B_i^* X_i + D_i^* C_i) \end{array}$$

where

$$X_i = \mathrm{Ric}\left[\begin{array}{cc} A_i - B_i R_i^{-1} D_i^* C_i & -B_i R_i^{-1} B_i^* \\ -C_i^* \tilde{R}_i^{-1} C_i & -(A_i - B_i R_i^{-1} D_i^* C_i)^* \end{array}\right].$$

2. Define a generalized system

$$G(s) = \left[\begin{array}{cc} \left[\begin{array}{c} M_1 \\ N_1 \end{array}\right] & \left[\begin{array}{c} M_2 \\ N_2 \end{array}\right] \\ -I & 0 \end{array}\right]$$

$$= \left[\begin{array}{cc|cc} A_1 + B_1 F_1 & 0 & B_1 R_1^{-1/2} & 0 \\ 0 & A_2 + B_2 F_2 & 0 & B_2 R_2^{-1/2} \\ \hline F_1 & F_2 & R_1^{-1/2} & R_2^{-1/2} \\ C_1 + D_1 F_1 & C_2 + D_2 F_2 & D_1 R_1^{-1/2} & D_2 R_2^{-1/2} \\ 0 & 0 & -I & 0 \end{array}\right].$$

3. Apply standard $\mathcal{H}_\infty$ algorithm to

$$\vec{\delta}_g(P_1, P_2) = \inf_{Q \in \mathcal{H}_\infty} \left\|\left[\begin{array}{c} M_1 \\ N_1 \end{array}\right] - \left[\begin{array}{c} M_2 \\ N_2 \end{array}\right] Q\right\|_\infty = \inf_{Q \in \mathcal{H}_\infty} \|\mathcal{F}_\ell(G, Q)\|_\infty.$$

Using the above procedure, it is easy to show that

$$\delta_g\left(\frac{1}{s}, \frac{1}{s+0.1}\right) = 0.0995,$$

which confirms that the two systems given at the beginning of this chapter are indeed close in the gap metric. This example shows an important feature about the gap metric (similarly, the ν-gap metric defined in the next section): The distance between two plants, as measured by the gap metric δ_g (or the ν-gap metric δ_ν), has very little to do with any difference between their open-loop behavior (indeed, there is no reason why it should). This point will be further illustrated by an example in the next section.

A lower bound for the gap metric can also be obtained easily without actually solving the corresponding $\mathcal{H}_\infty$ optimization. Let

$$\Phi = \begin{bmatrix} \tilde{M}_2 & \tilde{N}_2 \\ -\tilde{N}_2 & \tilde{M}_2 \end{bmatrix}.$$

Then $\Phi^\sim \Phi = \Phi \Phi^\sim = I$ and

$$\vec{\delta}_g(P_1, P_2) = \inf_{Q \in \mathcal{H}_\infty} \left\| \begin{bmatrix} \tilde{M}_2 & \tilde{N}_2 \\ -\tilde{N}_2 & \tilde{M}_2 \end{bmatrix} \left\{ \begin{bmatrix} M_1 \\ N_1 \end{bmatrix} - \begin{bmatrix} M_2 \\ N_2 \end{bmatrix} Q \right\} \right\|_\infty$$

$$= \inf_{Q \in \mathcal{H}_\infty} \left\| \begin{bmatrix} \tilde{M}_2 M_1 + \tilde{N}_2 N_1 - Q \\ -\tilde{N}_2 M_1 + \tilde{M}_2 N_1 \end{bmatrix} \right\|_\infty \geq \|\Psi(P_1, P_2)\|_\infty$$

where

$$\Psi(P_1, P_2) := -\tilde{N}_2 M_1 + \tilde{M}_2 N_1 = \begin{bmatrix} \tilde{M}_2 & \tilde{N}_2 \end{bmatrix} \begin{bmatrix} 0 & I \\ -I & 0 \end{bmatrix} \begin{bmatrix} M_1 \\ N_1 \end{bmatrix}. \quad (17.1)$$

It will be seen in the next section that $\|\Psi(P_1, P_2)\|_\infty$ is related to the ν-gap metric. The above lower bound may actually be achieved. Consider, for example,

$$P_1 = \frac{k_1}{s+1}, \quad P_2 = \frac{k_2}{s+1}.$$

Then it is easy to verify that $P_i = N_i/M_i$, $i = 1, 2$, with

$$N_i = \frac{k_i}{s + \sqrt{1 + k_i^2}}, \quad M_i = \frac{s+1}{s + \sqrt{1 + k_i^2}},$$

are normalized coprime factorizations and it can be further shown, as in Georgiou and Smith [1990], that

$$\delta_g(P_1, P_2) = \|\Psi(P_1, P_2)\|_\infty = \begin{cases} \dfrac{|k_1 - k_2|}{|k_1| + |k_2|}, & \text{if } |k_1 k_2| > 1; \\[2ex] \dfrac{|k_1 - k_2|}{\sqrt{(1 + k_1^2)(1 + k_2^2)}}, & \text{if } |k_1 k_2| \leq 1. \end{cases}$$

An immediate consequence of Theorem 17.1 is the connection between the uncertainties in the gap metric and the uncertainties characterized by the normalized coprime factors. The following corollary states that a ball of uncertainty in the directed gap is equivalent to a ball of uncertainty in the normalized coprime factors.

17.1. Gap Metric

Corollary 17.2 *Let P have a normalized coprime factorization $P = NM^{-1}$. Then for all $0 < b \leq 1$,*

$$\left\{P_1 : \vec{\delta}_g(P, P_1) < b\right\}$$

$$= \left\{P_1 : P_1 = (N + \Delta_N)(M + \Delta_M)^{-1}, \Delta_N, \Delta_M \in \mathcal{H}_\infty, \left\|\begin{bmatrix} \Delta_N \\ \Delta_M \end{bmatrix}\right\|_\infty < b\right\}.$$

Proof. Suppose $\vec{\delta}_g(P, P_1) < b$ and let $P_1 = N_1 M_1^{-1}$ be a normalized right coprime factorization. Then there exists a $Q \in \mathcal{H}_\infty$ such that

$$\left\|\begin{bmatrix} M \\ N \end{bmatrix} - \begin{bmatrix} M_1 \\ N_1 \end{bmatrix} Q\right\|_\infty < b.$$

Define

$$\begin{bmatrix} \Delta_M \\ \Delta_N \end{bmatrix} := \begin{bmatrix} M_1 \\ N_1 \end{bmatrix} Q - \begin{bmatrix} M \\ N \end{bmatrix} \in \mathcal{H}_\infty.$$

Then $\left\|\begin{bmatrix} \Delta_M \\ \Delta_N \end{bmatrix}\right\|_\infty < b$ and $P_1 = (N_1 Q)(M_1 Q)^{-1} = (N + \Delta_N)(M + \Delta_M)^{-1}$.

To show the converse, note that $P_1 = (N + \Delta_N)(M + \Delta_M)^{-1}$ and there exists a $\tilde{Q}^{-1} \in \mathcal{H}_\infty$ such that $P_1 = \left\{(N + \Delta_N)\tilde{Q}\right\}\left\{(M + \Delta_M)\tilde{Q}\right\}^{-1}$ is a normalized right coprime factorization. Hence by definition, $\vec{\delta}_g(P, P_1)$ can be computed as

$$\vec{\delta}_g(P, P_1) = \inf_Q \left\|\begin{bmatrix} M \\ N \end{bmatrix} - \begin{bmatrix} M + \Delta_M \\ N + \Delta_N \end{bmatrix} \tilde{Q} Q\right\|_\infty \leq \left\|\begin{bmatrix} M \\ N \end{bmatrix} - \begin{bmatrix} M + \Delta_M \\ N + \Delta_N \end{bmatrix}\right\|_\infty < b$$

where the first inequality follows by taking $Q = \tilde{Q}^{-1} \in \mathcal{H}_\infty$. □

The following is a list of useful properties of the gap metric shown by Georgiou and Smith [1990].

- If $\delta_g(P_1, P_2) < 1$, then $\delta_g(P_1, P_2) = \vec{\delta}_g(P_1, P_2) = \vec{\delta}_g(P_2, P_1)$.

- If $b \leq \lambda(P) := \inf_{\text{Re } s > 0} \sigma\left(\begin{bmatrix} M(s) \\ N(s) \end{bmatrix}\right)$, then

$$\left\{P_1 : \vec{\delta}(P, P_1) < b\right\} = \left\{P_1 : \delta(P, P_1) < b\right\}.$$

Recall that

$$b_{\text{obt}}(P) := \left\{\inf_{K \text{ stabilizing}} \left\|\begin{bmatrix} I \\ K \end{bmatrix}(I+PK)^{-1}\begin{bmatrix} I & P \end{bmatrix}\right\|_\infty\right\}^{-1}$$

$$= \sqrt{1 - \lambda_{\max}(YQ)} = \sqrt{1 - \left\|\begin{bmatrix} \tilde{N} & \tilde{M} \end{bmatrix}\right\|_H^2}$$

and

$$b_{P,K} := \left\|\begin{bmatrix} I \\ K \end{bmatrix}(I+PK)^{-1}\begin{bmatrix} I & P \end{bmatrix}\right\|_\infty^{-1} = \left\|\begin{bmatrix} I \\ P \end{bmatrix}(I+KP)^{-1}\begin{bmatrix} I & K \end{bmatrix}\right\|_\infty^{-1}.$$

The following results were shown by Qiu and Davison [1992a].

Theorem 17.3 *Suppose the feedback system with the pair (P_0, K_0) is stable. Let $\mathcal{P} := \{P: \delta_g(P, P_0) < r_1\}$ and $\mathcal{K} := \{K: \delta_g(K, K_0) < r_2\}$. Then*

(a) The feedback system with the pair (P, K) is also stable for all $P \in \mathcal{P}$ and $K \in \mathcal{K}$ if and only if

$$\arcsin b_{P_0, K_0} \geq \arcsin r_1 + \arcsin r_2.$$

(b) The worst possible performance resulting from these sets of plants and controllers is given by

$$\inf_{P \in \mathcal{P},\, K \in \mathcal{K}} \arcsin b_{P,K} = \arcsin b_{P_0, K_0} - \arcsin r_1 - \arcsin r_2.$$

The sufficiency part of the theorem follows from Theorem 17.8 in the next section. Note that the theorem is still true if one of the uncertainty balls is taken as closed ball. In particular, one can take either $r_1 = 0$ or $r_2 = 0$.

Example 17.1 Consider

$$P_1 = \frac{s-1}{s+1}, \quad P_2 = \frac{2s-1}{s+1}.$$

Then $P_1 = N_1/M_1$ and $P_2 = N_2/M_2$ with

$$N_1 = \frac{1}{\sqrt{2}}\frac{s-1}{s+1}, \quad M_1 = \frac{1}{\sqrt{2}}, \quad N_2 = \frac{2s-1}{\sqrt{5}s+\sqrt{2}}, \quad M_2 = \frac{s+1}{\sqrt{5}s+\sqrt{2}}$$

are normalized coprime factorizations. It is easy to show that

$$\delta_g(P_1, P_2) = 1/3 > \|\Psi(P_1, P_2)\|_\infty = \sup_\omega \frac{|\omega|}{\sqrt{10\omega^2+4}} = \frac{1}{\sqrt{10}},$$

17.1. Gap Metric

which implies that any controller K that stabilizes P_1 and achieves only $b_{P_1,K} > 1/3$ will actually stabilize P_2 by Theorem 17.3. The following MATLAB command can be used to compute the gap:

$$\gg \delta_g(\mathbf{P_1},\mathbf{P_2}) = \text{gap}(\mathbf{P_1},\mathbf{P_2},\text{tol})$$

Next, note that $b_{\text{obt}}(P_1) = 1/\sqrt{2}$ and the optimal controller achieving $b_{\text{obt}}(P_1)$ is $K_{\text{obt}} = 0$. There must be a plant P with $\delta_\nu(P_1, P) = b_{\text{obt}}(P_1) = 1/\sqrt{2}$ that can not be stabilized by $K_{\text{obt}} = 0$; that is, there must be an unstable plant P such that $\delta_\nu(P_1, P) = b_{\text{obt}}(P_1) = 1/\sqrt{2}$. A such P can be found using Corollary 17.2:

$$\{P: \delta_g(P_1, P) \leq b_{\text{obt}}(P_1)\}$$

$$= \left\{ P: P = \frac{N_1 + \Delta_N}{M_1 + \Delta_M},\ \Delta_N, \Delta_M \in \mathcal{H}_\infty,\ \left\| \begin{bmatrix} \Delta_N \\ \Delta_M \end{bmatrix} \right\|_\infty \leq b_{\text{obt}}(P_1) \right\}.$$

that is, there must be $\Delta_N, \Delta_M \in \mathcal{H}_\infty$, $\left\| \begin{bmatrix} \Delta_N \\ \Delta_M \end{bmatrix} \right\|_\infty = b_{\text{obt}}(P_1)$ such that

$$P = \frac{N_1 + \Delta_N}{M_1 + \Delta_M}$$

is unstable. Let

$$\Delta_N = 0, \quad \Delta_M = \frac{1}{\sqrt{2}} \frac{s-1}{s+1}.$$

Then

$$P = \frac{N_1 + \Delta_N}{M_1 + \Delta_M} = \frac{s-1}{2s}, \quad \delta_\nu(P_1, P) = b_{\text{obt}}(P_1) = 1/\sqrt{2}.$$

Example 17.2 We shall now consider the following question: Given an uncertain plant

$$P(s) = \frac{k}{s-1}, \quad k \in [k_1, k_2],$$

(a) Find the best nominal design model $P_0 = \dfrac{k_0}{s-1}$ in the sense

$$\inf_{k_0 \in [k_1, k_2]} \sup_{k \in [k_1, k_2]} \delta_g(P, P_0).$$

(b) Let k_1 be fixed and k_2 be variable. Find the k_0 so that the largest family of the plant P can be guaranteed to be stabilized a priori by any controller satisfying $b_{P_0,K} = b_{\text{obt}}(P_0)$.

For simplicity, suppose $k_1 \geq 1$. It can be shown that $\delta_g(P, P_0) = \frac{|k_0 - k|}{k_0 + k}$. Then the optimal k_0 for question (a) satisfies

$$\frac{k_0 - k_1}{k_0 + k_1} = \frac{k_2 - k_0}{k_2 + k_0};$$

that is, $k_0 = \sqrt{k_1 k_2}$ and

$$\inf_{k_0 \in [k_1, k_2]} \sup_{k \in [k_1, k_2]} \delta_g(P, P_0) = \frac{\sqrt{k_2} - \sqrt{k_1}}{\sqrt{k_2} + \sqrt{k_1}}.$$

To answer question (b), we note that by Theorem 17.3, a family of plants satisfying $\delta_g(P, P_0) \leq r$ with $P_0 = k_0/(s+1)$ is stabilizable a priori by any controller satisfying $b_{P_0, K} = b_{\text{obt}}(P_0)$ if, and only if, $r < b_{P_0, K}$. Since $P_0 = N_0/M_0$ with

$$N_0 = \frac{k_0}{s + \sqrt{1 + k_0^2}}, \quad M_0 = \frac{s - 1}{s + \sqrt{1 + k_0^2}}$$

is a normalized coprime factorization, it is easy to show that

$$\left\| \begin{bmatrix} N_0 \\ M_0 \end{bmatrix} \right\|_H = \frac{\sqrt{k_0^2 + (1 - \sqrt{1 + k_0^2})^2}}{2\sqrt{1 + k_0^2}}$$

and

$$b_{\text{obt}}(P_0) = \sqrt{\frac{1}{2} \left(1 + \frac{1}{\sqrt{1 + k_0^2}} \right)}.$$

Hence we need to find a k_0 such that

$$b_{\text{obt}}(P_0) \geq \max \left\{ \frac{k_0 - k_1}{k_0 + k_1}, \frac{k_2 - k_0}{k_2 + k_0} \right\};$$

that is,

$$\sqrt{\frac{1}{2} \left(1 + \frac{1}{\sqrt{1 + k_0^2}} \right)} \geq \max \left\{ \frac{k_0 - k_1}{k_0 + k_1}, \frac{k_2 - k_0}{k_2 + k_0} \right\}$$

for a largest possible k_2. The optimal k_0 is given by the solution of the equation:

$$\sqrt{\frac{1}{2} \left(1 + \frac{1}{\sqrt{1 + k_0^2}} \right)} = \frac{k_0 - k_1}{k_0 + k_1}$$

and the largest $k_2 = k_0^2/k_1$. For example, if $k_1 = 1$, then $k_0 = 7.147$ and $k_2 = 51.0793$.

In general, given a family of plant P, it is not easy to see how to choose a best nominal model P_0 such that (a) or (b) is true. This is still a very important open question.

17.2 ν-Gap Metric

The shortfall of the gap metric is that it is not easily related to the frequency response of the system. On the other hand, the ν-gap metric to be introduced in this section has a clear frequency domain interpolation and can, in general, be computed from frequency response. The presentation given in this section, Sections 17.3, 17.4, and 17.5 are based on Vinnicombe [1993a, 1993b], to which readers are referred for further detailed discussions.

To define the new metric, we shall first review a basic concept from the complex analysis.

Definition 17.1 *Let $g(s)$ be a scalar transfer function and let Γ denote a Nyquist contour indented around the right of any imaginary axis poles of $g(s)$, as shown in Figure 17.1. Then the winding number of $g(s)$ with respect to this contour, denoted by* wno(g), *is the number of counterclockwise encirclements around the origin by $g(s)$ evaluated on the Nyquist contour Γ. (A clockwise encirclement counts as a negative encirclement.)*

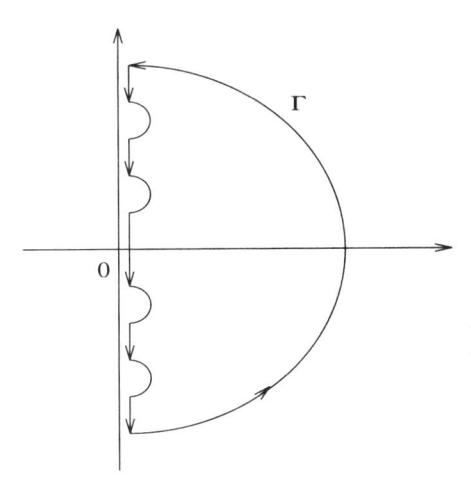

Figure 17.1: The Nyquist contour

The following argument principle is standard and can be found from any complex analysis book.

Lemma 17.4 (The Argument Principle) *Let Γ be a closed contour in the complex plane. Let $f(s)$ be a function analytic along the contour; that is, $f(s)$ has no poles on Γ. Assume $f(s)$ has Z zeros and P poles inside Γ. Then $f(s)$ evaluated along the contour Γ once in an anticlockwise direction will make $Z - P$ anticlockwise encirclements of the origin.*

Let $G(s)$ be a matrix (or scalar) transfer matrix. We shall denote $\eta(G)$ and $\eta_0(G)$, respectively, the number of open right-half plane and imaginary axis poles of $G(s)$.

The winding number has the following properties:

Lemma 17.5 *Let g and h be biproper rational scalar transfer functions and let F be a square transfer matrix. Then*

(a) $\mathrm{wno}(gh) = \mathrm{wno}(g) + \mathrm{wno}(h)$;

(b) $\mathrm{wno}(g) = \eta(g^{-1}) - \eta(g)$;

(c) $\mathrm{wno}(g^\sim) = -\mathrm{wno}(g) - \eta_0(g^{-1}) + \eta_0(g)$;

(d) $\mathrm{wno}(1+g) = 0$ *if* $g \in \mathcal{RL}_\infty$ *and* $\|g\|_\infty < 1$;

(e) $\mathrm{wno}\det(I+F) = 0$ *if* $F \in \mathcal{RL}_\infty$ *and* $\|F\|_\infty < 1$.

Proof. Part (a) is obvious by the argument principle. To show part (b), note that by the argument principle $\mathrm{wno}(g)$ equals the excess of the number of open right-half plane zeros of g over the number of open right-half plane poles of g; that is, $\mathrm{wno}(g) = \eta(g^{-1}) - \eta(g)$, since the number of right-half plane zeros of g is the number of right-half plane poles of g^{-1}. Next, suppose the order of g in part (c) is n. Then $\eta(g^\sim) = n - \eta(g) - \eta_0(g)$ and $\eta\left[(g^\sim)^{-1}\right] = n - \eta(g^{-1}) - \eta_0(g^{-1})$, which gives $\mathrm{wno}(g^\sim) = \eta\left[(g^\sim)^{-1}\right] - \eta(g^\sim) = \eta(g) - \eta(g^{-1}) - \eta_0(g^{-1}) + \eta_0(g) = -\mathrm{wno}(g) - \eta_0(g^{-1}) + \eta_0(g)$. Part (d) follows from the fact that $1 + \mathrm{Re}\, g(j\omega) > 0$, $\forall \omega$ since $\|g\|_\infty < 1$. Finally, part (e) follows from part (d) and $\det(I+F) = \prod_{i=1}^m (1 + \lambda_i(F))$ with $|\lambda_i(F)| < 1$. □

Example 17.3 Let

$$g_1 = \frac{1.2(s+3)}{s-5}, \quad g_2 = \frac{s-1}{s-2}, \quad g_3 = \frac{2(s-1)(s-2)}{(s+3)(s+4)}, \quad g_4 = \frac{(s-1)(s+3)}{(s-2)(s-4)}.$$

Figure 17.2 shows the functions, g_1, g_2, g_3, and g_4, evaluated on the Nyquist contour Γ. Clearly, we have

$$\mathrm{wno}(g_1) = -1, \quad \mathrm{wno}(g_2) = 0, \quad \mathrm{wno}(g_3) = 2, \quad \mathrm{wno}(g_4) = -1$$

and they are consistent with the results computed from using Lemma 17.5.

The ν-gap metric introduced in Vinnicombe [1993a, 1993b] is defined as follows:

17.2. ν-Gap Metric

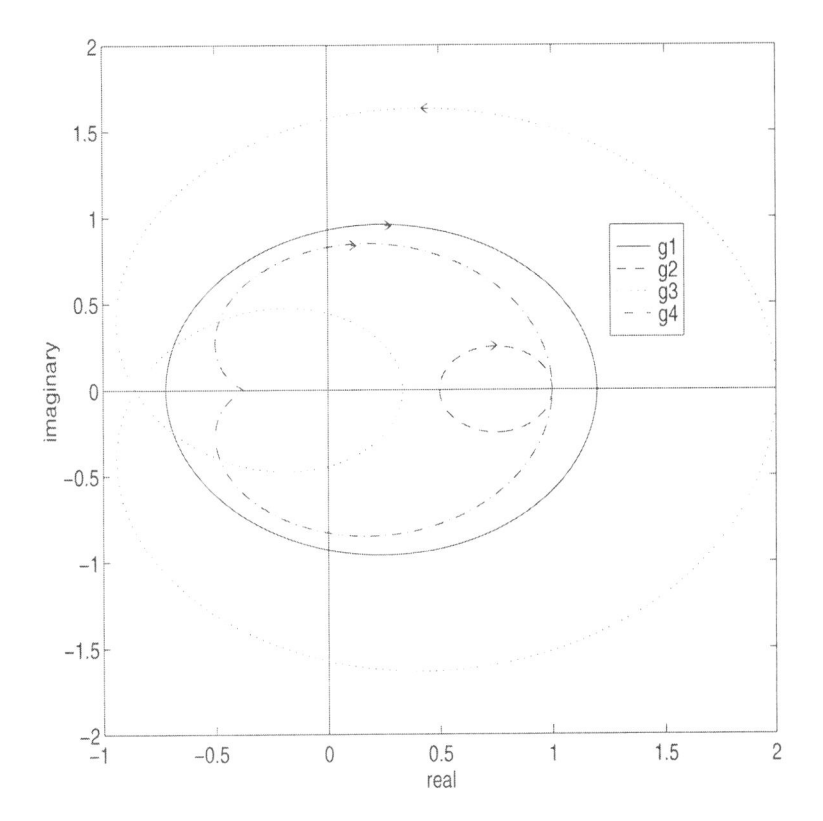

Figure 17.2: g_1, g_2, g_3, and g_4 evaluated on Γ

Definition 17.2 *The ν-gap metric is defined as*

$$\delta_\nu(P_1, P_2) = \begin{cases} \|\Psi(P_1, P_2)\|_\infty, & \text{if } \det \Theta(j\omega) \neq 0 \ \forall \omega \\ & \text{and wno} \det \Theta(s) = 0, \\ 1, & \text{otherwise} \end{cases}$$

where $\Theta(s) := \tilde{N}_2^\sim N_1 + \tilde{M}_2^\sim M_1$ and $\Psi(P_1, P_2) := -\tilde{N}_2 M_1 + \tilde{M}_2 N_1$.

Note that it can be shown as in Vinnicombe [1993a] that

$$\delta_\nu(P_1, P_2) = \delta_\nu(P_2, P_1) = \delta_\nu(P_1^T, P_2^T)$$

and δ_ν is indeed a metric (a proof of this fact is quite complex).

Computing $\delta_\nu(P_1, P_2)$: Let

$$P_1 = \left[\begin{array}{c|c} A_1 & B_1 \\ \hline C_1 & D_1 \end{array}\right], \quad P_2 = \left[\begin{array}{c|c} A_2 & B_2 \\ \hline C_2 & D_2 \end{array}\right].$$

1. Let $P_i = N_i M_i^{-1}$, $i = 1, 2$ be normalized right coprime factorizations. Then

$$\left[\begin{array}{c} M_i \\ N_i \end{array}\right] = \left[\begin{array}{c|c} A_i + B_i F_i & B_i R_i^{-1/2} \\ \hline F_i & R_i^{-1/2} \\ C_i + D_i F_i & D_i R_i^{-1/2} \end{array}\right], \quad \begin{array}{l} R_i = I + D_i^* D_i \\ \tilde{R}_i = I + D_i D_i^* \\ F_i = -R_i^{-1}(B_i^* X_i + D_i^* C_i) \end{array}$$

where

$$X_i = \text{Ric}\left[\begin{array}{cc} A_i - B_i R_i^{-1} D_i^* C_i & -B_i R_i^{-1} B_i^* \\ -C_i^* \tilde{R}_i^{-1} C_i & -(A_i - B_i R_i^{-1} D_i^* C_i)^* \end{array}\right].$$

2. Compute the zeros of

$$\Theta(s) := N_2^\sim N_1 + M_2^\sim M_1 = \left[\begin{array}{c} M_2 \\ N_2 \end{array}\right]^\sim \left[\begin{array}{c} M_1 \\ N_1 \end{array}\right].$$

Let n_0 = number of imaginary axis zeros of Θ, n_+ = number of open right-half plane zeros of Θ, and n = number of open right-half plane poles of Θ. Then wno $\det(N_2^\sim N_1 + M_2^\sim M_1) = n_+ - n$.

3. If either $n_0 \neq 0$ or $n_+ \neq n$, $\delta_\nu(P_1, P_2) = 1$. Otherwise, $\delta_\nu(P_1, P_2) = \|\Psi(P_1, P_2)\|_\infty$ with $\Psi(P_1, P_2) = -\tilde{N}_2 M_1 + \tilde{M}_2 N_1$:

$$\left[\begin{array}{c} \tilde{M}_i \\ \tilde{N}_i \end{array}\right] = \left[\begin{array}{c|cc} A_i + L_i C_i & L_i & B_i + L_i D_i \\ \hline \tilde{R}_i^{-1/2} C_i & \tilde{R}_i^{-1/2} & \tilde{R}_i^{-1/2} D_i \end{array}\right]$$

$$L_i = -(B_i D_i^* + Y_i C_i^*)\tilde{R}_i^{-1}$$

where

$$Y_i = \text{Ric}\left[\begin{array}{cc} (A_i - B_i D_i^* \tilde{R}_i^{-1} C_i)^* & -C_i^* \tilde{R}_i^{-1} C_i \\ -B_i R_i^{-1} B_i^* & -(A_i - B_i D_i^* \tilde{R}_i^{-1} C_i) \end{array}\right].$$

The MATLAB command **nugap** can be used to carry out the preceding computation:

$$\gg \delta_\nu(\mathbf{P_1}, \mathbf{P_2}) = \mathbf{nugap}(\mathbf{P_1}, \mathbf{P_2}, \mathbf{tol})$$

where tol is the computational tolerance.

17.2. ν-Gap Metric

Example 17.4 Consider, for example, $P_1 = 1$ and $P_2 = \dfrac{1}{s}$. Then

$$M_1 = N_1 = \frac{1}{\sqrt{2}}, \quad M_2 = \frac{s}{s+1}, \quad N_2 = \frac{1}{s+1}.$$

Hence

$$\Theta(s) = \frac{1}{\sqrt{2}} \frac{1-s}{1-s} = \frac{1}{\sqrt{2}}, \quad \Psi(P_1, P_2) = \frac{1}{\sqrt{2}} \frac{s-1}{s+1},$$

and $\delta_\nu(P_1, P_2) = \frac{1}{\sqrt{2}}$. (Note that Θ has no poles or zeros!)

The ν-gap metric can also be computed directly from the system transfer matrices without first finding the normalized coprime factorizations.

Theorem 17.6 *The ν-gap metric can be defined as*

$$\delta_\nu(P_1, P_2) = \begin{cases} \|\Psi(P_1, P_2)\|_\infty, & \text{if } \det(I + P_2^\sim P_1) \neq 0 \; \forall \omega \text{ and} \\ & \text{wno} \det(I + P_2^\sim P_1) + \eta(P_1) - \eta(P_2) - \eta_0(P_2) = 0, \\ 1, & \text{otherwise} \end{cases}$$

where $\Psi(P_1, P_2)$ can be written as

$$\Psi(P_1, P_2) = (I + P_2 P_2^\sim)^{-1/2}(P_1 - P_2)(I + P_1^\sim P_1)^{-1/2}.$$

Proof. Since the number of unstable zeros of M_1 (M_2) is equal to the number of unstable poles of P_1 (P_2), and

$$N_2^\sim N_1 + M_2^\sim M_1 = M_2^\sim (I + P_2^\sim P_1) M_1,$$

we have

$$\text{wno} \det(N_2^\sim N_1 + M_2^\sim M_1) = \text{wno} \det \{M_2^\sim (I + P_2^\sim P_1) M_1\}$$
$$= \text{wno} \det M_2^\sim + \text{wno} \det(I + P_2^\sim P_1) + \text{wno} \det M_1.$$

Note that $\text{wno} \det M_1 = \eta(P_1)$, $\text{wno} \det M_2^\sim = -\text{wno} \det M_2 - \eta_0(M_2^{-1}) = -\eta(P_2) - \eta_0(P_2)$, and

$$\text{wno} \det(N_2^\sim N_1 + M_2^\sim M_1) = -\eta(P_2) - \eta_0(P_2) + \text{wno} \det(I + P_2^\sim P_1) + \eta(P_1).$$

Furthermore,

$$\det(N_2^\sim N_1 + M_2^\sim M_1) \neq 0, \; \forall \omega \iff \det(I + P_2^\sim P_1) \neq 0, \; \forall \omega.$$

The theorem follows by noting that
$$\Psi(P_1, P_2) = (I + P_2 P_2^\sim)^{-1/2}(P_1 - P_2)(I + P_1^\sim P_1)^{-1/2}$$
since $\Psi(P_1, P_2) = -\tilde{N}_2 M_1 + \tilde{M}_2 N_1 = \tilde{M}_2(P_1 - P_2)M_1$ and
$$\tilde{M}_2 \tilde{M}_2^\sim = (I + P_2 P_2^\sim)^{-1}, \quad M_1 M_1^\sim = (I + P_1^\sim P_1)^{-1}.$$

□

This alternative formula is useful when doing the hand calculation or when computing from the frequency response of the plants since it does not need to compute the normalized coprime factorizations.

Example 17.5 Consider two plants $P_1 = 1$ and $P_2 = 1/s$. Then wno $\det(1 + P_2^\sim P_1) =$ wno$[(s-1)/s] = 1$, as shown in Figure 17.3(a), and wno $\det(1 + P_2^\sim P_1) + \eta(P_1) - \eta(P_2) - \eta_0(P_2) = 0$. On the other hand, wno $\det(1 + P_1^\sim P_2) + \eta(P_2) - \eta(P_1) =$ wno $(s+1)/s = 0$, as shown in Figure 17.3(b).

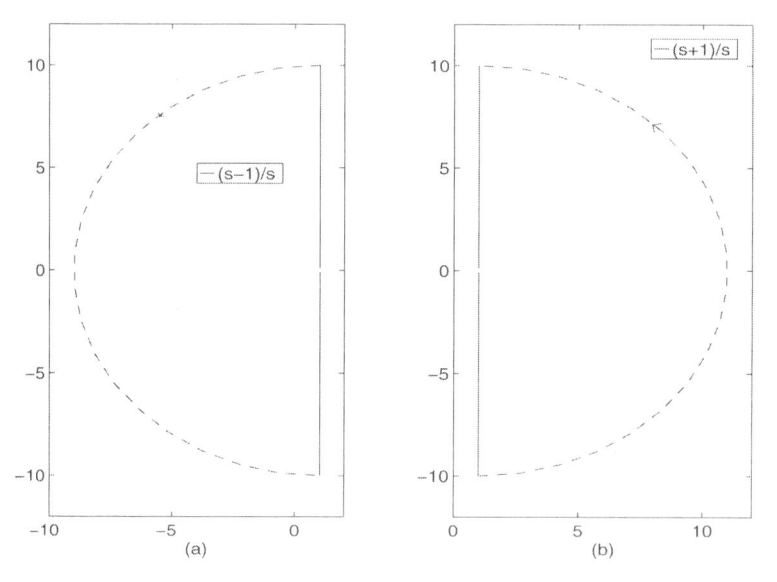

Figure 17.3: $\dfrac{s-1}{s}$ and $\dfrac{s+1}{s}$ evaluated on Γ

Similar to the gap metric, it is shown by Vinnicombe [1993a, 1993b] that the ν-gap metric can also be characterized as an optimization problem (however, we shall not use it for computation).

17.2. ν-Gap Metric

Theorem 17.7 *Let $P_1 = N_1 M_1^{-1}$ and $P_2 = N_2 M_2^{-1}$ be normalized right coprime factorizations. Then*

$$\delta_\nu(P_1, P_2) = \inf_{\substack{Q, Q^{-1} \in \mathcal{L}_\infty \\ \text{wno } \det(Q) = 0}} \left\| \begin{bmatrix} M_1 \\ N_1 \end{bmatrix} - \begin{bmatrix} M_2 \\ N_2 \end{bmatrix} Q \right\|_\infty.$$

Moreover, $\delta_g(P_1, P_2) b_{\text{obt}}(P_1) \le \delta_\nu(P_1, P_2) \le \delta_g(P_1, P_2)$.

It is now easy to see that

$$\{P : \delta_\nu(P_0, P) < r\}$$
$$\supset \left\{ P = (N_0 + \Delta_N)(M_0 + \Delta_M)^{-1} : \begin{bmatrix} \Delta_N \\ \Delta_M \end{bmatrix} \in \mathcal{H}_\infty, \left\| \begin{bmatrix} \Delta_N \\ \Delta_M \end{bmatrix} \right\|_\infty < r \right\}.$$

Define

$$\frac{1}{b_{P,K}(\omega)} := \bar\sigma\left(\begin{bmatrix} I \\ K(j\omega) \end{bmatrix} (I + P(j\omega)K(j\omega))^{-1} \begin{bmatrix} I & P(j\omega) \end{bmatrix} \right)$$

and

$$\psi(P_1(j\omega), P_2(j\omega)) = \bar\sigma\left(\Psi(P_1(j\omega), P_2(j\omega)) \right).$$

The following theorem states that robust stability can be checked using the frequency-by-frequency test.

Theorem 17.8 *Suppose (P_0, K) is stable and $\delta_\nu(P_0, P_1) < 1$. Then (P_1, K) is stable if*

$$b_{P_0, K}(\omega) > \psi(P_0(j\omega), P_1(j\omega)), \quad \forall \omega.$$

Moreover,

$$\arcsin b_{P_1, K}(\omega) \ge \arcsin b_{P_0, K}(\omega) - \arcsin \psi(P_0(j\omega), P_1(j\omega)), \quad \forall \omega$$

and

$$\arcsin b_{P_1, K} \ge \arcsin b_{P_0, K} - \arcsin \delta_\nu(P_0, P_1).$$

Proof. Let $P_1 = \tilde M_1^{-1} \tilde N_1$, $P_0 = N_0 M_0^{-1} = \tilde M_0^{-1} \tilde N_0$ and $K = UV^{-1}$ be normalized coprime factorizations, respectively. Then

$$\frac{1}{b_{P_1, K}(\omega)} = \bar\sigma\left(\begin{bmatrix} V \\ U \end{bmatrix} (\tilde M_1 V + \tilde N_1 U)^{-1} \begin{bmatrix} \tilde M_1 & \tilde N_1 \end{bmatrix} \right) = \bar\sigma\left((\tilde M_1 V + \tilde N_1 U)^{-1} \right).$$

That is,

$$b_{P_1, K}(\omega) = \underline\sigma(\tilde M_1 V + \tilde N_1 U) = \underline\sigma\left(\begin{bmatrix} \tilde M_1 & \tilde N_1 \end{bmatrix} \begin{bmatrix} V \\ U \end{bmatrix} \right).$$

Similarly,
$$b_{P_0,K}(\omega) = \underline{\sigma}(\tilde{M}_0 V + \tilde{N}_0 U) = \underline{\sigma}\left(\begin{bmatrix} \tilde{M}_0 & \tilde{N}_0 \end{bmatrix} \begin{bmatrix} V \\ U \end{bmatrix}\right).$$

Note that
$$\psi(P_0(j\omega), P_1(j\omega)) = \overline{\sigma}\left(\begin{bmatrix} \tilde{M}_1 & \tilde{N}_1 \end{bmatrix} \begin{bmatrix} N_0 \\ -M_0 \end{bmatrix}\right)$$

$$\begin{bmatrix} N_0 & \tilde{M}_0^\sim \\ -M_0 & \tilde{N}_0^\sim \end{bmatrix}^\sim \begin{bmatrix} N_0 & \tilde{M}_0^\sim \\ -M_0 & \tilde{N}_0^\sim \end{bmatrix} = I.$$

To simplify the derivation, define

$$G_0 = \begin{bmatrix} N_0 \\ -M_0 \end{bmatrix}, \quad \tilde{G}_0 = \begin{bmatrix} \tilde{M}_0 & \tilde{N}_0 \end{bmatrix}, \quad \tilde{G}_1 = \begin{bmatrix} \tilde{M}_1 & \tilde{N}_1 \end{bmatrix}, \quad F = \begin{bmatrix} V \\ U \end{bmatrix}.$$

Then
$$\psi(P_0, P_1) = \overline{\sigma}(\tilde{G}_1 G_0), \quad b_{P_0,K}(\omega) = \underline{\sigma}(\tilde{G}_0 F), \quad b_{P_1,K}(\omega) = \underline{\sigma}(\tilde{G}_1 F)$$

and
$$\begin{bmatrix} G_0 & \tilde{G}_0^\sim \end{bmatrix}^\sim \begin{bmatrix} G_0 & \tilde{G}_0^\sim \end{bmatrix} = I \Longrightarrow \begin{bmatrix} G_0 & \tilde{G}_0^\sim \end{bmatrix} \begin{bmatrix} G_0 & \tilde{G}_0^\sim \end{bmatrix}^\sim = I.$$

That is,
$$G_0 G_0^\sim + \tilde{G}_0^\sim \tilde{G}_0 = I.$$

Note that
$$I = \tilde{G}_1 \tilde{G}_1^\sim = \tilde{G}_1 (G_0 G_0^\sim + \tilde{G}_0^\sim \tilde{G}_0) \tilde{G}_1^\sim = (\tilde{G}_1 G_0)(\tilde{G}_1 G_0)^\sim + (\tilde{G}_1 \tilde{G}_0^\sim)(\tilde{G}_1 \tilde{G}_0^\sim)^\sim.$$

Hence
$$\underline{\sigma}^2(\tilde{G}_1 \tilde{G}_0^\sim) = 1 - \overline{\sigma}^2(\tilde{G}_1 G_0).$$

Similarly,
$$I = F^\sim F = F^\sim (G_0 G_0^\sim + \tilde{G}_0^\sim \tilde{G}_0) F = (G_0^\sim F)^\sim (G_0^\sim F) + (\tilde{G}_0 F)^\sim (\tilde{G}_0 F)$$

$$\Longrightarrow \overline{\sigma}^2(G_0^\sim F) = 1 - \underline{\sigma}^2(\tilde{G}_0 F).$$

By the assumption, $\psi(P_0, P_1) < b_{P_0,K}(\omega)$; that is,
$$\overline{\sigma}(\tilde{G}_1 G_0) < \underline{\sigma}(\tilde{G}_0 F), \quad \forall \omega$$

and
$$\overline{\sigma}(G_0^\sim F) = \sqrt{1 - \underline{\sigma}^2(\tilde{G}_0 F)} < \sqrt{1 - \overline{\sigma}^2(\tilde{G}_1 G_0)} = \underline{\sigma}(\tilde{G}_1 \tilde{G}_0^\sim).$$

Hence
$$\overline{\sigma}(\tilde{G}_1 G_0)\overline{\sigma}(G_0^\sim F) < \underline{\sigma}(\tilde{G}_1 \tilde{G}_0^\sim)\underline{\sigma}(\tilde{G}_0 F);$$

17.2. ν-Gap Metric

that is,

$$\overline{\sigma}(\tilde{G}_1 G_0 \tilde{G}_0^\sim F) < \underline{\sigma}(\tilde{G}_1 \tilde{G}_0^\sim \tilde{G}_0 F), \quad \forall \, \omega$$

$$\implies \left\|(\tilde{G}_1 \tilde{G}_0^\sim \tilde{G}_0 F)^{-1}(\tilde{G}_1 G_0 \tilde{G}_0^\sim F)\right\|_\infty < 1.$$

Now

$$\tilde{G}_1 F = \tilde{G}_1(\tilde{G}_0^\sim \tilde{G}_0 + G_0 \tilde{G}_0^\sim) F = (\tilde{G}_1 \tilde{G}_0^\sim \tilde{G}_0 F) + (\tilde{G}_1 G_0 \tilde{G}_0^\sim F)$$

$$= (\tilde{G}_1 \tilde{G}_0^\sim \tilde{G}_0 F) \left(I + (\tilde{G}_1 \tilde{G}_0^\sim \tilde{G}_0 F)^{-1}(\tilde{G}_1 G_0 \tilde{G}_0^\sim F) \right).$$

By Lemma 17.5,

$$\text{wno } \det(\tilde{G}_1 F) = \text{wno } \det(\tilde{G}_1 \tilde{G}_0^\sim \tilde{G}_0 F) = \text{wno } \det(\tilde{G}_1 \tilde{G}_0^\sim) + \text{wno } \det(\tilde{G}_0 F).$$

Since (P_0, K) is stable $\implies (\tilde{G}_0 F)^{-1} \in \mathcal{H}_\infty \implies \eta((\tilde{G}_0 F)^{-1}) = 0$

$$\implies \text{wno } \det(\tilde{G}_0 F) := \eta((\tilde{G}_0 F)^{-1}) - \eta(\tilde{G}_0 F) = 0.$$

Next, note that

$$P_0^T = (\tilde{N}_0^T)(\tilde{M}_0^T)^{-1}, \quad P_1^T = (\tilde{N}_1^T)(\tilde{M}_1^T)^{-1}$$

and $\delta_\nu(P_0^T, P_1^T) = \delta_\nu(P_0, P_1) < 1$; then, by definition of $\delta_\nu(P_0^T, P_1^T)$,

$$\text{wno } \det((\tilde{N}_0^T)^\sim(\tilde{N}_1^T) + (\tilde{M}_0^T)^\sim(\tilde{M}_1^T)) = \text{wno } \det(\tilde{G}_1 \tilde{G}_0^\sim)^T = \text{wno } \det(\tilde{G}_1 \tilde{G}_0^\sim) = 0.$$

Hence wno $\det(\tilde{G}_1 F) = 0$, but wno $\det(\tilde{G}_1 F) := \eta((\tilde{G}_1 F)^{-1}) - \eta(\tilde{G}_1 F) = \eta((\tilde{G}_1 F)^{-1})$ since $\eta(\tilde{G}_1 F) = 0$, so $\eta((\tilde{G}_1 F)^{-1}) = 0$; that is, (P_1, K) is stable.

Finally, note that

$$\tilde{G}_1 F = \tilde{G}_1(\tilde{G}_0^\sim \tilde{G}_0 + G_0 \tilde{G}_0^\sim) F = (\tilde{G}_1 \tilde{G}_0^\sim)(\tilde{G}_0 F) + (\tilde{G}_1 G_0)(\tilde{G}_0^\sim F)$$

and

$$\underline{\sigma}(\tilde{G}_1 F) \geq \underline{\sigma}(\tilde{G}_1 \tilde{G}_0^\sim)\underline{\sigma}(\tilde{G}_0 F) - \overline{\sigma}(\tilde{G}_1 G_0)\overline{\sigma}(\tilde{G}_0^\sim F)$$

$$= \sqrt{1 - \overline{\sigma}^2(\tilde{G}_1 G_0)}\underline{\sigma}(\tilde{G}_0 F) - \overline{\sigma}(\tilde{G}_1 G_0)\sqrt{1 - \underline{\sigma}^2(\tilde{G}_0 F)}$$

$$= \sin(\arcsin \underline{\sigma}(\tilde{G}_0 F) - \arcsin \overline{\sigma}(\tilde{G}_1 G_0))$$

$$= \sin(\arcsin b_{P_0,K}(\omega) - \arcsin \psi(P_0(j\omega), P_1(j\omega)))$$

and, consequently,

$$\arcsin b_{P_1,K}(\omega) \geq \arcsin b_{P_0,K}(\omega) - \arcsin \psi(P_0(j\omega), P_1(j\omega))$$

and

$$\inf_\omega \arcsin b_{P_1,K}(\omega) \geq \inf_\omega \arcsin b_{P_0,K}(\omega) - \sup_\omega \arcsin \psi(P_0(j\omega), P_1(j\omega)).$$

That is, $\arcsin b_{P_1,K} \geq \arcsin b_{P_0,K} - \arcsin \delta_\nu(P_0, P_1)$. □

The significance of the preceding theorem can be illustrated using Figure 17.4. It is clear from the figure that $\delta_\nu(P_0, P_1) > b_{P_0,K}$. Thus a frequency-independent stability test cannot conclude that a stabilizing controller K for P_0 will stabilize P_1. However, the frequency-dependent test in the preceding theorem shows that K stabilizes both P_0 and P_1 since $b_{P_0,K}(\omega) > \psi(P_0(j\omega), P_1(j\omega))$ for all ω. Furthermore,

$$b_{P_1,K} \geq \inf_\omega \sin\left(\arcsin b_{P_0,K}(\omega) - \arcsin \psi(P_0, P_1)\right) > 0.$$

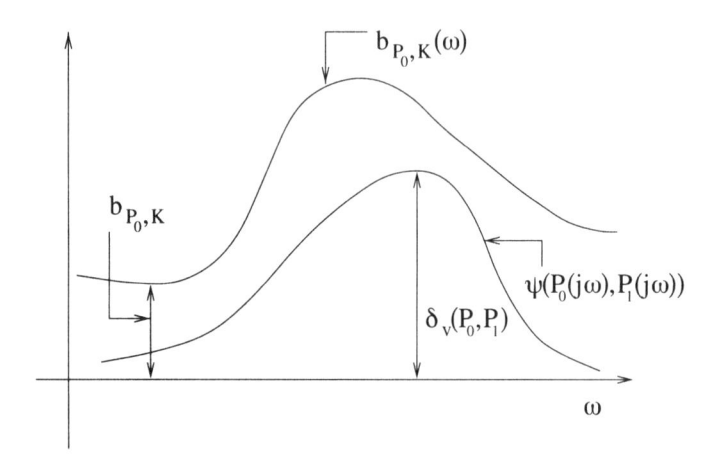

Figure 17.4: K stabilizes both P_0 and P_1 since $b_{P_0,K}(\omega) > \psi(P_0, P_1)$ for all ω

The following theorem is one of the main results on the ν-gap metric.

Theorem 17.9 *Let P_0 be a nominal plant and $\beta \leq \alpha < b_{\text{obt}}(P_0)$.*

(i) For a given controller K,

$$\arcsin b_{P,K} > \arcsin \alpha - \arcsin \beta$$

for all P satisfying $\delta_\nu(P_0, P) \leq \beta$ if and only if $b_{P_0,K} > \alpha$.

(ii) For a given plant P,

$$\arcsin b_{P,K} > \arcsin \alpha - \arcsin \beta$$

for all K satisfying $b_{P_0,K} > \alpha$ if and only if $\delta_\nu(P_0, P) \leq \beta$.

Proof. The sufficiency follows essentially from Theorem 17.8. The necessity proof is harder, see Vinnicombe [1993a, 1993b] for details. □

The preceding theorem shows that any plant at a distance less than β from the nominal will be stabilized by any controller stabilizing the nominal with a stability

17.2. ν-Gap Metric

margin of β. Furthermore, any plant at a distance greater than β from the nominal will be *destabilized* by *some* controller that stabilizes the nominal with a stability margin of at least β.

Similarly, one can consider the system robust performance with simultaneous perturbations on the plant and controller.

Theorem 17.10 *Suppose the feedback system with the pair (P_0, K_0) is stable. Then*

$$\arcsin b_{P,K} \geq \arcsin b_{P_0,K_0} - \arcsin \delta_\nu(P_0, P) - \arcsin \delta_\nu(K_0, K)$$

for any P and K.

Proof. Use the fact that $b_{P,K} = b_{K,P}$ and apply Theorem 17.8 to get

$$\arcsin b_{P,K} \geq \arcsin b_{P_0,K} - \arcsin \delta_\nu(P_0, P).$$

Dually, we have

$$\arcsin b_{P_0,K} \geq \arcsin b_{P_0,K_0} - \arcsin \delta_\nu(K_0, K).$$

Hence the result follows. $\square$

Example 17.6 Consider again the following example, studied in Vinnicombe [1993b], with

$$P_1 = \frac{s-1}{s+1}, \quad P_2 = \frac{2s-1}{s+1}$$

and note that

$$1 + P_2^\sim P_1 = 1 + \frac{-2s-1}{-s+1}\frac{s-1}{s+1} = \frac{3s+2}{s+1}.$$

Then

$$1 + P_2^\sim(j\omega)P_1(j\omega) \neq 0, \quad \forall \omega, \quad \text{wno } \det(I + P_2^\sim P_1) + \eta(P_1) - \eta(P_2) = 0$$

and

$$\delta_\nu(P_1, P_2) = \|\Psi(P_1, P_2)\|_\infty = \sup_\omega \frac{|P_1 - P_2|}{\sqrt{1+|P_1|^2}\sqrt{1+|P_2|^2}} = \sup_\omega \frac{|\omega|}{\sqrt{10\omega^2+4}} = \frac{1}{\sqrt{10}}.$$

This implies that any controller K that stabilizes P_1 and achieves only $b_{P_1,K} > 1/\sqrt{10}$ will actually stabilize P_2. This result is clearly less conservative than that of using the gap metric. Furthermore, there exists a controller such that $b_{P_1,K} = 1/\sqrt{10}$ that destabilizes P_2. Such a controller is $K = -1/2$, which results in a closed-loop system with P_2 illposed.

Example 17.7 Consider the following example taken from Vinnicombe [1993b]:

$$P_1 = \frac{100}{2s+1}, \quad P_2 = \frac{100}{2s-1}, \quad P_3 = \frac{100}{(s+1)^2}.$$

P_1 and P_2 have very different open-loop characteristics—one is stable, the other unstable. However, it is easy to show that

$$\delta_\nu(P_1, P_2) = \delta_g(P_1, P_2) = 0.02, \quad \delta_\nu(P_1, P_3) = \delta_g(P_1, P_3) = 0.8988,$$

$$\delta_\nu(P_2, P_3) = \delta_g(P_2, P_3) = 0.8941,$$

which show that P_1 and P_2 are very close while P_1 and P_3 (or P_2 and P_3) are quite far away. It is not surprising that any reasonable controller for P_1 will do well for P_2 but not necessarily for P_3. The closed-loop step responses under unity feedback,

$$K_1 = 1,$$

are shown in Figure 17.5.

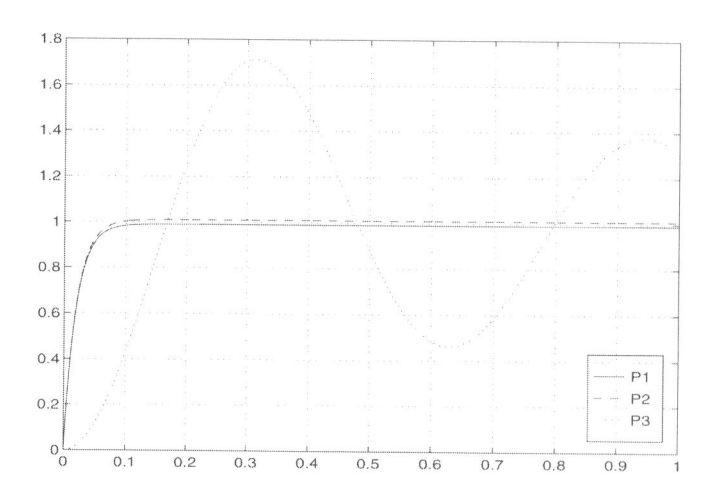

Figure 17.5: Closed-loop step responses with $K_1 = 1$

The corresponding stability margins for the closed-loop systems with P_1 and P_2 are

$$b_{P_1, K_1} = 0.7071, \quad \text{and} \quad b_{P_2, K_1} = 0.7,$$

respectively, which are very close to their maximally possible margins,

$$b_{\text{obt}}(P_1) = 0.7106, \quad \text{and} \quad b_{\text{obt}}(P_2) = 0.7036$$

17.2. ν-Gap Metric

(in fact, the optimal controllers for P_1 and P_2 are $K = 0.99$ and $K = 1.01$, respectively). While the stability margin for the closed-loop system with P_3 is

$$b_{P_3, K_1} = 0.0995,$$

which is far away from its optimal value, $b_{\text{obt}}(P_3) = 0.4307$, and results in poor performance of the closed loop. In fact, it is not hard to find a controller that will perform well for both P_1 and P_2 but will destabilize P_3.

Of course, this does not necessarily mean that all controllers performing reasonably well with P_1 and P_2 will do badly with P_3, merely that some do — the unit feedback being an example. It may be harder to find a controller that will perform reasonably well with all three plants; the maximally stabilizing controller of P_3,

$$K_3 = \frac{2.0954s + 10.8184}{s + 23.2649},$$

is a such controller, which gives

$$b_{P_1, K_3} = 0.4307, \quad b_{P_2, K_3} = 0.4126, \quad \text{and} \quad b_{P_3, K_3} = 0.4307.$$

The step responses under this control law are shown in Figure 17.6.

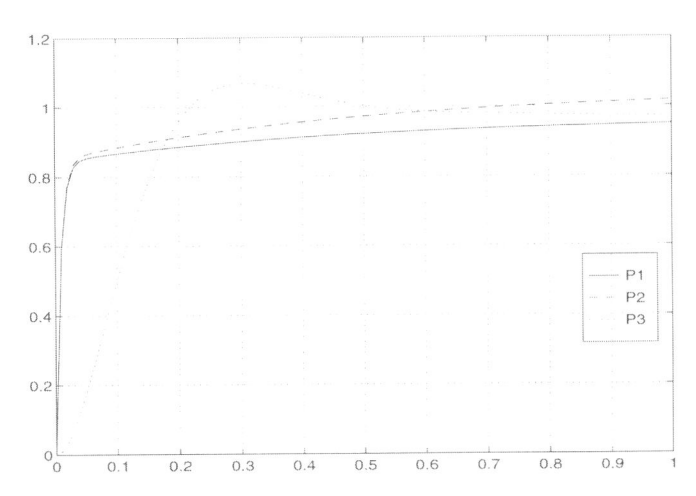

Figure 17.6: Closed-loop step responses with $K_3 = \dfrac{2.0954s + 10.8184}{s + 23.2649}$

17.3 Geometric Interpretation of ν-Gap Metric

The most salient feature of the ν-gap metric is that it can be computed pointwise in frequency domain:

$$\delta_\nu(P_1, P_2) = \sup_\omega \psi(P_1(j\omega),\ P_2(j\omega))$$

provided the winding number condition is satisfied. (For a more extensive coverage of material presented in this section and the next two sections, readers are encouraged to consult the original references by Vinnicombe [1992a, 1992b, 1993a, 1993b].)

In particular, for a single-input single-output system,

$$\psi(P_1(j\omega),\ P_2(j\omega)) = \frac{|P_1(j\omega) - P_2(j\omega)|}{\sqrt{1 + |P_1(j\omega)|^2}\ \sqrt{1 + |P_2(j\omega)|^2}}. \tag{17.2}$$

This function has the interpretation of being the chordal distance between $P_1(j\omega)$ and $P_2(j\omega)$. To illustrate this, consider the Riemann sphere, which is a unit sphere tangent at its "south pole" to the complex plant at its origin shown in Figure 17.7.

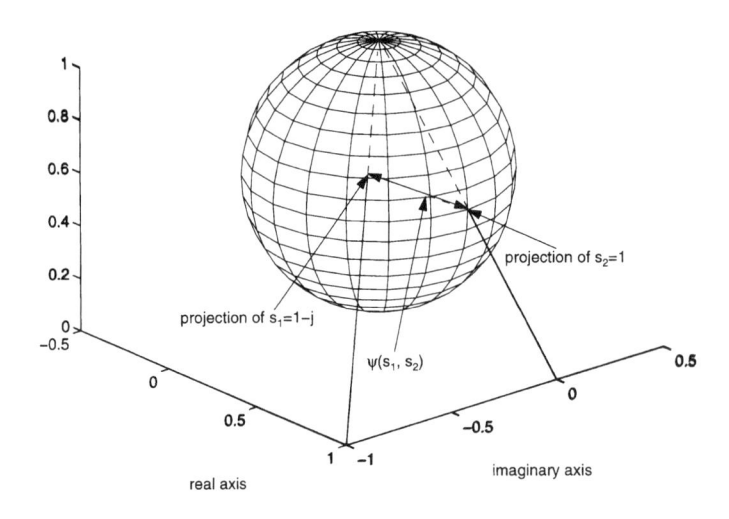

Figure 17.7: Projection onto the Riemann sphere

A point s_1 (e.g., $s_1 = 1-j$) in the complex plane is stereographically projected on the Riemann sphere by connecting the "north pole" to s_1 and determining the intersection of this straight line with the Riemann sphere, resulting in the projection, q_1, of s_1. The

17.3. Geometric Interpretation of ν-Gap Metric

coordinates of q_1 are

$$x_1 = \frac{\Re s_1}{1+|s_1|^2}, \quad y_1 = \frac{\Im s_1}{1+|s_1|^2}, \quad z_1 = \frac{|s_1|^2}{1+|s_1|^2}.$$

Thus, the north pole represents the point at infinity and the unit circle is projected onto the "equator." The chordal distance between two points, s_1 and s_2, is the Euclidean distance between their stereographical projections, q_1 and q_2:

$$d(s_1, s_2) = \sqrt{(x_1-x_2)^2 + (y_1-y_2)^2 + (z_1-z_2)^2} = \frac{|s_1-s_2|}{\sqrt{1+|s_1|^2}\sqrt{1+|s_2|^2}}.$$

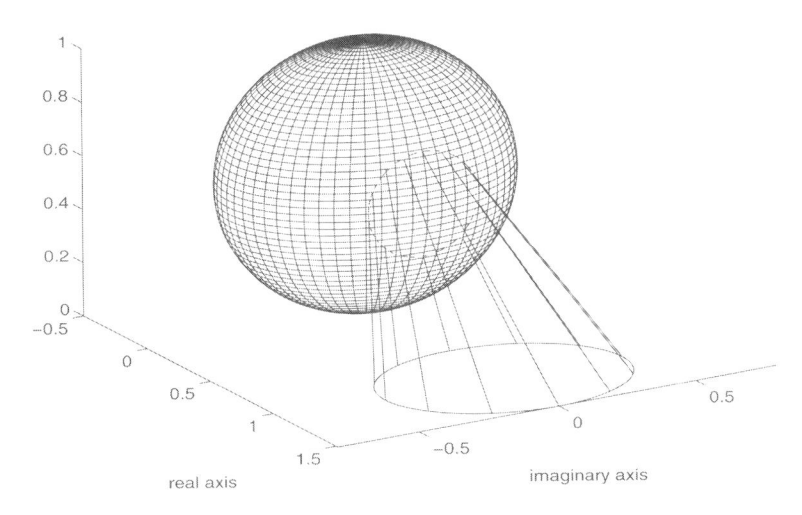

Figure 17.8: Projection of a disk on the Nyquist diagram onto the Riemann sphere

Now consider a circle of chordal radius r centered at $P_0(j\omega_0)$ on the Riemann sphere for some frequency ω_0; that is,

$$\frac{|P(j\omega_0) - P_0(j\omega_0)|}{\sqrt{1+|P(j\omega_0)|^2}\sqrt{1+|P_0(j\omega_0)|^2}} = r.$$

Let $P(j\omega_0) = R + jI$ and $P_0(j\omega_0) = R_0 + jI_0$. Then it is easy to show that

$$\left(R - \frac{R_0}{1-\alpha}\right)^2 + \left(I - \frac{I_0}{1-\alpha}\right)^2 = \frac{\alpha(1+|P_0|^2-\alpha)}{(1-\alpha)^2}, \quad \text{if } \alpha \neq 1$$

where $\alpha = r^2(1+|P_0|^2)$. This means that a ball of uncertainty on the ν-gap metric corresponds to a (large) ball of uncertainty on the Nyquist diagram. Figure 17.8 shows a circle of chordal radius 0.2 centered at the stereographical projection of $P_0(j\omega_0) = 1$ and the corresponding circle on the Nyquist diagram.

Figure 17.9 and Figure 17.10 illustrate the uncertainty on the Nyquist diagram corresponding to the balls of uncertainty on the Riemann sphere centered at p_0 with chordal radius 0.2. For example, an uncertainty of 0.2 at $|p_0(j\omega_0)| = 1$ for some ω_0 (i.e., $\delta_\nu(p_0, p) \leq 0.2$) implies that $0.661 \leq |p(j\omega_0)| \leq 1.513$ and the phase difference between p_0 and p is no more than $23.0739°$ at ω_0.

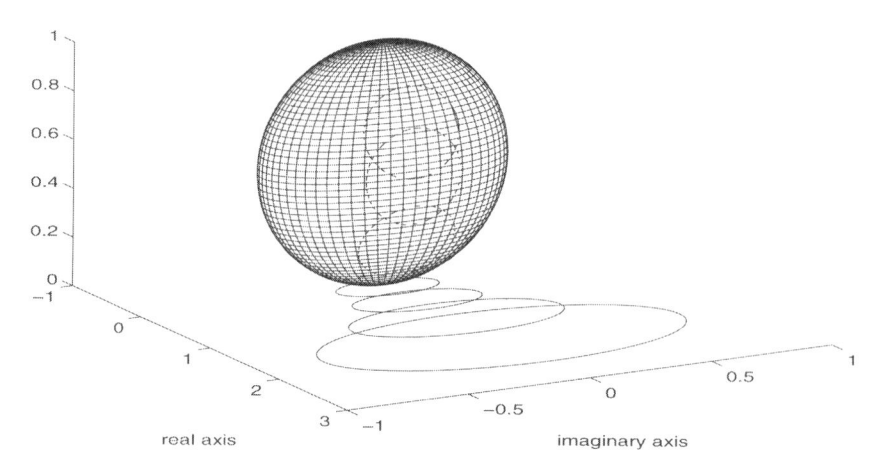

Figure 17.9: Uncertainty on the Riemann sphere and the corresponding uncertainty on the Nyquist diagram

Note that $\|\Psi(P_1, P_2)\|_\infty$ on its own without the winding number condition is useless for the study of feedback systems. This is illustrated through the following example.

Example 17.8 Consider

$$P_1 = 1, \quad P_2 = \frac{s - 1 - \epsilon}{s - 1}.$$

It is clear that P_2 becomes increasingly difficult to stabilize as $\epsilon \to 0$ due to the near unstable pole/zero cancellation. In fact, any stabilizing controller for P_1 will destabilize all P_2 for ϵ sufficiently small. This is confirmed by noting that $b_{\text{obt}}(P_1) = 1$, $b_{\text{obt}}(P_2) \approx \epsilon/2$, and

$$\delta_g(P_1, P_2) = \delta_\nu(P_1, P_2) = 1, \quad \epsilon \geq -2.$$

However, $\|\Psi(P_1, P_2)\|_\infty = \frac{|\epsilon|}{\sqrt{4 + 4\epsilon + 2\epsilon^2}} \approx \frac{\epsilon}{2}$ in itself fails to indicate the difficulty of the problem.

17.4. Extended Loop-Shaping Design

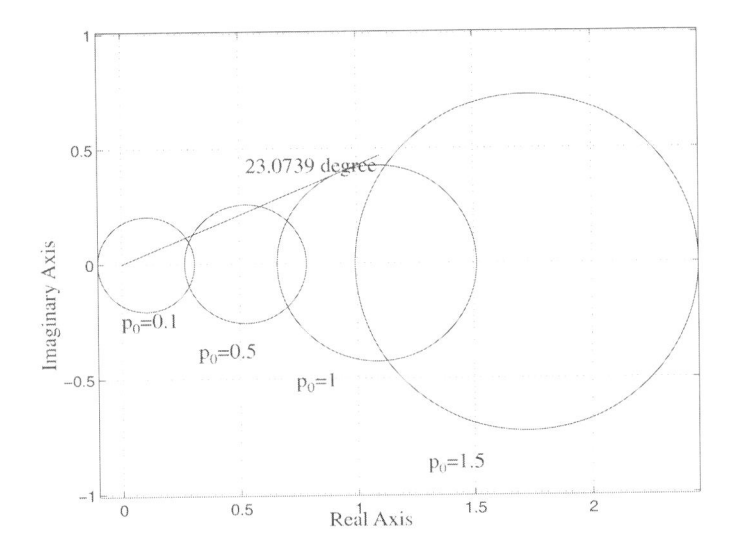

Figure 17.10: Uncertainty on the Nyquist diagram corresponding to the balls of uncertainty on the Riemann sphere centered at p_0 with chordal radius 0.2

17.4 Extended Loop-Shaping Design

Let $\mathcal{P}$ be a family of parametric uncertainty systems and let $P_0 \in \mathcal{P}$ be a nominal design model. We are interested in finding a controller so that we have the largest possible robust stability margin; that is,
$$\sup_K \inf_{P \in \mathcal{P}} b_{P,K}.$$

Note that by Theorem 17.8, for any $P_1 \in \mathcal{P}$, we have
$$\arcsin b_{P_1,K}(\omega) \geq \arcsin b_{P_0,K}(\omega) - \arcsin \psi(P_0(j\omega), P_1(j\omega)), \quad \forall \omega.$$

Now suppose we need $\inf_{P \in \mathcal{P}} b_{P,K} > \alpha$. Then it is sufficient to have
$$\arcsin b_{P_0,K}(\omega) - \arcsin \psi(P_0(j\omega), P_1(j\omega)) > \arcsin \alpha, \quad \forall \omega, \ P_1 \subset \mathcal{P};$$
that is,
$$b_{P_0,K}(\omega) > \sin \left(\arcsin \psi(P_0(j\omega), P_1(j\omega)) + \arcsin \alpha \right), \quad \forall \omega, \ P_1 \in \mathcal{P}.$$

Let $W(s) \in \mathcal{H}_\infty$ be such that
$$|W(j\omega)| \geq \sin \left(\arcsin \psi(P_0(j\omega), P_1(j\omega)) + \arcsin \alpha \right), \quad \forall \omega, \ P_1 \in \mathcal{P}.$$

Then it is sufficient to guarantee

$$\frac{|W(j\omega)|}{b_{P_0,K}(\omega)} < 1.$$

Let $P_0 = \tilde{M}_0^{-1}\tilde{N}_0$ be a normalized left coprime factorization and note that

$$\frac{1}{b_{P_0,K}(\omega)} := \bar{\sigma}\left(\begin{bmatrix} I \\ K(j\omega) \end{bmatrix}(I + P_0(j\omega)K(j\omega))^{-1}\tilde{M}_0^{-1}(j\omega)\right).$$

Then it is sufficient to find a controller so that

$$\left\|\begin{bmatrix} I \\ K \end{bmatrix}(I + P_0 K)^{-1}\tilde{M}_0^{-1}W\right\|_\infty < 1.$$

The process can be iterated to find the largest possible α.

Combining the preceding robust stabilization idea and the $\mathcal{H}_\infty$ loop-shaping in Chapter 16, one can devise an extended loop-shaping design procedure as follows. (A more advanced loop-shaping procedure can be found in Vinnicombe [1993b].)

Design Procedure:

Let $\mathcal{P}$ be a family of parametric uncertain systems and let P_0 be a nominal model.

(a) Loop-Shaping: The singular values of the nominal plant are shaped, using a precompensator W_1 and/or a postcompensator W_2, to give a desired open-loop shape. The nominal plant P_0 and the shaping functions W_1, W_2 are combined to form the shaped plant, P_s, where $P_s = W_2 P_0 W_1$. We assume that W_1 and W_2 are such that P_s contains no hidden modes.

(b) Compute *frequency-by-frequency*:

$$f(\omega) = \sup_{P \in \mathcal{P}} \psi(P_s(j\omega), W_2(j\omega)P(j\omega)W_1(j\omega)).$$

Set $\alpha = 0$.

(b) Fit a stable and minimum phase rational transfer function $W(s)$ so that

$$|W(j\omega)| \geq \sin(\arcsin f(\omega) + \arcsin \alpha) \quad \forall \omega.$$

(c) Find a K_∞ such that

$$\beta := \inf_{K_\infty}\left\|\begin{bmatrix} I \\ K_\infty \end{bmatrix}(I + P_0 K_\infty)^{-1}\tilde{M}_0^{-1}W\right\|_\infty.$$

(d) If $\beta \approx 1$, stop and the final controller is $K = W_1 K_\infty W_2$. If $\beta \ll 1$, increase α and go back to (b). If $\beta \gg 1$, decrease α and go back to (b).

17.5 Controller Order Reduction

The controller order-reduction procedure described in Chapter 15 can, of course, be applied to the loop-shaping controller design in Chapter 16 and the gap or ν-gap metric optimization here. However, the controller order-reduction in the loop-shaping controller design, or gap metric, or ν-gap metric optimization is especially simple. The following theorem follows immediately from Theorems 17.7 and 17.10.

Theorem 17.11 *Let P_0 be a nominal plant and K_0 be a stabilizing controller such that $b_{P_0,K_0} \leq b_{\text{opt}}(P_0)$. Let $K_0 = UV^{-1}$ be a normalized coprime factorization and let $\hat{U}, \hat{V} \in \mathcal{RH}_\infty$ be such that*

$$\left\| \begin{bmatrix} U \\ V \end{bmatrix} - \begin{bmatrix} \hat{U} \\ \hat{V} \end{bmatrix} \right\|_\infty \leq \varepsilon.$$

Then $K := \hat{U}\hat{V}^{-1}$ stabilizes P_0 if $\varepsilon < b_{P_0,K_0}$. Furthermore,

$$\arcsin b_{P,K} \geq \arcsin b_{P_0,K_0} - \arcsin \varepsilon - \arcsin \beta$$

for all $\{P : \delta_\nu(P, P_0) \leq \beta\}$.

Hence to reduce the controller order one only needs to approximate the normalized coprime factors of the controller. An algorithm for finding the best approximation is also presented in Vinnicombe [1993a, 1993b].

17.6 Notes and References

This chapter is based on Georgiou and Smith [1990] and Vinnicombe [1993a, 1993b]. Early studies of the gap metric can be found in Zames and El-Sakkary [1980] and El-Sakkary [1985]. The pointwise gap metric was introduced by Qiu and Davision [1992b].

17.7 Problems

Problem 17.1 *Calculate the gap $\delta_g(P_i, P_j)$ with*

$$P_1 = \frac{1}{s+1}, \quad P_2 = \frac{1}{s-1}, \quad P_3 = \frac{s+2}{(s+1)^2}, \quad P_4 = \frac{s-2}{(s+1)^2}, \quad P_5 = \frac{1}{(s+1)^2}.$$

Problem 17.2 *Let $P = \frac{10}{\tau s+1}, \tau \in [1,3]$ and let $P_0 = \frac{10}{\tau_0 s+1}$. Find the optimal $\tau_0 \in [1,3]$ minimizing*

$$\min_{\tau_o} \max_{\tau} \delta_g(P, P_0).$$

Problem 17.3 *Repeat Problems 17.1–17.2 for the ν-gap metric.*

Chapter 18

Miscellaneous Topics

This chapter considers two somewhat different problems. The first section gives a brief introduction into the problem of model validation and the second section considers the mixed real and complex μ analysis and synthesis.

18.1 Model Validation

A key to the success of the robust control theory developed in this book is to have appropriate descriptions of the uncertain system (whether it is an additive uncertainty model or a general linear fractional model). Then an important question is how one can decide if a model description is appropriate (i.e., how to validate a model).

For simplicity of presentation, we have chosen to present the discrete time model validation in this section with the understanding that a continuous time problem can be approximated by fast sampling. Suppose we have modeled a set of uncertain dynamical systems by

$$\mathbf{\Delta} := \{\Delta : \Delta \in \mathcal{H}_\infty, \ \|\Delta\|_\infty \leq 1\}$$

where the $\mathcal{H}_\infty$ norm of a discrete time system $\Delta \in \mathcal{H}_\infty$ is defined as $\|\Delta(z)\|_\infty = \sup_{|z|>1} \overline{\sigma}(\Delta(z))$. In order to verify whether this model assumption is correct, some experimental data are collected. For example, let the input to the system be the sequence $u = (u_0, u_1, \ldots, u_{l-1})$ and the output $y = (y_0, y_1, \ldots, y_{l-1})$. A natural question is whether these data are consistent with our modeling assumption. In other words, does there exist a model $\Delta \in \mathbf{\Delta}$ such that the output of the Δ for the period of $t = 0, 1, \ldots, l-1$ is exactly $y = (y_0, y_1, \ldots, y_{l-1})$ with the input $u = (u_0, u_1, \ldots, u_{l-1})$? If there does not exist a such Δ, then the model is invalidated. If there exists a such Δ, however, it does not mean that the model is validated but it only means that the model is not validated by this set of data and it may be invalidated by another set of data in the future. Hence it is actually more accurate to say our validation procedure is model invalidation.

Let Δ be a stable, causal, linear, time-invariant system with transfer matrix

$$\Delta(z) = h_0 + h_1 z^{-1} + h_2 z^{-2} + \cdots$$

where $h_i, i = 0, 1, \ldots$ are the matrix Markov parameters. Suppose we have applied the input sequence $u = (u_0, u_1, \ldots, u_{l-1})$ to the system and collected the output for the period $t = 0, 1, \ldots, \ell - 1$, $y = (y_0, y_1, \ldots, y_{l-1})$. Then the input and output sequences are related by a Toeplitz matrix:

$$\begin{bmatrix} y_0 \\ y_1 \\ \vdots \\ y_{l-1} \end{bmatrix} = \begin{bmatrix} h_0 & 0 & \cdots & 0 \\ h_1 & h_0 & \ddots & 0 \\ \vdots & \vdots & \ddots & 0 \\ h_{l-1} & h_{l-2} & \cdots & h_0 \end{bmatrix} \begin{bmatrix} u_0 \\ u_1 \\ \vdots \\ u_{l-1} \end{bmatrix}.$$

This equation shows that, for $u_0 \neq 0$ and SISO Δ, the inputs and outputs uniquely determine the first ℓ Markov parameters of the transfer function $\Delta(z)$. The model is validated (or more accurately not invalidated) if the remaining Markov parameters can be chosen so that $\Delta(z) \in \boldsymbol{\Delta}$. The existence of such a choice is the classical tangential Carathéodory-Fejér interpolation problem, for which a solution to the MIMO case can be found in Foias and Frazho [1990, page 195]. We shall state this result in the following theorem. But we shall define some notation first.

Let $(v_0, v_1, \ldots, v_{\ell-1}, v_\ell, v_{\ell+1}, \ldots)$ be a sequence and let π_ℓ denote the truncation operator such that

$$\pi_\ell(v_0, v_1, \ldots, v_{\ell-1}, v_\ell, v_{\ell+1}, \ldots) = (v_0, v_1, \ldots, v_{\ell-1}).$$

Let $v = (v_0, v_1, \ldots, v_{\ell-1})$ be a sequence of vectors and denote

$$T_v := \begin{bmatrix} v_0 & 0 & \cdots & 0 \\ v_1 & v_0 & \ddots & 0 \\ \vdots & \vdots & \ddots & 0 \\ v_{l-1} & v_{l-2} & \cdots & v_0 \end{bmatrix}.$$

Theorem 18.1 *Given* $u = (u_0, u_1, \ldots, u_{l-1})$ *and* $y = (y_0, y_1, \ldots, y_{l-1})$, *there exists a* $\Delta \in \mathcal{H}_\infty$, $\|\Delta\|_\infty \leq 1$ *such that*

$$y = \pi_\ell \Delta u$$

if and only if $T_y^* T_y \leq T_u^* T_u$.

Note that the output of Δ after time $t = \ell - 1$ is irrelevant to the test. The condition $T_y^* T_y \leq T_u^* T_u$ is equivalent to

$$\sum_{j=1}^i \|y_j\|^2 \leq \sum_{j=1}^i \|u_j\|^2, i = 0, 1, \ldots, \ell - 1$$

18.1. Model Validation

or
$$\|\pi_i y\|_2 \leq \|\pi_i u\|_2, \, i = 0, 1, \ldots, \ell - 1,$$

which is obviously necessary. In fact, the last condition holds for stable, linear, time-varying operator Δ with $\sup_{u \neq 0} \frac{\|\Delta u\|_2}{\|u\|_2} \leq 1$; see Poolla et al. [1994]. Note that if $u_0 \neq 0$, then T_u is of full column rank and the condition can also be written as $\bar{\sigma}\left(T_y(T_u^* T_u)^{-\frac{1}{2}}\right) \leq 1$.

Using the above theorem, we can derive solutions to some model validation problems easily. For example, consider a set of additive models shown in Figure 18.1.

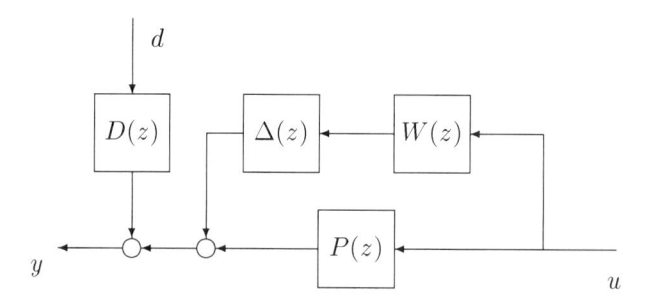

Figure 18.1: Model validation for additive uncertainty

In this case,
$$y = (P + \Delta W)u + Dd, \quad \|\Delta\|_\infty \leq 1$$

where $P(z), W(z), D(z)$ and $\Delta(z)$ are causal, linear, time-invariant systems (but not necessarily stable). The disturbance d is assumed to come from a convex set, $d \in \mathcal{D}_{\text{convex}}$; for example, $\mathcal{D}_{\text{convex}} = \{d : d \in \ell_2[0, \infty), \|d\|_2 \leq 1\}$. For simplicity, we shall also assume that $W(\infty)$ is of full column rank. Let
$$D(z) = D_0 + D_1 z^{-1} + D_2 z^{-2} + \cdots.$$

Theorem 18.2 *Given a set of input-output data $u_{\text{expt}} = (u_0, u_1, \ldots, u_{\ell-1})$ with $u_0 \neq 0$, $y_{\text{expt}} = (y_0, y_1, \ldots, y_{\ell-1})$ for the additive perturbed uncertainty system with an additive disturbance $d \in \mathcal{D}_{\text{convex}}$, where $\mathcal{D}_{\text{convex}}$ is a convex set, let*
$$\hat{u} = (\hat{u}_0, \hat{u}_1, \ldots, \hat{u}_{\ell-1}) = \pi_\ell(W u_{\text{expt}})$$
$$\hat{y} = (\hat{y}_0, \hat{y}_1, \ldots, \hat{y}_{\ell-1}) = y_{\text{expt}} - \pi_\ell P u_{\text{expt}}.$$

Then there exists a $\Delta \in \mathcal{H}_\infty$, $\|\Delta\|_\infty \leq 1$ such that
$$y_{\text{expt}} = \pi_\ell\left((P + \Delta W)u_{\text{expt}} + Dd\right)$$

for some $d \in \mathcal{D}_{\text{convex}}$ if and only if there exists a $d = (d_0, d_1, \ldots, d_{l-1}) \in \pi_\ell \mathcal{D}_{\text{convex}}$ such that
$$\bar{\sigma}\left[(T_{\hat{y}} - T_D T_d)(T_{\hat{u}}^* T_{\hat{u}})^{-1/2}\right] \leq 1$$

where
$$T_D := \begin{bmatrix} D_0 & 0 & \cdots & 0 \\ D_1 & D_0 & \ddots & 0 \\ \vdots & \vdots & \ddots & 0 \\ D_{l-1} & D_{l-2} & \cdots & D_0 \end{bmatrix}.$$

Proof. Note that the system input-output equation can be written as
$$(y - Pu) - Dd = \Delta(Wu).$$

Since P, W, D, and Δ are causal, linear, and time invariant, we have $\pi_\ell Dd = \pi_\ell D \pi_\ell d$, $\pi_\ell(y - Pu) = y_{\text{expt}} - \pi_\ell P \pi_\ell u = y_{\text{expt}} - \pi_\ell P u_{\text{expt}}$ and $\pi_\ell W u = \pi_\ell W \pi_\ell u = \pi_\ell W u_{\text{expt}}$. Denote
$$\hat{d} = (\hat{d}_0, \hat{d}_1, \ldots, \hat{d}_{\ell-1}) = \pi_\ell(Dd).$$

Then it is easy to show that
$$\begin{bmatrix} \hat{d}_0 \\ \hat{d}_1 \\ \vdots \\ \hat{d}_{\ell-1} \end{bmatrix} = T_D \begin{bmatrix} d_0 \\ d_1 \\ \vdots \\ d_{\ell-1} \end{bmatrix}$$

and $T_{\hat{d}} = T_D T_d$. Now note that
$$T_{\pi_\ell(y-Pu-Dd)} = T_{\pi_\ell(y-Pu)} - T_{\pi_\ell(Dd)} = T_{\hat{y}} - T_D T_d, \quad T_{\pi_\ell W u} = T_{\hat{u}}$$

and $\pi_\ell \Delta W u = \pi_\ell \Delta \pi_\ell (Wu)$ since Δ is causal. Applying Theorem 18.1, there exists a $\Delta \in \mathcal{H}_\infty$, $\|\Delta\|_\infty \leq 1$ such that
$$\pi_\ell[(y - Pu) - Dd] = \pi_\ell \Delta(Wu) = \pi_\ell \Delta \pi_\ell(Wu)$$

if and only if
$$(T_{\hat{y}} - T_D T_d)^*(T_{\hat{y}} - T_D T_d) \leq T_{\hat{u}}^* T_{\hat{u}}$$

which is equivalent to
$$\bar{\sigma}\left[(T_{\hat{y}} - T_D T_d)(T_{\hat{u}}^* T_{\hat{u}})^{-\frac{1}{2}}\right] \leq 1.$$

Note that $T_{\hat{u}}$ is of full column rank since $W(\infty)$ is of full column rank and $u_0 \neq 0$, which implies $\hat{u}_0 \neq 0$. $\square$

Note that
$$\inf_{d \in \mathcal{D}_{\text{convex}}} \bar{\sigma}\left[(T_{\hat{y}} - T_D T_d)(T_{\hat{u}}^* T_{\hat{u}})^{-\frac{1}{2}}\right] \leq 1$$

is a convex problem and can be checked numerically.

Many other classes of model validation problems can be solved analogously. For example, consider a coprime factor model validation problem with

$$y = (M + \Delta_M W_M)^{-1}(N + \Delta_N W_N)u + d$$

where M, N, W_M, and W_N are causal, linear, time-invariant systems, and $\Delta_M, \Delta_N \in \mathcal{H}_\infty$, $\left\| \begin{bmatrix} \Delta_M & \Delta_N \end{bmatrix} \right\|_\infty \leq 1$. Then the problem can be solved by multiplying $M + \Delta_M W_M$ from the left of the system equation and rewriting the system equation as

$$(My - Nu - Md) = \begin{bmatrix} \Delta_M & \Delta_N \end{bmatrix} \begin{bmatrix} W_M(d-y) \\ W_N u \end{bmatrix}.$$

The model validation of a general LFT uncertainty system is considered in Davis [1995] and Chen and Wang [1996]. For continuous time model validation, see Rangan and Poolla [1996] and Smith and Dullerud [1996].

18.2 Mixed μ Analysis and Synthesis

In Chapter 10, we considered analysis and synthesis of systems with *complex* uncertainties. However, in practice, many systems involve parametric uncertainties that are real (for example, the uncertainty about a spring constant in a mechanical system). In this case, one has to cover this real parameter variation with a complex disk in order to use the complex μ analysis and synthesis tools, which usually results in a conservative solution. In this section, we shall consider briefly the analysis and synthesis problems with possibly both real parametric and complex uncertainties.

The mixed real and complex μ involves three types of blocks: *repeated real scalar*, *repeated complex scalar*, and *full* blocks. Three nonnegative integers, S_r, S_c, and F, represent the number of *repeated real scalar* blocks, the number of *repeated complex scalar* blocks, and the number of *full* blocks, and they satisfy

$$\sum_{i=1}^{S_r} k_i + \sum_{i=1}^{S_c} r_i + \sum_{j=1}^{F} m_j = n.$$

The ith repeated real scalar block is $k_i \times k_i$, the jth repeated complex scalar block is $r_j \times r_j$, and the ℓth full block is $m_\ell \times m_\ell$. The admissible set of uncertainties $\mathbf{\Delta} \subset \mathbb{C}^{n \times n}$ is defined as

$$\begin{aligned}\mathbf{\Delta} = \{&\mathrm{diag}\ [\phi_1 I_{k_1}, \ldots, \phi_{s_r} I_{k_{s_r}}, \delta_1 I_{r_1}, \ldots, \delta_{s_c} I_{r_{s_c}}, \\ &\Delta_1, \ldots, \Delta_F] : \phi_i \in \mathbb{R},\ \delta_j \in \mathbb{C},\ \Delta_\ell \in \mathbb{C}^{m_\ell \times m_\ell}\}. \end{aligned} \quad (18.1)$$

The mixed μ is defined in the same way as for the complex μ: Let $M \in \mathbb{C}^{n \times n}$; then

$$\mu_{\mathbf{\Delta}}(M) := (\min\{\overline{\sigma}(\Delta) : \Delta \in \mathbf{\Delta}, \det(I - M\Delta) = 0\})^{-1} \quad (18.2)$$

unless no $\Delta \in \mathbf{\Delta}$ makes $I - M\Delta$ singular, in which case $\mu_{\mathbf{\Delta}}(M) := 0$. Or, equivalently,

$$\frac{1}{\mu_{\mathbf{\Delta}}(M)} := \inf\{\alpha : \det(I - \alpha M\Delta) = 0, \ \overline{\sigma}(\Delta) \leq 1, \ \Delta \in \mathbf{\Delta}\}.$$

Let $\rho_R(M)$ be the real spectral radius (i.e., the largest magnitude of the real eigenvalues of M). For example, if a 4×4 matrix M has eigenvalues $1 \pm j3, -2, 1$, then $\rho(M) = |1 + j3| = \sqrt{10}$ and $\rho_R(M) = |-2| = 2$. It is easy to see that

$$\mu_{\mathbf{\Delta}}(M) = \max_{\Delta \in \mathbf{B\Delta}} \rho_R(M\Delta)$$

where $\mathbf{B\Delta} := \{\Delta : \Delta \in \mathbf{\Delta}, \ \overline{\sigma}(\Delta) \leq 1\}$. Note that $\max_{\Delta \in \mathbf{B\Delta}} \rho_R(M\Delta) = \max_{\Delta \in \mathbf{B\Delta}} \rho(M\Delta)$ if $s_r = 0$. [This should not be confused with the fact that, for a given matrix $\Delta \in \mathbf{B\Delta}$ and M, $\rho_R(M\Delta)$ may not be equal to $\rho(M\Delta)$. For example, $M = 2e^{j\frac{\pi}{4}}$ and $\Delta = 1$; then $\rho(M\Delta) = 2$ but $\rho_R(M\Delta) = 0$ since M has no real eigenvalues. However, one can choose another $\Delta_1 = e^{-j\frac{\pi}{4}}$ such that $\rho_R(M\Delta_1) = 2 = \rho(M\Delta)$.]

Define

$$\mathcal{Q} = \{\Delta \in \mathbf{\Delta} : \ \phi_i \in [-1, 1], \ |\delta_i| = 1, \ \Delta_i \Delta_i^* = I_{m_i}\}$$

$$\mathcal{D} = \left\{ \begin{array}{l} \text{diag}\left[\tilde{D}_1, \ldots, \tilde{D}_{s_r}, D_1, \ldots, D_{s_c}, d_1 I_{m_1}, \ldots, d_{F-1} I_{m_{F-1}}, I_{m_F}\right] : \\ \tilde{D}_i \in \mathbb{C}^{k_i \times k_i}, \ \tilde{D}_i = \tilde{D}_i^* > 0, \ D_i \in \mathbb{C}^{r_i \times r_i}, \ D_i = D_i^* > 0, \ d_j \in \mathbb{R}, d_j > 0 \end{array} \right\}.$$

$$\mathcal{G} = \left\{\text{diag}[G_1, \ldots, G_{s_r}, 0, \ldots, 0] : \ G_i = G_i^* \in \mathbb{C}^{k_i \times k_i}\right\}.$$

It was shown in Young [1993] that

$$\mu_{\mathbf{\Delta}}(M) = \max_{Q \in \mathcal{Q}} \rho_R(QM).$$

Note that the above maximization is not necessarily achieved on the vertices for the real parameters; hence one must search over the entire interval for each real parameter. Again this maximization problem can have many local maximums and a power algorithm has been developed in Young [1993] to compute a lower bound.

It should also be noted that even though the complex μ (i.e., $s_r = 0$) is a continuous function of the data, the mixed μ (i.e., $s_r \neq 0$) may only be upper semicontinuous; see Packard and Pandey [1993]. It was also shown in Braatz et al. [1994] and Toker and Özbay [1995] that the computation of μ is a NP hard problem, which means that it may not be computable in a polynomial time. Of course, it should not be interpreted as every μ problem will not be solvable in a polynomial time; merely some might not.

Obviously, the upper bound for the complex μ can be applied for the mixed μ when the intervals of the real parameters are covered by complex disks. However, a better bound can be obtained for the mixed μ by exploiting the phase information of the real parameters. To motivate the improved bound for the mixed μ, we consider again the upper bound for the complex μ problem. It is known that

$$\mu_{\mathbf{\Delta}}(M) \leq \inf_{D \in \mathcal{D}} \overline{\sigma}(DMD^{-1}).$$

18.2. Mixed μ Analysis and Synthesis

This bound can be reformulated using linear matrix inequalities by noting the following:

$$\bar{\sigma}(DMD^{-1}) \leq \beta \iff (DMD^{-1})^*DMD^{-1} \leq \beta^2 I \iff M^*D^*DM - \beta^2 D^*D \leq 0.$$

Since D is nonsingular and $D^*D \in \mathcal{D}$, we have

$$\mu_\Delta(M) \leq \inf_{D \in \mathcal{D}} \min_\beta \left\{ \beta : M^*DM - \beta^2 D \leq 0 \right\}.$$

The following upper bound for the mixed μ was derived by Fan, Tits, and Doyle [1991] and reformulated in the current form by Young [1993].

Theorem 18.3 *Let $M \in \mathbb{C}^{n \times n}$ and $\Delta \in \mathbf{\Delta}$. Then*

$$\mu_\Delta(M) \leq \inf_{D \in \mathcal{D}, G \in \mathcal{G}} \min_\beta \left\{ \beta : M^*DM + j(GM - M^*G) - \beta^2 D \leq 0 \right\}.$$

Proof. Suppose we have a $Q \in \mathcal{Q}$ such that QM has a real eigenvalue $\lambda \in \mathbb{R}$. Then there is a vector $x \in \mathbb{C}^n$ such that

$$QMx = \lambda x.$$

Let $D \in \mathcal{D}$. Then $D^{\frac{1}{2}} \in \mathcal{D}$, $D^{\frac{1}{2}}Q = QD^{\frac{1}{2}}$ and

$$D^{\frac{1}{2}}QMx = QD^{\frac{1}{2}}Mx = \lambda D^{\frac{1}{2}}x.$$

Since $\bar{\sigma}(Q) \leq 1$, it follows that

$$\lambda^2 \left\| D^{\frac{1}{2}}x \right\|^2 = \left\| QD^{\frac{1}{2}}Mx \right\|^2 \leq \left\| D^{\frac{1}{2}}Mx \right\|^2.$$

Hence

$$x^*(M^*DM - \lambda^2 D)x \geq 0.$$

Next, let $G \in \mathcal{G}$ and note that $Q^*G = QG = GQ$; then

$$x^*GMx = \left(\frac{1}{\lambda}QMx\right)^* GMx = \frac{1}{\lambda}x^*M^*Q^*GMx = \frac{1}{\lambda}x^*M^*QGMx$$

$$= \frac{1}{\lambda}x^*M^*GQMx = \frac{1}{\lambda}x^*M^*G(QMx) = x^*M^*Gx.$$

That is,

$$x^*(GM - M^*G)x = 0.$$

Note that $j(GM - M^*G)$ is a Hermitian matrix, so it follows that for such x

$$x^*(M^*DM + j(GM - M^*G) - \lambda^2 D)x \geq 0.$$

It is now easy to see that if we have $D \in \mathcal{D}$, $G \in \mathcal{G}$ and $0 \leq \beta \in \mathbb{R}$ such that

$$M^*DM + j(GM - M^*G) - \beta^2 D \leq 0$$

MISCELLANEOUS TOPICS

then $|\lambda| \leq \beta$, and hence $\mu_{\boldsymbol{\Delta}}(M) \leq \beta$. □

This upper bound has an interesting interpretation: covering the uncertainties on the real axis using possibly off-axis disks. To illustrate, let $M \in \mathbb{C}$ be a scalar and $\Delta \in [-1, 1]$. We can cover this real interval using a disk as shown in Figure 18.2.

The off-axis disk can be expressed as

$$j\frac{G}{\beta} + \sqrt{1 + \left(\frac{G}{\beta}\right)^2}\, \tilde{\Delta}, \quad \tilde{\Delta} \in \mathbb{C}, \; |\tilde{\Delta}| \leq 1.$$

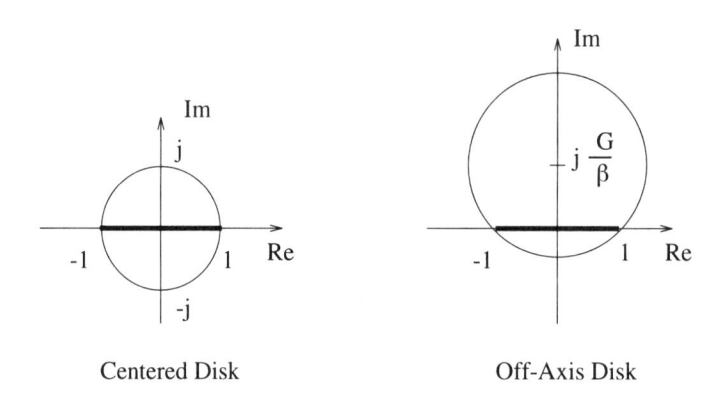

Figure 18.2: Covering real parameters with disks

Hence $1 - \Delta\frac{M}{\beta} \neq 0$ for all $\Delta \in [-1, 1]$ is guaranteed if

$$1 - \left(j\frac{G}{\beta} + \sqrt{1 + \left(\frac{G}{\beta}\right)^2}\, \tilde{\Delta}\right)\frac{M}{\beta} \neq 0, \quad \tilde{\Delta} \in \mathbb{C}, \; |\tilde{\Delta}| \leq 1$$

$$\iff 1 - \frac{\sqrt{1 + \left(\frac{G}{\beta}\right)^2}\, \frac{M}{\beta}}{1 - j\frac{G}{\beta}\frac{M}{\beta}} \tilde{\Delta} \neq 0, \quad \tilde{\Delta} \in \mathbb{C}, \; |\tilde{\Delta}| \leq 1$$

$$\iff \left(\frac{\sqrt{1 + \left(\frac{G}{\beta}\right)^2}\, \frac{M}{\beta}}{1 - j\frac{G}{\beta}\frac{M}{\beta}}\right)^* \left(\frac{\sqrt{1 + \left(\frac{G}{\beta}\right)^2}\, \frac{M}{\beta}}{1 - j\frac{G}{\beta}\frac{M}{\beta}}\right) \leq 1$$

$$\iff \frac{M^*}{\beta}\frac{M}{\beta} + j\left(\frac{G}{\beta}\frac{M}{\beta} - \frac{M^*}{\beta}\frac{G}{\beta}\right) - 1 \leq 0$$

$$\iff M^*M + j(GM - M^*G) - \beta^2 \leq 0.$$

18.2. Mixed μ Analysis and Synthesis

The scaling G allows one to exploit the phase information about the real parameters so that a better upper bound can be obtained. We shall demonstrate this further using a simple example.

Example 18.1 Let
$$G(s) = \frac{s^2 + 2s + 1}{s^3 + s^2 + 2s + 1}.$$

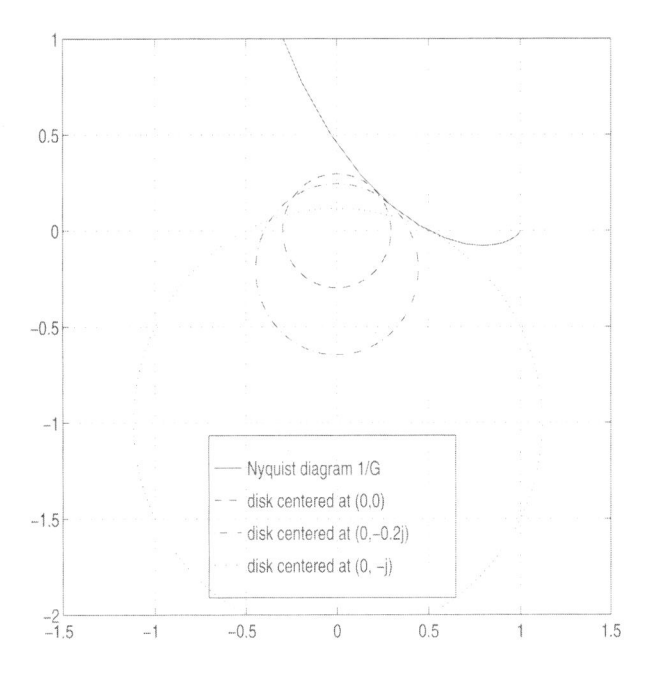

Figure 18.3: Computing the real stability margin by covering with disks

We are interested in finding the largest k such that $1 + \Delta G(s)$ has no zero in the right-half plane for all $\Delta \in [-k, k]$. Of course, the largest k can be found very easily by using well-known stability test, which gives

$$k_{\max} = \left(\sup_\omega \mu_\Delta(G(j\omega))\right)^{-1} = \left(\sup_\omega \max_{\phi \in [-1,1]} \rho_R(\phi G(j\omega))\right)^{-1}$$

$$= \left(\sup_\omega \{|G(j\omega)| : \Im G(j\omega) = 0\}\right)^{-1} = \inf_\omega \left\{\frac{1}{|G(j\omega)|} : \Im G(j\omega) = 0\right\} = 0.5.$$

Now we use the complex covering idea to find the best possible k. Note that we only need to find the smallest $|\Delta|$ so that $1 + \Delta G(j\omega_0) = 0$ for some ω_0 or, equivalently, $\Delta + 1/G(j\omega_0) = 0$. The frequency response of $1/G$ and the disks covering an interval $[-k, k]$ are shown in Figure 18.3. It is clear that a centered disk would give $k = 1/\|G\|_\infty = 0.2970$ and an off-axis disk centered at $(0, -0.2j)$ would give $k = 0.3984$ while an off-axis disk centered at $(0, -j)$ would give the exactly value $k = 0.5$.

The following alternative characterization of the upper bound is useful in the mixed μ synthesis.

Theorem 18.4 *Given $\beta > 0$, there exist $D \in \mathcal{D}$ and $G \in \mathcal{G}$ such that*

$$M^*DM + j(GM - M^*G) - \beta^2 D \leq 0$$

if and only if there are $D_1 \in \mathcal{D}$ and $G_1 \in \mathcal{G}$ such that

$$\bar{\sigma}\left(\left(\frac{D_1 M D_1^{-1}}{\beta} - jG_1\right)(I + G_1^2)^{-\frac{1}{2}}\right) \leq 1.$$

Proof. Let $D = D_1^2$ and $G = \beta D_1 G_1 D_1$. Then

$$M^*DM + j(GM - M^*G) - \beta^2 D \leq 0$$

$$\iff M^*D_1^2 M + j(\beta D_1 G_1 D_1 M - \beta M^* D_1 G_1 D_1) - \beta^2 D_1^2 \leq 0$$

$$\iff (D_1 M D_1^{-1})^*(D_1 M D_1^{-1}) + j(\beta G_1 D_1 M D_1^{-1} - \beta(D_1 M D_1^{-1})^* G_1) - \beta^2 I \leq 0$$

$$\iff \left(\frac{D_1 M D_1^{-1}}{\beta} - jG_1\right)^*\left(\frac{D_1 M D_1^{-1}}{\beta} - jG_1\right) - (I + G_1^2) \leq 0$$

$$\iff \bar{\sigma}\left[\left(\frac{D_1 M D_1^{-1}}{\beta} - jG_1\right)(I + G_1^2)^{-\frac{1}{2}}\right] \leq 1.$$

$\square$

Similarly, the following corollary can be shown.

Corollary 18.5 $\mu_\Delta(M) \leq r\beta$ *if there are $D_1 \in \mathcal{D}$ and $G_1 \in \mathcal{G}$ such that*

$$\bar{\sigma}\left(\left(\frac{D_1 M D_1^{-1}}{\beta} - jG_1\right)(I + G_1^2)^{-\frac{1}{2}}\right) \leq r \leq 1.$$

18.2. Mixed μ Analysis and Synthesis

Proof. This follows by noting that

$$\bar{\sigma}\left(\left(\frac{D_1 M D_1^{-1}}{\beta} - jG_1\right)(I+G_1^2)^{-\frac{1}{2}}\right) \leq r \leq 1$$

$$\implies \left(\frac{D_1 M D_1^{-1}}{r\beta} - j\frac{G_1}{r}\right)^* \left(\frac{D_1 M D_1^{-1}}{r\beta} - j\frac{G_1}{r}\right) \leq I + G_1^2 \leq I + \left(\frac{G_1}{r}\right)^2.$$

Let $G_2 = \dfrac{G_1}{r} \in \mathcal{G}$. Then

$$\left(\frac{D_1 M D_1^{-1}}{r\beta} - jG_2\right)^* \left(\frac{D_1 M D_1^{-1}}{r\beta} - jG_2\right) \leq I + G_2^2$$

$$\implies \bar{\sigma}\left(\left(\frac{D_1 M D_1^{-1}}{r\beta} - jG_2\right)(I+G_2^2)^{-\frac{1}{2}}\right) \leq 1$$

$$\implies \mu_\Delta(M) \leq r\beta.$$

$\square$

Note that this corollary is not necessarily true if $r > 1$. It is fairly easy to check that the well-posedness condition, main loop theorem, robust stability, and robust performance theorems for the mixed μ setup are exactly the same as the ones for complex μ problems.

We are now in the position to consider the synthesis problem with mixed uncertainties. Consider again the general system diagram in Figure 18.4. By the robust performance condition, we need to find a stabilizing controller K so that

$$\min_K \sup_\omega \mu_\Delta\left(\mathcal{F}_\ell(P,K)\right) \leq \beta.$$

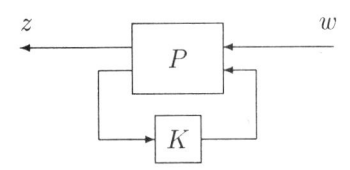

Figure 18.4: Synthesis framework

By Theorems 18.3 and 18.4, $\mu_\Delta\left(\mathcal{F}_\ell(P(j\omega), K(j\omega))\right) \leq \beta$, $\forall \omega$ if there are frequency-dependent scaling matrices $D_\omega \in \mathcal{D}$ and $G_\omega \in \mathcal{G}$ such that

$$\sup_\omega \bar{\sigma}\left[\left(\frac{D_\omega\left(\mathcal{F}_\ell(P(j\omega), K(j\omega))\right)D_\omega^{-1}}{\beta} - jG_\omega\right)(I+G_\omega^2)^{-\frac{1}{2}}\right] \leq 1, \quad \forall\omega.$$

Similar to the complex μ synthesis, we can now describe a mixed μ synthesis procedure that involves $D, G - K$ iterations.

$D, G - K$ Iteration:

(1) Let K be a stabilizing controller. Find initial estimates of the scaling matrices $D_\omega \in \mathcal{D}$, $G_\omega \in \mathcal{G}$ and a scalar $\beta_1 > 0$ such that

$$\sup_\omega \bar{\sigma}\left[\left(\frac{D_\omega\left(\mathcal{F}_\ell(P(j\omega), K(j\omega))\right)D_\omega^{-1}}{\beta_1} - jG_\omega\right)(I + G_\omega^2)^{-\frac{1}{2}}\right] \leq 1, \quad \forall \omega.$$

Obviously, one may start with $D_\omega = I$, $G_\omega = 0$, and a large $\beta_1 > 0$.

(2) Fit the frequency response matrices D_ω and jG_ω with $D(s)$ and $G(s)$ so that

$$D(j\omega) \approx D_\omega, \quad G(j\omega) \approx jG_\omega, \quad \forall \omega.$$

Then for $s = j\omega$

$$\sup_\omega \bar{\sigma}\left(\left(\frac{D_\omega\left(\mathcal{F}_\ell(P(j\omega), K(j\omega))\right)D_\omega^{-1}}{\beta_1} - jG_\omega\right)(I + G_\omega^2)^{-\frac{1}{2}}\right)$$

$$\approx \sup_\omega \bar{\sigma}\left[\left(\frac{D(s)\left(\mathcal{F}_\ell(P(s), K(s))\right)D^{-1}(s)}{\beta_1} - G(s)\right)(I + G^\sim(s)G(s))^{-\frac{1}{2}}\right].$$

(3) Let $D(s)$ be factorized as

$$D(s) = D_{ap}(s)D_{\min}(s), \quad D_{ap}^\sim(s)D_{ap}(s) = I, \quad D_{\min}(s), D_{\min}^{-1}(s) \in \mathcal{H}_\infty.$$

That is, D_{ap} is an all-pass and $D_{\min}$ is a stable and minimum phase transfer matrix. Find a normalized right coprime factorization

$$D_{ap}^\sim(s)G(s)D_{ap}(s) = G_N G_M^{-1}, \quad G_N, \; G_M \in \mathcal{H}_\infty$$

such that

$$G_M^\sim G_M + G_N^\sim G_N = I.$$

Then

$$G_M^{-1} D_{ap}^\sim (I + G^\sim G)^{-1} D_{ap} (G_M^{-1})^\sim = I$$

and, for each frequency $s = j\omega$, we have

$$\bar{\sigma}\left[\left(\frac{D(s)\left(\mathcal{F}_\ell(P(s), K(s))\right)D^{-1}(s)}{\beta_1} - G(s)\right)(I + G^\sim(s)G(s))^{-\frac{1}{2}}\right]$$

$$= \bar{\sigma}\left[\left(\frac{D_{\min}\left(\mathcal{F}_\ell(P, K)\right)D_{\min}^{-1}}{\beta_1} - D_{ap}^\sim G D_{ap}\right)D_{ap}^\sim(I + G^\sim G)^{-\frac{1}{2}}\right]$$

$$= \bar{\sigma}\left[\left(\frac{D_{\min}\left(\mathcal{F}_\ell(P,K)\right)D_{\min}^{-1}}{\beta_1} - G_N G_M^{-1}\right) D_{ap}^\sim (I + G^\sim G)^{-\frac{1}{2}}\right]$$

$$= \bar{\sigma}\left[\left(\frac{D_{\min}\left(\mathcal{F}_\ell(P,K)\right)D_{\min}^{-1}G_M}{\beta_1} - G_N\right) G_M^{-1} D_{ap}^\sim (I + G^\sim G)^{-\frac{1}{2}}\right]$$

$$= \bar{\sigma}\left[\frac{D_{\min}\left(\mathcal{F}_\ell(P,K)\right)D_{\min}^{-1}G_M}{\beta_1} - G_N\right].$$

(4) Define

$$P_a = \begin{bmatrix} D_{\min}(s) & \\ & I \end{bmatrix} P(s) \begin{bmatrix} D_{\min}^{-1}(s)G_M(s) & \\ & I \end{bmatrix} - \beta_1 \begin{bmatrix} G_N \\ 0 \end{bmatrix}$$

and find a controller K_{new} minimizing $\|\mathcal{F}_\ell(P_a, K)\|_\infty$.

(5) Compute a new β_1 as

$$\beta_1 = \sup_\omega \inf_{\tilde{D}_\omega \in \mathcal{D}, \tilde{G}_\omega \in \mathcal{G}} \{\beta(\omega): \ \Gamma \leq 1\}$$

where

$$\Gamma := \bar{\sigma}\left[\left(\frac{\tilde{D}_\omega \mathcal{F}_\ell(P, K_{\text{new}})\tilde{D}_\omega^{-1}}{\beta(\omega)} - j\tilde{G}_\omega\right)(I + \tilde{G}_\omega^2)^{-\frac{1}{2}}\right].$$

(6) Find $\hat{D}_\omega$ and $\hat{G}_\omega$ such that

$$\inf_{\hat{D}_\omega \in \mathcal{D}, \hat{G}_\omega \in \mathcal{G}} \bar{\sigma}\left[\left(\frac{\hat{D}_\omega \mathcal{F}_\ell(P, K_{\text{new}})\hat{D}_\omega^{-1}}{\beta_1} - j\hat{G}_\omega\right)(I + \hat{G}_\omega^2)^{-\frac{1}{2}}\right].$$

(7) Compare the new scaling matrices $\hat{D}_\omega$ and $\hat{G}_\omega$ with the previous estimates D_ω and G_ω. Stop if they are close, else replace D_ω, G_ω and K with $\hat{D}_\omega$, $\hat{G}_\omega$ and K_{new}, respectively, and go back to step (2).

18.3 Notes and References

The model validation problems are discussed in Smith and Doyle [1992] in the frequency domain; in Poolla, Khargonekar, Tikku, Krause, Nagpal [1994 in the discrete time domain (on which Section 18.1 is based); and in Rangan and Poolla [1996] and Smith and Dullerud [1996] in the continuous time domain. See also Davis [1995] and Chen and Wang [1996]. The mixed μ problems are discussed in detail in Young [1993] (on which Section 18.2 is based), Fan, Tits, and Doyle [1991], Packard and Pandey [1993], and references therein.

18.4 Problems

Problem 18.1 Write a MATLAB program for the additive model validation problem and try it on a simple experiment in your laboratory.

Bibliography

[1] Al-Saggaf, U. M. and G. F. Franklin (1987). "An error bound for a discrete reduced order model of a linear multivariable system," *IEEE Trans. Automat. Contr.*, Vol. AC-32, pp. 815-819.

[2] Al-Saggaf, U. M. and G. F. Franklin (1988). "Model reduction via balanced realizations: an extension and frequency weighting techniques," *IEEE Trans. Automat. Contr.*, Vol. AC-33, No. 7, pp. 687-692.

[3] Anderson, B. D. O. (1967). "An algebraic solution to the spectral factorization problem," *IEEE Trans. Automat. Contr.*, Vol. AC-12, pp. 410-414.

[4] Anderson, B. D. O. (1993). "Bode Prize Lecture (Control design: moving from theory to practice)," *IEEE Control Systems*, Vol. 13, No.4, pp. 16-25.

[5] Anderson, B. D. O. and Y. Liu (1989). "Controller reduction: concepts and approaches," *IEEE Trans. Automat. Contr.*, Vol. AC-34, No. 8, pp. 802-812.

[6] Anderson, B. D. O. and J. B. Moore (1989). *Optimal Control: Linear Quadratic Methods.* Prentice Hall, Englewood Cliffs, New Jersey.

[7] Anderson, B. D. O. and S. Vongpanitlerd (1973). *Network Analysis and Synthesis: A Modern System Theory Approach.* Prentice Hall, Englewood Cliffs, New Jersey.

[8] Arnold, W. F. and A. J. Laub (1984). "Generalized Eigenproblem algorithms and software for algebraic Riccati equations," *Proceedings of the IEEE*, Vol. 72, No. 12, pp. 1746-1754.

[9] Balas, G. (1990). *Robust Control of Flexible Structures: Theory and Experiments*, PhD. Thesis, California Institute of Technology.

[10] Balas, G., J. C. Doyle, K. Glover, A. Packard, and R. Smith (1994). *μ-Analysis and Synthesis Toolbox*, MUSYN Inc. and The MathWorks, Inc.

[11] Başar, T. and P. Bernhard (1991). $\mathcal{H}_\infty$-*Optimal Control and Related Minimax Design Problems: A Dynamic Game Approach.* Systems and Control: Foundations and Applications. Birkhäuser, Boston.

[12] Bernstein, D. S. and W. M. Haddard (1989). "LQG control with an $\mathcal{H}_\infty$ performance bound: A Riccati equation approach," *IEEE Trans. Automat. Contr.*, Vol. AC-34, pp. 293-305.

[13] Bode, H. W. (1945). *Network Analysis and Feedback Amplifier Design*, Van Nostrand, Princeton.

[14] Boyd, S., V. Balakrishnan, and P. Kabamba (1989). "A bisection method for computing the $\mathcal{H}_\infty$ norm of a transfer matrix and related problems," *Math. Control, Signals, and Systems*, Vol. 2, No. 3, pp. 207-220.

[15] Boyd, S. and C. Barratt (1991). *Linear Controller Design -Limits of Performance*, Prentice Hall, Englewood Cliffs, New Jersey.

[16] Boyd, S. and C. A. Desoer (1985). "Subharmonic functions and performance bounds in linear time-invariant feedback systems," *IMA J. Math. Contr. and Info.*, Vol. 2, pp. 153-170.

[17] Boyd, S., L. El Ghaoui, E. Feron, and V. Balakrishnan (1994). *Linear Matrix Inequalities in Systems and Control Theory*, SIAM, Philadelphia.

[18] Braatz, R. D., P. M. Young, J. C. Doyle, and M. Morari (1994). "Computational complexity of μ calculation," *IEEE Trans. Automat. Contr.*, Vol. 39, No. 5, pp. 1000-1002.

[19] Brogan, W. L. (1991). *Modern Control Theory*, 3rd ed., Prentice Hall, Englewood Cliffs, New Jersey.

[20] Bruinsma, N. A. and M. Steinbuch (1990). "A fast algorithm to compute the $\mathcal{H}_\infty$-norm of a transfer function matrix", *Systems and Control Letters*, Vol. 14, pp. 287-293.

[21] Bryson Jr., A. E. and Y-C. Ho (1975). *Applied Optimal Control*, Hemisphere Publishing Corporation.

[22] Chen, C. T. (1984). *Linear System Theory and Design*, Holt, Rinehart and Winston.

[23] Chen, J. (1995). "Sensitivity integral relations and design trade-offs in linear multivariable feedback systems," *IEEE Trans. Auto. Contr.*, Vol. 40, No. 10, pp. 1700-1716.

[24] Chen, J. and S. Wang (1996). "Validation of linear fractional uncertain models: solutions via matrix inequalities," *IEEE Trans. Auto. Contr.*, Vol. 41, No. 6, pp. 844-849.

[25] Chen, T. and B. A. Francis (1992). "$\mathcal{H}_2$-optimal sampled-data control," *IEEE Trans. Automat. Contr.*, Vol. 36, No. 4, pp. 387-397.

[26] Chen, X. and K. Zhou (1996). "On the algebraic approach to $\mathcal{H}_\infty$ control," *1996 IFAC World Congress*, San Francisco, California, pp. 505-510.

[27] Chilali, M. and P. Gahinet (1996). "$\mathcal{H}_\infty$ design with pole placement constraints: an LMI approach," *IEEE Trans. Automat. Contr.*, Vol. 41, No. 3, pp. 358-367.

[28] Christian, L. and J. Freudenberg (1994). "Limits on achievable robustness against coprime factor uncertainty," *Automatica*, Vol. 30, No. 11, pp. 1693-1702.

[29] Chu, C. C. (1985). *$\mathcal{H}_\infty$ optimization and robust multivariable control*. Ph.D. thesis, University of Minnesota.

[30] Dahleh, M. and I. J. Diaz-Bobillo (1995). *Control of Uncertain Systems: A Linear Programming Approach*, Prentice Hall, Englewood Cliffs, New Jersey.

[31] Daniel, R. W., B. Kouvaritakis, and H. Latchman (1986). "Principal direction alignment: a geometric framework for the complete solution to the μ problem," *IEE Proceedings*, Vol. 133, Part D, No. 2, pp. 45-56.

[32] Davis, R. A. (1995). *Model Validation for Robust Control*, Ph.D. dissertation, University of Cambridge, UK.

[33] Desoer, C. A., R. W. Liu, J. Murray, and R. Saeks (1980). "Feedback system design: the fractional representation approach to analysis and synthesis," *IEEE Trans. Automat. Contr.*, Vol. AC-25, No. 6, pp. 399-412.

[34] Desoer, C. A. and M. Vidyasagar (1975). *Feedback Systems: Input-Output Properties*, Academic Press, New York.

[35] Doyle, J. C. (1978). "Guaranteed Margins for LQG Regulators," *IEEE Trans. Automat. Contr.*, Vol. AC-23, No. 4, pp. 756-757.

[36] Doyle, J.C. (1982). "Analysis of feedback systems with structured uncertainties," *IEE Proceedings*, Part D, Vol.133, pp. 45-56.

[37] Doyle, J. C. (1984). "Lecture notes in advances in multivariable control," *ONR/Honeywell Workshop*, Minneapolis.

[38] Doyle, J. C. (1985). "Structured uncertainty in control system design", Proc. IEEE Conf. Dec. Contr., Ft. Lauderdale.

[39] Doyle, J. C., B. Francis, and A. Tannenbaum (1992). *Feedback Control Theory*, Macmillan Publishing Company.

[40] Doyle, J. C., K. Glover, P.P . Khargonekar, B. A. Francis (1989). "State-space solutions to standard $\mathcal{H}_2$ and $\mathcal{H}_\infty$ control problems," *IEEE Trans. Automat. Contr.*, Vol. AC-34, No. 8, pp. 831-847. Also see 1988 American Control Conference, Atlanta, June 1988.

[41] Doyle, J., K. Lenz, and A. Packard (1986). "Design examples using μ synthesis: space shuttle lateral axis FCS during reentry," *IEEE Conf. Dec. Contr.*, December 1986, pp. 2218-2223.

[42] Doyle, J. C., A. Packard and K. Zhou (1991). "Review of LFTs, LMIs and μ," *Proc. IEEE Conf. Dec. Contr.*, England, pp. 1227-1232

[43] Doyle, J. C., R. S. Smith, and D. F. Enns (1987). "Control of plants with input saturation nonlinearities," *American Control Conference*, pp. 1034-1039.

[44] Doyle, J. C. and G. Stein (1981). "Multivariable feedback design: Concepts for a classical/modern synthesis," *IEEE Trans. Automat. Contr.*, Vol. AC-26, pp. 4-16, Feb. 1981.

[45] Doyle, J. C., J. Wall and G. Stein (1982). "Performance and robustness analysis for structured uncertainty," in *Proc. IEEE Conf. Dec. Contr.*, pp. 629-636.

[46] Doyle, J. C., K. Zhou, K. Glover, and B. Bodenheimer (1994). "Mixed $\mathcal{H}_2$ and $\mathcal{H}_\infty$ performance objectives II: optimal control," *IEEE Trans. Automat. Contr.*, Vol. 39, No. 8, pp. 1575-1587.

[47] El-Sakkary, A. (1985). "The gap metric: robustness of stabilization of feedback systems," *IEEE Trans. Automat. Contr.*, Vol. 30, pp. 240-247.

[48] Enns, D. (1984a). *Model Reduction for Control System Design*, Ph.D. dissertation, Department of Aeronautics and Astronautics, Stanford University, Stanford, California.

[49] Enns, D. (1984b). "Model reduction with balanced realizations: an error bound and a frequency weighted generalization," *Proc. Conf. Dec. Contr.*, Las Vegas, Nevada.

[50] Fan, M. K. H. and A. L. Tits (1986). "Characterization and efficient computation of the structured singular value," *IEEE Trans. Automat. Contr.*, Vol. AC-31, No. 8, pp. 734-743.

[51] Fan, M. K. H. and A. L. Tits, and J. C. Doyle (1991). "Robustness in the presence of mixed parametric uncertainty and unmodeled dynamics," *IEEE Trans. Automat. Contr.*, Vol. AC-36, No. 1, pp. 25-38.

[52] Foias, C. and A. E. Frazho (1990). *The commutant Lifting Approach to Interpolation Problems*, Birkhäuser.

[53] Francis, B. A. (1987). *A course in $\mathcal{H}_\infty$ control theory*, Lecture Notes in Control and Information Sciences, Vol. 88, Springer-Verlag.

[54] Francis, B. A. and J. C. Doyle (1987). "Linear control theory with an $\mathcal{H}_\infty$ optimality criterion," *SIAM J. Contr. Optimiz.*, Vol. 25, pp. 815-844.

[55] Franklin, G. F., J. D. Powell, and M. L. Workman (1990), *Digital Control of Dynamic Systems*, 2nd Ed, Addison-Wesley Publishing company.

[56] Freudenberg, J. S. (1986). "The general structured singular problem with two blocks of uncertainty." *Proc. 1986 Allerton Conf.*, Monticello, Illinois.

[57] Freudenberg, J. S. and D. P. Looze (1988). *Frequency Domain Properties of Scalar and Multivariable Feedback Systems*, Lecture Notes in Contr. and Info. Science, Vol. 104, Springer-Verlag, Berlin.

[58] Gahinet, P. (1996). "Explicit controller formulas for LMI-based $\mathcal{H}_\infty$ synthesis," *Automatica*, Vol. 32, No. 7, pp. 1007-1014.

[59] Gahinet, P. and P. Apkarian (1994). "A linear matrix inequality approach to $\mathcal{H}_\infty$ control," *Int. J. Robust and Nonlinear Control*, Vol. 4, pp. 421-448.

[60] Garnett, J. B. (1981). *Bounded Analytic Functions*, Academic Press.

[61] Georgiou, T. T. (1988). "On the computation of the gap metric," *Systems and Control Letters*, Vol. 11, pp. 253-257.

[62] Georgiou, T. T. and M. C. Smith (1990). "Optimal robustness in the gap metric," *IEEE Trans. Automat. Contr.*, Vol. AC-35, No. 6, pp. 673-686.

[63] Georgiou, T. T. and M. C. Smith (1992). "Robust stabilization in the gap metric: controller design for distributed plants," *IEEE Trans. Automat. Contr.*, Vol. AC-37, No. 8, pp. 1133-1143.

[64] Glover, K. (1984). "All optimal Hankel-norm approximations of linear multivariable systems and their $\mathcal{L}_\infty$-error bounds," *Int. J. Contr.*, Vol. 39, pp. 1115-1193, 1984.

[65] Glover, K. (1986a). "Robust stabilization of linear multivariable systems: Relations to approximation," *Int. J. Contr.*, Vol. 43, No. 3, pp. 741-766.

[66] Glover, K. (1986b). "Multiplicative approximation of linear multivariable systems with L_∞ error bounds, " *Proc. Amer. Contr. Conf.*, Seattle, WA, pp. 1705-1709.

[67] Glover, K. and J. Doyle (1988). "State-space formulae for all stabilizing controllers that satisfy an $\mathcal{H}_\infty$ norm bound and relations to risk sensitivity," *Systems and Control Letters*, Vol. 11, pp. 167-172.

[68] Glover, K. and J. C. Doyle (1989). "A state space approach to $\mathcal{H}_\infty$ optimal control," in *Three Decades of Mathematical Systems Theory: A Collection of Surveys at the Occasion of the 50th Birthday of Jan C. Willems*, H. Nijmeijer and J. M. Schumacher (Eds.), Springer-Verlag, Lecture Notes in Control and Information Sciences, Vol. 135, 1989.

[69] Glover, K., D. J. N. Limebeer, J. C. Doyle, E. M. Kasenally, and M. G. Safonov (1991). "A characterization of all solutions to the four block general distance problem," *SIAM J. Contr. Optimiz.*, Vol. 29, No. 2, pp. 283-324.

[70] Glover, K., D. J. N. Limebeer, and Y. S. Hung (1992). "A structured approximation problem with applications to frequency weighted model reduction, " *IEEE Trans. Automat. Contr.*, Vol. AC-37, No. 4, pp. 447-465.

[71] Glover, K. and D. McFarlane (1989). "Robust stabilization of normalized coprime factor plant descriptions with $\mathcal{H}_\infty$ bounded uncertainty," *IEEE Trans. Automat. Contr.*, Vol. 34, No. 8, pp. 821-830.

[72] Glover, K. and D. Mustafa (1989). "Derivation of the maximum entropy $\mathcal{H}_\infty$ controller and a state space formula for its entropy," *Int. J. Contr.*, Vol. 50, pp. 899.

[73] Goddard, P. J. and K. Glover (1993). "Controller reduction: weights for stability and performance preservation," *Proceedings of the 32nd Conference on Decision and Control*, San Antonio, Texas, December 1993, pp. 2903-2908.

[74] Goddard, P. J. and K. Glover (1994). "Performance preserving frequency weighted controller approximation: a coprime factorization approach," *Proceedings of the 33nd Conference on Decision and Control*, Orlando, Florida, December 1994.

[75] Gohberg, I., and S. Goldberg (1981). *Basic Operator Theory.* Birkhauser, Boston.

[76] Gohberg, I., P. Lancaster, and L. Rodman (1986). "On Hermitian solutions of the symmetric algebraic Riccati equation," *SIAM J. Contr. Optimiz.*, Vol. 24, No. 6, November 1986, pp. 1323-1334.

[77] Golub, G. H., and C. F. Van Loan (1983). *Matrix Computations*, Johns Hopkind Univ. Press, Baltimore.

[78] Green, M. (1988). "A relative error bound for balanced stochastic truncation," *IEEE Trans. Automat. Contr.*, Vol. AC-33, No. 10, pp. 961-965.

[79] Green, M. (1992). "$\mathcal{H}_\infty$ controller synthesis by J-lossless coprime factorization," *SIAM J. Contr. Optimiz.*, Vol. 30, No. 3, pp. 522-547.

[80] Green, M., K. Glover, D. J. N. Limebeer, and J. C. Doyle (1988), "A J-spectral factorization approach to H_∞ control," *SIAM J. Contr. Optimiz.*, Vol. 28, pp. 1350-1371.

[81] Green, M. and D. J. N. Limebeer (1995). *Linear Robust Control*, Prentice Hall, Englewood Cliffs, New Jersey.

[82] Hinrichsen, D., and A. J. Pritchard (1990). "An improved error estimate for reduced-order models of discrete time system," *IEEE Trans. Automat. Contr.*, AC-35, pp. 317-320.

BIBLIOGRAPHY

[83] Hoffman, K. (1962). *Banach Spaces of Analytic Functions*, Prentice Hall, Englewood Cliffs, New Jersey.

[84] Horn, R. A. and C. R. Johnson (1990). *Matrix Analysis*, Cambridge University Press, first published in 1985.

[85] Horn, R. A. and C. R. Johnson (1991). *Topics in Matrix Analysis*, Cambridge University Press.

[86] Horowitz, I. M. (1963). *Synthesis of Feedback Systems*, Academic Press, London.

[87] Hung, Y. S. (1989). "$\mathcal{H}_\infty$-optimal control–Part I: model matching, – Part II: solution for controllers," *Int. J. Contr.*, Vol. 49, pp. 1291-1359.

[88] Hung, Y. S. and K. Glover (1986). "Optimal Hankel-norm approximation of stable systems with first order stable weighting functions," *Systems and Control Letters*, Vol. 7, pp. 165-172.

[89] Hyde, R. A. and K. Glover (1993). "The application of scheduled $\mathcal{H}_\infty$ controllers to a VSTOL aircraft," *IEEE Trans. Automat. Contr.*, Vol. 38, No. 7, pp. 1021-1039.

[90] Jonckheere, E. and L. M. Silverman (1983). "A new set of invariants for linear systems - applications to reduced order compensator design," *IEEE Transactions on Automatic Control*, Vol 28, No. 10, pp. 953-964.

[91] Kailath, T. (1980). *Linear Systems*, Prentice Hall, Englewood Cliffs, New Jersey.

[92] Kalman, R. E. (1964). "When is a control system optimal?" *ASME Trans. Series D: J. Basic Engineering*, Vol. 86, pp. 1-10.

[93] Kalman, R. E. and R. S. Bucy (1960). "New results in linear filtering and prediction theory," *ASME Trans. Series D: J. Basic Engineering*, Vol. 83, pp. 95-108.

[94] Khargonekar, P. P., I. R. Petersen, and M. A. Rotea (1988). "$\mathcal{H}_\infty$- optimal control with state-feedback," *IEEE Trans. Automat. Contr.*, Vol. AC-33, pp. 786-788.

[95] Khargonekar, P. P., I. R. Petersen, and K. Zhou (1990). "Robust stabilization and $\mathcal{H}_\infty$-optimal control," *IEEE Trans. Automat. Contr.*, Vol. 35, No. 3, pp. 356-361.

[96] Khargonekar, P. P. and E. Sontag (1982). "On the relation between stable matrix fraction factorizations and regulable realizations of linear systems over rings," *IEEE Trans. Automat. Contr.*, Vol. 27, pp. 627-638.

[97] Khargonekar, P. P. and A. Tannenbaum (1985). "Noneuclidean metrics and the robust stabilization of systems with parameter uncertainty," *IEEE Trans. Automat. Contr.*, Vol. AC-30, pp. 1005-1013.

[98] Kimura, H. (1997). *Chain-Scattering Approach to $\mathcal{H}_\infty$-Control*, Birkhäuser, Boston.

[99] Kwakernaak, H. and R. Sivan (1972). *Linear Optimal Control Systems*, Wiley-Interscience, New York.

[100] Kucera, V. (1972). "A contribution to matrix quadratic equations," *IEEE Trans. Automat. Contr.*, AC-17, No. 3, 344-347.

[101] Lancaster, P. and L. Rodman (1995). *Algebraic Riccati Equations*. Oxford University Press, Oxford.

[102] Lancaster, P. and M. Tismenetsky (1985). *The Theory of Matrices: with applications*, 2nd ed., Academic Press.

[103] Latham, G. A. and B. D. O. Anderson (1986). "Frequency weighted optimal Hankel-norm approximation of stable transfer function," *Systems and Control Letters*, Vol. 5, pp. 229-236.

[104] Laub, S. J. (1980). "Computation of 'balancing' transformations," *Proc. 1980 JACC*, Session FA8-E.

[105] Lenz, K. E., P. P. Khargonekar, and John C. Doyle (1987). "Controller order reduction with guaranteed stability and performance, " *American Control Conference*, pp. 1697-1698.

[106] Limebeer, D. J. N., B. D. O. Anderson, P. P. Khargonekar, and M. Green (1992). "A game theoretic approach to $\mathcal{H}_\infty$ control for time varying systems," *SIAM J. Contr. Optimiz.*, Vol. 30, No. 2, pp. 262-283.

[107] Liu, K. Z., T. Mita, and R. Kawtani (1990). "Parameterization of state feedback $\mathcal{H}_\infty$ controllers," *Int. J. Contr.*, Vol. 51, No. 3, pp. 535-551.

[108] Liu, Y. and B. D. O. Anderson (1986). "Controller reduction via stable factorization and balancing," *Int. J. Contr.*, Vol. 44, pp. 507-531.

[109] Liu, Y. and B. D. O. Anderson (1989). "Singular perturbation approximation of balanced systems," *Int. J. Contr.*, Vol. 50, No. 4, pp. 1379-1405.

[110] Liu, Y. and B. D. O. Anderson (1990). "Frequency weighted controller reduction methods and loop transfer recovery," *Automatica*, Vol. 26, No. 3, pp. 487-497.

[111] Liu, Y., B. D. O. Anderson, and U. Ly (1990). "Coprime factorization controller reduction with Bezout identity induced frequency weighting," *Automatica*, Vol. 26, No. 2, pp. 233-249.

[112] Lu, W. M., K. Zhou, and J. C. Doyle (1996). "Stabilization of uncertain linear systems and linear multidimensional systems," *IEEE Trans. on Automatic Control*, vol. 41, No. 1, pp. 50-65.

BIBLIOGRAPHY

[113] Luenberger, D. G. (1971). "An introduction to observers," *IEEE Trans. Automat. Contr.*, Vol. 16, No. 6, pp. 596-602.

[114] Maciejowski, J. M. (1989). *Multivariable Feedback Design*, Addison-Wesley, Reading, Massachusetts.

[115] Mageirou, E. F. and Y. C. Ho (1977). "Decentralized stabilization via game theoretic methods," *Automatica*, Vol. 13, pp. 393-399.

[116] Martensson, K. (1971). "On the matrix Riccati equation," *Information Sciences*, Vol. 3, pp. 17-49.

[117] McFarlane, D. C., K. Glover, and M. Vidyasagar (1990). "Reduced-order controller design using coprime factor model reduction," *IEEE Trans. Automat. Contr.*, Vol. 35, No. 3, pp. 369-373.

[118] McFarlane, D. C. and K. Glover (1990). *Robust Controller Design Using Normalized Coprime Factor Plant Descriptions*, Vol. 138, Lecture Notes in Control and Information Sciences, Springer-Verlag.

[119] McFarlane, D. C. and K. Glover (1992). "A loop shaping design procedure using $\mathcal{H}_\infty$ synthesis," *IEEE Trans. Automat. Contr.*, Vol. 37, No. 6, pp. 759-769.

[120] Meyer, D. G. (1990). "Fractional balanced reduction: model reduction via fractional representation," *IEEE Trans. Automat. Contr.*, Vol. 35, No. 12, pp. 1341-1345.

[121] Meyer, D. G. and G. Franklin (1987). "A connection between normalized coprime factorizations and linear quadratic regulator theory," *IEEE Trans. Automat. Contr.*, Vol. 32, pp. 227-228.

[122] Molinari, B. P. (1975). "The stabilizing solution of the discrete algebraic Riccati equation," *IEEE Trans. Automat. Contr.*, June 1975, pp. 396-399.

[123] Moore, B. C. (1981). "Principal component analysis in linear systems: controllability, observability, and model reduction," *IEEE Trans. Automat. Contr.*, Vol. 26, No. 2, pp. 17-32.

[124] Moore, J. B., K. Glover, and A. Telford (1990). "All stabilizing controllers as frequency shaped state estimate feed back," *IEEE Trans. Auto. Contr.*, Vol. AC-35, pp. 203-208.

[125] Morari, M. and E. Zafiriou (1989). *Robust Process Control*. Prentice Hall, Englewood Cliffs, New Jersey.

[126] Mullis, C. T. and R. A. Roberts (1976). "Synthesis of minimum roundoff noise fixed point digital filters," *IEEE Trans. Circuits Syst.*, Vol. 23, pp. 551-562.

[127] Mustafa, D. (1989). "Relations between maximum entropy / $\mathcal{H}_\infty$ control and combined $\mathcal{H}_\infty/LQG$ control," *Systems and Control Letters,* Vol. 12, No. 3, p. 193.

[128] Mustafa, D. and K. Glover (1990). *Minimum Entropy $\mathcal{H}_\infty$ Control,* Lecture Notes in Control and Information Sciences, Springer-Verlag.

[129] Mustafa, D. and K. Glover (1991). "Controller reduction by $\mathcal{H}_\infty$-balanced truncation," *IEEE Trans. Automat. Contr.,* Vol. AC-36, No. 6, pp. 669-682.

[130] Naylor, A. W., and G. R. Sell (1982). *Linear Operator Theory in Engineering and Science,* Springer-Verlag.

[131] Nagpal, K. M. and P. P. Khargonekar (1991). "Filtering and smoothing in an $\mathcal{H}_\infty$ setting," *IEEE Trans. Automat. Contr.,* Vol. AC-36, No. 2, pp. 152-166.

[132] Nehari, Z. (1957). "On bounded bilinear forms," *Annals of Mathematics,* Vol. 15, No. 1, pp. 153-162.

[133] Nett, C. N., C. A. Jacobson and N. J. Balas (1984). "A connection between state-space and doubly coprime fractional representations," *IEEE Trans. Automat. Contr.,* Vol. AC-29, pp. 831-832.

[134] Nett, C. N. and J. A. Uthgenannt (1988). "An explicit formula and an optimal weight for the 2-block structured singular value interaction measure," *Automatica,* Vol. 24, No. 2, pp. 261-265.

[135] Packard, A. (1991). *Notes on μ_Δ.* Unpublished lecture notes.

[136] Packard, A. (1994). "Gain scheduling via linear fractional transformations." *Systems and Control Letters,* Vol. 22, pp. 79-92.

[137] Packard, A. and J. C. Doyle (1988a). "Structured singular value with repeated scalar blocks," *Proc. American Control Conference,* Atlanta.

[138] Packard, A. and J. C. Doyle (1988b). *Robust Control of Multivariable and Large Scale Systems,* Final Technical Report for Air Force Office of Scientific Research.

[139] Packard, A. and J. C. Doyle (1993), "The complex structured singular value," *Automatica,* Vol. 29, pp. 71-109.

[140] Packard, A. and P. Pandey (1993). "Continuity properties of the real/complex structured singular value," *IEEE Trans. Automat. Contr.,* Vol. 38, No. 3, pp. 415-428.

[141] Packard, A., K. Zhou, P. Pandey, J. Leonhardson, and G. Balas (1992). "Optimal, constant I/O similarity scaling for full information and state feedback control problems," *Systems and Control Letters,* Vol. 19, No. 4, pp. 271-280.

BIBLIOGRAPHY

[142] Paganini, F. (1995). "Necessary and sufficient conditions for robust $\mathcal{H}_2$ performance," *Proc. IEEE Conf. dec. Contr.*, New Orleans, Louisiana, pp. 1970-1975.

[143] Paganini, F. (1996). "A set-based approach for white noise modeling," *IEEE Trans. Automat. Contr.*, Vol. AC-41, No. 10, pp. 1453-1465.

[144] Pernebo, L. and L. M. Silverman (1982). "Model reduction via balanced state space Representation," *IEEE Trans. Automat. Contr.*, Vol. AC-27, No. 2, pp. 382-387.

[145] Petersen, I. R. (1987). "Disturbance attenuation and $\mathcal{H}_\infty$ optimization: a design method based on the algebraic Riccati equation," *IEEE Trans. Automat. Contr.*, Vol. AC-32, pp. 427-429.

[146] Poolla, K., P. P. Khargonekar, A. Tikku, J. Krause and K. Nagpal (1994). "A time domain approach to model validation," *IEEE Trans. Automat. Contr.*, Vol. 39, No. 5, pp. 951-959.

[147] Poolla, K. and A. Tikku (1995). "Robust performance against time-varying structured perturbations," *IEEE Trans. Automat. Contr.*, Vol. 40, No. 9, pp. 1589-1602.

[148] Qiu, L. and E. J. Davison (1992a). "Feedback stability under simultaneous gap metric uncertainties in plant and controller," *Systems and Control Letters*, Vol. 18, pp. 9-22.

[149] Qiu, L. and E. J. Davison (1992b). "Pointwise gap metrics on transfer matrices," *IEEE Trans. Automat. Contr.*, Vol. 37, No. 6, pp. 741-758.

[150] Qiu, L. (1995). "On the robustness of symmetric systems," *Proc. IEEE Conf. Dec. Contr.*, New Orleans, Louisiana, pp. 2659-2660.

[151] Qiu, L., B. Bernhardsson, A. Rantzer, E. J. Davison, P. M. Young, and J. C. Doyle (1995). "A formula for computation of the real stability Radius," *Automatica*, Vol. 31, No. 6, pp. 879-890.

[152] Ran, A. C. M. and R. Vreugdenhil (1988). "Existence and comparison theorems for algebraic Riccati equations for continuous- and discrete-time systems," *Linear Algebra and its Applications*, Vol. 99, pp. 63-83.

[153] Redheffer, R. M. (1959). "Inequalities for a matrix Riccati equation," *Journal of Mathematics and Mechanics*, Vol. 8, No. 3.

[154] Redheffer, R. M. (1960). "On a certain linear fractional transformation," *J. Math. and Physics*, Vol. 39, pp. 269-286.

[155] Rangan, S. and K. Poolla (1996). "Time domain validation for sampled-data uncertainty models," *IEEE Trans. Automat. Contr.*, Vol. 41, No. 7, pp. 980-991.

[156] Rosenbrock, H. H. (1974). *Computer-Aided Control System Design*. Academic Press, London.

[157] Safonov, M. G. (1980). *Stability and Robustness of Multivariable Feedback Systems*, MIT Press.

[158] Safonov, M. G. (1982). "Stability margins of diagonally perturbed multivariable feedback systems," *Proc. IEE, Part D*, Vol. 129, No. 6, pp. 251-256.

[159] Safonov, M. G. (1984). "Stability of interconnected systems having slope-bounded nonlinearities," *6th Int. Conf. on Analysis and Optimization of Systems*, Nice, France.

[160] Safonov, M. G. and J. C. Doyle (1984). "Minimizing conservativeness of robustness singular values," in *Multivariable Control*, S.G. Tzafestas, Ed., Reidel, New York.

[161] Safonov, M. G. and D. J. N. Limebeer, and R. Y. Chiang (1990). "Simplifying the $\mathcal{H}_\infty$ Theory via Loop Shifting, matrix pencil and descriptor concepts," *Int. J. Contr.*, Vol 50, pp. 2467-2488.

[162] Sampei, M., T. Mita, and M. Nakamichi (1990). "An algebraic approach to $\mathcal{H}_\infty$ output feedback control problems," *Systems and Control Letters*, Vol. 14, pp. 13-24.

[163] Scherer, C. (1992). "$\mathcal{H}_\infty$ control for plants with zeros on the imaginary axis," *SIAM J. Contr. Optimiz.*, Vol. 30, No. 1, pp. 123-142.

[164] Scherer, C. (1992). "$\mathcal{H}_\infty$ optimization without assumptions on finite or infinite zeros," *SIAM J. Contr. Optimiz.*, Vol. 30, No. 1, pp. 143-166.

[165] Seron, M. M., J. H. Braslavsky, and G. C. Goodwin (1997). *Fundamental Limitations in Filtering and Control*, Springer, Berlin.

[166] Shamma, J. (1994). "Robust stability with time-varying structured uncertainty," *IEEE Trans. Automat. Contr.*, Vol. 39, No. 4, pp. 714-724.

[167] Skogestad, S. and I. Postlethwaite (1996). *Multivariable Feedback Control*, John Wiley & Sons, New York.

[168] Smith, R. S. and J. C. Doyle (1992). "Model validation: a connection between robust control and identification," *IEEE Trans. Automat. Contr.*, Vol. 37, No. 7, pp. 942-952.

[169] Smith, R. S. and G. Dullerud (1996). "Continuous-time control model validation using finite experimental data," *IEEE Trans. Automat. Contr.*, Vol. 41, No. 8, pp. 1094-1105.

[170] Stein, G. and J. C. Doyle (1991). "Beyond singular values and loop shapes," *AIAA Journal of Guidance and Control*, Vol. 14, No. 1, pp. 5-16.

[171] Stoorvogel, A. A. (1992). *The $\mathcal{H}_\infty$ Control Problem: A State Space Approach*, Prentice Hall, Englewood Cliffs, New Jersey.

[172] Tadmor, G. (1990). "Worst-case design in the time domain: the maximum principle and the standard $\mathcal{H}_\infty$ problem," *Mathematics of Control, Systems and Signal Processing*, pp. 301-324.

[173] Tits, A. L. (1995). "On the small mu theorem," *Proc. American Control Conference*, Seattle, Washington, pp. 2870-2872.

[174] Tits, A. L. and M. K. H. Fan (1995). "On the small μ theorem," *Automatica*, Vol. 31, No. 8, pp. 1199-1201.

[175] Toker, O. and H. Özbay (1995). "On the NP-hardness of the purely complex μ computation, analysis/synthesis, and some related problems in multidimensional systems," *Proc. American Control Conference*, Seattle, Washington, pp. 447-451.

[176] Ushida, S. and H. Kimura (1996). "A counterexample to Mustafa-Glovers' monotonicity conjecture," *Systems and Control Letters*, Vol. 28, No. 3, pp. 129-137.

[177] Van Dooren, P. (1981). "A generalized eigenvalue approach for solving Riccati equations," *SIAM J. Sci. Stat. Comput.*, Vol. 2, No. 2, pp. 121-135.

[178] Vidyasagar, M. (1984). "The graph metric for unstable plants and robustness estimates for feedback stability," *IEEE Trans. Automat. Contr.*, Vol. 29, pp. 403-417.

[179] Vidyasagar, M. (1985). *Control System Synthesis: A Factorization Approach*. MIT Press, Cambridge, Massachusetts.

[180] Vidyasagar, M. (1988). "Normalized coprime factorizations for non-strictly proper systems," *IEEE Trans. Automat. Contr.*, Vol. 33, No. 3, pp. 300-301.

[181] Vidyasagar, M. and H. Kimura (1986). "Robust controllers for uncertain linear multivariable systems," *Automatica*, Vol. 22, pp. 85-94.

[182] Vinnicombe, G. (1992a). "On the frequency response interpretation on an indexed $\mathcal{L}_2$-gap metric," *Proceedings of the American Control Conference*, Chicago, Illinois, pp. 1133-1137.

[183] Vinnicombe, G. (1992b). "Robust Design in the graph topology: a benchmark example," *Proceedings of the American Control Conference*, Chicago, Illinois, pp. 2063-2064.

[184] Vinnicombe, G. (1993a). "Frequency domain uncertainty and the graph topology," *IEEE Trans. Automat. Contr.*, Vol. 38, No. 9, pp. 1371-1383.

[185] Vinnicombe, G. (1993b). *Measuring the Robustness of Feedback Systems*, Ph.D. dissertation, Cambridge University.

[186] Whittle, P. (1990). *Risk-sensitive Optimal Control*, John Wiley and Sons, New York.

[187] Willems, J.C. (1971). "Least Squares Stationary Optimal Control and the Algebraic Riccati Equation", *IEEE Trans. Automat. Contr.*, Vol. AC-16, No. 6, pp. 621-634.

[188] Willems, J. C. (1981). "Almost invariant subspaces: an approach to high gain feedback design – Part I: almost controlled invariant subspaces," *IEEE Trans. Automat. Contr.*, Vol. AC-26, pp. 235-252.

[189] Willems, J. C., A. Kitapci and L. M. Silverman (1986). "Singular optimal control: a geometric approach," *SIAM J. Contr. Optimiz.*, Vol. 24, pp. 323-337.

[190] Wimmer, H. K. (1985). "Monotonicity of Maximal solutions of algebraic Riccati equations," *Systems and Control Letters*, Vol. 5, April 1985, pp. 317-319.

[191] Wonham, W. M. (1985). *Linear Multivariable Control: A Geometric Approach*, third edition, Springer-Verlag, New York.

[192] Yang, X. H. and A. Packard (1995). "A low order controller design method," *Proc. IEEE Conf. Dec. Contr.*, New Orleans, Louisiana, pp. 3068-3073.

[193] Youla, D. C. and J. J. Bongiorno, Jr. (1985). "A feedback theory of two-degree-of-freedom optimal Wiener-Hopf design," *IEEE Trans. Automat. Contr.*, Vol. AC-30, No. 7, pp. 652-665.

[194] Youla, D. C., H. A. Jabr, and J. J. Bongiorno (1976a). "Modern Wiener-Hopf design of optimal controllers: part I," *IEEE Trans. Automat. Contr.*, Vol. AC-21, pp. 3-13.

[195] Youla, D. C., H. A. Jabr, and J. J. Bongiorno (1976b). "Modern Wiener-Hopf design of optimal controllers: part II," *IEEE Trans. Automat. Contr.*, Vol. AC-21, pp. 319-338.

[196] Youla, D. C., H. A. Jabr, and C. N. Lu (1974). "Single-loop feedback stabilization of linear multivariable dynamical plants," *Automatica*, Vol. 10, pp. 159-173.

[197] Youla, D. C. and M. Saito (1967). "Interpolation with positive-real functions," *Journal of The Franklin Institute*, Vol. 284, No. 2, pp. 77-108.

[198] Young, N. (1988). *An Introduction to Hilbert Space*, Cambridge University Press.

[199] Young, P. M. (1993). *Robustness with Parametric and Dynamic Uncertainty*, Ph.D. thesis, California Institute of Technology.

[200] Zames, G. (1966). "On the input-output stability of nonlinear time-varying feedback systems, parts I and II," *IEEE Trans. Automat. Contr.*, Vol. AC-11, pp. 228 and 465.

[201] Zames, G. (1981). "Feedback and optimal sensitivity: model reference transformations, multiplicative seminorms, and approximate inverses," *IEEE Trans. Automat. Contr.*, Vol. AC-26, pp. 301-320.

[202] Zames, G. and A. K. El-Sakkary (1980). "Unstable systems and feedback: The gap metric," *Proc. Allerton Conf.*, pp. 380-385.

[203] Zhang, C. and M. Fu (1996). "A revisit to the gain and phase margins of linear quadratic regulators," *IEEE Trans. Automat. Contr.*, Vol. 41, No. 10, pp. 1527-1530.

[204] Zhou, K. (1992). "On the parameterization of $\mathcal{H}_\infty$ controllers," *IEEE Trans. Automat. Contr.*, Vol. 37, No. 9, pp. 1442-1445.

[205] Zhou, K. (1995). "Frequency weighted $\mathcal{L}_\infty$ norm and optimal Hankel norm model reduction," *IEEE Trans. Automat. Contr.*, vol. 40, no. 10, pp. 1687-1699, October 1995.

[206] Zhou K. and J. Chen (1995). "Performance bounds for coprime factor controller reductions," *Systems and Control Letters*, Vol. 26, No. 2, pp. 119-127.

[207] Zhou, K., J. C. Doyle and K. Glover (1996). *Robust and Optimal Control*. Prentice Hall, Upper Saddle River, New Jersey.

[208] Zhou, K. and P. P. Khargonekar (1988). "An algebraic Riccati equation approach to $\mathcal{H}_\infty$ optimization," *Systems and Control Letters*, Vol. 11, pp. 85-92.

[209] Zhou, K., K. Glover, B. Bodenheimer, and J. Doyle (1994). "Mixed $\mathcal{H}_2$ and $\mathcal{H}_\infty$ performance objectives I: robust performance analysis," *IEEE Trans. Automat. Contr.*, Vol. 39, No. 8, pp. 1564-1574.

[210] Zhou, T. and H. Kimura (1992). "Minimal $\mathcal{H}_\infty$-norm of transfer functions consistent with prescribed finite input-output data," in *Proc. SICE' 92*, Kumamotp, Japan, pp. 1079-1082.

[211] Zhu, S. Q. (1989). "Graph topology and gap topology for unstable systems," *IEEE Trans. Automat. Contr.*, Vol. 34, No. 8, pp. 848-855.

Index

additive approximation, 105
additive uncertainty, 131, 142
admissible controller, 222
algebraic Riccati equation, 6, 233
    complementarity property, 234
    solutions, 15
    stability property, 234
    stabilizing solution, 234
all-pass dilation, 120
all-pass function, 245
Al-Saggaf, U. M., 126
analytic function, 47
Anderson, B. D. O., 312
Argument Principle, 357
Arnold, W. F., 246

Balakrishnan, V., 62, 299
balanced model reduction, 99
    additive, 105
    error bound, 119
    frequency-weighted, 124
    multiplicative, 125
    relative, 125
    stability, 117
balanced realization, 4, 37, 105, 107
Balas, G., 216
Balas, N. J, 77
basis, 11
Bezout identity, 71
bisection algorithm, 57
Bode, H. W., 102
Bode integral, 81
Bode's gain and phase relation, 94
Bongiorno, J. J., 231

bounded real lemma, 7, 238
Boyd, S., 50, 62, 102, 299
Braatz, R. D., 382
Braslavsky, J. H., 102
Brogan, W., 24, 41
Bruinsma, N. A., 62

Cauchy-Schwarz inequality, 46
Chen, C. T., 41
Chen, J., 102, 344, 381, 389
Chen. T., 54
Chen, X., 299
Chilali, M., 299
Christian, L., 341
co-inner function, 245
complementary inner function, 246
complementary sensitivity, 82
conjugate system, 34
controllability, 3, 27
    Gramian, 3, 53, 106
    matrix, 29
controllable canonical form, 36
controller parameterization, 221
controller reduction, 8, 305
coprime factor uncertainty, 144, 315
coprime factorization, 4, 71, 228, 315
    normalized, 318

Davis, R. A., 381, 389
Davison, E. J., 341, 354, 375
design limitation, 4, 81
design model, 129
design tradeoff, 81

Desoer, C. A., 50, 62, 102, 139, 158, 231
detectability, 27, 31
D-G-K iteration, 388
D-K iteration, 214
directed gap, 350
dom(Ric), 234
double coprime factorization, 71
Doyle, J. C., 3, 77, 102, 152, 180, 192, 194, 206, 216, 231, 246, 267, 299, 312, 383, 389
dual system, 34
Dullerud, G., 381, 389

eigenvalue, 12
eigenvector, 12, 234
El Ghaoui, L., 299
El-Sakkary, A., 341, 349, 375
Enns, D., 126, 312
entropy, 286
error bound, 4

$\mathcal{F}_\ell$, 165
$\mathcal{F}_u$, 166
Fan, M. K. H., 201, 217, 383, 389
Feron, E., 299
feedback, 65
filtering, 297
Foias, C., 378
Fourier transform, 49
Francis, B. A., 54, 77, 102, 231, 299
Franklin, G. F., 126
Frazho, A. E., 378
frequency weighting, 85
frequency-weighted balanced reduction, 8, 124
Freudenberg, J. S., 102, 341, 348
Frobenius norm, 18

Gahinet, P., 299
gain, 94
gap metric, 9, 154, 341, 349
generalized eigenvector, 15, 234

Georgiou, T. T., 334, 341, 350, 375
Gilbert's realization, 37
Glover, K., 3, 126, 246, 299, 309, 312, 325, 341, 344
Goddard, P. J., 309, 312
Gohberg, I., 62
Goldberg, I., 62
Golub, G. H., 24
Goodwin, G. C., 102
Gramian, 106, 110
graph metric, 341
graph topology, 341
Green, M., 126

Hamiltonian matrix, 56, 233
Hankel singular value, 110
Hardy spaces, 48
Hermitian matrix, 13
$\mathcal{H}_\infty$ control, 2, 7, 269
    loop shaping, 315
    singular problem, 294
$\mathcal{H}_\infty$ filtering, 8, 297
$\mathcal{H}_\infty$ norm, 2, 55
$\mathcal{H}_\infty$ optimal controller, 282, 286
$\mathcal{H}_\infty$ performance, 85, 88
$\mathcal{H}_\infty$ space, 45, 47, 50
$\mathcal{H}_\infty^-$ space, 50
Hilbert space, 45
Hinrichsen, D., 126
Horn, R. A., 24
Horowitz, I. M., 102, 341
$\mathcal{H}_2$ norm, 53
$\mathcal{H}_2$ optimal control, 253
$\mathcal{H}_2$ performance, 85, 87
$\mathcal{H}_2$ space, 45, 47
$\mathcal{H}_2$ stability margin, 265
$\mathcal{H}_2^\perp$ space, 48
Hung, Y. S., 126
Hurwitz, 29
Hyde, R. A., 341

image, 12

induced norm, 17
inner function, 245
inner product, 46
inner-outer factorization, 248
input sensitivity, 82
integral control, 8, 294
internal model principle, 294
internal stability, 4, 68
invariant subspace, 15
invariant zero, 39, 242
inverse of a transfer function, 35

Jacobson, V. A., 77
Johnson, C. R., 24
Jonckheere, E., 347

Kabamba, P., 62
Kailath, T., 41, 71
kernel, 12
Khargonekar, P. P., 299, 312, 379, 389
Kimura, H., 302
Kitapci, A., 294
Krause, J., 379, 389
Kwakernaak, H., 267

Lancaster, P., 24, 246
Laub, A. J., 246
Lebesgue measure, 46
left coprime factorization, 71
Lenz, K., 312
$\mathcal{L}_\infty$ norm, 55
$\mathcal{L}_\infty$ space, 50
Limebeer, D. J. N., 126
linear combination, 11
linear fractional transformation (LFT), 2, 5, 163
linear matrix inequality (LMI), 239, 277
Liu, Y., 312
loop gain, 83
loop shaping, 9, 315, 325
loop transfer matrix, 82
Looze, D. P., 102, 348

LQG stability margin, 265
LQG/LTR, 102
LQR porblem, 255
LQR stability margin, 259
$\mathcal{L}_2$ norm, 53
$\mathcal{L}_2$ space, 48
$\mathcal{L}_2(-\infty, \infty)$ space, 47
Lu, W. M., 231
Ly, U., 312
Lyapunov equation, 13, 53, 106

main loop theorem, 197
Martensson, K., 246
matrix
    Hermitian, 13
    inequality, 239
    inversion formulas, 13
    norm, 16
    square root of a, 23
maximum modulus theorem, 47
McFarlane, D. C., 312, 325, 341, 344
minimal realization, 35, 109
minimum entropy controller, 286
Mita, T., 299
mixed μ, 9, 381
modal controllability, 31
modal observability, 31
model invalidation, 377
model reduction, 105
model uncertainty, 65, 129
model validation, 9, 377
Moore, B. C., 126
Moore, J. B., 231
μ, 5, 183
    lower bound, 192
    synthesis, 213
    upper bound, 192
Mullis, C. T., 126
multiplication operator, 50
multiplicative approximation, 125
multiplicative uncertainty, 131, 143
Mustafa, D., 312

Nagpal, K. M., 299, 379, 389
Nakamichi, M., 299
Naylor, A. W., 62
Nett, C. N., 77
nominal performance (NP), 137
nominal stability (NS), 137
nonminimum phase zero, 81
norm, 16
normal rank, 38
normalized coprime factorization, 8, 154
    loop shaping, 325
ν-gap metric, 9, 154, 349
null space, 12

observability, 3, 27
    Gramian, 3, 53, 106
observable canonical form, 36
observable mode, 31
observer, 31
observer-based controller, 31
optimality of $\mathcal{H}_\infty$ controller, 282
orthogonal complement, 12
orthogonal matrix, 12
output sensitivity, 82
Özbay, H., 382

Packard, A., 180, 192, 216, 299, 314, 382, 389
Pandey, P., 217, 382, 389
Parseval's relations, 49
PBH (Popov-Belevitch-Hautus) tests, 31
performance limitation, 81
Pernebo, L., 126
phase, 94
plant condition number, 150
Poisson integral, 81, 335, 348
pole, 38
Poolla, K., 379, 381, 389
positive (semi-)definite matrix, 23
positive real, 247
Postlethwaite, I., 102
Pritchard, A. J., 126

Qiu, L., 220, 341, 354, 375
quadratic performance, 253

Ran, A. C. M., 299
Rangan, S., 381, 389
range, 12
real μ, 381
real spectral radius, 13
realization, 35
    balanced, 110
    input normal, 113
    minimal, 35
    output normal, 113
Redheffer star product, 178
reduced-order controller, 8
regulator problem, 253
relative approximation, 125
return difference, 82
$\mathcal{RH}_\infty$ space, 50
$\mathcal{RH}_\infty^-$ space, 50
$\mathcal{RH}_2$ space, 48
$\mathcal{RH}_2^\perp$ space, 48
Riccati equation, 233
Riccati operator, 234
right coprime factorization, 71
Roberts, R. A., 126
robust performance, 137
    $\mathcal{H}_2$ performance, 147
    $\mathcal{H}_\infty$ performance, 147, 197
    structured, 202
robust stability (RS), 5, 137
    structured, 200
robust stabilization, 315
Rodman, L., 246

Safonov, M. G., 132
Saito, M., 215
Sampei, M., 299
Schur complement, 14
Sell, G. R., 62
sensitivity function, 82
Seron, M. M., 102

Silverman, L. M., 126, 347
singular value decomposition (SVD), 19, 51
singular $\mathcal{H}_\infty$ problem, 294
singular vector, 20
Sivan, R., 267
skewed performance specification, 150
Skogestad, S., 102
small gain theorem, 3, 129, 137
Smith, M. C., 334, 341, 375
Smith, R. S., 381, 389
span, 11
spectral radius, 12
stability, 27
    internal, 68
    margin, 265
stabilizability, 27
stabilizable, 29
stabilization, 221
stabilizing controller, 6, 221
stable invariant subspace, 15, 234
star product, 178
Stein, G., 102, 152, 217, 267
Steinbuch, M., 62
strictly positive real, 247
Stoorvogel, A. A., 294, 299
structured singular value, 5, 183
    lower bound, 192
    upper bound, 192
structured uncertainty, 5, 183
Sylvester equation, 13

Tannenbaum, A., 102
Tikku, A., 379, 389
Tismenetsky, M., 24
Tits, A. L., 201, 217, 383, 389
Toker, O., 382
trace, 12
tradeoff, 81

uncertainty, 1, 65, 129
    state space, 171

unstructured, 5, 129
unitary matrix, 12
Ushida, S., 302

Van Dooren, P., 246
Van Loan, C. F., 24
Vidyasagar, M., 62, 77, 139, 158, 231, 341, 349
Vinnicombe, G., 312, 341, 349, 366, 375
Vreugdenhil, R., 299

Wall, J., 158
Wang, S., 381, 389
weighted model reduction, 124
weighting function, 4, 85, 89
well-posedness, 66, 167
Willems, J. C., 246
winding number, 357
Wonham, W. M., 41

Yang, X. H., 314
Youla, D. C., 215, 221, 231
Youla parameterization, 224, 228
Young, P. M., 217, 382, 389

Zames, G., 9, 132, 158, 349, 375
zero, 3, 38
Zhou, K., 3, 126, 180, 217, 246, 299, 344
Zhu, S. Q., 341